THE AGE OF MIGRATION

About the Authors

Hein de Haas, PhD, is Professor of Sociology at the University of Amsterdam, The Netherlands. He is a founding member and former director of the International Migration Institute (IMI) at the University of Oxford, United Kingdom, and now directs the IMI at its current home at the University of Amsterdam. Dr. de Haas is also Professor in Migration and Development at Maastricht University/United Nations University–MERIT. His research focuses on the linkages between migration and broader processes of social transformation and development in origin and destination countries.

Stephen Castles, DPhil, was Honorary Professor of Sociology at the University of Sydney, Australia, before retiring in 2017, and served as the first director of the IMI at the University of Oxford, United Kingdom. His research has focused on international migration dynamics, global governance, migration and development, and migration trends in Africa, Asia, and Europe. Dr. Castles's books and articles have been translated into many languages and have been highly influential in the development of international migration studies.

Mark J. Miller, PhD, is Emma Smith Morris Professor Emeritus of Political Science and International Relations at the University of Delaware. He served as editor of the *International Migration Review* from 1983 to 2005. Dr. Miller has conducted research and consulted widely on comparative immigration and refugee policies, global migration, and migration and security. He is a recipient of the Francis Alison Award, the highest honor extended to faculty by the University of Delaware.

THE AGE OF MIGRATION

International Population Movements in the Modern World

SIXTH EDITION

Hein de Haas
Stephen Castles
and
Mark J. Miller

THE GUILFORD PRESS
New York   London

First published in English under the title The Age of Migration, 6th Edition by Hein de Haas, Stephen Castles, Mark J. Miller by Springer Nature Limited Copyright © 2020, Hein de Haas, Stephen Castles, Mark J. Miller

This edition has been published under license from Springer Nature Limited. The authors have asserted their right to be identified as the author of this Work. For copyright reasons this edition is not for sale outside of the United States and its dependencies, Canada, Mexico, and the Philippines.

Published in the United States of America by
The Guilford Press
A Division of Guilford Publications, Inc.
370 Seventh Avenue, New York, NY 10001
www.guilford.com

Last digit is print number: 9 8 7 6 5 4 3 2 1

Library of Congress Cataloging-in-Publication Data is available from The Guilford Press.

ISBN 978–1–4625–4289–5 (paperback)

This book is printed on paper suitable for recycling and made from fully managed and sustained forest sources. Logging, pulping and manufacturing processes are expected to conform to the environmental regulations of the country of origin.

Printed in the United Kingdom by Bell and Bain Ltd, Glasgow

Contents

List of Figures

List of Tables

List of Maps

List of Boxes

Preface to the Sixth Edition

The Age of Migration provides an accessible introduction to the study of global migrations and their consequences for society. Originally published in 1993, it is designed to combine theoretical knowledge with up-to-date information on migration flows and their implications for states as well as people everywhere. International migration has become a major theme for public debate, and *The Age of Migration* is widely used by policy makers, scholars and journalists. It is recommended as a textbook in politics and social science all over the world.

Hein de Haas joined Stephen Castles and Mark J. Miller as an author of the fifth edition and for this, new and expanded sixth edition, he has taken on the lead role in preparing the manuscript. It has been thoroughly revised and updated. As with previous editions, the sixth edition is essentially a new book. Its revised structure now comprises three thematic clusters. After the introductory chapter, Chapters 2, 3, 4 and 5 are concerned with categories, theories and the history of migration and ethnic diversity. Chapters 6, 7, 8 and 9 present overviews of migration in all major world regions, while Chapters 10, 11, 12, 13 and 14 are devoted to the politics and policies of migration and the effects of migration upon destination and origin societies.

Since its inception, each successive edition of *The Age of Migration* has improved the balance between coverage of Western and non-Western regions and perspectives on migration. This is done with a particular eye to avoid a simplistic casting of a world divided in a 'migrant sending' Global South and a 'migrant receiving' Global North. Such divisions are less and less tenable, both because of shifts in global migration patterns as well as a growing realization that such an understanding of migration was colonial, ahistorical and inaccurate in the first place. Even more than previous editions, this edition aims to provide a broad historical comparative vision that represents an awareness that most countries have dealt with substantial immigration and emigration.

The Age of Migration stresses that, despite regional differences, contemporary migration is shaped by similar forces of social and economic transformation affecting societies and people around the world. To facilitate comparison, the regional chapters that analyse migration trends in Europe (Chapter 6), the Americas (Chapter 7), the Asia-Pacific region (Chapter 8) and Africa and the Middle East (Chapter 9) use a uniform periodization. Continuing the trend of the previous editions to improve regional coverage of the book, the sixth edition of *The Age of Migration* contains a systematic assessment of migration trends in the former Soviet Union and Eastern Europe, which were underrepresented in previous editions.

It is therefore the first edition providing a global coverage of migration trends, migration debates and migration policies. To make the regional chapters more interesting for the general public and students (and not only for those with a particular interest in those regions), the chapters discuss how regional migration and policy trends illustrate more general themes and theories addressed in the introductory and theoretical chapters. Furthermore, the analyses draw on cutting-edge empirical evidence on migration and issues from across the social sciences (ranging from anthropology to economics).

Reflecting the rapid growth of empirical evidence over the past decades, *The Age of Migration* therefore provides a state-of-the-art overview of the field of migration studies.

This edition uses new global data on migration and migration policies that have become available since the last edition, such as through the World Bank's *Global Bilateral Migration Database* and recent updates of this global data on migrant populations by the UN's Population Division, as well as Oxford University's DEMIG (Determinants of International Migration) project. These advances in data have drastically increased the capacity for analysing long-term trends in global and regional migration patterns as well as the evolution of migration policies. Even though empirical evidence and data are still biased towards Western countries, recent advances in data and research have enabled a more balanced coverage of migration issues that steers away from Western-centric accounts. New data have been incorporated in the various new graphs and tables analysing migration and policy trends throughout the book. Maps, which were redrawn for the fourth edition, have been maintained—and new world maps have been added.

The sixth edition examines recent events and emerging trends anew. Labour migration to new industrial economies in North America, Europe, the Gulf, East Asia and the Pacific has been growing fast, while violent conflicts are leading to large-scale movements of displaced people, especially in less-developed regions. Improvements in transport and communication facilitate temporary, circular and repeated movements. New types of mobility are emerging as increasing numbers of people move for education, marriage or retirement, or in search of new lifestyles. Demographic and economic changes in immigration countries are raising awareness of future demand for migrant labour, while, at the same time, public concern about ethnic diversity is leading to measures to increase social cohesion, for instance through 'integration contracts' and citizenship tests.

Much has changed in the world since the publication of the first edition, yet the book's central argument remains the same. International population movements are reforging states and societies around the world in ways that affect bilateral and regional relations, economic restructuring, security, national identity and sovereignty. As a key dynamic within globalization, *The Age of Migration* highlights the need to see migration as an intrinsic part of broader economic and social change, and contributing to further transformations of the international political order. However, what sovereign states do in the realm of migration policies continues to matter a great deal. The notion of open borders remains elusive even within regional integration frameworks, except for European citizens circulating within the European Union. Sovereign states have a legitimate desire to control migration. The real question is *how* to achieve more effective migration policies, that benefit both migrants and societies involved. In order to identify what such policies might look like, it is important to achieve a better understanding of the causes and consequences of migration, which is what this book aims to achieve.

The authors thank the following for help in preparing the sixth edition. Hein de Haas is indebted to Sonja Fransen, Agnieszka Kubal, Lea Müller-Funk, Katharina Natter, Kerilyn Schewel, Hélène Thiollet and Simona Vezzoli for providing valuable feedback on drafts of various chapters. Special thanks go out to Alejandro Olayo-Méndez for his careful reading of several chapters. He is also grateful to Sarah Salehi who did essential work on compiling migration data and preparation of the bibliography and glossary, to Siebert Wielstra for vital assistance in preparing the final book manuscript, and to his mother, Annie de Haas-de Jonge, for providing the space and patience to write.

Hein de Haas expresses gratitude to the European Research Council (ERC), which has enabled him to do essential background research as part of a Starting Grant to the DEMIG (Determinants of International Migration) project at the University of Oxford funded under the European Commission's Seventh Framework Programme (FP7/2007–2013, ERC Grant Agreement 240940) as well as a Consolidator Grant to the MADE (Migration as Development) project at the University of Amsterdam funded under the European Commission's Horizon 2020 Programme (H2020/2015–2020, ERC Grant Agreement 648496).

We would like to thank our commissioning editor, Andrew Malvern, as well as Peter Atkinson and Aine Flaherty at Red Globe Press for giving a great deal of support, encouragement and invaluable feedback and encouragement on the sixth edition. The authors wish to acknowledge the many valuable criticisms on this and earlier editions from reviewers and colleagues, although it was not possible to respond to them all.

Hein de Haas would particularly like to thank Dovelyn Mendoza for reading the entire manuscript, her honest criticism and providing many suggestions which have greatly improved the quality, nuance and the overall balance of the book. Besides her intellectual engagement with the book, he is grateful for her encouragement to soldier on, and, above all, her care, companionship and unremitting love. He also thanks Selma, Dalila, Edgar and Stefano for injecting so much optimism, energy and joy.

Stephen Castles would like to thank the many friends and colleagues who contributed to the previous editions of *The Age of Migration*. Special thanks go to Ellie Vasta for her intellectual engagement with the contents of this book and her critique and input during the revision process for the sixth edition.

Mark Miller wishes to thank James Miller and Barb Ford for their help in his contribution to the sixth edition.

HEIN de HAAS
STEPHEN CASTLES
MARK J. MILLER

Note on Migration Statistics

When studying migration and minorities it is vital to use statistical data, but it is also important to be aware of the limitations of such data. Statistics are collected in different ways, using different methods and different definitions by authorities of various countries. These can even vary between different agencies within a single country.

A key point is the difference between *flow* and *stock* figures. The *flow* of migrants is the number of migrants who enter a country (*inflow, entries* or *immigration*) in a given period (usually a year), or who leave the country (*outflow, departures* or *emigration*). The balance between these figures is known as *net migration*. The *stock* (or population) of migrants is the number of migrants present in a country on a specific date. Flow figures are useful for understanding trends in mobility, while stock figures on the size of migrant populations help us to examine the long-term impact of migration on a given population.

Until recently, figures on immigrants in 'classical immigration countries' (the US, Canada, Australia and New Zealand) were mainly based on the criterion of a person being *foreign-born* (or *overseas-born*), while data for many European immigration countries were mainly based on the criterion of a person being a *foreign national* (or *foreign resident, foreign citizen, foreigner* or *alien*). The foreign born include persons who have become *naturalized*, that is, who have taken on the nationality (or citizenship) of the receiving country. The category excludes children born to immigrants in the receiving country (the *second generation*) if they are citizens of that country. The term 'foreign nationals' excludes those who have taken on the nationality of the receiving country, but includes children born to immigrants who retain their parents' nationality (see OECD, 2006: 260–261).

The two ways of looking at the concept of immigrants reflect the perceptions and laws of different types of immigration countries. However, with longer settlement and recognition of the need to improve integration of long-term immigrants and their descendants, laws on nationality and ideas on its significance are changing. Many countries now provide figures for both the foreign-born and foreign nationals. These figures cannot be aggregated, so we will use both types in the book, as appropriate. In addition, some countries now provide data on children born to immigrant parents, on ethnicity, on race, or on combinations of these. When using statistics it is therefore very important to be aware of the definition of terms (which should always be given clearly in presenting data), the significance of different concepts and the purpose of the specific statistics (for detailed discussion see OECD, 2006, Statistical Annexe).

If data on both foreign-born and foreign citizens are available, the book has preferred to use data on foreign-born, because this allows for better comparisons across countries. Unless mentioned otherwise, the migration statistics mentioned in the text and graphs of this book are based on global migrant stock data from the *Trends in International Migrant Stock: The 2017 Revision* database from the United Nations (POP/DB/MIG/Stock/Rev.2017) compiled by the Population Division of the Department of Economic and

Social Affairs of the United Nations (UNDESA 2017). This database contains estimates of country-of-birth specific population data for all countries in the world for 1990, 2000, 2010 and 2017. For 1960, 1970 and 1980 we used data from the World Bank's Global Bilateral Migration Database. We compiled this data to provide overviews of migration trends and to draw the various migration graphs. We also used a range of other data sources, particularly the OECD, national sources and the DEMIG database.

The Age of Migration Website

The Age of Migration is accompanied by a companion website featuring freely accessible resources, including:

- Additional case studies;
- PowerPoint slides;
- Updates to cover important developments that affect the text;
- Weblinks;
- Further resources for both students and instructors.

The website is found at www.age-of-migration.com.

List of Abbreviations

ASEAN	Association of South East Asian Nations
BfA	*Bundesanstalt für Arbeit* (Federal Labour Office)
CAA	The Cuban Adjustment Act
CAP	Common Agricultural Policy
CEE	Central and Eastern Europe
CIS	Commonwealth of Independent States
DACA	Deferred Action for Childhood Arrivals
EC	European Community, predecessor of the EU
EEC	European Economic Community
ECOWAS	Economic Community of West African States
EU	European Union
EU25	The 25 member states of the EU from May 2004 to December 2006
EU27	The 27 member states of the EU from January 2007 to June 2013
EU28	The 28 member states of the EU since July 2013
FDI	Foreign Direct Investment
FRG	Federal Republic of Germany
GCC	Gulf Cooperation Council
GCM	Global Compact for Safe, Orderly and Regular Migration
GFMD	Global Forum on Migration and Development
IDPs	Internally displaced persons
ILO	International Labour Office
IMF	International Monetary Fund
INS	Immigration and Naturalization Service
IOM	International Organization for Migration
IRCA	Immigration Reform and Control Act
MERCOSUR	Mercado Común del Sur
NAFTA	North American Free Trade Agreement
NELM	New economics of labour migration
NGO	Non-Governmental Organization
ODA	Official development assistance
OECD	Organization for Economic Cooperation and Development
SGI	Société générale d'immigration (France)
UAE	United Arab Emirates
UK	United Kingdom
UKIP	United Kingdom Independence Party
UN	United Nations
UNDESA	United Nations Department of Economic and Social Affairs
UNDP	United Nations Development Programme

UNHCR	United Nations High Commissioner for Refugees
UNICEF	United Nations Children's Fund
UNRWA	United Nations Relief and Works Agency for Palestine Refugees in the Near East
WFP	World Food Programme
WTO	World Trade Organization

1 Introduction

Migration raises high hopes and deep fears: hopes for the *migrants* themselves, for whom migration often embodies the promise of a better future. At the same time, migration can be a dangerous undertaking, and every year thousands die in attempts to cross borders. Family and friends are often left behind in uncertainty. If a migrant fails to find a job or is expelled, it can mean the loss of all family savings. However, if successful, migration can mean a stable source of family income, decent housing, the ability to cure an illness, resources to set up a business and the opportunity for children to study.

In receiving societies, migration is equally met with ambiguity. Settler societies, nascent empires and bustling economies have generally welcomed immigrants, as they fill labour shortages, boost population growth and stimulate businesses and trade. However, particularly in times of economic crisis and conflict, immigrants are often the first to be blamed for problems, and face discrimination, *racism* and sometimes violence. This particularly applies to migrants who look, behave or believe differently than majority populations.

Migration and the resulting ethnic and racial diversity are amongst the most emotive subjects in contemporary societies. While global migration rates have remained relatively stable over the past half a century, the political salience of migration has increased. For origin societies, the departure of people raise concern about a *brain drain*, but it also creates the hope that the money and knowledge migrants obtain abroad can foster development back home. For destination societies, the arrival of migrant groups can fundamentally change the social, cultural and political fabric of societies, particularly in the longer run.

This became apparent during the US presidential election of 2016. During the election campaign, Donald J. Trump promised voters that he would build a border wall to prevent Mexican immigration. Trump stoked up fear of Mexican immigrants by saying "They are bringing drugs. They are bringing crime. They are rapists". At campaign rallies, Trump also tapped into anti-Islam sentiment, linking Muslim immigration to terrorism and expressing a desire for a Muslim registry. In January 2017 the Trump administration introduced a controversial ban on the entry of passport holders from seven predominantly Muslim countries. Although this policy met social and legal resistance, this reflected a campaign promise of a 'Muslim ban', based on reducing perceived security risks and curbing refugee migration to the US.

In Europe, the growing political salience of migration is reflected in the rise of anti-immigrant and anti-Islam parties and a subsequent move to the right of the entire political spectrum on migration and diversity issues (see Davis 2012). Parties like the Front National in France, the Lega Nord in Italy, and the Freedom Parties of Austria and the Netherlands have been established features of the political landscape for over two decades now. Although in most countries such parties have not been able to gain majorities, they are perceived as a major electoral threat by established parties, and their influence on debates is therefore larger than their voting share may suggest, as they tempt rival parties to adopt similar positions on immigration and diversity in order to retain voters.

In Europe, fears of mass migration came to a boiling point in 2015, when more than one million refugees and asylum seekers from Syria and elsewhere crossed the Mediterranean Sea. Concerns about immigration also played a central important role in the 2016 *Brexit* referendum, with 52 per cent of the voters supporting leaving the European Union (EU). In the lead-up to the vote, Nigel Farage's United Kingdom Independence Party (UKIP) stoked up fears of mass immigration. Anti-immigration parties from across Europe have argued that free intra-EU mobility undermines national sovereignty, and that abolishing free movement – or leaving the EU – is the only way to regain control over what is portrayed as unfettered migration.

The increasing ethnic and cultural diversity of immigrant-receiving societies creates dilemmas for societies and governments in finding ways to respond to these changes. Young people of immigrant background are protesting against discrimination and exclusion from the societies in which they had grown up – and often been born – and claim their right to equal opportunities in obtaining jobs, education and practising their religion. Some politicians and elements of the media shift the blame to the migrants themselves, claiming that they fail to integrate by deliberately maintaining distinct cultures and religions, and that immigration has become a threat to security and social cohesion.

Migration is an inherently divisive political issue. There is little evidence that migrants take away jobs or that migration is the fundamental cause of deteriorating working conditions, welfare provisions and public services. In fact, most evidence suggests that migration has positive impacts on overall growth, innovation and the vitality of economies and societies. For the most part, the growth of diversity and *transnationalism* is seen as a beneficial process, because it can help overcome the violence and destructiveness that characterized the era of nationalism – this was for instance a major motive behind the creation of the EU.

On the other hand, the benefits of migration are not equally distributed across members of destination societies. Businesses and high-income groups tend to reap the primary benefits from the labour and services delivered by migrants. Lower income groups, who have often experienced a deterioration of working conditions, real wages and social security as a result of economic deregulation and *globalization*, enjoy few, if any, direct economic benefits from migration, while they are often most directly confronted with the social and cultural change that migration is bringing about. Some politicians are therefore tempted to rally support by blaming the most vulnerable members of society – migrant workers and asylum seekers – for problems not of their making.

Beyond the usual allegations that migrants take away jobs and benefits, and undercut wages, migration has been increasingly been linked to security concerns. On the one hand, this reflects genuine worries about the involvement of small fractions of immigrant and immigrant-origin populations in extremist violence and terrorism. International migration is sometimes directly or indirectly linked to conflict. Events such as 9/11 (the 2001 attacks on the World Trade Center in New York and the Pentagon in Washington, DC) as well as the attacks by Islamist radicals in Europe and elsewhere involved immigrants or their offspring. This can lead to useful debates about how to counter such violence and how to prevent the marginalization and radicalization of immigrant populations.

It becomes more problematic when such attacks are used to stoke up xenophobia for political gain, by representing migration as a fundamental threat to the security, identity and cultural integrity of destination societies. This has led to the frequent portrayal of migrants and asylum seekers as criminals, rapists and terrorists, or as 'foreign hordes' plotting a takeover by bringing in foreign cultures and religions such as Islam. Anti-immigrant sentiment and migrant scapegoating by politicians and opinion makers have created a climate where far-right and racist attacks have flared up. It is in this political climate that extreme-right violence has proliferated, such as the attacks on New Zealand mosques in 2019, with a danger of provoking counterreactions by Islamist radicals. In this way, violence by white *supremacists* and Islamist radicals can feed into each other in a dangerous vicious circle.

This is by no means a uniquely Western phenomenon. As migration is globalizing, the cultural, social and economic changes that inevitably result from the arrival and settlement of large groups of migrants are deeply affecting societies around the world, often leading to polarization between social and economic groups opposing and favouring large-scale immigration. Oil economies in the Gulf region have the highest immigration rates in the world, and their economies have become structurally dependent on foreign labour. Lack of worker rights, prohibition of unions and fear of deportation leave migrant workers no choice other than to accept exploitative conditions. In Japan and Korea too, politicians often express fears of loss of ethnic homogeneity through immigration. The government of multiracial Malaysia has regularly blamed immigrants for crime and other social problems, and announced 'crackdowns' against irregular migrants whenever there are economic slowdowns. African countries have among the most restrictive immigration regimes in the world, and racist attacks and mass expulsions have regularly occurred in countries such as South Africa, Nigeria and Libya.

Migration is such a politically divisive issue because it is directly linked to issues of national identity and sovereignty. However, as migrants stay longer they become an increasingly permanent feature of societies, while economies have come to increasingly depend on continuous inflows of lower and higher migrant labour. Time and again, this has compelled governments to come to terms with such new realities by creating facilities for the legalization, integration and naturalization of migrants. As settlement takes place and migrants claim their place as new members of society, this is almost bound to create political tension and, sometimes, conflict.

Migration in an age of globalization

This book is about contemporary migrations and the way they are changing societies. The perspective is international: large-scale movements of people arise from processes of global integration. Migrations are not isolated phenomena: movements of commodities, capital and ideas almost always give rise to movements of people, and vice versa. Global cultural interchange, facilitated by improved transport and the proliferation of print and electronic media, can also increase migration aspirations by diffusing images and information about life and opportunities in other places. International migration therefore ranks as one of the most important factors in global change – both as a manifestation and a further cause of such change.

There are several reasons to expect the age of migration to endure: increasing levels of education and specialization combined with the growing complexity of labour markets will continue to generate demand for all sorts of lower- and higher-skilled migrant labour; inequalities in wealth and job opportunities will continue to motivate people to move in search of better living standards; while violent conflict and political oppression in some countries is likely to fuel future refugee movements.

Migration is not just – or even mainly – a reaction to difficult conditions at home: it is primarily driven by the search for better opportunities and preferred lifestyles elsewhere. Some migrants experience abuse or exploitation, but most benefit and are able to improve their long-term life perspectives through migrating. Conditions are sometimes tough for migrants but are often preferable to limited opportunities at home – otherwise migration would not continue.

According to the Population Division of the United Nations, the global number of international migrants (defined as people living outside their native country for at least a year) has grown from about 93 million in 1960 to 170 million in 2000 and from there further to an estimated 258 million in 2017. Although this seems a staggering increase, in relative terms international migration has remained remarkably stable, fluctuating around levels of around 3 per cent of the world population (see Figure 1.1).

These facts challenge popular narratives of rapidly accelerating migration, as the number of international migrants has grown at a roughly equal pace with overall global population since 1960. Some researchers have argued that this percentage was actually higher in the late nineteenth century, during the heyday of trans-Atlantic migration between Europe and America. For instance, the approximately 48 million Europeans that left the continent between 1846 and 1924 represented about 12 per cent of the European population in 1900. In the same period, about 17 million people left the British Isles, equal to 41 per cent of Britain's population in 1900 (Massey 1988: 381).

Although international migration has thus not increased in relative terms, falling costs of travel and infrastructure improvements have increased non-migratory forms of mobility such as tourism, business trips and commuting. Another important change has been increasing of long-distance migration between world-regions and the growing

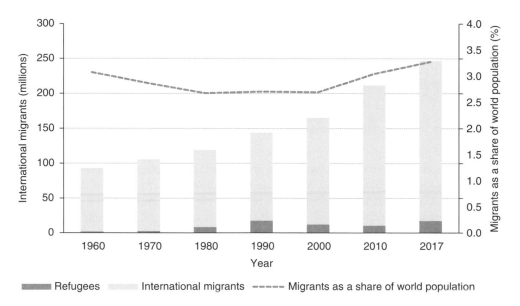

Figure 1.1   International migrants and refugees, as a percentage of world population, 1960–2017

Sources: Authors' calculations based on the *Global Bilateral Migration Database* (World Bank) (1960–1980 data) and UNPD (2017) (1990–2017 data)

share of non-Europeans in global migrant populations. These trends have increased the diversity of immigrant populations in terms of *ethnicity*, culture, religion, language and education.

The vast majority of people remain in their countries of birth. Only 3 per cent of the world population are international migrants, so 97 per cent stay at home. Yet many more people move within countries. Internal migration (often in the form or rural–urban movement) is far higher than international migration, especially in large and populous countries like China, India, Indonesia, Brazil and Nigeria. It is impossible to know exact numbers of internal migrants, but they are likely to represent at least 80 per cent of all migrants in the world (UNDP 2009). The number of internal migrants in China alone has been estimated at levels of 250 million (Li and Wang 2015), which roughly equal the worldwide number of international migrants. Although this book focuses on international migration, internal and international mobility are closely interlinked and driven by the same development processes, and the book will therefore refer to internal migration where relevant.

In addition, the impact of international migration is considerably larger than such percentages suggest, particularly in origin communities and in destination cities where migrants tend to concentrate. The departure of migrants has considerable consequences for origin communities. Money sent home by migrants allows families to significantly improve living standards, keep children at school and to invest in local businesses. Under unfavourable circumstances, however, the departure of people can also further undermine prospects for growth and change in remittance-dependent and migration-obsessed communities. In destination countries, migrants concentrate in certain urban areas, or in areas of intensive horticulture, where the social, economic and cultural impacts of migration can be life-changing, either positively or negatively. Migration thus affects not only the migrants themselves but also origin and destination societies as a whole. There can be few people in either industrial or developing countries today who do not have a personal experience of migration or its effects (Map 1.1).

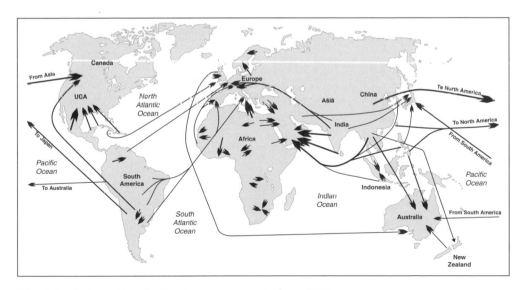

Map 1.1   International migratory movements from 1973

Note: The size of the arrowheads gives an approximate indication of the volume of flows. Exact figures are often unavailable

The growth and diversity of international migration

Reflecting broader patterns of globalization and labour market dynamics, international migrants have increasingly concentrated in particular regions and a relatively limited set of prime destination countries (see Czaika and de Haas 2014). Map 1.1 gives a very rough idea of the major migratory flows since 1973. Figure 1.2 shows the evolution of migrant populations in major world regions between 1960 and 2017. The data shows that migrant numbers in the industrial regions of Western Europe, North America and the Middle East have been growing fast. Immigrant populations been growing at a much slower pace in Africa, Eastern Europe, Central Asia and Latin America.

Figure 1.3 examines migrant populations as a share of the total population of different world regions. The relative magnitude of immigration is highest in North America and the Middle East, where in 2017 immigrants represent 15.3 per cent of the total population, and Western Europe, where this share is 12.7 per cent. By contrast, immigrants have represented a *declining* share of populations in Africa, Asia-Pacific and Latin America.

Table 1.1 displays the 25 most important countries of origin and destination in the world. With an estimated 44.5 million immigrants in 2017, the US is by far the most important migration destination in the world. Saudi Arabia, Germany and Russia come next with immigrant populations of around 12 million. Other important destinations are the UK, the United Arab Emirates (UAE), France, Canada and Australia. Many developing countries, including South Africa, Côte d'Ivoire, Thailand, Pakistan, Afghanistan and Iran are home to significant immigrant populations. Measured as share of their population, Gulf countries such as Saudi Arabia, UAE, Kuwait, Oman, as well as 'global city states' such as Singapore and Hong Kong have the highest immigration levels.

India, Mexico and Russia are the most important origin countries of migrants, followed by China, Bangladesh, Syria, Pakistan, Ukraine, the Philippines and the UK. In relative terms, the world's most prominent sources of migrant labour, such as Mexico, the Philippines, Morocco and Poland, have between 5 and 12 per cent of their population living abroad. Such percentages can be much higher in small countries, island states in countries affected by warfare, such as Somalia, Syria and amongst Palestinians.

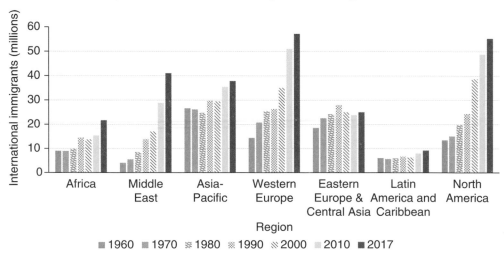

Figure 1.2   International immigrants by region, 1960–2017

Source: Calculations based on Global Bilateral Migration Database and United Nations Population Division

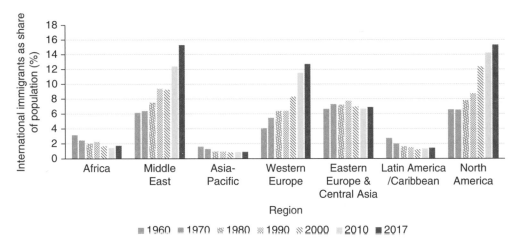

Figure 1.3 International immigrants as a share of the population by region, 1960–2017
Source: Calculations based on Global Bilateral Migration Database and United Nations Population Division

The data also shows that prominent immigration countries such as Russia, United Kingdom and Germany often have high numbers of their own citizens living abroad. This exemplifies that globalization and economic development often go along with high levels of immigration *and* emigration. It is difficult to crunch countries into categories of 'sending' and 'receiving' countries: many countries are both. And most people move *within* regions. These facts expose the flawed nature of popular views that represent contemporary global migration as a massive move or 'exodus' (Collier 2013) from the global 'South' to the global 'North'.

Some of those who move are *forced migrants*: people compelled to flee their homes and seek refuge elsewhere. The reasons for flight include political or ethnic violence or persecution, development projects like large dams, wildlife conservation projects, or natural disasters like floods, hurricanes or earthquakes. Although the vast majority of forced migrants move over short distances, some cross borders. There is no evidence of a long-term increase in refugee numbers, as levels rather go up and down depending on the outbreak and end of wars. The number of registered international refugees decreased from 17.8 million to 8.7 million between 1992 and 2005, mainly because of a decline in the number of conflicts. After 2005, the number rebounded to 17.2 million in 2016, primarily as a result of war in Syria, but it may also reflect the statistical inclusion of refugees, particularly in Africa, who were previously not accounted for.

Most forced migrants remain in the poorest areas of the world, either within their countries or in neighbouring countries. Many refugees prefer to stay close to home, and of those wishing to move farther, only a minority of refugees have the resources to achieve this goal. In fact, the biggest victims of war, oppression and environmental disaster are those who cannot flee and become trapped in life-endangering situations. Since 1990, refugees have counted for between 7 and 10 per cent of the global migrant population (Fransen and de Haas 2019; Hatton 2009) (see also Figure 1.1). In 2017 refugees represented about 10 per cent of the global migrant population, thus about 0.3 per cent of the total world population. These shares vary greatly across regions. The highest percentages can be found in the Africa and the Middle East, where about one quarter of international migrants are refugees. These percentages are lower and generally declining in most other world regions.

Table 1.1   Major immigration and emigration countries, 2017

Destination country	Immigrants	Share of population (%)	Origin country	Emigrants	Share of population (%)
US	44,525,900	13.7	India	16,588,000	1.2
Saudi Arabia	12,185,000	37.0	Mexico	12,965,000	10.0
Germany	12,165,000	14.8	Russian Fed.	10,636,000	7.4
Russia	11,652,000	8.1	China	9,962,000	0.7
UK	8,842,000	13.4	Bangladesh	7,500,000	4.6
United Arab Em.	8,313,000	88.4	Syria	6,864,000	37.6
France	7,903,000	12.2	Pakistan	5,979,000	3.0
Canada	7,861,000	21.5	Ukraine	5,942,000	13.4
Australia	7,036,000	28.8	Philippines	5,681,000	5.4
Spain	5,947,000	12.8	UK	4,921,000	7.4
Italy	5,907,000	10.0	Afghanistan	4,826,000	13.6
India	5,189,000	0.4	Poland	4,701,000	12.3
Ukraine	4,964,000	11.2	Indonesia	4,234,000	1.6
Turkey	4,882,000	6.0	Germany	4,208,000	5.1
South Africa	4,037,000	7.1	Kazakhstan	4,074,000	22.4
Kazakhstan	3,635,000	20.0	Palestine	3,804,000	77.3
Thailand	3,589,000	5.2	Romania	3,579,000	18.2
Pakistan	3,398,000	1.7	Turkey	3,419,000	4.2
Jordan	3,234,000	33.3	Egypt	3,413,000	3.5
Kuwait	3,123,000	75.5	Italy	3,029,000	5.1
Hong Kong	2,883,000	39.1	US	3,017,000	0.9
Malaysia	2,704,000	8.5	Morocco	2,899,000	8.1
Iran	2,699,000	3.3	Myanmar	2,895,000	5.4
Singapore	2,623,000	46.0	Colombia	2,736,000	5.6
Switzerland	2,506,000	29.6	Viet Nam	2,727,000	2.9
Japan	2,321,000	1.8	Rep. of Korea	2,478,000	4.9
Côte d'Ivoire	2,197,000	9.0	Portugal	2,267,000	21.9
Argentina	2,165,000	4.9	France	2,207,000	3.4
Oman	2,073,000	44.7	Uzbekistan	1,992,000	6.2
Netherlands	2,057,000	12.1	Somalia	1,988,000	13.5

Source: United Nations Population Division, 2017 estimates, including refugees

The number of 'refugee-producing' countries has shown a declining trend and refugees represent a relatively small share of all migrants. The idea of a global 'refugee crisis' has no basis in fact, at least not from a Western perspective. About 85 per cent of all refugees live in developing countries (UNHCR 2017). Countries such as Turkey, Pakistan, Lebanon,

Source: Getty Images/NurPhoto

Photo 1.1   Syrian refugees in Istanbul, Turkey, in August 2016

Iran, Ethiopia and Jordan currently host the largest refugee populations. In 2018, Turkey hosted more than 3.6 million Syrian refugees, equivalent to about 4.4 per cent of its 82 million population. In the same year, almost 1 million Syrian refugees lived in Lebanon, on a total population of 6 million. Western societies, by contrast, receive a comparatively low number of refugees, and current numbers are anything but unprecedented.

Trends and patterns of global migration

Throughout the world, long-standing migratory patterns are persisting in new forms, while new movements are developing in response to economic, political and cultural change, and violent conflicts. Since the end of the Second World War, the main trends and patterns of migration have been:

1. The *globalization of migration*: This is the tendency for more and more countries to be significantly affected by international migration. Immigration countries tend to receive migrants from an increasingly diverse array of origin countries, so that most immigration countries have entrants from a broad spectrum of economic, social and cultural backgrounds. A growing number of lower income countries have been entering the global migration stage, along with processes of social transformation and economic development that have motivated and enabled a growing number of people to migrate. While the share of international migrants as a percentage of the world population has remained rather constant, there has been a strong growth in inter-continental migration. Long-distance migration between major world regions has increased fast. While in 1960 movements between continents migration represented 38 per cent of global migration, this share had risen to 55 per cent in 2017. As part of

this process, migrants from an increasingly diverse pool of origin countries have been concentrating in a relatively small pool of prime destination countries where economic power and employment opportunities are concentrated (see Czaika and de Haas 2014). Some researchers have claimed that this has led to unprecedented patterns of 'superdiversity' in destination cities (see Vertovec 2007).

2. The *changing direction of dominant migration flows*: Since their 'discovery' of the Americas, Europeans have been moving outward to conquer, colonize and settle in foreign lands. As a result of decolonization, rapid economic growth and demographic changes, these patterns were reversed after World War II. Europe transformed from a continent of colonizers and emigrants to a destination region for an increasingly diverse array of origin countries. As part of this 'global migration reversal', Europeans represent a declining share of immigrants in classical immigration countries such as the US, Canada, Australia and New Zealand. In 1960 more than three quarters of all international migrants who moved to another world region were Europeans. This proportion had shrunk to 22 per cent by 2017. By contrast, the share of Latin Americans, and in particular Asians, in long-distance, inter-regional migration has increased.

3. *The emergence of new migration destinations*. Particularly since the 1973 Oil Shock, the oil-rich Gulf region emerged as a global magnet for migrant workers from Asia and also Africa, hosting about 28 million workers in 2017. In absolute numbers, the Gulf is now the third most important migration destination after North America and Western Europe. In addition, new migration destinations also emerged in the industrial cores of East Asia (Japan, South Korea) while industrializing countries in Southeast Asia (Thailand, Malaysia, Singapore) have attracted increasing numbers of mainly regional migrants from countries such as the Philippines, Indonesia, Vietnam, Myanmar and Bangladesh.

4. The *proliferation of migration transitions*: This occurs when traditional lands of emigration become lands of immigration. Growing transit migration is often the prelude to becoming predominantly immigration countries. States as diverse as Poland, Spain, Morocco, Mexico, the Dominican Republic, Turkey and South Korea have been experiencing various stages and forms of a migration transition. But many other countries, for example in Latin America and sub-Saharan Africa, have experienced *reverse migration transitions* as they transformed from immigration to emigration countries and the diversity of their population decreases rather than increases.

5. The *feminization of labour migration*: In the past many labour migrations were male-dominated, and women were often dealt with under the category of family reunion, even if they did take up employment. Although the share of female migrants as a share of total migration has remained stable at levels of around 46 per cent for decades (de Haas *et al.* 2019a), the feminization of migration primarily pertains to the increasing participation of women in labour migration. Today women workers form the majority in movements as diverse as those of Cape Verdeans to Italy, Ecuadorians to Spain, Ethiopians to the Middle East, Thai to Japan, Myanmar to Thailand, Indonesians to Malaysia, and Filipinas to the Middle East and Europe. However, because of the high presence of female migrant workers in largely *informal sectors* such as personal care and cleaning, women are often still a less visible part of the workforce compared to men.

6. The *politicization and securitization of migration*: Domestic politics, bilateral and regional relationships and national security policies of *states* around the world are increasingly affected by international migration. Communities of migrants and their

Source: Kerilyn Schewel

Photo 1.2   A return migrant's store, paid with money she earned in Saudi Arabia. Ziway, Ethiopia, 2018

descendants are demanding a place in destination societies, frequently sparking tensions and intense debate on *citizenship*, diversity and identity. Particularly since the end of the Cold War, this has also gone along with the securitization of migration, which is the tendency by some politicians and media to portray migration as a fundamental threat to the security and cultural integrity of destination societies. On the other hand, as part of a 'global race for talent', governments around the world have tried to facilitate the immigration of skilled workers, investors and students. The arrival of asylum seekers typically sparks debates amongst humanitarian and religious groups advocating their need for protections, and groups who may see their large-scale arrival as a threat to native workers, security and the welfare state.

The challenges of international migration

Migration has gained increasing political salience over the past decades. That is why we have called this book *The Age of Migration*. While movements of people across borders have shaped societies since time immemorial, what is distinctive in recent decades is thus their centrality to (1) domestic and (2) international politics. This does not imply that migration is something new – indeed, human beings have always moved in search of new opportunities or to escape conflict and oppression. However, migration took on a new, more global, character with the beginnings of European expansion from the sixteenth century and, particularly, the Industrial Revolution from the nineteenth century, which set in motion a massive transfer of population from rural to urban areas both within and across borders.

Since the 1950s Europe has transformed from a source of colonizers and settlers to a global destination region for migrants from an increasingly diverse array of non-Western origin countries. Decreasing migration out of Europe went along with an increasing share of Asians, Latin Americans and, to a lesser extent, Africans in global migrant populations. This would have fundamental repercussions for migration to traditional countries of European settlement in the Americas, Australia and New Zealand. The increasing share of non-European, non-white and often non-Christian migrants in fast-growing immigrant populations in Europe and North America has sparked significant unease and fierce political debate between liberals welcoming increasing diversity, and conservative voices claiming that 'too much' diversity forms a threat to the cultural integrity and social cohesion of destination societies. Likewise, migration has also become a politically contentious issue in new migrant destinations in the Gulf, Africa and Asia.

While global migration rates have not increased, more migrants move over large distances. Furthermore communication and travel have become easier as a result of new transport and communication technologies. This has enabled migrants to remain in almost constant touch with families and friends back home and to travel back and forth more often, and to maintain multiple and transnational identities, which can challenge traditional ideas of national identity. International migration has thus become a central dynamic within globalization. While people have always moved, an increasing number of low- and middle-income countries have become integrated in global migration systems centred around old destinations in North America and Russia and more recent destinations in Western Europe, the Gulf and East Asia. This globalization of migration has confronted societies with unprecedented levels of diversity.

Immigration poses two major challenges for domestic and international politics. First, the arrival and settlement of migrants and the resulting increase in diversity have challenged dominant concepts of nation states and have sparked intense debate about identity, belonging and integration. The second defining feature of the age of migration is the challenge that migration poses to the sovereignty of states, specifically to their ability to regulate movements of people across their borders in a globalizing world. The central tension is the following: while more effective regulation of migration would benefit from improved international cooperation, governments are often unwilling to give up national sovereignty on vital issues around migration and citizenship.

International migration in global governance

States have always struggled to control migration, but these challenges seem particularly significant in an era where broad trends of globalization and economic deregulation seem to run counter to the wish to regulate the arrival and stay of foreigners. While most governments have abolished the exit controls of the past, efforts to regulate *immigration* are at an all-time high and involve intensive bilateral, regional and international diplomacy. Although the majority of migrants move within the law, a significant share of migrants cross borders in irregular ways. Paradoxically, irregular migration is often a *consequence* of tighter control measures, which have blocked earlier forms of spontaneous and more circular mobility.

The experiences with large-scale migration such as from Mexico to the US and Turkey and Morocco to the EU have shown that ill-conceived migration restrictions can be counterproductive by interrupting circulation, encouraging permanent settlement and encouraging undocumented migration (de Haas *et al.* 2019; Massey and Pren 2012). This exposes the significant challenge that the effective regulation of migration represents for governments. Policy trends seem contradictory: on the one hand, politicians cling to national sovereignty, with such slogans as 'British jobs for British workers'. On the other hand, politicians are sensitive to businesses lobbying to let more migrant workers in or to turn a blind eye to illegal employment practices, and often have limited legal and practical tools to curb immigration of family members and refugees.

The complexity and fragmentation of power require governments to cooperate with other organizations and institutions, both public and private, foreign and domestic. An important manifestation of global governance is the expansion of consultative processes within regional unions such as the EU, the Economic Community of West African States (ECOWAS), the Association of South East Asian Nations (ASEAN) or the South-American Mercado Común del Sur (Mercosur). The implementation of free travel protocols in such unions has boosted intra-regional mobility and commerce, although the adoption and implementation of the right of establishment (including work) and residency has not been achieved except for the EU, where it also remains an important point of political contention.

Between 1945 and the 1980s, many governments in Western Europe and North America did not see international migration as a central political issue. In developing countries, the picture was more mixed. For labour-exporting countries such as the Philippines, Egypt and Morocco, organized emigration was seen as a key political 'safety valve' to alleviate discontent, decrease poverty and to generate remittances. In most developing countries, however, governments were often more concerned about internal migration from rural areas to large cities.

This situation began to change in the late 1980s, when migration governance started to become an increasingly prominent topic in international fora. The Organization for Economic Cooperation and Development (OECD) – which regroups the wealthiest democracies in the world – convened its first international conference on international migration in 1986 (OECD 1987). This was partly related to processes of regional economic integration, which also had a migration component. As most European Community (EC, the predecessor of the EU) countries started to remove their internal boundaries with the signature of the *Schengen* Agreement in 1985 and its full implementation in 1995, they became increasingly concerned about the joint control of external borders as well as the alignment of visa and immigration policies.

The adoption of the 1990 Convention on the Rights of Migrant Workers and Their Families by the UN General Assembly brought into sharp relief global tensions and differences surrounding international migration. It did not come into force until 2003, and major immigration countries refused to sign the convention. By October 2018 it had been ratified by just 54 of the UN's 193 states, virtually all of them countries of emigration. Many emigration countries, including India, China, Brazil and Ethiopia have not signed the Convention.

The opposed interests between major origin and destination countries have been a central obstacle standing in the way of achieving improved international migration governance. Destination countries are generally unwilling to concede national sovereignty that would force them to adopt more liberal immigration regimes. Governments of origin countries have often limited success, or interest in, defending the rights of migrant workers. This can be partly because they have a weak negotiation positions vis-à-vis more powerful destination countries, because they fear migrants' political activism from abroad, or simply because ruling elites are not very much concerned about the significant exploitation, extortion and abuse by state agents and employers that migrant workers often have to endure.

Globalization has coincided with the strengthening of global institutions: the World Trade Organization (WTO) for trade, the International Monetary Fund (IMF) for finance, the World Bank for economic development, and so on. But the will to cooperate has not been as strong in the migration field. There are international bodies with specific tasks – such as the United Nations High Commissioner for Refugees (UNHCR) for refugees, the International Labour Office (ILO) and the International Organization for Migration (IOM) – but no institution has the responsibility, capacity and authority to bring about significant change.

The UN General Assembly held its first High-Level Dialogue on International Migration and Development in 2006. The Secretary General's report on this meeting recommended a forum for UN member states to discuss migration and development issues. The Global Forum on Migration and Development (GMFD) has met annually since, although its role has been purely advisory, and it is hard to see any concrete results of these meetings. In 2016 the General Assembly of the UN decided to develop a Global Compact for Safe, Orderly and Regular Migration (GCM), which was adopted in 2018 in Marrakech, Morocco. Although the Global Compact is non-legally binding, several countries, including the US and Hungary, pulled out of the GCM before it was even adopted. Despite the lip service paid to lofty goals of improved global migration governance, the key obstacle for achieving a higher degree of effective international collaboration on migration remains that governments are generally reluctant to surrender national sovereignty on migration controls to supra-national bodies.

Aims and argument of the book

The Age of Migration sets out to provide an understanding of the contemporary global dynamics of migration and of the consequences for societies, migrants and non-migrants everywhere. It provides an interdisciplinary introduction to the subject of international migration and the emergence of increasingly diverse societies. This will help readers to put more detailed accounts of specific *migratory processes* in context. The book argues that international migration is a central dynamic in globalization and is recasting states and societies in distinctive and powerful ways.

Building upon the latest insights from research and cutting-edge data, this book offers information on

- Long- and short-term trends and patterns of global migration;
- Theories on the causes and continuation of migration;
- Impacts of migration on destination and origin societies;
- Migration experiences and migrant identities;
- The implications of migration for states and politics;
- The evolution and effectiveness of migration policies.

This book provides a synthesis of theoretical and empirical insights from all disciplines studying migration – ranging from anthropology and history to sociology, geography, political science and economics. *The Age of Migration* aims to present these insights in ways that are accessible to readers from diverse academic or professional backgrounds, and will accentuate the contribution of research from all disciplines to our common understanding of migration. This is based on a conviction that different disciplines and theories provide different views on migration, which are often complementary and help us to develop a richer and more critical view by helping to look at the same issues from different angles. We therefore hope that the evidence presented in the book provides much-needed nuance in a polarized debate in which, all too often, 'pro-' and 'anti-' migration voices exaggerate the downsides and upsides of migration.

We can summarize the main arguments of this book as follows:

- Migration is an intrinsic – and therefore inevitable – part of broader processes of global change and development;
- Labour demand in destination societies is the main driver of international migration;
- While states play a key role in initiating migration, once set in motion, migration processes tend to gain their own momentum;
- Migration is a partly autonomous process that will almost inevitably go along with some degree of permanent settlement;
- Migration is neither the cause of, nor a panacea for, structural socioeconomic problems in destination and origin societies;
- While governments have a legitimate desire to control migration, ill-considered migration policies often have counterproductive effects.

Our first objective is to describe and explain contemporary international migration. The second objective is to explain how migrant settlement is bringing about increased ethnic diversity and how it affects broader social, cultural and political change in destination *and* origin societies. Understanding these changes is the precondition for effective political action to deal with problems, tensions *and* opportunities linked to migration and ethnic diversity. The third objective is to link the two analyses, by examining the complex interactions between migration and broader processes of change in origin and destination societies. There are large bodies of empirical and theoretical work on both themes. However, the two are often inadequately linked. The linkages can best be understood by analysing the migratory process in its totality.

Structure of the book

The Age of Migration is structured as follows. A first group of chapters (2–5) provide the theoretical and historical background necessary to understand contemporary global trends. Chapter 2 examines the main categories and typologies that researchers, the media and politicians use to describe and analyse migration. It will provide definitions and familiarize readers with important terms that will appear throughout the text, and it will highlight the dangers of uncritically adopting categories and discourses of migration that provide distorted and biased views of migration.

Chapter 3 will discuss theories and concepts that are useful to explain the causes and continuation of migration. It will show the need to understand migration as an intrinsic part of broader processes of development and social transformation. It will argue that a combination of insights from various theories contribute to obtain a richer, more profound, understanding of migration processes. For instance, the analysis of theories will help to understand the paradox that development in poor countries often leads to more, *instead of less*, migration and why middle-income countries such as Mexico, Turkey, Morocco and the Philippines tend to have the highest emigration levels. Labour market theories will help us to understand why low-skilled migration continues even in times of high domestic unemployment. Other theories will show the need to analyse migration as a partly autonomous social process – because, once started, migration tends to gain its own momentum through the formation of *migrant networks* and other feedback processes. This helps to explain why migration often continues despite governments' attempts to stop more people from coming in, and why perhaps well-intended, but ill-considered policies can become counterproductive.

Chapter 4 focuses on the social and political issues in destination countries arising from ethnic and cultural diversity resulting from the arrival of migrants. The chapter will argue how immigration often results in the establishment of (semi-) permanent migrant communities, despite frequent attempts of governments to prevent this. The chapter will discuss the various modes of migrant incorporation or integration, ranging from assimilation to systematic racism, segregation and minority formation, or between forms of multiculturalism. It will show how experiences of incorporation are dependent on factors such as class, gender and state ideologies.

Chapter 5 describes the history of international migration from early modern times until 1945. This chapter illustrates the necessity of adopting a long-term view in order to achieve a fundamental understanding of contemporary migration processes, to understand what is new, and what is rather a continuation of previous trends and patterns. The chapter will particularly raise awareness about the key roles that states have played in shaping contemporary world migration such as through warfare, colonization, slavery and recruitment of workers. Such patterns are often reproduced until the present day because of the cultural, commercial and social links created by colonial and imperial interventions.

A second group of chapters (6–9) provide a more detailed analysis of migration trends in the different world regions since 1945. Chapter 6 will focus on migration in Europe, including migration trends in the former Soviet Union and its successor countries. Chapter 7 analyses migration trends in North, Middle and South America and the Caribbean. Chapter 8 focuses on migration in the Asia-Pacific region, and Chapter 9 will analyse migration trends in Africa and the Middle East. Because of the limited space, these chapters are inevitably broad-brush. The aim is not to be exhaustive, but to analyse how general trends in the global political economy are manifested in different migratory outcomes in various regions and countries. This is based on the conviction that, notwithstanding the complexity and diversity of migration, contemporary migration is shaped by similar forces of political and economic transformation affecting societies and people around the world.

To reflect this argument, and to facilitate comparison, the regional chapters will apply a similar periodization. Four periods will be distinguished, signifying main shifts in the international political economy:

- The period between 1945 and 1973 was characterized by strong economic growth, decolonization and conflict related to the *Cold War*. It is also known as the second era of globalization of trade and finance. Different from the first era of globalization between 1870 and 1914, this period was characterized by high government intervention in economic systems and labour markets as well as strong controls of capital flows. In Western countries migration regimes were liberalized, coinciding with high levels of post-colonial and (predominantly Mediterranean) labour migration to Western Europe, increasing migration from Mexico and the Caribbean to North America as well as conflict-related migrations in Africa and Asia partly linked to the Cold War and post-colonial state formation. While some developing countries embarked upon 'labour export' policies, others attempted to prevent a brain drain. Authoritarianism and economic stagnation led to the demise of South America as a global migration destination;

- The 1973 Oil Shock marked a period of global economic restructuring and the emergence of a new international division of labour, which would last until 1989. The economic boom following the Oil Shock heralded the rise of the Gulf region as a major destination for migrant workers from the Middle East, Asia and Africa. Economic stagnation, austerity and relocation of industrial production to low-wage countries, particularly in Asia, temporarily decreased demand for lower-skilled workers in Western countries, although migration continued partly through family migration. Rapid economic growth in East Asian countries such as Japan, South Korea and Malaysia started to attract migrants from poorer Asian countries, while Asian migration to the Gulf and North America increased fast;

- The fall of the Berlin Wall 1989 and the subsequent collapse of Communist regimes marked the start of the third era of 'neoliberal' globalization. The end of the Cold War heralded a period of market triumphalism, economic deregulation and accelerated globalization of trade and finance. While the political turmoil around the Cold War and the concomitant disintegration of the former Soviet Union and destabilization of 'strong states' in Africa and Asia led to a temporary spike in refugee migration, the removal of exit restrictions boosted East-West labour migration and circulation from and between former Communist countries. A combination of economic growth, demographic ageing and labour market deregulation increased the demand for lower- and high-skilled migrant labour in in North America, Europe and East Asia, while the Gulf consolidated its position as a prime migration destination for migrant workers from South and Southeast Asia;

- Since 2008, the Great Recession heralded a period of economic uncertainty, fiscal crises and political polarization, particularly in Europe and North America, accentuating the rise of new economic powers in Asia and elsewhere. Intensified global competition for skilled workers and fee-paying students coincided with a diffusion of point-systems to make immigration serve economic needs, fuelling migration from an increasingly diverse array of origin countries. Fast-growing East and Southeast Asian and Gulf economies consolidated their position as global migration destinations while countries like Turkey, Russia, South Africa and Brazil strengthened their position as regional migration destinations. Popular discontent with the consequences of neoliberal globalization and growing inequality was mobilized by populist leaders stoking up xenophobia. This contributed to the Brexit vote and election of Trump in 2016 but also to growing opposition against free trade and calls for more taxation of the rich and increased controls on capital flows (see Piketty 2014).

Each of the chapters will evaluate the repercussions of these global shifts for different world regions. While the regions differ significantly, the analyses will also stress instances where patterns are repeated, and reveal the extent to which the challenges posed by migration to the sovereignty of states are often rather similar across various historical and regional contexts.

A third group of chapters (10–14) elaborate on a number of key themes and challenges represented by migration. Chapter 10 shows how migration has been central to processes of modern state formation while simultaneously being a challenge for national identity and state sovereignty. It explains the complex fields of national and international politics out of which immigration and emigration policies arise. It will also discuss attempts to achieve improved international governance of migration as well as refugee movements.

Chapter 11 analyses how migration policies have evolved over the post-1945 era, and analyses their effectiveness. It will give an overview of the migration policy toolbox: the various instruments governments use to control the arrival, integration and return of migrants. The chapter challenges the idea that migration policies have become more restrictive despite rhetoric suggesting the contrary, and shows how modern migration policies have increasingly been about selection rather than numbers. The chapter will discuss the effectiveness of migration policies, the unintended effects of migration restrictions, and the significant trade-off and dilemmas this creates for policy makers and societies.

Chapter 12 considers the labour market position of migrant workers and the meaning of migration for the economies of destination countries. The analysis illustrates the large extent to which migration is driven by the intrinsic and therefore largely inevitable demand for migrant labour in industrial and post-industrial societies. The analysis shows that migrants have become a structural feature of the lower- and higher-skilled workforce in high- and middle-income countries around the world. It highlights the continued relevance of recruitment as a driving and facilitating force for international migration. The chapter also discusses the key role of migration in labour market restructuring and the emergence of neoliberal economies based on employment practices such as sub-contracting, temporary employment and informal-sector work.

Chapter 13 examines the social position of immigrants in destination societies, looking at such factors as legal status, social policy, formation of ethnic communities, racism, citizenship and national identity. It will address crucial questions around immigrant integration and the factors explaining why some migrations lead to the formation of ethnic minorities suffering from various forms of isolation, racism and marginalization, and why many other migrants achieve high levels of socio-economic mobility, often (but not always) combined with rapid cultural assimilation and high levels of intermarriage.

Chapter 14 discusses how emigration affects processes of development and social transformation in origin countries. In particular, this chapter will examine the debate on the development impacts of migration, such as between those arguing that migration leads to a *brain drain* versus those arguing that migration can be a *brain gain* through remittances and the knowledge that migrants send back. It will examine the factors and conditions that explain why the development impact of migration seems more negative in some cases, and more positive in others. The chapter will show that although it is unlikely to solve more structural development problems, migration has considerable benefits for migrants and their families that provide a strong motivating force for people to migrate.

Chapter 15 sums up the arguments of the book, reviews current trends in global migration and speculates on the future of global migration. It discusses the dilemmas faced by governments and people in attempting to find appropriate responses to the challenges of an increasingly mobile world, and point to some of the major obstacles blocking the way to better international cooperation on migration issues.

Guide to Further Reading

There are too many books on international migration to list here. Many important works are referred to in the guide to further reading sections found at the end of each chapter. A wide range of relevant literature is listed in the Bibliography. *Migration and Development* (1997) by Ronald Skeldon is an excellent introduction for readers aiming to develop an understanding of migration as part of broader development processes. *Worlds in Motion* (1998) by Douglas Massey and his colleagues gives a comprehensive overview of migration theories and migration trends around the world. Major books that focus on particular regions will be mentioned in other chapters.

Important information on all aspects of international migration is provided by several specialized journals. *International Migration Review* (New York: Center for Migration Studies) was established in 1964 and provides excellent comparative information. *International Migration* (IOM, Geneva) is also a valuable comparative source. *Population and Development Review* is a prominent journal on population studies with many contributions on migration, while *Demography* is a prominent interdisciplinary journal publishing predominantly quantitative analyses of migration. Some journals, which formerly concentrated on Europe, are becoming more global in focus. These include the *Journal of Ethnic and Migration Studies*, the *Revue Européenne des Migrations Internationales*, *Race and Class* and *Ethnic and Racial Studies*. A journal concerned with transnational issues is *Global Networks*. Recently many new migration journals have appeared, including *Migration Studies and Comparative Migration Studies*.

Several international organizations provide comparative information on migrations. The most useful is the OECD's annual *International Migration Outlook*. The IOM has published its *World Migration Report* annually since 2000.

Several internet sites are concerned with issues of migration and ethnic diversity. A few of the most significant ones are listed here. Since they are in turn linked with many other websites, this list should provide a starting point for further exploration:

African Centre for Migration & Society (ACMS), University of the Witwatersrand: www.wits.ac.za/acms

Center for Migration Studies, New York: www.cmsny.org

Centre for Migration Studies, University of Ghana: www.cms.ug.edu.gh

Centre on Migration, Policy and Society (COMPAS), University of Oxford: www.compas.ox.ac.uk

Gulf Labour Markets, Migration, and Population (GLMM) Programme: www.gulfmigration.org

International Centre for Migration Policy Development (ICMPD), Vienna: www.icmpd.org

International Migration Institute (IMI), University of Amsterdam: www.migrationinstitute.org

International Migration, Integration and Social Cohesion (IMISCOE): www.imiscoe.org

▶

International Network on Migration and Development, Autonomous University
    of Zacatecas: www.migracionydesarrollo.org
International Organization for Migration (IOM): www.iom.int
Migration Policy Institute (MPI), Washington DC: www.migrationpolicy.org
Migration Matters: www.migrationmatters.me
Migration Observatory, University of Oxford: www.migrationobservatory.ox.ac.uk
Migration Policy Centre (MPC), European University Institute, Florence:
    www.migrationpolicycentre.eu
Migration Research Center at Koç University (MiReKoc), Istanbul:
    www.mirekoc.ku.edu.tr
Refugee Studies Centre (RSC), University of Oxford: www.rsc.ox.ac.uk
Sussex Centre for Migration Research: www.sussex.ac.uk/migration
The Hugo Observatory on Environment, Migration, Politics:
    www.labos.ulg.ac.be/hugo
United Nations High Commissioner for Refugees (UNHCR): www.unhcr.org
United Nations Population Division:
    www.un.org/en/development/desa/population/migration

Extra resources can be found at: **www.age-of-migration.com**

2 Categories of Migration

Categories are essential tools for understanding migration. They help us to make sense of complex social processes, to see patterns and to compare. However, the uncritical use of categories can also be a source of confusion and distortion. Common misperceptions about migration start with the language and categories that politicians, media and researchers use to describe different types of migration and migrants. A particular problem is the uncritical adoption of legal and policy categories to describe migration. These categories are not always meaningful, and can stand in the way of achieving a better understanding of migration processes. Language matters. After all, categories and terms shape the way we perceive the world around us. For instance, it matters whether the same group of people is described as 'illegal aliens', 'irregular migrants' or 'undocumented workers'.

It is important to make a distinction between (1) analytical categories mainly by researchers and some policy makers, (2) administrative categories used by governments and states, and (3) discursive categories mainly used by politicians and media. Often, the distinction is not clear, and this is an important source of confusion. In many ways, states have shaped the way people talk about migration. This is evident in the term migration itself. Migration can be defined as a *change of residency across administrative borders*. Whether these borders are municipalities, provinces, departments, federal or national states, the feature of migration is that people move across such borders to live in another administrative unit. The key issue is that states draw such borders, and that the ways in which states and their judicial institutions define and categorize migration matters a great deal to the daily lives of migrants, the rights they can obtain as well as their participation in the social and economic life of destination societies.

This is why attempts to define what is a migrant from an 'objective' or non-state perspective are somehow elusive, as the very concept of migration largely reflects the need of modern administrative systems to define who is resident, who has to pay taxes, who has the right to remain, and, last but not least, who can become a citizen. Because the crossing of (internal and international) administrative borders has real implications for people's lives, it therefore seems sensible to use this as a basis for defining migration. It matters a lot whether migrants cross international borders or not, or whether migrants have a visa or residence permit or not.

Yet our understanding of migration becomes obstructed if we uncritically view migration solely through political state-defined categories, without evaluating their value in migrants' daily lives, perceptions and plans. For instance, governments often persist in framing lower-skilled immigrants as guestworker or 'temporary workers' even if they have been living and working in destination countries for several decades. Categories of migrants can implicitly convey judgements linked to nationality, race, profession and class, as well as motives, trustworthiness, rights and the overall 'desirability' of migrants. For instance, in Europe lower-skilled workers are seen as 'migrant workers', but higher-skilled workers often designated as 'expats'. Although higher-skilled workers are migrants, many observers, and also higher-skilled workers themselves, may resist the migrant label.

In the same vein, it matters whether politicians and media designate people fleeing warn-torn countries as 'refugees' and 'asylum seekers', or, alternatively, as 'illegal migrants', potential 'terrorists', 'criminals' or even 'rapists'.

Besides categories describing migrants, caution is also warranted with regard to the categories used to describe and analyse migration *processes*. In recent years, there has been a tendency to cast migration from Africa to Europe – in an alleged response to poverty, violence, population growth and climate change – in terms of 'flows', 'waves' and 'tides' (see also Christian Aid 2007). This has fuelled perceptions that this phenom-enon is taking on the proportions of an exodus, that have, however, no basis in evidence. Governments and organizations can use categories to reframe migration for political pur-poses. An example is the rebranding of the hundreds of thousands of Sudanese refugees and other migrants living in Egypt as 'transit migrants' (Roman 2006). Such a process of relabelling is inherently political, because it implies that *de facto* settlers are only there temporarily and that they are somehow expected to leave.

These examples show the extent to which migration categories can become subject to political games. Neither language nor categories are neutral, and therefore deserve to be assessed critically with regard to their usefulness to describe migrants and migration processes. The ways such discursive categories shape our understanding in implicit but extremely powerful ways matters greatly for the ways governments treats migrants, but the appropriate and critical use of categories is also essential to achieve a better, non-ideological understanding of migration processes. To this purpose, the remainder of this chapter will review common categories of migrants and migration, and assess their usefulness for understanding of migration processes. The purpose is not to come to a final 'choice' of terminology, but to understand advantages and disadvantages of different categories, and to define key terms as much as possible (Figure 2.1).

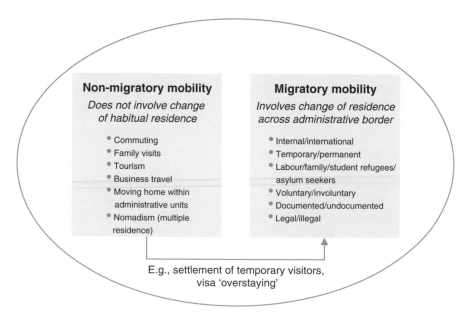

Figure 2.1   Different forms of migratory and non-migratory human mobility

Defining migration and human mobility

Migration has only become a meaningful term since humanity switched from the itinerant and nomadic lifestyles of hunter–gatherers to agrarian lifestyles starting around 12,500–10,000 years ago in various world regions including Mesopotamia, Ethiopia, West Africa, Mesoamerica and New Guinea. The domestication of plants and the 'invention of agriculture' allowed people to settle down permanently in a process also known as *sedentarization*. Before this process, people had no fixed residence as hunter–gatherer lifestyles were inherently mobile. It was only when people started to settle down in permanent settlements that the concept of migration became relevant, as migration implies a change in residence.

The invention of agriculture (also known as the 'Neolithic Revolution') also allowed the emergence of states, which had a vital interest in controlling resident populations for political, military and economic purposes, particularly to extract benefits from populations through taxation, slavery and other forms of bonded labour and to prevent rebellion. For instance, the Roman Empire saw a strategic interest in censuses to control and tax people. For modern national states with their fixed borders, controlling population and defining migration has become even more important, partly because of the increased social, economic and political rights associated with modern citizenship.

It is important to differentiate between *migratory* and *non-migratory* mobility in the context of sedentary lifestyles and modern statehood (see Figure 2.1). *Human mobility* refers to all forms of human movement outside of their direct living place and social environment (house, village or neighbourhood), irrespective of the distance and time-period implied, or whether this involves the crossing of administrative borders. Non-migratory and migratory mobility are subsets of human mobility. *Migratory mobility* equals *migration* and involves the change of residence across administrative borders. The movement within the same neighbourhood, town or municipality thus does not qualify as migration from an administrative point of view, although from a sociological point of view we should not ignore that, for the people involved, even the change of a block or street may have great sociological or cultural impact, for instance if it means living amongst a different ethnic, religious or class group. In Northern Ireland, it can mean a change from a Catholic to a Protestant environment, or in the case of Belgium, from a French- to a Flemish-speaking environment.

Non-migratory mobility comprises all forms of mobility that do not qualify as migration. Major categories include commuting, shopping and tourism, family visits (including by migrants) and business-related mobility. The distinction between migratory and non-migratory mobility is essential in understanding how industrialization and globalization have affected human mobility. While global levels of migration (as a proportion of the world population) have not significantly increased over recent decades, non-migratory mobility, most notably commuting, tourism and business travel, have increased rapidly. Improved infrastructure and transport have enlarged the distances over which people can commute on a daily basis. As a form of non-migratory mobility, commuting has thus potentially decreased the need to migrate (see Zelinsky 1971). Also, the outsourcing of industries and services to low-wage countries may increase the need for various kind of business trips but have potentially decreased the demand for low-skilled migrant workers.

In practice, governments, statistical agencies and relevant international organizations such as the United Nations Population Division (which compiles global migration data)

Source: Getty Images/Dan Kitwood

Photo 2.1　A view of the 'Peace Wall' separating the Catholic and Protestant communities in March 2017, Belfast, United Kingdom

use various definitions of migration. The main difference relates to the amount of time a migrant has to spend in the new destination to count as a migrant. Most governments put this limit somewhere between 3 and 12 months. For instance, tourist visas in many countries, including in the European Schengen zone, are valid for a maximum period of three months. If people want to stay longer, they usually have to apply for a visa extension or a residence permit, after which their status changes to migrants. In situations where it is difficult to obtain a visa extension or a residence or work permit, some *de facto* migrants may prefer to prolong their stay as tourists, through a so-called *visa run* of leaving and re-entering their de facto country of residence, often on the same day.

In countries using population registers, such as in much of continental Europe, people are often counted as migrants the moment or a few months after they register as such. In countries that mainly rely on censuses to keep track of their populations, such as the UK and the US, the administrative visibility of migrants often takes longer to materialize. However, most governments will consider people migrants if they have stayed for longer than one year.

Despite differences in measurement and time criteria applied, from a conceptual point of view the defining features of migration are the (1) change in residence and the (2) crossing of an administrative border. This border can be an internal border, such as between municipalities, counties, provinces or states, or international, between states. When migrants move within countries, they are called *internal migrants* or, alternatively, the term *domestic migrants* is used. If migration involves the crossing of a border between states, it is usually seen as *international migration*. The UN Population Division defines an international migrant as 'any person who changes his or her country of usual residence', irrespective of the reason for migration or the legal status of migrants. It also makes a distinction between *short-term* or *temporary migration*, covering movements with a duration between 3 and 12 months, and *long-term* or *permanent migration*, referring to a change of country of residence for a duration of one year or more. However, different governments attribute very different meanings to permanent and temporary migration.

Objectively speaking, a migrant is a person who is living in another country, state, province or municipality than where she or he was born. This includes students, 'expats' and other higher-skilled workers. Therefore, terms like second or third generation 'migrants' to officially designate the descendants of migrants, which is fairly common in Western Europe for instance, sound like a contradiction in terms. After all, these groups are natives, and often citizens, of destination countries. The uncritical use of such terms can be part of *othering* discourses that marks these groups as not fully part of destination societies based on race, appearance, religion or other aspects of ethnicity.

As Chapter 4 will argue, such 'other-definition' of migrant-origin groups by dominant groups of destination countries can encourage the formation of ethnic minorities, to the point that also these minorities also continue to define themselves as 'migrants'. The continued designation of migrants' offspring as 'migrants' or otherwise distinct ethnic groups is not necessarily always negative. One can also argue that a lack of designation, identification and measurement of ethnic minority groups implies a refusal by governments to recognize their existence and identity, their experiences of discrimination, racism and exclusion, and their rights to emancipation. This has for instance been an issue in France, where dominant republican ideologies (see Chapter 4) preclude the collection of statistics for migrant groups once they have attained French citizenship.

In classical immigration countries such as the US, the usage of second generation migrants can be seen as a personal affront by descendants of migrants. At the same time, immigrants, their descendants and ethnic minorities in the US can take a lot of pride in 'hyphenated identities' (such as 'Mexican-American', 'Filipino-American' or 'African-American'), whilst this is rarely seen as conflicting with citizenship and their self-identification as fully belonging to the American nation. In other countries, descendants of migrants may proudly call themselves 'migrants' as a form of conscious self-identification with emancipation struggles. These examples exemplify the large extent to which migration terminology is seldom neutral, but intertwined with ideology, politics and struggles for recognition.

Internal and international migration

The distinction between internal and international migration seems clear, but can be blurred in practice. For instance, migration within large and ethnically diverse countries such as Brazil, Nigeria, the Democratic Republic of Congo, India or China can involve a much more radical change in terms of language, culture and lifestyle than migration between, say, the UK and the US, Morocco and Algeria, Azerbaijan and Iran, or Ukraine and Russia. Also, international migration is not necessarily more difficult than internal migration in terms of needing permits and visa. International migration is nominally free between states of regional blocs such as the European Union (EU), the Economic Community of West African States (ECOWAS) and Australia and New Zealand. Several states including India, China, Nigeria and the former Soviet Union have subjected internal migration to controls or permit systems (Torpey 2007).

This does not mean that the distinction between internal and international migration is not useful. On average, the crossing of an international border involves higher costs and a bigger change in culture and social environments, as well as higher psychosocial strains put upon migrants in terms of suffering from separation from their loved ones and the difficulty of adapting to a new climate, diet, customs, bureaucratic systems, language and culture and the feeling of being uprooted this often involves. However, it is important to be aware that the distinction between the two can be blurred in both legal and social terms.

Concrete experiences of internal and international migrants are not necessarily different and some forms of migration do not fit easily in either category. It is for instance a tricky question whether to consider intra-EU migrants as internal or international. In fact, they seem to constitute an in-between category. The arbitrariness of the distinction between internal and international migration is also shown by the migrants' transfer of status when borders are created or shifted. For instance, after the disintegration of the Soviet Union, Ukrainians in Russia and Russians in Ukraine became international migrants overnight without moving at all.

That is why increases in the total number of international migrants may not reflect increasing human mobility, but rather a redefinition of people's legal status from internal to international migrants. The number of independent countries has increased from about 74 in 1946 to 195 in 2018, mainly as a result of decolonization and the concomitant disintegration of European empires including the Soviet Union, as well as the further splitting up of countries such as Yugoslavia and Sudan. The increase in global migrant numbers may therefore seem partially artificial. On the other hand, such redefinition of status and the associated loss of citizenship can have very real consequences in terms of rights and experiences of people, such as the discrimination Russians living in former Soviet republics have experienced in periods of post-Soviet nationalistic fervour (see Chapter 6). Such 'othering' can equally target non-migrant ethnic and religious minorities who do not fit in the new nationalistic mould.

Another reason why it is often so difficult to make a sharp distinction between internal and international migration is that both are manifestations of the same more general processes of social transformation and economic development. For instance, industrialization and economic growth typically go hand in hand with rural-to-urban migration. While most rural-to-urban migration is usually contained within the borders of one country, a smaller or substantial part may spill over international borders, particularly if countries are small in population and size, and bigger cities are located in other countries. This partly explains why large, populous countries such as China, India, Nigeria, Brazil and the United States have relatively low rates of emigration expressed in proportions of their populations. Ambitious people living in smaller countries like Moldova and Lesotho or in island states like Cape Verde, Samoa and Tonga are more likely to have to leave their country to achieve your life goals.

Moreover, internal and international migration are often functionally related. For instance, international migration can be preceded by internal, rural-to-urban migration, as in towns and cities where people may save money, obtain the knowledge and get the aspirations allowing them to move abroad. International migration can also lead to internal migration. For instance, in Morocco, international migrants invested *remittances* in small enterprises located in towns and cities rather than in their village of origin, which can lead to resettlement of their families in urban areas (Berriane 1997; see also King and Skeldon 2010). International migration can also trigger *replacement migration*, with the jobs left by emigrants in cities being filled up by internal rural-to-urban migrants, or by migrants from poorer countries.

Given these conceptual and functional links, it is unfortunate that research and debates on international and internal migration have evolved largely in isolation of each other. There is much to be gained from connecting, comparing and integrating research on migration within and across borders. For instance, research comparing internal migration in large countries such as Brazil, China, India and the US could generate key insights into how migration evolves in the absence of border restrictions. Rigid categorizations based on legal definitions can stand in the way of gaining a more comprehensive understanding of the nature and drivers of migration processes which transcends such conceptual borders.

Temporary and permanent migration

Governments often make a distinction between temporary and permanent migration. This can be part of political discourses that try to assuage public worries about high immigration. For instance, in the 1960s and 1970s politicians in Germany, the Netherlands and Belgium often stressed that Mediterranean *guestworkers* were temporary workers who would return after a few years. Largely in the same vein, leaders of Gulf states also stress that immigration is only temporary, even though migration is often gaining a more long-term character than they officially proclaim. Also governments of traditional immigration countries such as Australia and the US have expanded their temporary immigration programmes, even though a substantial proportion of migrants end up staying through the various pathways to permanency offered to migrants who are able to find a job.

Processes of labelling particular forms of migration as 'temporary' and 'permanent' are partly political and ideological in nature. First of all, different states and their judicial systems attach different meanings to these labels. In countries that ran immigration programmes based on permanent inflows of entire families – such as Canada, Australia and New Zealand in the post-Second World War decades – most migrations that do not fit into that definition are seen as temporary. In most European countries, temporary migration is generally associated with workers coming on temporary contracts generally lasting a few months to a few years. In fact, most countries in the world lack permanent immigration programmes, with the notable exception of policies that states grant permanent residency or instant citizenship to 'co-ethnics' upon their arrival as part of nation-building policies – such as Jews in Israel, *Aussiedler* (ethnic Germans living in the former Soviet Union) in Germany, and ethnic Turks in Turkey.

The distinction between temporary and more permanent forms of migration primarily reflects the preoccupation of states with protecting their sovereignty in selectively granting rights to particular groups of 'deserving' migrants, particularly as regards the right to work, social services and citizenship. While this desire is innate to nation states and modern democracies, efforts to 'predestine' migrants are often at odds with the more complex and fluid realities on the ground, in which migrants whose stay was thought to be temporary end up staying permanently, and migrants who came to stay permanently return after a few years. Because circumstances change, migrants' plans change as well. This highlights the need to conceptualize migration as a *process* rather than a one-off move between place or country A and B, in which people may move back and forth between A and B, or make a secondary move to place C.

It is therefore difficult to know whether and at which point a migration is temporary or permanent, as there can be a wide gap between intentions of governments as well as migrants and eventual outcomes. However, the political and ideological framing of migrants as 'temporary' or 'refugees' can have very real consequences. For instance, refugees are frequently seen as temporary migrants, as guests, who are supposed to return once conflict has ended, even if such refugee populations have lived in their adopted homelands for decades, new generations have grown up and their return has become very unlikely. However, in other cases refugees are welcome as heroes, particularly if they fled hostile regimes, and their right to stay is generally not questioned, such as the Vietnamese 'boat people' who were seen as fleeing communism and were granted asylum in the West, or Cubans fleeing to the US under the Castro regime. This again exemplifies that terminology is often not neutral and subject to political framing.

Origin and destination

'Host' and 'home' countries

Another set of migration categories are pairs that pertain to the countries, regions or places migrants come from and go to, most notably 'home–host', 'sending–receiving' and 'origin–destination'. An uncritical use of these terms frequently creates biases in the analysis of migration processes. First, the common 'home–host' terminology can reflect guestworker visions according to which migrants are 'guests', who are welcomed (or tolerated), but are also expected to leave, and, thus, cannot claim rights. On the one hand, 'host' has a positive connotation of hospitality and the duty to treat guests well. On the other hand, if migrants stay longer, the position of 'guest' can become a problem if it implies long-term denial of rights linked to the idea that migrants should go 'home'.

The 'home'–'host' dichotomy becomes particularly problematic if it reveals ideas or ideologies in which the notion of 'home' is unequivocal and unchangeable. Such worldviews are problematic because of their essentialist nature. *Essentialism* is the idea that every social unit has a number of fixed characteristics that are inextricably linked to its identity. This is problematic for understanding a social *process* like migration because in reality people's characteristics such as knowledge, ideas, skills as well as identities, perceptions and aspirations typically change over time. Numerous studies have shown that migrants' (and non-migrants' alike) identities and concomitant feelings of belonging and home are socially constructed, and tend to change over life courses (De Bree *et al.* 2010; Ghorashi 2005). Moreover, migrants often foster multiple, transnational belongings that defy the idea of a singular belonging to one nation or state.

This challenges essentialist ideas that there are natural links between people, culture and territory (de Bree 2007; Pedersen 2003). In other words, what initially was experienced as the host country, often becomes the new home country over time, and many migrants can feel at home in different places at the same time. The home-host terminology can therefore come down to a top-down imposition of identities on migrants without investigating their own changing intentions, experiences and feelings. Under certain circumstances, such framing can become part of discriminatory practices and racist discourses which frame migrants and their descendants as people whose 'home' is somewhere else and can therefore be deprived of human rights or deported.

'Sending' and 'receiving' countries

Although seemingly less moralistic and essentialist than the home-host dichotomy, the 'sending-receiving' dichotomy can be problematic because of the assumed passivity of migrants. The suggestion that migrants are 'sent' by their governments conveys the implicit denial of people's *agency*. Human agency can be defined as people's capacity to make their own choices and to impose these choices on the world. Although human agency is always limited by structural constraints, such as access to resources, time and policy constraints, the vast majority of migrants and their families exercise some level of agency in making migration decisions and organizing and paying for their journeys. For instance, efforts to recruit migrant labour would fail if people were not willing to sign up. Even in the case of refugee migration, people need some level of agency in terms of access to resources in order to be able to migrate.

The common use of 'sending–receiving' and also 'home–host' terminology reveal a perception of the world in which states can and should organize and decide on who can leave and come. It also reflects the moral ideal or 'orderly' migration advocated by governments and organizations like the IOM. However, this bias generally ignores the perspective and experiences of migrants themselves. On the other hand, the notion of 'sending' and 'receiving' states cannot be altogether dismissed if we consider the important role governments have often played in shaping migration patterns. There are many historical antecedents to this, such as the way in which governments of European countries such as Spain, Portugal, Italy and the Netherlands often took a very active role in planned emigration policies until the 1960s. This involved the establishment of special offices that played an active role in selecting and assisting prospective migrants to the Americas, Australia and New Zealand. The Gulf states and Israel have also played a central role in the planning and regulation of migration. In recent decades, governments of countries like the Philippines, Ethiopia and Mexico have often played an active role in selecting and preparing emigrants as part of national development strategies. Such organized migration often happened in the context of recruitment agreements that 'sending' states have signed with 'receiving' countries.

However, in these cases people's agency plays an important role too. Also in the case of official recruitment and assisted emigration, people often try to circumvent bureaucratic rules and controls through migrating spontaneously, generally through chain migration in which previously migrated friends and family played an important role in helping with acquiring documents, buying travel tickets, identifying, paying smugglers, and finding work and housing upon arrival. Employers often see such spontaneous migration and recruitment through networks of family and friends as an efficient and flexible way to quickly react to fluctuations in labour demand. Governments therefore often either support or turn a blind eye to such practices. The terminology of 'sending' and 'receiving' countries therefore reflects a state bias that can conceal how migration is a partly autonomous social process through the agency that migrants deploy.

The other potential problem of this terminology is that it can reinforce dichotomous world views according to which the world is divided into 'receiving' (generally imagined to be situated in the 'global North') and 'sending' states (generally imagined to be situated in the 'global South'). Such views are simplistic because much migration occurs within world regions, between and among lower- and middle-income countries. Many countries have significant levels of emigration and immigration, a reality which does not fit within simple binary schemes. While from the perspective of migrants, certain countries can indeed be considered as 'host' or 'home' or 'sending' or receiving', it is more problematic to adopt a worldview that crunches entire countries into a neat (post-colonial) world order between 'sending South' and a 'receiving North' (see Bakewell 2009).

'Origin' and 'destination' countries

At first sight, the 'origin-destination' country dichotomy seems less problematic. The other advantage is that we can also apply it to places, areas and regions, and not only countries. There is thus a less strong state bias in this terminology. However, the term 'origin' can also have essentialist overtones, because of the implicit assumption that each person has a single, unchangeable 'origin'. Although most migrants will probably see their country in which they were born and grew up as their 'origin' (or 'home') country, other migrants and minorities will base their constructed 'origin' and 'home' countries in much more ancestral terms, such as this is the case amongst many Jewish, Armenian and Lebanese (see Cohen 1997; Leichtman 2005). Although 'destination' sounds more straightforward as a category, field studies have clearly shown that destinations often change *en route*. For instance, some migrants from sub-Saharan countries who initially saw Morocco as a 'transit country' choose to settle in Morocco as second-best option after they failed or did not venture to cross the Mediterranean (de Haas 2008).

This exemplifies that all terms come with their problems and biases. This does not mean that we should not use these terms – on the contrary, we need categories to make sense of the social world around us and create some order in our thinking. Rather, it shows the importance to always remain critical with regards to the applicability and appropriateness of terminology.

Migration motives: Labour, family, study, business

Another set of migration categories relate to the motive for migrating. This is usually matched by a legal entry channel as part of migration regimes. The most common of these are *labour*, *family*, *student* and *business* or *investor migrants*. Importantly, while such categories are useful to know people's *prime* motive for migrating, they are often less useful to analyse the main cause of migration, which we need to find in more structural processes of development and change. For instance, knowing that many Indonesians move to Malaysia, or Mexicans to the US, to work does not tell us anything about the socioeconomic processes that created a demand for migrant labour or the circumstances that led to people leaving. A second complication is that migrants often have multiple motives for migrating, and that these may also change over time. For instance, many family migrants end up working, and an increasing number of governments try to retain student migrants by giving them special job seeker visas.

Labour migration, which is also known as economic migration, is the most important prime cause of migration. For labour migration, a frequent distinction is made between low- and high-skilled migrants. Such dichotomous distinctions can be deceiving, as in reality migrants can be found across all skill levels. Second, the term 'low-skilled' seems to devalue all sorts of manual and service labour that still requires considerable skills and physical abilities, and that often come with major responsibilities, for instance in care work for children, the sick and the elderly. Third, the jobs that migrants take often do not match their education and skills because of difficulties to get diplomas and skills recognized, language problems and discrimination. This phenomenon, which is also known as *brain waste* (although this term is problematic in its own terms because of its utilitarian overtones) can cause migrants to work under their educational levels.

Family migration is often a direct consequence of the decision of labour migrants to settle, which can set in motion further *chain migration* with social networks reducing the costs and risks of migration. In some cases, there is a gap between the legal entry channel and the main migration motive. This often happens when there is a demand for migrant labour but no sufficient legal migration channels to match such demand. People may then start to use other legal channels for migration, such as the family or asylum channels, even though their prime migration motive may be work. Such mismatch between labour market dynamics and legal migration channels may also encourage irregular migration. Family migration is often a popular legal entry category, as it is difficult for governments, particularly of democratic states, to deny migrants the right to live with family members for humanitarian and legal reasons. A further distinction can be made between primary and secondary family reunion. In the case of *primary family reunion*, a migrant is joined by spouses and children living in origin countries. In the case of *secondary family reunion*, migration follows new unions between migrants' offspring and spouses in origin countries.

Forced migration: Refugees and asylum seekers

A special legal migration category applies to people fleeing violence and oppression. According to the 1951 UN Convention Relating to the Status of Refugees, a *refugee* is a person who,

> owing to a well-founded fear of being persecuted for reasons of race, religion, nationality, membership of a particular social group, or political opinion, is outside the country of his nationality, and is unable to or, owing to such fear, is unwilling to avail himself of the protection of that country.

The term refugee is distinct from, but often confused with, asylum seeker. An *asylum seeker* is a person who has applied for refugee status and is still awaiting a decision on her or his recognition as a refugee. Such refugee status determination processes are usually carried out by destination country governments, although in other cases – particularly where state institutions are weak – these procedures are carried out by the United Nations High Commissioner for Refugees (UNHCR) and the United Nations Relief and Works Agency for Palestine Refugees in the Near East (UNWRA). The UNHCR was established in 1950 in the aftermath of the Second World War to help millions of European refugees and displaced people to find a new home. Since then, its mission has expanded to operations around the world. Most countries in the world have signed the UN refugee convention and often enshrined important elements of it in national law (see Chapter 10).

According to the UN Refugee Convention, it is a fundamental right of people to cross international borders to seek protection from violence and persecution. People thus have the right to leave their country and cross borders in search of protection.

Terms like 'illegal asylum seekers', which are sometimes used in political and media discourses – can therefore not be applied to asylum seekers. An essential part of international refugee law is therefore the principle of *non-refoulement*, which protects asylum seekers from return to countries where they may fear persecution. Return is only possible if refugee status determination procedures have established that an asylum request is not founded.

The UN Refugee Convention and its 1967 Protocol and its 1979 Handbook provide a universal refugee definition and incorporate the basic rights and obligations of refugees. They also lay down principles for, but do not regulate the actual determination of refugee status. This is why procedures and recognition rates vary hugely across countries. The majority of refugees in the world have not been recognized on an individual level, but on a so-called *prima facie* basis, in which groups of people receive refugee status on the basis of apparent, objective circumstances in the origin country. A prima facie approach is common in the large-scale arrival of refugee groups where individual status determination is impractical or impossible.

The legal basis on which refugees gain access to residency status is a humanitarian one and is enshrined into international humanitarian law. This is a very different legal basis compared to other migrant categories, the regulation of which is the sole prerogative of sovereign states or regional unions such as the EU. Several activists and scholars therefore oppose the application of the term migrant to refugees. They so do to emphasize the different legal basis of refugee protection and to avoid confusion between these categories, out of fear that such conflation of terms would undermine the case for refugee protection, particularly through the inaccurate portrayal of refugees as 'illegal migrants' in media and politics.

This book will consider refugees as a particular category of migrants, with a particular sets of rights. Besides the fact that refugees are objectively migrants, another reason for doing so is that, over time, many refugees end up working, setting up businesses, reunifying their families at the destinations and starting to form communities just as other migrants do. Refugee migration is thus a manifestation of human mobility that is related to, but distinct from, other forms of migration. Acknowledgement of this reality does not necessarily undermine refugees' right to protection. However, this highlights again that migration terminology is often subject to political games and power struggles, and the care that is required in the conscious choice of particular categories.

In more recent decades, a special category of *internally displaced persons* (IDPs) has gained increasing recognition as a group of 'internal refugees' in need of protection. According to the United Nations Guiding Principles on Internal Displacement, IDPs are

> persons or groups of persons who have been forced or obliged to flee or to leave their homes or places of habitual residence, in particular as a result of or in order to avoid the effects of armed conflict, situations of generalized violence, violations of human rights or natural or human-made disasters, and who have not crossed an internationally recognized state border.

This definition does not give any special rights to IDPs because, in principle, they have the same rights and guarantees as other citizens of their country. Contrary to the UN Refugee Convention, the UN Guiding Principles on Internal Displacement are not legally binding, although they have received increased recognition, particularly within the African Union through the 2009 Convention for the Protection and Assistance of Internally Displaced Persons in Africa.

Refugee migration is often cast as *forced migration* while other forms of migration are generally seen as 'voluntary'. However, virtually all migrants face some level of constraints limiting their agency, such as access to money and border restrictions. At the same time, refugees fleeing persecution and violence still need some agency in the form of access to resources *in order to be able to flee*. That is why the most vulnerable populations are often unable to leave if economic crises, droughts, or famine occur, while others may become immobilized by conflict (Lubkemann 2008).

Migration is thus forced when staying is not an option. As such, the forced character of their migration does not therefore preclude that refugees have agency. Refugees exercise their agency as far as possible in the face of appalling circumstances, for instance in making decisions about where to go. The fact that refugees deploy considerable resources in order to be able to flee, and that international migrants are therefore generally not among the poorest, contrasts with political and moralistic discourses that unilaterally portray refugees as victims. Refugees are forced migrants because they have no choice to stay at their homes. The distinction between forced and voluntary migration is primarily driven by the (legitimate) interest of states in classifying migrants and to make decisions about who deserves refugee protection. The label 'forced migrant' therefore also applies to people who are forced to leave their homes because of development projects such as roads, dams and factories (see below).

Illegal, unauthorized, irregular, undocumented

Another range of categories reflects attempts to characterize the legal status of individual migrants and, at the more conceptual level, forms of migration. The most familiar one is the distinction between legal and illegal migration. The distinction between *illegal entry* and *illegal stay* is essential. Although public debates tend to be focused on illegal entry, the biggest source of illegal stay is migrants who entered legally but who have *overstayed* the duration of their visa or residence permit. This is the main reason why building walls and fences alone will not stop situations of illegality.

This category of illegality, which reflects governmental views, had been criticized on legal and social-scientific grounds. According to classical jurisprudence, a person cannot be illegal (see Kubal 2013). Acts can be illegal. People cannot. The related moral argument is that 'nobody is illegal', that it is unacceptable to characterize human beings as 'illegal'. Humanitarian organizations defending the rights of migrants therefore typically reject this label (Collyer and de Haas 2012; Jordan and Düvell 2002; Van Liempt 2007). The category of 'illegality' can become dangerously broad, and is often stretched to those who intend to make an asylum claim, but have not yet done so (see also Kubal 2013).

There are also social-scientific reasons for criticizing the use of the 'illegal' category to describe migrants. First, what governments see as 'illegal' is not necessarily a socially 'illicit' form of behaviour, and may be tolerated or even encouraged in migrants' origin communities (Van Liempt 2007). Another critique is that the legal/illegal distinction is not as clear-cut as it seems. Irregular entry or overstaying visas are often put under the same umbrella as much more ambiguous situations. For instance, in many countries, immigrants without residency rights can access legal employment, can obtain documents such as drivers' licences, and pay taxes, with governments often turning a blind eye towards their legal or semi-legal employment. In some countries, governments even acknowledge this ambiguity by registering undocumented migrants, such with the *padrón municipal* (municipal register) in Spain or 'pink cards' in Thailand (Mendoza 2018). Migrants often occupy various in-between statuses between full legality and full illegality. They therefore often find themselves in legal limbo zones (Kubal 2013; Ruhs and Anderson 2010). In addition, their status often shifts over time, with migrants moving in and out of situations of illegality.

However, the counter-argument is that the legal status of migrants does matter to their lives and decisions, and the term should therefore not be avoided but rather used more carefully (Black 2003; Collyer and de Haas 2012). In general, the term seems more problematic in reference to individual people. Referring to individual *people* as 'illegal' seems neither accurate, nor desirable, nor analytically useful. However, the term *illegal migration* can therefore still be used to refer to a migratory phenomenon at the more general level. As a phenomenon, illegal migration from Mexico to the US does exist as long as people cross borders without permission. However, it seems less appropriate to impose the label onto individual people as a fixed category.

To overcome problems with the term 'illegal', researchers and journalists have used alternative terms, such as 'irregular', 'undocumented' and 'unauthorized' migration. These terms are useful, but can have their own problems. With regards to *irregular migration*, one could argue that in several cases, such as migration from Mexico to the US, or from Zimbabwe to South Africa, irregular migration has become the norm rather than the exception. On the other hand, if we interpret 'irregular' as migration taking place outside the regulatory norms of states, it can still be an appropriate term, although it should thus not be applied to asylum seekers.

The categories of *unauthorized migration* and *undocumented migrant* (derived from the French *sans papiers*, which means 'without papers') seem more objective and, importantly, value-neutral terms. On the other hand, these terms also cannot entirely avoid some of the problems as encountered with 'illegal migration'. 'Unauthorized' reveals a state bias, and can be more easily applied to the phenomenon of migration in general than for individual migrant (as a human being, a person cannot be unauthorized, but their movement or stay can), for which reason the term *undocumented* seems to be preferable. On the other hand, migrants without a document permitting their entry or stay often do possess a range of documents such as drivers' licences, registration documents, insurance documents and tax files. In many countries, migrants can work legally and pay taxes even though their stay is unauthorized. This exemplifies that there are no 'perfect' categories, and that each category comes with its own advantages and disadvantages. The most important lesson is to be always critical and careful in the use of categories.

Smuggling and trafficking

In political and media representations, irregular migration is often associated with smuggling and trafficking. However, it is important to be careful, as asylum seekers who use smugglers to cross borders cannot be classified as illegal migrants: according to international and national law they have the legal right to seek protection, and in this sense their migration is 'regular'. The terms of smuggling and trafficking are often confounded, but need to be clearly distinguished as they mean quite different things. *Smuggling* refers to the use of paid or unpaid *migration intermediaries* to cross borders without authorization. According to the UN Protocol Against the Smuggling of Migrants by Land, Sea and Air, migrant smuggling is defined as

> the procurement, in order to obtain, directly or indirectly, a financial or other material benefit, of the illegal entry of a person into a State Party of which the person is not a national or permanent resident.

However, this definition is too narrow, as smuggling can also happen voluntarily out of humanitarian motives. One example are the resistance movements smuggling Jews out of Nazi occupied territory during the Second World War. Research has shown that smugglers can also be friends, family or former migrants, who usually, but not always, receive payment for their migration assistance (Van Liempt 2007). Smuggling is essentially a form of service provision, which flourishes when borders are tightly controlled but causes of migration, such as labour demand in destination countries and conflict in origin countries, persist. Clients of smugglers include not only economic migrants, but also refugees unable to make an asylum claim because restrictive border rules prevent them from entering countries of potential asylum (Gibney 2000).

If dependence on smugglers and other migration facilitators becomes very high, this can evolve in exploitative forms of migration which are also known as trafficking. The UN Protocol to Prevent, Suppress and Punish Trafficking in Persons defines *human trafficking* as

> the recruitment, transportation, transfer, harbouring or receipt of persons, by means of the threat or use of force or other forms of coercion, of abduction, of fraud, of deception, of the abuse of power or of a position of vulnerability or of the giving or receiving of payments or benefits to achieve the consent of a person having control over another person, for the purpose of exploitation. Exploitation shall include, at a minimum, the exploitation or the prostitution of others or other forms of sexual exploitation, forced labour or services, slavery or practices similar to slavery, servitude or the removal of organs.

Human trafficking involves a strong element of exploitation, which sets it apart from smuggling. Smuggling is consensual and resembles a business transaction, or can be a voluntary service, and involves the movement across borders. As the definition shows, human trafficking is not necessarily related to migration. Smuggling is frequently mislabelled as trafficking, giving the misleading impression of migrants being forced on journeys, while in fact they engage into a business transaction.

It is safe to say that vast majority of illegal border crossings involve smuggling, not trafficking. It is sometimes difficult to make a sharp distinction between smuggling and trafficking, as elements of exploitation and abuse may emerge during transit or at destination, even if migrants' original decisions to move were voluntary. But even in many trafficking situations, there is often some degree of consent and knowledge on the part of the migrants. Initial consent does not invalidate a trafficking claim, as smuggling can lead to trafficking over time.

The risks and exploitation experienced by migrants who hire smugglers to cross borders does not mean they are trafficked. Border enforcement has resulted in an increasing death toll as smugglers use longer and more dangerous routes. However, research has shown that migrants are willing to take significant risks to reach their destination (Mbaye 2014). By disrupting the modus operandi and traditional routes of smugglers, increased border enforcement has contributed to rising fees and increased the risks of crossing borders. Underscoring the blurred line between smuggling and trafficking, insolvent migrants have, upon arrival, been forced into a condition of debt bondage until they reimburse the smuggler. Although they left as voluntarily migrants using smugglers, people can *become* trafficking victims along the journey or during their stay.

Anderson and Andrijasevic (2008) argued that the lack of definitional clarity and the constant slippage and confusion between terms like 'trafficking', 'illegal immigration' and 'forced prostitution' divert attention from the role of the state in tolerating and contributing to poor work and vulnerable workers, and that the language of trafficking and the portrayal of people involved as victims of criminals is part of attempts to depoliticize migration and struggles over rights and citizenship. The existence of trafficking and other forms of exploitation does not preclude the fact that victims of trafficking have agency and see long-term benefits of staying in exploitative patronage relations – this can for instance explain why Nigerian sex workers in Italy who are 'liberated' by authorities and deported back to Nigeria, often return in order to finish their contracts (see Carling 2006).

The difficult – and inconvenient – reality is that many migrants voluntarily engage in exploitative relations with employers, smugglers and other 'intermediaries' because they usually see a long-term benefit in such a relation. For instance, over the past centuries, many Europeans and Asians would only agree to become indentured workers based on the knowledge that they would be free once they had served their contracts or paid off their migration debts (see Chapter 5). The existence of a certain level of consent amongst bonded labourers does not justify exploitation, nor does exploitation preclude a certain level of consent and agency. As with other categories, it does not mean that we cannot use such terms, but that they should always be carefully defined and that we should be suspicious of ideological and simplistic uses of such terms.

Climate refugees: The construction of a migration threat

As the cases of 'illegal migration' and 'trafficking' illustrate, the deliberate or uncritical use of particular categories by politicians and media can have a distorting effect on perceptions of migration phenomena. Often such categories are embedded in discourses that give misleading representation of the magnitude, nature and causes of migration processes. This is illustrated by recent discourses that portray climate change as a major cause of massive 'South–North' movement of climate refugees. While these discourses sound intuitive, and have gained the status of 'truth' in many circles, they are in fact highly deceiving (Foresight 2011; Gemenne 2011, see Box 2.1).

Concerns about climate change-induced migration have emerged in the context of debates on global warming and the inability of states to take effective action to mitigate it through. Global warming is the most pressing issue facing humanity, and the lack of willingness of states and the international community to address it effectively – particularly through reducing of carbon emissions – is a valid source of major public concern and protest.

Box 2.1 Climate change and migration

Research evidence challenges the popular idea that climate change will lead to mass migration. A major study involving scientists from around the world (Black *et al.* 2011; Foresight 2011) showed that most estimates of huge numbers of environmental/ climate change migrants are methodologically flawed (see particularly Gemenne 2011). Migration is a multi-causal phenomenon, which precludes the drawing of direct links with (climate- and non climate-related) environmental change (Foresight 2011).

First, migration is likely to continue regardless of the environment, because it is driven by powerful economic, political and social processes (see Box 3.1). This makes it difficult to directly attribute internal migration to environmental factors. For instance, this challenges the popular idea that much migration within Bangladesh is an 'obvious example' of mass displacement due to the sea-level rise (Findlay and Geddes 2011). In fact, many people migrate into areas of greater environmental vulnerability, such as fertile deltas and cities built on floodplains, because of improved livelihood opportunities they can expect to find there.

Second, environmental degradation or natural disasters need to go hand in hand with political instability, poverty, violent conflict and corruption to create conditions for large-scale displacement (Black 2001). For instance, if people get displaced or die as a result of natural disaster, this is not just the direct consequence of the disaster, but also reflects the inability of governments to help people to cope with such stresses, such as by building flood defences, timely evacuation efforts and building regulations.

Third, the most vulnerable are often deprived of the *capabilities* to move at all. When Hurricane Katrina hit New Orleans in 2005, many of the car-less poor got trapped in the city, and African Americans were therefore were overrepresented amongst those who died (Gemenne 2010). In the same vein, when people are impoverished by such factors as drought, they often lack the resources to move, trapping them in situations of extreme vulnerability. In fact, migration is often an effective adaptation to increase resilience by diversifying and increasing family income (Foresight 2011; Gemenne and Blocher 2017).

Detailed studies from Africa failed to find a simple causal link between environmental stress and migration (Jónsson 2010). In Malawi, droughts decreased rural out-migration (Lewin *et al.* 2012). In Burkina Faso, droughts increased short-distance migration between villages, but reduced international moves to Côte d'Ivoire (Henry *et al.* 2004). Also, an analysis of global data from 1960 to 2000 found no direct effect of climatic factors on international migration (Beine and Parsons 2015).

Yet the absence of the displaced millions by climate change migration fearmongers is no reason for complacency. The forecasted acceleration of climate change is likely to have severe effects on production, livelihoods and human security. Populations likely to be affected most are in mega-cities built only a few metres above sea-level (Hugo 2013). Climate change will create serious challenges for humanity. However, using the spectre of mass migration to make the case for urgent action on reducing CO_2 emissions is a clear example of 'being right for the wrong reason', with potentially harmful credibility implications for organizations using these argument. More importantly, though, it overlooks the fact that the implications of environmental adversity are likely to be most severe for the most vulnerable populations lacking the means to move.

Source: Getty Images/Handout

Photo 2.2   When Hurricane Katrina hit New Orleans in August 2005, the most severely affected populations were the poorest who lacked cars and other resources that would enable them to move out

However, to link this issue with the spectre of mass migration is a dangerous exercise based on myth rather than fact. Environmentalists have claimed that the effects of global warming, especially on sea-levels and rainfall patterns, will lead to massive population displacements. In a publication entitled 'Environmental Exodus', the influential biodiversity specialist Norman Myers drew a direct, simplistic link between environmental change and large-scale migration, arguing that there would already be 25 million 'environmental refugees' which would further increase to 200 million by 2050 (Myers and Kent 1995).

The typical approach of environmentalists has been to map climate-change-induced developments (such as sea-level rise, drought or desertification) onto settlement patterns to predict future human displacement. For instance, if climate change models predicted a sea-level rise of (say) 50 centimetres, it would be possible to map all coastal areas affected by this and work out how many people lived in such areas. The assumption then was that all these people would have to move (see Myers and Kent 1995). Some have put forward scenarios of mass displacements as a cause of future global insecurity (Homer-Dixon and Percival 1996), while certain NGOs even escalated forecasts of future population displacements up to one billion by 2050 (Christian Aid 2007).

Such forecasts have turned out to be highly speculative and scientifically unsound (Gemenne 2011). Researchers have argued that migration is driven by many factors, and can rarely be reduced to the effects of just one form of change, such as climate change (Black 2001; Castles 2002; Foresight 2011). It is not possible to make a direct link between climate change, environmental stress and large-scale migration. Box 2.1

summarizes evidence from recent research, but there are five main reasons to be sceptical on the idea that climate change will lead to mass migration.

First, people can use various adaptation strategies, such as flood defences, changes in livelihoods or short-distance mobility to cope with environmental stresses. Second, in cases of floods, tsunamis and other environmental havoc, the vast majority of people move over short distances, such as to the next neighbourhood, village or town. Third, most of such displacements tend to be temporary, because most people wish to return home as soon as possible. Fourth, most people living in the poorer countries of the world do not have the resources to move over large distances. Fifth, processes of impoverishment influenced by environmental stresses can actually deprive vulnerable people of the means to travel and migrate over large distances, and they might find themselves therefore trapped at home.

It is therefore unlikely that climate change will lead to mass migration from 'South' to 'North'. Even though the methodologies of studies on which previous estimates were based are problematic, they are still often quoted mainly because the idea that climate change will lead to mass migration serves powerful political agendas on the left and right. In fact, they misuse the topic of climate change and migration to give a 'human face to climate change' (Gemenne 2011: 225; see also Klepp 2017). Urging governments to 'do something', NGOs and humanitarian organizations use alarmist narratives turning climate change, and especially climate change migration, into a security topic (Klepp 2017).

By drawing a simplistic, direct causal link between climate change and migration, the 'climate refugee' discourse depoliticizes the migration of vulnerable people (see also Bettini 2013). *Depoliticization* refers to discursive strategies that remove the political dimension from a social issue. Political issues affect the vulnerability of people and their resilience with environmental and other stresses. For instance, poverty, poor housing and weak governmental services explain why the suffering and death toll is so much higher when a hurricane hits a country like Haiti compared to the US.

Politicians have frequently depoliticized social issues by shifting the blame to exogenous or environmental or climatic factors 'beyond their control'. For instance, in Morocco politicians and bureaucrats routinely blame 'drought' and 'desertification' to explain a whole range of problems in rural areas, from low agrarian productivity to the *rural exodus*. In fact, the crisis narrative of desertification has been invoked to facilitate and justify policies that have disadvantaged nomadic groups and deprived them of their mobility freedoms by forcing them to settle down (Davis 2005). Governments may also use 'the climate' as an excuse to displace people, for instance in the case of discourses around sea level rise in Pacific islands. In the Maldives the government has recycled older, highly controversial, proposals for the resettlement of its population dispersed over 200 islands onto 10–15 islands. The main motive has always been economic, because the government finds it too costly to provide services and resources to dispersed populations. However, in recent years the same ideas are gaining renewed leverage by being couched in environmental and 'sea level rise' terms (Kothari 2014).

A simplistic view of the relation between environmental factors and migration can distract the attention away from the political causes of much forced migration. In fact, apart from conflicts and persecution, development projects (such as dams, mining, airports, industrial areas and middle-class housing complexes) and wildlife conservation are a major cause of displacement. Development-induced displacement is actually the largest single form of forced migration, leading to the internal displacement of an estimated 10–15 million people per year, mainly affecting disempowered groups such as indigenous peoples, other ethnic minorities and slum-dwellers (Cernea and McDowell 2000; Tan 2020).

Climate change mitigation can become a cause of displacement in itself. In China, hydropower, irrigation and water transfer projects are an integral part of climate change mitigation and adaptation strategies, but also displace a large number of people (Tan 2020). Wildlife conservation and other environmental protection projects are estimated to prompt the displacement – or forced settlement (the *involuntary immobilization,* or *sedentarization*) in case of pastoralist and nomadic people – and the loss of land and property for hundreds of thousand of people each year (Chatty and Colchester 2002). Displacees tend to be amongst the most vunerable people, unable to defend themselves and they often get barely compensated for the loss of livelihood.

These examples expose the importance to remain aware of the deployment of categories, concepts and discourses by political actors that, ironically, serve to try to depoliticize social problems. By deploying alarmist rhetoric around future waves of 'climate refugees', media, politicians, UN agencies, non-governmental organizations (NGOs) and humanitarian organizations have turned climate change into an immediate security threat linked to migration. This ignores evidence showing that climate change is unlikely to cause mass migration, and draws the attention away from the political and economic causes of people's vulnerability to environmental (and other) stresses. Once more, this illustrates how the use of particular categories can become part of political struggles to impose certain world views or to advance particular interests.

Conclusion

This chapter outlined the main concepts and categories used in describing and analyzing migration. On the one hand, categories are indispensable to make sense of the complex and inherently 'messy' social processes such as migration. Categories help to structure thinking, define phenomena and distinguish patterns and regularities. In other words, social categories are essential devices helping to generalize and to 'see the wood for the trees'. On the other hand, the chapter highlighted the need to be careful and conscious in choosing and using particular categories. Language is not innocent, but reflects largely unconcious biases or conscious choices. Some categories oversimplify migration and can actually stand in the way of achieving a deeper understanding of migration as a complex social process. Other categories reflect state and legal perspectives and are infused with moral or ideological meanings. In fact, all categories come with their own problems.

This is not necessarily problematic. Governments need categories and clear legal definitions in order to implement policies, and decide which migrants can access certain rights – this is crucial for refugee policies for instance. The categories states use to implement policies do matter a great deal for the daily lives of migrants. The issue is more to be aware of the biases that the use of particular categories can create, and to critically assess the extent to which such categories are relevant and useful to explain the lived experiences and perceptions of migrants themselves. It is also important to be aware of the blurred distinction between dichotomous categories, such as between internal and international migration, permanent and temporary migration, and forced and voluntary migration. This does not necessarily invalidate the categories as such, but is reason to be always conscious of the biases and simplifications their uncritical use involves.

The examples of 'trafficking' and 'climate refugees' also showed the danger of the unreflective use of migration categories which are part of political and ideological narratives and discourses that deliberately ignore evidence, actively distort the truth and thus obstruct a better understanding of migration processes. Although they are often shrouded in quasi-scientific language, such narratives often buy into simplistic and outdated models to explain migration and ignore the accumulated insights about the nature and drivers of migration processes generated by decades of migration scholarship. Besides an understanding of categories, a good knowledge of theories is essential for obtaining insights into the nature, patterns and causes of migration processes, as well as to strengthen capacity for critical and independent scrutiny of the ways in which politicians and media discuss and frame migration issues. To provide such a solid basis for analysing migration processes, the following chapter will therefore give an overview of the most important migration theories.

Guide to Further Reading

Ruhs and Anderson (2010) and Kubal (2013) provide useful discussions on the blurred lines between regular and irregular migration and situations of 'semi-legality' and 'semi-compliance'. Bakewell (2000) provides an insightful discussion of the frequent gap between government discourses and the lived experiences of forced migrants. Van Liempt and Sersli (2013) and Brachet (2018) provide critical perspectives on the ambiguous category of 'smugglers', ranging from 'migration facilitators' (by migrants) to 'criminals' (by states). A special issue of the Annals of the American Academy of Political and Social Science, *Migrant Smuggling as a Collective Strategy and Insurance Policy* gives an overview on research evidence on smuggling from around the world (Zhang *et al.* 2018). Anderson and Andrijasevic (2008) provide a critical discussion of the politicization of trafficking. For environmental change, climate and migration, a key reading is the Foresight Report. All papers along with the main report are available at www.bis.gov.uk/foresight, while Black *et al.* (2011) provides a useful summary. Gemenne (2011) reviews critically reviews the various methodologies used to estimate the number of displaced by environmental change. Gemenne and Blocher (2017) discuss how migration can serve as an *adaptation* to climate change.

Extra resources can be found at: **www.age-of-migration.com**

3 Theories of Migration

Migration is hardly ever a simple individual action in which a person decides to move in search of better life-chances, pulls up his or her roots in the place of origin and quickly becomes assimilated in a new country. Much more often migration and settlement are long-term processes that will be played out for the rest of the migrant's life, and affect subsequent generations too. Migration can even transcend death: members of some migrant groups arrange for their bodies to be taken back for burial in their native soil (see Tribalat 1995: 109–111). Migration is often a collective action, arising out of social, economic and political change and affecting entire communities and societies in both origin and destination areas. Moreover, the experience of migration and of living in another country often leads to modification of original plans, so that migrants' intentions at the time of departure are poor predictors of actual behaviour.

Conventional wisdom holds that migration is driven by geographical differences in income, employment and other opportunities. However, the paradox is that economic and human development in poor societies tends to initially *increase* movement. Improved access to education and information, social capital and financial resources typically increase people's aspirations and capabilities to migrate (see de Haas 2014a). Development typically expands people's access to material resources, social networks and knowledge. As noted earlier, this explains why most migrants do not move from the poorest to the wealthiest countries, why the poorest countries tend to experience lower levels of long-distance emigration and why industrializing societies have the highest levels of internal and international mobility.

However, if it is true that development and global inequality boost migration, it is difficult to understand why the volume of international migration as a share of the world population has remained remarkably stable at levels of around three per cent over the past decades. Such paradoxes show that the relation between migration and broader processes of development is intrinsically complex, and that patterns and trends of real-world migration often defy intuition. In order to achieve a better understanding of the nature and causes of migration processes, this chapter reviews the main insights offered by various migration theories. This will help to understand the more descriptive accounts of migration, settlement and minority formation from around the globe in later chapters. However, the reader may prefer to read those first and come back to the theory later.

Since the late nineteenth century, various social science disciplines have developed several theories that aim to understand the nature and causes of migration processes. These theories differ in their assumptions, thematic focus and level of analysis, ranging from global accounts of shifting migration patterns to theories of migrants' transnational identities. Often, theoretical and disciplinary divides are artificial. For instance, it does not seem very useful to develop separate theories for internal and international migration. Although international migration is more often (albeit not always) subject to control by states and internal migration is generally (albeit not always) free, both forms of migration are driven by the same processes of social, economic and political change and the two are often closely linked.

It is equally debatable whether it is useful to develop separate theories for different categories of migrants, such as for 'forced' and 'voluntary' or for refugee, family or economic migration. Motives for migrating are often manifold. For instance, migrants who primarily move for economic reasons may also flee political oppression. The other way around, political oppression often goes along with economic exclusion. It is thus difficult to strictly separate economic from social, cultural and political causes of migration.

To gain a deeper understanding of migration processes, it is important to see migration

- On the macro-level, as an intrinsic part of broader processes of development and social transformation rather than 'a problem to be solved' or a temporary reaction to geographical inequalities or 'disequilibria';
- On the micro-level, as a function of (1) capabilities and (2) aspirations to migrate within a given set of constraints – instead of some sort of automated reaction to, or linear function of, 'push' and 'pull' factors. This helps to understand the complex, and often counterintuitive, ways in which macro-level processes of social transformation and development shape migration processes.

These two central theoretical premises will guide the analysis in this chapter and throughout the book. For this review of theories, it is useful to make a basic distinction between theories on the (1) causes of migration processes, and theories on the impacts of migration for (2) destination and (3) origin societies. This chapter will focus on the first set of concepts and theories, while Chapters 4 and 14 will focus on the second and third set of concepts and theories. However, it is important to link theories on causes and consequences of migration in order to develop an understanding of migration as a dynamic process which is in constant interaction with broader change processes in destination and origin societies. This book uses the term 'migration studies' in the widest sense, to embrace both bodies of investigation.

In looking at the causes of migration, it is useful to also make a further distinction between

- Macro- and micro-level theories on the *causes* of migration; and
- Meso-level theories on the *continuation* of migration which focus on feedback mechanisms such as migrant networks that explain why migration processes can gain their own momentum and become partly self-perpetuating (see Massey *et al.* 1993).

Any migratory movement can be seen as the result of interacting macro- and micro-structures. Macro-structures refer to large-scale institutional factors, such as the political economy of the world market, labour market dynamics, interstate relationships, and efforts by the states of origin and destination countries to control migration. Micro-structures embrace the practices, family ties and beliefs of migrants themselves. These two levels are linked by a number of meso-level social mechanisms and structures: examples of these include migrant networks, immigrant communities, business sectors catering to migrants and the migration industry. Such social structures tend to facilitate further migration within established migration corridors.

Explaining the migratory process

The concept of the *migratory process* sums up the complex sets of factors and interactions which shape migration. Migration is a process which affects every dimension of social existence, and which develops its own complex 'internal' dynamics. The great majority of people in the world (around 97 per cent) may not be classified as international migrants, yet their communities and way of life are often affected by migration. The changes are generally much bigger for the migrants themselves.

Research on migration is interdisciplinary: sociology, anthropology, political science, history, economics, geography, demography, psychology, cultural studies, law, archaeology and the humanities are all relevant (see also Brettell and Hollifield 2014). Within each discipline a variety of approaches exist, based on differences in theory and methods. For instance, researchers who base their work on quantitative analysis of large data-sets (such as censuses or surveys) will ask different questions and get different results from those who do qualitative studies of small groups based on open interviews or participant observation. Those who examine the role of migrant labour within the world economy using long-term historical approaches will again reach different conclusions. Each of these methods has its place, as long as it lays no claim to be the only correct one.

This chapter will not review migration theories along disciplinary lines. This is done on purpose, because such distinctions are often artificial and can obstruct a more comprehensive understanding of migratory processes. Different disciplines and theories provide different views on migration, which are more often complementary than mutually exclusive. After all, most disciplines – ranging from anthropology to economics – are part of the family of the *social sciences*, and each of these disciplines as well as theories and methodologies should be seen as different ways or angles of looking at the same social realities rather than exclusive truth statements.

An early contribution to migration studies consisted of two articles by the nineteenth-century geographer Ravenstein (1885; 1889), in which he formulated his 'laws of migration'. Ravenstein saw migration as an inseparable part of economic development, and he asserted that the major causes of migration were economic. Lee (1966) argued that migration decisions are determined by 'plus' and 'minus' factors in areas of origin and destination; intervening obstacles (such as distance, physical barriers, immigration laws and so on); and personal factors. Ravenstein and Lee provide many basic insights that are still valid, such as that migration in one direction tends to generate movements in the opposite direction and that migration often takes place in clear spatial patterns linking particular destination and origins.

Most migration theories can be grouped together into two main *paradigms*, following a more general division in social sciences between 'functionalist' and 'historical–structural' theoretical paradigms. Functionalist social theory tends to see society as a system, a collection of interdependent parts (individuals, actors), somehow analogous to the functioning of an organism, in which an inherent tendency toward equilibrium exists. Functionalist migration theory generally treats migration as a positive phenomenon, as an 'optimization' mechanism serving the interests of most people, increasing productivity and contributing to greater equality within and between societies.

Rooted in neo-Marxist political economy, historical–structural theories primarily see migration as an exploitation mechanism. They emphasize how social, economic, cultural and political structures constrain and direct the behaviour of individuals in ways that generally do not lead to greater equilibrium, but rather reinforce such disequilibria, unless governments intervene to redistribute resources. They emphasize that economic and political power is unequally distributed, and that cultural beliefs (such as religion and tradition) and social practices tend to reproduce such structural inequalities. They see migration as providing a cheap, exploitable labour force, which mainly serves the interests of the wealthy in receiving areas, causes a 'brain drain' in origin areas, and therefore reinforces social and geographical inequalities.

Functionalist theories

Push–pull models

A particularly popular analytical framework is what is often referred to as the 'push–pull' model (Passaris 1989). Push–pull models identify economic, environmental and demographic factors which are assumed to push people out of places of origin and pull them into destination places. 'Push factors' usually include population growth and population density, lack of economic opportunities and political repression, while 'pull factors' usually include demand for labour, availability of land, economic opportunities and political freedoms. In this logic, 'gravity' models developed by geographers from the early twentieth century were derived from Newton's law of gravity and predict the volume of migration between places and countries as a more or linear function of distance, population size and economic opportunities in destination and origin areas.

At first sight, the push–pull framework seems attractive because of its apparent ability to incorporate all major factors affecting migration decision-making (Bauer and Zimmermann 1998: 103). However, in reality push–pull models are inadequate to explain migration, since they are purely descriptive models enumerating factors which are assumed to play 'some' role in migration in a relative arbitrary manner, without specifying their role and interactions. As Skeldon put it:

> The disadvantage with the push-pull model is that … it is never entirely clear how the various factors combine together to cause population movement. We are left with a list of factors, all of which can clearly contribute to migration, but which lack a framework to bring them together in an explanatory system … The push–pull theory is but a platitude at best. (Skeldon 1990: 125–126)

The same conditions that make some people leave, make others stay or attract people from other places. Push–pull factors models therefore have difficulties explaining why many countries and regions simultaneously experience substantial immigration and emigration, why migrants would return, or why most people do not migrate at all. People have different perceptions, preferences and ambitions, and therefore react in different ways to the same circumstances. As Lee (1966) argued there are many factors that retain people in origin areas. Push–pull models also tend to feed into dichotomous, stereotypical worldviews in which countries in the 'Global South' are unilaterally portrayed as pools of poverty, violence and misery from which everybody wants to leave to go to the beacons of prosperity and progress in the 'Global North'.

The way they are usually applied, push–pull models are often also deterministic by assuming that demographic, environmental, political and economic factors 'cause' migration, without taking account of the role of other factors. For instance, population growth, resource scarcity or environmental adversity in rural areas do not necessarily result in migration, because 'population pressure' can also encourage innovation (such as the introduction of irrigation, terraces or fertilizers), enabling peasants and farmers to maintain or even increase productivity (see Boserup 1965). Furthermore, scarcity and impoverishment can actually impede migration if people cannot afford the costs and risk of migrating (Henry *et al.* 2004). Because conflict and impoverishment tend to immobilize people, the 'involuntarily immobile' are therefore the greatest victims of economic, political and environmental havoc (Carling 2002; Lubkemann 2008) (see also Box 2.1).

Environmental or demographic factors should not be isolated from other social, economic, political and institutional factors affecting people's living standards and aspirations. For instance, while Eastern European countries have very low fertility and low or negative population growth, they have experienced large-scale emigration. At the same time, the Gulf countries have combined high fertility with low emigration and very high immigration. Improved education and media exposure may increase feelings of *relative deprivation*, and may give rise to higher aspirations and, therefore, increased migration, even if local conditions and opportunities have improved. As long as aspirations rise faster than local opportunities, this will result in increasing migration. This explains the paradox of development-driven emigration booms.

Neoclassical migration theory

As part of the functionalist paradigm, neoclassical migration theory assumes that social forces tend towards equilibrium. Rooted in modernization theory (Rostow 1960), it sees migration as a constituent part of the whole development process, by which surplus labour in the rural sector supplies the workforce for the urban industrial economy (Lewis 1954; Todaro 1969: 139). Neoclassical theory sees migration primarily as a function of geographical differences in the supply and demand for labour. The resulting wage differentials encourage workers to move from low-wage, labour-surplus areas to high-wage, labour-scarce areas.

At the micro-level, neoclassical theory views migrants as individual, rational actors, who decide to move on the basis of a cost–benefit calculation, maximizing their income. Migrants are expected to go where they can be the most productive and can earn the highest wages. In this context, Borjas (1989; 1990) proposed the idea of an international immigration market, in which potential migrants base their choice of destination on individual, cost–benefit calculations.

At the macro-level, neoclassical theory views migration as a process which optimizes the allocation of production factors. Migration will make labour less scarce at the destination and scarcer in origin areas. Capital is expected to move in the opposite direction, in search for cheaper labour, such as through outsourcing of industrial production. Neo-classical theory suggests that this process will eventually result in convergence between wages (Harris and Todaro 1970; Lewis 1954; Ranis and Fei 1961; Schiff 1994; Todaro and Maruszko 1987). In the long run, migration should therefore help to make wages and conditions in sending and receiving countries more equal, lowering the incentives for migrating.

Neoclassical migration theory was advanced by Todaro (1969) and Harris and Todaro (1970) to explain rural–urban migration in developing countries but has also been applied to international migration (c.f. Borjas 1989; Todaro and Maruszko 1987).

Harris and Todaro tried to explain the continuation of rural-to-urban labour migration in developing countries despite high unemployment in cities. They argued that it is necessary to extend the wage differential approach by adjusting the 'expected' rural–urban income differential for the probability of finding an urban job (Todaro 1969: 138). Migration would continue as long as income differences remain high enough to outweigh the risk of becoming unemployed (Todaro 1969: 147). Later, this model was refined (Bauer and Zimmermann 1998: 97) to include other factors, such as the financial and social costs of migration.

Human capital theory

An alternative but complementary approach was proposed by Sjaastad (1962), who applied human capital theory to migration. He viewed migration as an investment that increases the productivity of *human capital* – such as knowledge and skills. People vary in terms of personal skills, knowledge, physical abilities, age and gender, so there will also be differences in the extent to which they can expect to gain from migrating. People decide to invest in migration, in the same way as they might invest in education, and they are expected to migrate if the additional lifetime benefits (primarily derived from higher wages) in the destination are greater than the costs incurred through migrating (Chiswick 2000). Differences in such expected 'returns on investments' partly explain why the young and the higher skilled tend to migrate more. Sjaastad's model can be extended by also conceptualizing student migration as an investment in human capital.

Human capital theory helps to explain the 'selectivity' of migration (the phenomenon that migrants tend to come from particular sub-sections of origin populations) and helps to understand how migration is shaped by the structure of labour markets as well as differences in skills and income distributions in origin and destination societies. A related, important insight is that the geographical scope of labour markets generally increases with specialization levels. While lower skilled workers can often find jobs nearby, specialized workers often have to move more often to find a job that matches their skills and preferences. The growing structural complexity of labour markets and concomitant increases in specialization is an additional explanation for understanding why levels of migration and mobility tend to increase with development, increasing education and divisions of labour.

Critique of functionalist theories

Neoclassical and human capital theory are valuable in understanding the selective nature of migration. However, neoclassical theory can be criticized because of the unrealistic nature of its central assumptions. The first assumption is that people are rational actors who maximize income or 'utility' based on a systematic comparison of lifetime costs and benefits of remaining at home or moving to an infinite range of potential destinations. The second, related assumption is that potential migrants have perfect knowledge of wage levels and employment opportunities in destination regions. The third assumption is that (capital, insurance and other) markets are perfect and accessible for the poor. For instance, neoclassical models implicitly assume that (poor) people could borrow money from banks in order to finance migration. Yet in reality, banks often do not cater to non-elite groups lacking collateral. Because these assumptions are unrealistic, neoclassical theories are often incapable of explaining real-life migration patterns, particularly if migration occurs in conditions of poverty and high constraints.

Neither push–pull nor neoclassical theories have much room for human agency, which is the limited, but real ability of human beings to make independent choices. They portray human beings as socially isolated individuals who react in predictable, uniform and therefore rather passive ways to external factors, while in practice people's aspirations and capabilities to migrate depend on factors such as age, gender, knowledge, social contacts, preferences and perceptions of the outside world. These theories generally do not consider how migrants perceive their world and relate to their kin, friends and community members. As far as they deal with structural factors, such as states, government policies or recruitment practices, at all, neoclassical approaches see them as distortions of perfect markets which affect migration costs rather than as migration drivers *in their own right*.

Historical–structural theories

Structural constraints such as limited access to money, connections and information have proven to be crucial factors affecting migration decisions. Historians, anthropologists, sociologists and geographers have shown that state actors and factors such as historical ties, recruitment practices, structural inequality and past migrations have a deep impact on migrants' decisions and behaviour (Portes and Böröcz 1989). This explains why real-life migration patterns often deviate enormously from neoclassical predictions. Instead of a random process, migration is a strongly *patterned* process because structural factors such as social stratification, market access, power inequalities and cultural repertoires constrain people's individual choices, affecting their preferences and 'channelling' their decisions in very particular directions.

An alternative explanation of migration was provided in the 1970s and 1980s by what came to be called the *historical–structural approach*. Instead of an optimisation mechanism as in neo-classical theories, historical–structural theories tend to see migration as an exploitation mechanism. Historical–structural theories view the *control and exploitation of labour* by states and corporations as vital to the survival of the capitalist system. While neoclassical and other 'functionalist' theories reduce state regulation to one of the 'intermediate' factors influencing migration costs, historical–structural theories see states, multinational corporations and employment agencies as drivers of migration processes in their own right.

Historical–structural theories see migration as one of the manifestations of capitalism, imperialism and the unequal terms of trade between developed and underdeveloped countries (Massey *et al.* 1998: 34–41). Migration is seen mainly as a way of mobilizing cheap labour for capital. They stress that the availability and control of migrant labour is both a legacy of colonialism and the result of war and structural international inequalities (see Cohen 1987). Functionalist migration theories were developed to explain migrations which are seen as largely spontaneous, voluntary and unconstrained, like most internal migrations, or the migrations from Europe to the US before 1914. Historical–structural accounts of migration stress the key role of large-scale recruitment of labour, whether of indentured Indian workers by the British for the railways in East Africa, Mexicans for the US agribusiness, or Turks and Moroccans for the factories and mines of Germany, France and the Netherlands, in shaping contemporary migration systems.

Photo 3.1 Italian emigrant family on their way to America, departing from the central rail station in Milan, 1889

Source: Giuseppe Primoli/inv. 2478/A, Archivio Primoli, Fondazione Primoli

Historical–structural theories criticize neoclassical approaches by arguing that migrants do *not really* have a free choice because they are fundamentally constrained by structural forces. Within this perspective, inhabitants of rural areas are forced to move because traditional economic structures have been undermined as a result of their incorporation into the global capitalist economy and social transformations accompanying the mechanization of agriculture, concentration of landownership, and the indebtedness and dispossession of smallholder peasants. Through these processes, rural populations consisting of peasants, farm workers and craftspeople would have become increasingly deprived of their traditional livelihoods. These uprooted rural populations then become part of the urban proletariat to the benefit of employers relying on their cheap labour. From this perspective, businesses have a strong interest in high immigration, as this will create a disposable, vulnerable and cheap 'labour reserve' that can be hired and fired at will, while the socioeconomic costs of unemployment (during recessions) and marginalization of migrants are borne by migrants and the general population.

Historical–structural theory emphasizes that while economic and political power is unequally distributed, the capitalist economy has the tendency to *reinforce* these inequalities unless governments intervene through taxing the rich and redistributing resources to poor people and peripheral regions. Within this context, historical–structural theory sees migration as a way of mobilizing cheap labour for capital, which primarily serves to keep wages down and boost profits of businesses and economic growth in destination countries while depriving origin countries of valuable labour and skills through the 'brain drain' (see Chapter 14). In opposition to neoclassical theory, historical–structural theory therefore argues that migration increases geographical and class-based income gaps, exploiting the resources of poor countries and poor people to make the rich even richer, contributing to increased inequalities within and between societies (Castles and Kosack 1973; Cohen 1987; Sassen 1988).

Dependency and world systems theory

The intellectual roots of such analyses lay in Marxist political economy – especially in *dependency theory*, which became influential in the 1960s. Rooted in the Latin American experience with US political and economic hegemony, Andre Gunder Frank (1966; 1969)

argued that by draining poor countries of their resources, global capitalism contributed to the 'development of underdevelopment'. From this perspective, migration can be seen as one of the very causes of underdevelopment and growing global inequality, which is the opposite of neoclassical perspectives.

Reversing the idea that developing countries will 'catch up' with rich countries, dependency theory sees the underdevelopment of 'Third World' countries as a *result* of the exploitation of their resources, including labour, though colonial interference. In the postcolonial period such dependency was being perpetuated by unfair terms of trade with powerful developed economies and hegemonic relations (Baeck 1993; Frank 1969). From this perspective, selective immigration policies privileging the already privileged (the skilled and wealthy) and discriminating against the vulnerable (lower-skilled workers and the poor in general) can be seen as another manifestation of these unequal forms of trade.

In the 1970s and 1980s, a more comprehensive *world systems theory* developed (Amin 1974; Wallerstein 1974; 1980; 1984). It focused on the way 'peripheral' regions have been incorporated into a world economy controlled by 'core' capitalist nations. The incorporation of peripheral regions into the capitalist economy and concomitant growth of multinational corporations accelerated rural change and deprived peasants and rural workers of their livelihoods, leading to poverty, rural–urban migration, rapid urbanization and the growth of informal economies.

Dependency and world systems theory were at first mainly concerned with internal migration (Massey *et al.* 1998: 35), but from the mid-1970s, as the key role of migrant workers in wealthy economies became more obvious, world systems theorists began to analyse international labour migration as one of the ways in which relations of domination were forged between the core economies of capitalism and its underdeveloped periphery. From this perspective, migration and immigration regimes depriving migrant workers of rights and protection can be seen as reinforcing the effects of hegemony and control of world trade and investment in keeping the 'Third World' dependent on the 'First'.

Globalization theory

Dependency and world systems theories were precursors of the globalization theories that emerged in the 1990s. *Globalization* can be defined as 'the widening, deepening and speeding up of worldwide interconnectedness in all aspects of contemporary social life' (Held *et al.* 1999: 2). Globalization is manifest in a rapid increase in cross-border exchanges of all sorts, ranging from finance and trade to media products and ideas. Globalization is often portrayed primarily as an economic process associated with the upsurge in foreign direct investment (FDI) and the liberalization of cross-border flows of capital, goods and services, as well as the emergence of new international divisions of labour (Petras and Veltmayer 2000: 2).

However, globalization is not just about technological and economic change: it is also a deeply political process, often conceived in normative or ideological terms. Critics of globalization argue that it is not a natural or inevitable new world order, but rather the latest phase in the evolution of the capitalist world economy, which, since the fifteenth century, has expanded into every corner of the globe (Petras and

Veltmayer 2000). The current globalization paradigm emerged in the context of neoliberal ideologies – initiated in the 1980s by the Reagan administration in the US and the Thatcher government in the UK – designed to roll back the welfare states and decrease government intervention in labour and capital markets. This process accelerated after the fall of the Berlin Wall in 1989 – often seen as the start of the era of 'neoliberal globalization' and market triumphalism.

Globalization is therefore also as an ideology about how the world should be reshaped – summed up in the *Washington consensus*, a development ideology which stresses the importance of market liberalization, privatization and deregulation as development recipes (Gore 2000; Mitchell and Sparke 2016; Stiglitz 2002). International institutions, especially the International Monetary Fund (IMF), the World Bank and the World Trade Organization (WTO) are key instruments in efforts to impose this new neoliberal economic world order, for instance through *structural adjustment programmes*, onto poor and weaker states.

The effects of globalization on migration are more ambiguous than it may seem. It is often thought that globalization has spurred migration as a consequence of growing inequality and revolutions in transport and communication technology. However, such improvements have also increased the scope for trade and the outsourcing of production and services, which has arguably replaced some forms of migration. The opening of markets and transfer of industrial production to low-wage economies – like the *maquiladoras* of Mexico, export processing zones in North Africa, the offshore production areas of Southeast Asia or call-centres around the world – have given rise to a new international division of labour. The partial relocation of production to low-wage economies is often thought to have weakened the political left and trade unions in industrial countries and shored up authoritarian regimes in the 'South' (see Froebel *et al.* 1980).

Far from weakening the nation state, neoliberal globalization can be seen as a new form of imperialism, designed to reinforce the power of core states, their ruling classes and multinational corporations whose interests they serve (Hardt and Negri 2000; Petras and Veltmayer 2000; Weiss 1997). In this process, they support corrupt and authoritarian elites in peripheral countries who play a vital role in ensuring access to export markets and the reproduction of a docile workforce. Control of migration and differential treatment of various categories of migrants have become the basis for a new type of transnational class structure (Glick and Salazar 2013; Schewel 2019). While immigration restrictions often fail to curb migration as long as labour demand persists, they lead to an increase in irregular migration and the increased vulnerability of migrants in labour markets (see Castles *et al.* 2012).

Critique of historical–structural approaches

Historical–structural theories stress structural constraints and the limited extent to which migrants are free to make choices. This has led to the criticism that some historical–structural views largely rule out human agency by depicting migrants as victims of global capitalism who have no choice but to migrate in order to survive. Such deterministic views often do no justice to the diversity of migration and the fact that many people do make active choices and succeed in significantly improving their livelihoods through migrating. It would be just as unrealistic to depict all migrants as passive victims of capitalism as it would be to depict them as entirely rational and free actors who constantly make individual cost-benefit calculations, as functionalist theories do. Such extreme views actually do no justice to neo-Marxist theories that have paid ample attention to migrants' agency by emphasizing the role of migrant workers in trade unions and industrial disputes (Castles and Kosack 1973; Lever-Tracey and Quinlan 1988).

With their assumption that capitalism uprooted stable peasant societies, historical–structural views are often based on the 'myth of the immobile peasant' (see Skeldon 1997: 7–8), which is the implicit assumption that pre-modern societies consisted of isolated, stable, homogeneous and egalitarian peasant communities, in which migration was exceptional. Historical research has shown that peasant societies were mobile (de Haan 1999; Moch 1992). Skeldon (1997: 32) pointed out that the whole idea that the Industrial Revolution uprooted peasants from their stable communities was based on a romanticized elitist view of peasants. Views that capitalism has 'uprooted' peasants by ruining egalitarian, self-sufficient communities are often based on idealized views of the past, ignoring the fact that pre-modern societies were often characterized by high mortality, conflict, famines and epidemics as well as extreme inequalities, in which entire classes, castes, ethnic groups as well as women, serfs or slaves were often denied the most fundamental human freedoms.

For instance, Vecoli (1964) argued that the notion that Southern Italian peasants (*contadini*) living in the US were 'uprooted' from the Italian countryside (Handlin 1951) was based on the myth of the Italian village as a harmonious social entity based on solidarity, communality and neighbourliness. In reality, typical Italian peasants lived in dismal and highly exploitative conditions. For them, migration to the US did provide unprecedented opportunities. In such cases, migration can be an active choice and an opportunity to escape from the constraints put on them by 'traditional' societies. This makes it difficult to portray migrants unilaterally as victims of global capitalism. In the same way, it is difficult to portray employers unliterally as exploitative capitalists, since many employers value migrants' contribution and experience, seek to retain them by giving them promotion or helping them in to obtain legal status. While labour market exploitation is an important part of the migration experience, the danger of unilaterally subscribing to historical–structural theories is a failure to understand why many workers have a strong interest in migrating in spite of the discrimination and exploitation – whether by smugglers, bureaucrats, employers or co-workers – they frequently encounter.

Conceptualizing migratory agency

Both functionalist and historical–structural migration paradigms are too one-sided to fully explain migration processes. Neoclassical approaches neglect historical causes of movements and downplay the role of the state and structural constraints, while historical–structural approaches put too much emphasis on political and economic structures, often see the interests of capital as all-determining, and fail to explain why people see an interest in migrating and working abroad – even if they are exploited. Both sets of theories are helpful to understand various dimensions and manifestations of migration, but their common weakness is that they are top-down perspectives that largely rule out agency by portraying human beings as rather passive 'reactors' to macro-forces. Neither approach offers a meaningful, realistic conceptualization of migrants' agency within a set of broader constraints. We will therefore review a third set of theories, which help to explain migratory agency within a wider set of structural constraints, such as immigration restrictions, discrimination and inequality. While dual labour market theory focuses on the ways in which the structure of labour markets in destination countries perpetuates the demand for migrant labour, the new economics of labour migration and livelihood approaches provide additional important insights into why migrants often seem so eager to do seemingly unattractive jobs.

Dual labour market

Dual labour market theory helps us to understand how the demand for skilled immigrant labour is structurally embedded in modern capitalist economies while simultaneously explaining why migrants are highly motivated to do jobs that natives shun. Piore (1979) argued that international migration is caused by structural and chronic demand within advanced economies for lower skilled workers to carry out production tasks (for example, assembly line work or garment manufacture) and to staff service enterprises (catering, cleaning, care, etc.). This challenges the popular idea that wealthy nations mainly need high-skilled migrant workers. Changes in the economic and labour market structure of receiving countries drive the demand for particular labour skills. While demands of manufacturing industries in Europe and North America were met by inflows of manual workers until the early 1970s, the growing importance of the tertiary (service) sector has triggered a demand for both highly qualified and low-skilled workers over recent decades despite the declining importance of the secondary (industrial) sector.

Through outsourcing, international corporations can move the production process to cheap labour, or they can try to replace labour with machines, computers and robots. However, particularly in the service sector, construction and intensive agriculture, not all work processes can be mechanized, automated or outsourced. Domestic supply for such low-skilled labour has dramatically decreased because women have massively entered the formal labour market and youngsters continue education for much longer. Dual labour market theory shows the importance of institutional factors as well as race and gender in bringing about labour market segmentation. A division into primary and secondary labour markets emerges (Piore 1979) marked by a growing gulf between the highly paid core workers in finance, management and research, and the poorly paid workers, in unstable, precarious and often informal jobs, who service their needs (see also Sassen 2001). The growth of the secondary sector and informal employment have been reinforced through neoliberal reforms and the concomitant de-regularization of labour markets, which have put the middle class under pressure and have increased inequalities in income and, particularly, wealth (see Piketty 2014).

The workers in the primary labour market are positively selected on the basis of their degrees and formal skills, but also often through membership of majority ethnic groups, male gender and, in the case of migrants, regular legal status – which is facilitated by selective immigration policies discriminating in favour of the educated and wealthy. Conversely, workers in the secondary labour market are disadvantaged by lack of education – or, in the case of migrant workers, a lack of formal recognition of foreign degrees – as well as by gender, race, and uncertain or irregular legal status. The increasing labelling of precarious jobs – also known as 3D jobs (dirty, difficult and dangerous) – as low-status 'migrant jobs' further decreases their attractivity for native workers. Dual labour market theory is useful to understand how the irregular status of migrants may serve employers' interests by creating a vulnerable and docile workforce. In this perspective, politicians' xenophobic discourses not only fulfil a symbolic function in order to rally voters, but can also serve to legitimize exploitation of migrants on labour markets by providing a moral justification for depriving them of their basic rights.

Motivational issues and job status are a key elements of Piore's (1979) explanation of why migration continues even when domestic unemployment is high. Even when unemployed, native workers are often generally no longer willing to do these jobs because of the low status attached to work in jobs in sectors such as planting and harvesting, gardening, personal care, cleaning, garbage collection, bricklaying, hairdressing, dry-cleaning and ironing, dishwashing and catering work. In other cases, migrants do such work as self-employed workers, such as in Mexican gardeners in the US or Polish plumbers in the EU. Since migrants are motivated to do jobs that native workers are unwilling or unable to do, employers have increasingly relied on migrant workers to fill such gaps. As long as their migrants' social frame of reference is in origin communities, these jobs often represent significant progress in terms of salary, status and future prospects. This exemplifies the need to overcome the 'receiving country bias' by learning to also understand migration from an origin country perspective.

New economics and household theory approaches

The *new economics of labour migration* (NELM) emerged as a critical response to neoclassical migration theory. The economist Stark (1978; 1991) argued that, in the context of migration in and from the developing world, migration decisions are often not made by isolated individuals, but usually by families or households. NELM also highlights factors other than income maximization as influencing migration decision-making. First, NELM sees migration as *risk-sharing* behaviour of families or households. Such social groups may decide that one or more of their members should migrate, not primarily to obtain higher wages, but to diversify income sources in order to spread and minimize income risks (Stark and Levhari 1982), with the money remitted by migrants providing income insurance for households of origin. Importantly, NELM sees migration as an investment by families, who pool resources to enable the migration of one or more household member.

For instance, the addition of an extra source of income can make peasant households less vulnerable to environmental hazards such as droughts and floods. This risk-spreading motive is a powerful explanation for the occurrence of internal and international migration *even in the absence of wage differentials*. This helps us to explain the continuation of large-scale rural-to-urban migration in developing countries that has so frequently puzzled – and frustrated – policy makers. Notwithstanding the frequently challenging conditions in cities, rural-to-urban migration allows families to diversify their income besides improving their access to education, health care and economic opportunities.

Second, NELM sees migration as a family or household investment strategy to provide resources for investment in economic activities, such as the family farm or another small business. NELM examines households in the context of the imperfect credit (capital) and risk (insurance) markets that prevail in most developing countries (Lucas and Stark 1985; Stark and Levhari 1982; Taylor 1999). Such markets are often not accessible for non-elite groups. In particular through remittances, households can overcome such market constraints by generating capital to invest in economic activities and improve their welfare (Stark 1980). While international remittances usually receive most attention, internal remittances are also an important source of livelihood improvement (Housen *et al.* 2013).

Third, NELM argues that *relative deprivation* (or poverty), rather than *absolute poverty*, within origin communities are important migration-motivating factors. While the extremely poor are generally deprived of the capability to migrate over larger distances, the feeling of being less well-off than other community members can be a powerful incentive to migrate in order to attain a similar of higher socioeconomic status. This corroborates empirical evidence showing that although international inequality can obviously motivate people to migrate, it has limited explanatory power compared to the role of community-level income inequalities (see Czaika and de Haas 2012; de Haas *et al.* 2019a). This complements Piore's (1979) dual labour market theory, which argued that migrants are often motivated to do jobs that seem underpaid and unattractive to native workers (as long as the origin community remains their prime social reference group) as such work allows them to make huge progress *in comparison* to what they could have earned at home.

With NELM, economists began to address questions of household composition traditionally posed by anthropologists and sociologists (Lucas and Stark 1985: 901). NELM has therefore strong parallels with so-called 'livelihood approaches' that evolved from the late 1970s among geographers, anthropologists and sociologists conducting micro-research in developing countries. Questioning dependency theory, they observed that the poor cannot be reduced to passive victims of global capitalist forces but exert human agency by trying to actively improve their livelihoods despite the difficult conditions they live in (Lieten and Nieuwenhuys 1989).

This went along with the insight that – particularly in circumstances of uncertainty and economic hardship – people organize their livelihoods not individually (as neoclassical theories assume) but within wider social contexts. The household or family was often seen as the most appropriate unit of analysis, and migration as one of the main strategies which households employ to diversify and secure their livelihoods (McDowell and de Haan 1997). Rather than a response to emergencies and crises or a 'desperate flight from misery', field studies showed that migration is often a pro-active, *deliberate* decision to improve livelihoods and to reduce fluctuations in rural family incomes by making them less dependent on climatic vagaries and market shocks (de Haan *et al.* 2000: 28; McDowell and de Haan 1997: 18).

This shows that migration cannot be sufficiently explained by focusing on income differences alone. Factors such as social insecurity, income risk and inequality, difficult access to credit, insurance and product markets can also be important migration determinants. For instance, as Massey *et al.* (1987) point out, Mexican farmers may migrate to the US because, even though they have sufficient land, they lack the capital to make it productive. Migration can then become a mechanism to maintain or increase the productivity of their farms while working in the US. Household approaches seem particularly useful to explain migration within and from developing countries and also of disadvantaged social groups in wealthy countries, where the lack of socioeconomic security and the risk of falling into absolute poverty increase the importance of mutual help and risk sharing within families.

Household models have been criticized because they can obscure intra-household inequalities and conflicts of interest along the lines of gender, generation and age (de Haas and Fokkema 2010). It is thus important not to lose sight of intra-household power struggles, and the fact that families may disagree about migration decisions. For instance, some migrants leave without consent of knowledge of their family members. Instead of a move to help the family, migration can also be an individual strategy for rebellious youngsters to escape from asphyxiating social control, abuse and oppression within communities, wishing to cut ties with their families (see de Haas 2009). Conversely, male or female family members may feel forced to migrate as they are socially pressured to assume the role of breadwinner although they may not aspire to leave themselves, and suffer from the loneliness, separation and estrangement from loved ones that migration often involves.

Migration transition theories

Despite their fundamental differences, functionalist and historical–structural theories share an important central assumption, which is that migration is primarily an outgrowth of poverty, development 'disequilibria' and the resulting geographical income inequalities. This assumption, which is implicitly or explicitly based on push–pull models, informs the popular idea that growing inequalities associated to neoliberal globalization have spurred migration and that migration can be significantly curbed by reducing poverty and stimulating development in origin countries. However, as we have seen, this assumption is undermined by empirical observations that development in poor societies often increases emigration (see de Haas 2010c; Hatton and Williamson 1998; Skeldon 1997; Tapinos 1990, see Box 3.1). Prominent emigration countries such as Mexico, Turkey, Morocco and the Philippines are typically not amongst the poorest. Some of the theories discussed so far already help to explain part of this paradox. For instance, taken together, human capital theories and dual labour market theory show how specialization, divisions of labour and the growing structural complexity and segmentation of labour markets can increase various forms of migratory and non-migratory mobility. NELM and livelihoods approaches help us to understand how, instead of a desperate flight from poverty and misery, migration is a *resource* and part of a deliberate strategy of families to increase their long-term wellbeing.

The migration transition

A more comprehensive explanation of why development initially tends to increase emigration is provided by migration transition theories. These theories see migration as an *intrinsic* part of broader processes of development and social transformation associated with processes of modernization and industrialization (de Haas 2010b). Transition theories conceptualize how migration patterns tend to change over the course of this development process. They argue that development processes are generally associated with *increasing* levels of migratory and non-migratory mobility, but they also stress that this relation is complex and fundamentally *non-linear*.

This idea was initially developed by Zelinsky (1971), who linked the several phases of the demographic transition (from high to low fertility and mortality) and concomitant development processes (which he called the 'vital transition') to distinctive phases in a *mobility transition*. He argued that there has been a *general* expansion of individual mobility in modernizing societies, and that the specific character of migration processes changes over the course of this transition. While pre-modern societies would be characterized by limited circular migration, Zelinsky (1971) argued that all forms of internal and international migratory and non-migratory mobility increase in early transitional societies due to population growth, a decline in rural employment and rapid economic and technological development. This was the case in early nineteenth-century Britain, just as it was in late nineteenth-century Japan and Western Europe, in early twentieth century Italy, Korea in the 1970s and China in the 1980s and the 1990s.

From this perspective, migration is driven by pervasive processes of *social transformation* all around the world (see Polanyi 1944). It is necessary to seek the main migratory impacts of capitalist development in profound transformations in production structures and labour markets. Particularly since the Industrial Revolution, the fundamental socio-economic transformation has been that of the massive transfer of economic activities and population from rural to urban areas, going along with large-scale urbanization as well as increasing cross-border movement. This was the case in nineteenth- and twentieth-century Europe, but also currently in many African, Asian and Latin American countries.

For instance, in many developing countries, the post-1945 'green revolution' has involved the introduction of new strains of rice and other crops, which gave higher yields, but required big investments in fertilizers, pesticides and mechanization. This contributed to a concentration of land ownership in the hands of wealthy farmers while mechanization inevitably decreased rural employment. As a consequence, peasants often lost their livelihoods and agricultural workers their employment, and often migrated into burgeoning cities like São Paolo, Mexico City, Casablanca, Lagos, Johannesburg, Shanghai, Calcutta, Manila or Jakarta.

In late transitional societies, international emigration decreases with industrialization, declining population growth and rising wages, and falling rural-to-urban migration. As industrialization proceeds, labour supply declines and wage levels rise; as a result, emigration falls and immigration increases. In 'advanced societies' with low population growth, residential mobility, urban-to-urban migration and circular non-migratory mobility increase, after which countries transform into net immigration countries.

Figure 3.1 summarizes this non-linear relationship between development and levels of emigration: the migration transition. Recent studies based on global migration data confirmed that countries with medium levels of development generally have the highest emigration rates (de Haas 2010b; Clemens 2014) (see Box 3.1). Historical and contemporary experiences support the idea that development initially boosts emigration, but that beyond a certain level of development, societies transform from net emigration into net immigration countries (Massey 2000). The migration transition of Southern European countries such as Spain and Italy (since the 1970s), Asian countries such as Malaysia, Taiwan and South Korea (see DeWind *et al.* 2012), and more recently Turkey, Brazil and China seem to fit within this pattern (de Haas 2010b).

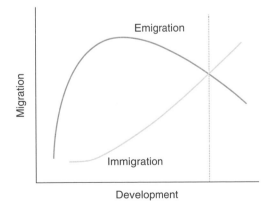

Figure 3.1   The migration transition
Source: de Haas, 2010c

Zelinsky (1971) already envisaged that technological advances do not necessarily increase migration, but can also remove the need to migrate such as through teleworking and forms of non-migratory mobility such as commuting, while rural-to-urban migration tends to slow down when urbanization reaches a certain saturation point. This can explain decreasing levels of internal migration in high-income countries, although they generally remain higher than in low-income societies. This highlights the ambiguous effects of technology on migration and may help to understand why global migration levels have not accelerated despite globalization processes.

Box 3.1 Will development stop migration?

It has often been argued that stimulating development is the only way to reduce unwanted migration. This is based on the underlying idea that much 'South–North' migration is driven by poverty. However, this ignores evidence that most migration neither occurs from the poorest countries nor from the poorest segments of the population. According to *migration transition theory* (Zelinsky 1971; Skeldon 1997), demographic shifts and economic development initially increase levels of internal and international migration.

Historical experiences seem to support the idea that societies go through migration transitions as part of broader development processes. In their study on large-scale European migration to North America between 1850 and 1913, Hatton and Williamson (1998) found that development initially boosts emigration. Migration was driven by the mass arrival of cohorts of young workers on the labour market, increasing incomes and a structural shift of labour out of agriculture towards the urban sector. The rapidly industrializing Northwestern European nations initially dominated migration to North America, with lesser developed Eastern and Southern European nations following suit only later.

However, it has long remained the subject of controversy whether these historical experiences can be generalized and apply to the contemporary world. Fortunately, recent advances in data have improved insights about the relationship between development and migration. In 2010, newly available global data on migrant populations enabled a first global assessment of the relationship between levels of development and migration (de Haas 2010c). Figure 3.2 shows how levels of emigration and immigration are related to development levels, as measured by the Human Development Index (HDI). The pattern for immigration is linear and intuitive: more developed countries attract more migrants. The relation between levels of human development and emigration is non-linear and counter-intuitive: middle-income countries tend to have the highest emigration levels.

Later studies using global data covering the 1960–2015 period confirmed that higher levels of economic and human development are initially associated with higher levels of emigration. (Clemens 2014; de Haas and Fransen 2018). Only when

▶

Labour frontiers, migration hierarchies and replacement migration

Skeldon (1990; 1997; 2012) elaborated on and amended Zelinsky's work and applied his model to actual global migration patterns. The core of his argument was that

> there is a relationship between the level of economic development, state formation and the patterns of population mobility. Very generally, we can say that where these are high, an integrated migration system exists consisting of global and local movements, whereas where they are low the migration systems are not integrated and mainly local (Skeldon 1997: 52).

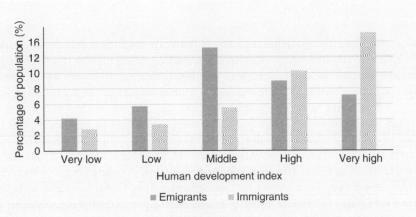

Figure 3.2   Immigrant and emigrant populations, by levels of development
Source: de Haas, 2010c

countries achieve upper-middle income status, such as has recently been the case with Mexico and Turkey, does emigration decrease alongside increasing immigration, leading to their transformation from net emigration to net immigration countries. Clemens (2014. 6) estimated that, on average, emigration starts to decrease if countries cross a wealth-threshold of per-capita GDP income levels of $7,000–8,000 (corrected for purchasing power parity), which is roughly the current GDP level of India, the Philippines and Morocco.

Development in low-income countries boosts migration because improvements in income, infrastructure and education typically increase people's capabilities and aspirations to migrate. International migration involves significant costs and risks which the poorest generally cannot afford, while education and access to information typically increases people's aspirations. Middle-income countries therefore tend to be the most migratory and international migrants predominantly come from relatively better-off sections of origin populations. Although these are averages that cannot be blindly applied to individual countries, it seems therefore very likely that any form of development in low-income countries such as in sub-Saharan Africa will lead to *more* emigration in the foreseeable future.

Skeldon distinguished the following global 'migration and development tiers':

* The old and new core countries (for example, Western Europe, North America, Japan, South Korea) characterized by net immigration;
* The *expanding core* (for example, Eastern China, Southern Africa, Eastern Europe) with high immigration *and* emigration as well as rural-to-urban migration;
* The labour frontier (for example, Mexico, Morocco, Egypt, Turkey, Philippines, Indonesia), which simultaneously experience high emigration and rural-to-urban migration; and
* The *resource niche* (for example, many sub-Saharan countries, parts of Central Asia, Andean countries), with variable, weaker forms of migration and low international migration (Skeldon 1997).

Skeldon emphasized the functional relations between migration and development tiers. For instance, the labour frontier countries generating migration to the US and EU tend to be middle-income, industrializing and urbanization countries such as Mexico, Turkey and Morocco going through profound social transformations. In such countries, rapid economic, educational and demographic transitions coincide with a temporary surplus of educated and ambitious young adults who are often prone to migrate for cultural and economic reasons. Such countries are also better connected to wealthy countries in terms of infrastructure and flows of information, capital, goods and tourists. This combination of factors is likely to foster the emergence of migration systems between such societies. From this perspective, as countries go through migration transitions and economic core areas are expanding, the labour frontier is also shifting outward geographically, expanding the 'catchment areas' from which migrant workers are coming, either spontaneously or through recruitment, while new core areas (former labour frontier countries) transform into destinations. In the case of Europe, for instance, since the 1960s and 1970s the labour frontier has shifted across the Mediterranean, from Southern Europe to the Maghreb and Turkey. In North America the labour frontier had recently expanded from Mexico to Central America, and future shifts are plausible.

From this perspective, it is possible to conceptualize global migration patterns as consisting of *multi-layered migration hierarchies*, with migrants from middle-income countries often moving to high-income countries, but with middle-income countries attracting migrants from poorer countries in their own right. This goes back to Ravenstein's (1885) original observation for Britain that migration from counties surrounding big cities such as London and Manchester leaves gaps in the rural population, which are subsequently filled by migrants from more remote districts. This results in complex geographical migration hierarchies, with labour gaps left by people leaving semi-peripheral areas to central areas being filled by migrants from even more peripheral areas. This can set in motion a chain of interdependent migrations, which is also known as *replacement migration*, with immigrants filling the places vacated by emigrants. Examples include migration from Ukraine and Nepal to Poland in partial reaction to large-scale Polish emigration to Western Europe, or the emergence of *global care chains*, which may for instance involve '(1) an older daughter from a poor family who cares for her siblings while (2) her mother works as a nanny caring for the children of a migrating nanny who, in turn, (3) cares for the child of a family in a rich country' (Hochschild 2000: 131; see also Parreñas 2000; Yeates 2004).

The migration hump, development stagnation and migration plateaus

The concept of migration transitions needs to be distinguished from the concept of *migration hump*. They are related, but conceptually distinct, concepts. Migration transition theory focuses on long-term associations between development and migration. The idea of the migration hump refers to short- to medium-term hikes in emigration in the wake of trade reforms and other economic shocks. Within the context of the North American Free Trade Agreement (NAFTA), Martin (1993) and Martin and Taylor (1996) argued that adjustment to new economic market conditions is never instantaneous and may lead to economic dislocations and rising unemployment. While the negative impacts of liberalization (particularly in previously protected sectors such as agriculture through imports of cheap US agrarian products) are often immediate, they argued that the expansion of production in sectors *potentially* favoured by trade reforms always takes time (Martin and Taylor 1996: 52).

For instance, free trade can drive peasants out of business who cannot compete with cheap imports of mass-produced agrarian products such as wheat. So, we can expect more short-term migration even if the long-term effects of free trade would be beneficial. In line with this argument, evidence suggests that for this reason NAFTA contributed to increasing migration from Mexico to the US in the first 15–20 years after the enactment of the trade agreement (Mahendra 2014b).

Migration hump theory can also be applied to fundamental political–economic reforms or shocks in other contexts, for instance to analyse the migratory consequences of post-communist reforms in Russia and Central and Eastern Europe. For instance, the economic dislocations generated by the sudden shift of political–economic regimes and market liberalization in Central and Eastern Europe after 1989 contributed an unprecedented emigration surge, particularly from countries which dismantled social security systems (Kureková 2013).

Unequal terms of trade, higher productivity and economics of scale in wealthy countries may structurally harm the competitiveness of poorer countries, leading to development stagnation. Under such circumstances, liberalization can lead to further concentration of economic activities in wealthy countries along with sustained migration of labourers to support them. This may result in a migration plateau of sustained out-migration (Martin and Taylor 1996) with may last for generations, which seems to have occurred in countries such as Ireland, Italy, Mexico, Morocco, Egypt and the Philippines. Migration transitions may also be reversed, if countries transform from net immigration into net emigration countries, as happened in Argentina and other South American nations over the second half of the twentieth century (see Chapter 7). This shows the continued relevance of historical–structural theories, as they show that 'development' is not inevitable and how global inequalities can sustain or worsen situations of underdevelopment.

This reveals that transition theories should not be blindly applied to predict the future. The more general danger of the migration transition and migration hump theories is to think that development and demographic change automatically leads to certain migration outcomes or that migration transitions are inevitable or irreversible. There is significant variation across countries in terms of the levels and composition of immigration and emigration over their development experiences. It also remains to be questioned whether and under which conditions the long-term effects of economic liberalization are beneficial for poorer population groups (see Rodrik 2011). More in general, whether countries will transform from emigration into immigration countries depends on many factors such as countries' positions in the global political economy trade as well as political reform needed to create the conditions for sustained and equitable development (Castles and Delgado Wise 2008; Nayar 1994).

The aspirations-capabilities model

Transition theories show how migration is an intrinsic part of broader processes of development and social transformation. They argue that the various demographic, economic and cultural transitions that industrializing, modernizing and urbanizing societies go through initially tend to boost levels of internal and international migration. But transition theories are less strong in explaining why people would actually be motivated to migrate when such development occurs. In order to reach a better understanding of how development processes shape people's migration behaviour, it is useful to conceptualize individual migration as a function of *capabilities* and *aspirations* to move (Carling 2002; de Haas 2003).

The economist and philosopher Amartya Sen defined human *capability* as the ability of human beings to lead lives they have reason to value, and to enhance the substantive choices (or 'freedoms') they have (Sen 1999). Although Sen did not analyse migration, his framework can be applied to migration to develop a richer understanding of human mobility (de Haas 2009; 2014a). From this perspective, human mobility enhances people's capabilities and, therefore, wellbeing for (1) *instrumental* (means to an end) and (2) *intrinsic* (directly wellbeing-enhancing) reasons. The instrumental dimension is related to the idea that migration is a resource that enables people to access better opportunities. The intrinsic dimension is the wellbeing derived from the awareness of having the freedom to explore new horizons, irrespective of the question whether people actually use such mobility freedoms (de Haas 2009; 2014a).

Income growth, improved education and improved communication and transport links increase people's capabilities to migrate over increasingly large distances. The same factors are also likely to change notions of the 'good life' (for instance, away from agrarian to urban lifestyles) and increase awareness about opportunities elsewhere. This typically increases *aspirations* to migrate as local opportunities no longer match changing cultural preferences and rising material aspirations. With development, both capabilities and aspirations to migrate can increase fast (see Figure 3.3), which explains that paradox that rapid development in low- and lower-middle income societies often coincides with increasing emigration (de Haas 2003; 2010c).

Aspirations are a function of people's general life aspirations and perceived geographical opportunity structures. If people have broader life aspirations that cannot be fulfilled at home, this often generates aspirations to migrate. For instance, education in rural areas tends to increase awareness of alternative, consumerist and urban lifestyles elsewhere – potentially leading to migration aspirations. As long as aspirations grow faster than the livelihood opportunities in origin communities, out-migration is likely to continue or even increase (de Haas 2003; 2014a). This for instance explains the paradox why emigration from rural Morocco has continued unabatedly despite significant improvements in living conditions over the past decades (de Haas 2003). In a study on Ethiopia, Schewel and Fransen (2018) showed that widening access to formal education, even at the primary level, tends to increase aspirations to leave. Development agendas aiming to keep people 'on the farm' (Rhoda 1983) by providing education, building roads or other services seem therefore based on flawed assumptions about the causes of migration.

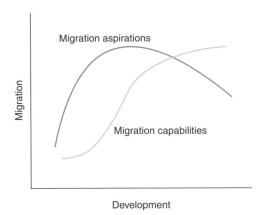

Figure 3.3   Hypothetical effect of development on capabilities and aspirations to migrate

Source: de Haas, 2010c

However, the relation between development and migration aspirations tends to be non-linear. While migration *capabilities* will further increase with development levels, migration *aspirations* are likely to decrease beyond a certain level of development, particularly when opportunity gaps with destination countries decrease significantly. However, for several reasons, migration and mobility in industrialized countries tend to remain higher than in pre-industrial, low-income societies. Besides transport and communication infrastructure, high levels of education and occupational specialization generate the migration of people who seek to match their particular skills and preferences to particular jobs. The specialization, segmentation and overall structural complexity of labour markets typically increase with education and economic development, and migration therefore remains an essential mechanism to make demand meet supply both domestically and internationally. Obviously, this labour market complexity argument is related to several elements of neo-classical theory, human capital theory, globalization theory and dual labour market theory reviewed earlier in this chapter. This shows how different theories can complement each other. Such factors help to understand why high emigration and immigration remain a structural feature of industrialized, wealthy societies.

Conceptualizing migration as a function of capabilities and aspirations to move within a given set of opportunity structures may also help to bridge certain rather dichotomous distinctions between migration categories (see Chapter 2). An example is the distinction between 'forced' and 'voluntary' migration. Rather than applying such dichotomous classifications, it seems more appropriate to conceive of a continuum running from low to high constraints under which migration occurs, in which all migrants have agency and deal with structural constraints, although to highly varying degrees (see de Haas 2009). It also shows the need to see migration as intrinsic well-being enhancing capability, resource or freedom, in itself.

Table 3.1 shows the various individual mobility types derived from the aspirations-capability perspective. This resulting perspective has the advantage of integrating voluntary and involuntary mobility as well as movement and non-movement within the same conceptual framework. From this perspective, people are only *voluntarily mobile* if they have both the aspirations and capabilities to move. Many people have the capabilities but do not aspire to move: they are the *voluntarily immobile*. Schewel (2019) proposed the category of *acquiescent immobility* to describe situations in which people are unable to migrate but also prefer to stay.

From a capabilities perspective, the term 'forced migration' may appear to be an oxymoron, because refugees need to have a certain level of agency (capabilities) in order to move. However, as already argued in Chapter 2, refugees are forced migrants *because they have no option to stay*. In situations of violent conflict, economic crisis, environmental

Table 3.1   **Aspirations and capabilities: Mobility types**

		Migration capabilities	
		Low	*High*
Migration aspirations	*High*	**Involuntary immobility** (Carling 2002) (feeling 'trapped')	**(Voluntary) mobility** (most forms of migration)
	Low	**Acquiescent immobility** (Schewel 2019)	**Voluntary immobility** *and* **Involuntary mobility** (e.g., refugees)

Source: Adapted from de Haas, 2014a

degradation or natural disaster, the most deprived are typically the ones 'forced to stay' – the involuntarily immobile cannot use migration as resource to cope with adversity and build resilience (see Lubkemann 2008). The most vulnerable often lack the resources to flee. In the same vein, restrictive migration policies can decrease capabilities to migrate among people who aspire to do so, and this can also create situations of *involuntary immobility* (Carling 2002).

It is only possible to speak about 'voluntariness' of mobility or immobility if there was a reasonable option to stay. We should therefore define human mobility freedoms not by the criterion of actual movement, but as 'people's capability (freedom) to choose where to live, *including the option to stay at home*'. The resulting view on migration does not presume either moving or staying as the norm, but acknowledges that they are the two sides of the same freedom-of-mobility coin (de Haas 2014a). Within this perspective, people can still enjoy migration capabilities without ever using them, because it adds to their sense of freedom. This may be one of the reasons why 'open borders' regimes sometimes lead to surprisingly low (permanent) emigration (see Vezzoli 2015); people who would otherwise have felt deprived of mobility options are not as obsessed with moving out at the first occasion as the involuntarily immobile tend to be.

The internal dynamics of migration processes

While the previous theories focused on the causes of migration and people's motives to migrate, another set of theories explain why migration processes, once started, tend to gain their own momentum over time (Massey *et al.* 1993). These theories focus on the social ties that are forged between origin and destination areas through reciprocal flows of people, information, ideas, money and goods. Most of these theories are interested in what motivates people and social groups to migrate, how they perceive the world and how they shape their identity during the migration process. These theories on the *internal dynamics of migration* analyse how migrants' agency leads to the emergence of intermediate, meso-level social structures through various feedback mechanisms, as well as the formation of transnational identities, and how this tends to sustain migration processes between particular areas and places. These theories underscore how, through their individual and collective agency, migrants can actively challenge and defy structural constraints such as poverty, social exclusion, oppression and migration restrictions.

Migration networks

Macro-structural factors such as warfare, colonialism, conquest, occupation, military service and labour recruitment, as well as shared culture, language and geographical proximity, often play a crucial role in the *initiation* of migration processes (Massey *et al.* 1998; Skeldon 1997). However, once a critical number of migrants have settled at the destination, other forces come into play. The choices made by recruiters or pioneer migrants influence the location choices of subsequent migrants. For instance, research on Mexican migrants in the 1970s showed that 90 per cent of those surveyed had obtained legal residence in the US through informal family and employer connections (Portes and Bach 1985).

Migration network theory explains how migrants create and maintain social ties with other migrants and with family and friends back home, and how this can lead to the emergence of social networks. Such networks are meso-level social structures which can facilitate further migration. The idea that migration is a path-dependent process with interpersonal relations shaping subsequent migration patterns is quite old (see Franz 1939; Lee 1966; Petersen 1958). Earlier scholars used the concept of 'chain migration' (Kenny 1962; Price 1963), which has been replaced with the term 'network migration' in recent research literature.

Migrant networks can be defined as sets of interpersonal ties that connect migrants, former migrants and non-migrants in origin and destination areas through bonds of kinship, friendship and shared community origin (Massey *et al.* 1993: 448). Migrant networks are a form of location-specific social capital that people draw upon to gain access to resources elsewhere (Massey *et al.* 1998). Bourdieu (1979; 1985) defined social capital as 'the aggregate of the actual or potential resources which are linked to the possession of a *durable network of* more or less institutionalized relationships of mutual acquaintance and recognition – or in other words, *to membership in a group*' (Bourdieu 1985: 248, emphasis in original French version).

Migrant networks tend to decrease the economic, social and psychological costs of migration. Migration can therefore be conceptualized as a diffusion process, in which

> expanding networks cause the costs of movement to fall and the probability of migration to rise; these trends feed off one another, and over time migration spreads outward to encompass all segments of society. This feedback occurs because the networks are created by the act of migration itself ... Once the number of network connections in an origin area reach a critical level, migration becomes self-perpetuating because migration itself creates the social structure to sustain it. (Massey 1990: 8)

Thus, besides (1) financial and (2) human capital, (3) social capital is a third resource affecting people's capabilities to migrate. Already settled migrants often function as 'bridgeheads' (Böcker 1994), reducing the risks and costs of subsequent migration and settlement by providing information, organizing travel, finding work and housing and assisting in adaptation to a new environment. Migrant groups develop their own social and economic infrastructure: places of worship, associations, shops, cafés, specialized services (such as lawyers and doctors speaking migrant languages) and other services.

The formation of a migrant community at one destination therefore increases the likelihood of more migration to the same place.

Migration intermediaries and the 'migration industry'

The internal dynamics of migration processes embrace non-migrants too: employers stimulate formal and informal recruitment and seek to retain capable workers. Krissman (2005) argued that definitions of migration networks should therefore include the employers who want new immigrant workers, as well as smugglers and other intermediaries that respond to this demand. Once a migratory movement is established, a variety of needs for special services arise. Researchers have often used the concept of *migration industry* when referring to such meso-level structures that go beyond migrant networks. The 'migration industry' would consist of employers, travel agents, recruiters, brokers, smugglers, humanitarian organizations, housing agents, immigration lawyers and other intermediaries who have a strong interest in the continuation of migration (see also Boswell and Geddes 2011: 39–41, 43–50).

Financial institutions have become part of the 'migration industry' as well, as many banks and other companies have established special transfer facilities for remittances. In a larger sense, organizations such as the IOM, the UNHCR and Frontex (the European Border and Coast Guard Agency), and private companies building and maintaining border control infrastructure, can also be seen as part of the 'migration industry'. For instance, soaring public expenditure on controlling the US–Mexican border has fuelled a military-industrial complex consisting of arms manufacturers, technology firms, (privatized) migrant detention centres, the military and state bureaucracies involved in deporting people (see Meissner *et al.* 2013). This in turn has created a huge parallel market for smugglers helping migrants across the border. In these ways, the migration control and migration facilitation industries can feed into and reinforce each other.

The term 'migration industry' has been contested because of its negative connotations, and its frequent association to illicit and profitable aspects of migration facilitation and control (Olayo-Méndez 2018). An alternative concept is 'migration intermediaries', which can be defined as actors that facilitate, and sometimes drive, migration within and across borders (Agunias 2009: 2). Although usually cast as exploitative, Agunias argues that such intermediaries can in fact be empowering for migrants:

> By providing information and extending critical services in many stages of migration and in places of origin, transit and destination, legitimate intermediaries build migrants' capabilities and expand their range of choice – the very essence of human development. (Agunias 2009: 2)

Yet their value is, in many cases, 'overshadowed by the costs they impose on migrants, from charging exorbitant fees to outright abuse of basic human rights' (Agunias 2009). There is rarely a bright line separating legitimate services and reasonable costs on one hand, and exploitative practices on the other (Agunias 2009: 60). In the same vein, the role of smugglers is often ambiguous. Contrary to politicians vilifying smugglers as 'unscrupulous' criminals, migrants often view smugglers as 'social bandits,' if not heroes,

Photo 3.2   UNHCR tents at Debaga refugee camp for refugees from Mosoul in February 2017, Debaga, Iraq

rather than as criminals (Kyle and Liang 2001: 1). Although deceit, abuse and extortion occur, most smugglers are service providers who have an interest in staying in business and therefore generally care about their reputation and have a certain interest in delivering (Pastore *et al.* 2006; Zhang *et al.* 2018) (see also Chapter 2).

Most literature on 'migration intermediaries' focuses on the role of recruiters, smugglers and other brokers, such as police officers or bureaucrats seeking to make money on the side by showing people loopholes in regulations or issuing false documents. But intermediaries can also include members of migrant communities such as shopkeepers, priests, teachers and other community leaders. It also includes humanitarian and religious organizations which are often active in organizing search and rescue operations, such as the involvement of Médecins sans Frontières in the Mediterranean Sea, or the network of shelters across Mexico that provide protection to Central American migrants on their way to the US (Olayo-Méndez 2018).

The development of dense networks of migration intermediaries is an inevitable extension of the social networks and transnational linkages which are part of the migration process. The cost and risk-reducing role of migration networks together with role of migration intermediaries have often frustrated governments in their efforts to control migration (Castles 2004) However, governments usually stimulate the growth of the 'migration industry' themselves. Even governments which initiate labour recruitment rarely provide all of the necessary services. While some countries use bilateral treaties, others, as the UK, have used private operators to contract guest-workers (GAO 2006: 21–23). Gulf and East Asian states (Japan, Taiwan and South

Korea) have made considerable use of brokers for migrant labour and (in South Korea) for marriage migration (Surak 2013). The role of the agents and brokers is vital: without them, many migrants would not have the information or contacts needed for successful migration.

Migration systems theory

While migration network theories focus on the role of social capital, *migration systems theory* looks at how migration is intrinsically linked to other forms of exchange, notably flows of goods, ideas and money; and how this changes the initial conditions under which migration takes place, both in origin and destination societies. The Nigerian geographer Mabogunje (1970), who pioneered migration systems theory, focused on the role of flows of information and new ideas (such as on the 'good life' and consumption patterns) in shaping migration systems within and across national borders.

A migration system can be defined as a 'set of places linked by flows and counter-flows of people, goods, services and information, which tend to facilitate further exchange, including migration, between the places'. Mabogunje stressed the importance of feedback mechanisms, through which information about the migrants' reception and progress at the destination is transmitted back to the place of origin. Favourable information would then encourage further migration and lead to situations of

> almost organized migratory flows from particular villages to particular cities...In many North-African cities, for instance, it is not uncommon for an entire district or craft occupation in a city to be dominated by permanent migrants from one or two villages. (Mabogunje 1970: 13)

Migration systems link people, families and communities over space. This encourages migration along certain spatial pathways or migration corridors, and discourages it along others: 'The end result is a set of relatively stable exchanges; yielding an identifiable geographical structure that persists across space and time' (Mabogunje 1970: 12). While Mabogunje focused on rural–urban migration within Africa, Kritz *et al. (1992)* and others applied this framework to international migration. International migration systems consist of countries – or rather places within different countries – that exchange relatively large numbers of migrants, and concomitant flows of goods, capital (remittances), ideas and information (see also Fawcett 1989; Gurak and Caces 1992). Migration systems can be conceptualized at various levels of analysis. In the South Pacific, West Africa or the Southern Cone of Latin America we can identify regional migration systems (Kritz *et al.* 1992). However, more distant regions may also be interlinked, such as the migration systems embracing the Caribbean, Western Europe and North America; or those linking Egypt, Sudan, Jordan and Yemen to the Gulf countries.

The key insight of migration systems theory is that one form of exchange between countries or places, such as trade, is likely to engender other forms of exchange, such as people, in both directions. Migratory movements generally arise from the existence of prior links between countries based on colonization, political influence, trade, investment or cultural ties. Thus migration from Mexico originated in the south-westward expansion of the US in the nineteenth century and the recruitment of Mexican workers by US employers in the twentieth century (Portes and Rumbaut 2006: 354–355), and both the Korean and the Vietnamese migrations to the US were consequences of US military involvement (Sassen 1988: 6–9) (see Chapter 7). The migrations from India, Pakistan and Bangladesh to Britain are inextricably linked to the British colonial presence on the Indian sub-continent (see Chapter 6). But it also works the other way around: large-scale migration between two countries tends to boost trade, capital flows, investment, travel and tourism between the same set of countries.

Cumulative causation

In a seminal paper, Massey (1990) reintroduced Myrdal's (1957) concept of *cumulative causation* to express 'the idea that migration induces changes in social and economic structures that make additional migration likely' (Massey 1990: 4–5). We can conceptualize such broader migration-affected changes in the communities and societies, as 'contextual feedback mechanisms' (de Haas 2010a), which, in their turn, often have a stimulating effect on subsequent migration. The money remitted by migrants is a good example. Remittances can increase income inequality in origin communities, which can subsequently increase feelings of relative deprivation and, hence, and in line with NELM theory (see above), migration aspirations among non-migrants. Relative deprivation and migration-facilitating network effects often reinforce each other, while remittances may also be used to pay for new journeys. While pioneer migrants are often relatively well-off, such feedback mechanisms can make migration more accessible for poorer groups and lead to a diffusion of migration within and across communities (de Haas 2010a; Jones 1998a; Massey 1990).

Information is not only instrumental in facilitating migration by increasing people's migratory capabilities, but new ideas and exposure to new lifestyles conveyed by migrants may also change people's cultural repertoires, preference and aspirations. Levitt (1998) coined the term *social remittances* to capture this flow of ideas, behavioural repertoires, identities and social capital from destination to origin communities (see also Levitt and Lamba-Nieves 2011). Such forms of cultural exchange can give rise to transnational and diasporic identities (see Chapter 4) that also tend to encourage migration in established migration corridors linking particular places and communities.

If migration becomes strongly associated with success, migrating can give rise to a *culture of migration* in which migration becomes the norm and staying home is associated with failure (de Haas 1998; Massey *et al.* 1993). Such migration-affected cultural change combined with social remittances can further strengthen migration aspirations. Other examples of contextual feedback include the formation of immigrant-specific economic niches in destination countries which create a specialized demand for co-ethnic workers (for example, Chinese cooks, Kosher or Halal butchers, musicians and clergymen). Table 3.2 summarizes the main contextual feedback mechanisms which have been identified in the literature. It also shows the extent to which social, economic and cultural transformation processes in origin and destination societies associated with migration are interrelated, highlighting the need to study them together.

Explaining migration system breakdown

Theories on migrant networks, migration intermediaries, migration systems and cumulative causation are useful to understand the crucial role of migrants' agency in creating meso-level social, cultural and economic structures which can make migration processes

Table 3.2   **Feedback mechanisms perpetuating migration processes**

Level	Domain		
	Social	*Economic*	*Cultural*
Intermediate (migrant group)	Migrant networks; migration intermediaries, 'Migration industry'	Remittance-financed migration	Transfers of migration-related ideas and information
Origin community (contextual)	Social stratification and relative deprivation	Income distribution, productivity and employment	Social remittances; cultures of migration
Destination community (contextual)	Patterns of clustering, integration and assimilation	Demand for migrant labour generated by clusters of migrant businesses and sectors where immigrants concentrate	Transnational identities, demand for marriage partners from origin countries

Source: Adapted from de Haas, 2010a

self-sustaining. However, these theories also have a number of weaknesses. First, they cannot explain why most initial migrations by pioneer migrants do *not* lead to the formation of migration networks and migration systems (de Haas 2010a). Second, through their focus on migration-facilitating mechanisms, they also have difficulties explaining the stagnation and weakening of migration systems over time. After all, In their circular logic, migration goes on *ad infinitum* (Böcker 1994; de Haas 2010a; Massey *et al.* 1998).

They do not specify under what general conditions migrant networks and migration systems weaken, or migration to *new* destinations occurs (de Haas 2010a). This shows the need for a more critical understanding of the role of social capital in migration processes (de Haas 2010a). Portes (1998) criticized one-sided, positive interpretations of social capital by researchers such as by Coleman (1988) and Putnam (2000) by arguing that strong social capital can also have negative implications, such as exclusion of non-group members and other outsiders, excessive social and material claims on successful group members and freedom-restricting pressures for social conformity (Portes 1998).

These 'downsides' of social capital can be applied to migration to understand non-formation and breakdown of migration networks (de Haas 2010a). Tight networks may be extremely useful in facilitating migration of group members but tend to exclude outsiders. Particular ethnic, religious or class groups can monopolize the access to migration opportunities, and thus exclude non-group members. This can explain the limited diffusion of migration within and across communities. Studies on Somali refugees and Moroccan migrants showed that constant claims on support from migrants by family and friends in origin communities, can lead to social distancing and a declining willingness to provide network assistance (de Haas 2010a; Lindley 2012). This may eventually lead to the breakdown of networks. Migration assistance does not automatically happen. After all, migrants have limited resources and might not always see the arrival of more immigrants as beneficial, particularly if they are perceived to compete for jobs, housing and other scarce resources. This can explain why settled migrants can evolve from being 'migration bridgeheads' to 'migration gatekeepers', who are hesitant or outright reluctant to assist prospective migrants (Böcker 1994; Collyer 2005), or who may therefore favour restrictive immigration policies, particularly when new arrivals are perceived to be a threat to group status.

Conclusion

This chapter has reviewed the most important migration theories. One central argument is that we should conceptualize migration as an *intrinsic* part of broader processes of development, social transformation and globalization rather than a 'problem to be solved'. Instead of reducing migration to more or less passive responses to poverty and geographical inequalities, as predicted by push–pull models, neoclassical and historical-structural theories, social transformation processes associated with industrial–capitalist development tend to boost migration by increasing people's capabilities and aspirations to move. This highlights the need for a socially embedded understanding of migration as a *normal* process, rather than more conventional views of migration as a largely temporary reaction (and solution) to development disequilibria (in functional-ist thinking) or even the sign (and cause) of development failure (in historical–structural thinking). A theoretical understanding of migration also shows the need for more nuanced views, that neither buy into simplistic optimistic or pessimistic views on migration, but that have an eye for the complexity of migration processes and the diversity of its causes and impacts.

A second argument is that migration processes have internal dynamics based on social networks and other contextual feedback mechanisms, which often give migration processes their own momentum. These internal dynamics are a testimony to the *agency* of migrants, leading to the formation of immigrant communities in receiving countries, the emergence of international networks and the rise of new transnational identities, which facilitate reverse flows of money (financial remittances) and ideas (social remittances) to origin societies. By lowering social, economic and psychological costs and risks, such feedback mechanisms can stimulate more migration between particular places and countries. In this way, migrants are frequently able to defy and circumvent government restrictions, often making migration notoriously difficult for governments to control.

A third argument is that the acknowledgment of migrants' agency should not obscure the real constraints migrants face. While theories on networks, migration intermediaries and migration systems help us to understand how migrants can actively overcome structural constraints, they cannot explain why not all migrants are eager to help others to come, how migrants can behave like gatekeepers instead of bridgeheads and why networks can disintegrate over time. It would also be naïve to assume that migration continues irrespective of changes in macro-level conditions such as political transformation, economic growth or labour market dynamics. Examination of historical and contemporary migrations (see Chapters 5–9) shows that origin and destination states continue to play a major role in initiating and shaping population movements.

In their seminal survey of migration theories, Massey *et al.* (1993) argued that the various theories operate at different levels of analysis and focus on different aspects of migration, but that they all provide important insights into migration. Insights from different theories can be useful to understand particular manifestations of migration occurring in particular contexts or at different levels of analysis. For instance, while neo-classical theories seem particularly useful to understand much migration of the highly educated within and among wealthy countries, historical–structural theories and the new economics of labour migration seem useful to understand migration taking place under conditions of high constraints, poverty and oppression. It may be possible to perceive irregular migration between Mexico and the US or between Morocco and the EU as part of a labour exploitation mechanism on the macro-structural level which mainly benefits employers, while at the same time acknowledging that, for the migrants themselves, it can be a rational strategy as remittances may enable them to significantly improve the living conditions of their families. This example shows the dangers of subscribing to one par-ticular train of theoretical thought in trying to understand migration.

It is regularly argued that attempts at theorizing migration are futile because migra-tion is such a diverse and complex process. This is not a very convincing argument, because, after all, almost all social processes are complex by nature. In fact, the argument can be turned around: We actually need theories and categories in order to make sense of complex and 'messy' social realities, as they help us to see the 'wood for the trees'. Furthermore, complexity does not imply that there are no patterns and that no regulari-ties can be discerned. One can even argue that migration is actually a strongly patterned social process, with most people migrating along a select number of specific geographical pathways as a result of networks and other migration system dynamics.

The different theoretical approaches lead to different ideas for migration policy. Neoclassical economists sometimes advocate 'open borders' and 'freedom of migration', believing that this will increase efficiency and lead to a global equalization of wage levels and opportunities in the long run. However, critics argue that immigration mainly serves the interest of capital by depressing wage levels – especially for low-skilled work – and undermining the bargaining power of trade unions. This is why left-wing parties and trade unions have historically opposed recruitment of guestworkers and other low-skilled workers, or demanded safeguards, such as equal pay and conditions so as not to under-mine the position of local workers. Historical–structural perspectives also argue that migration deprives poor countries of vital human resources through the *brain drain* (see Chapter 14). This is one of the reasons why, until recently, many countries have tried to limit or prevent emigration of the skilled.

The new economics and livelihood approaches explain migration occurring within and from developing countries because of economic insecurity, inequality (relative deprivation) and market failure. Within this perspective, policies concerned merely with controlling exit or entry are unlikely to succeed, but origin country governments could perhaps affect migration indirectly through progressive taxation and other poli-cies decreasing income inequality and increasing the access of the poor to insurance, credit, public health, education, state pensions and other social benefits (see also Massey *et al.* 1998: 27). As Kureková (2011) has shown for post-communist migration from Central and Eastern Europe, in middle- to high-income societies social spend-ing and increased social security may reduce forms of migration that are primarily driven by inequality and livelihood insecurity. However, as Mahendra (2014a) showed for Indonesia, cash transfers may increase internal migration in low-income societies through releasing resource constraints.

Dual labour market theory focuses on the demand-side, emphasizing that migration is driven by a chronic demand for immigrant labour that is structurally embedded in modern capitalist economies. Strong employer demand for cheap labour that is easy to control and exploit (such as undocumented migrants) creates unregulated markets for migrant labour and opportunities for smugglers and recruitment agents, which, in combination with the migration-facilitating role of networks, can defy immigration restrictions. Governments could counteract undocumented migration mainly by increasing labour market regulation, improving workers' protection and thereby removing incentives for employing irregular or temporary workers. However, this may push up labour cost and thus render unviable various businesses in agriculture, food processing and labour-intensive services.

Network, migration systems and dual labour markets theories help to explain why migration processes, once set in motion, often gain their own momentum, often frustrating states' efforts to control migration (see Chapter 11). This seems particularly true for liberal democracies, which have limited legal means to prevent migration and settlement, in particular of family migrants and asylum seekers. World systems, globalization and, particularly, transition theories argue that labour market dynamics and development processes drive migration, and that it is very difficult to significantly affect long-term migration trends, unless states and multinational organizations introduce rather radical changes in their political and economic systems. The key lesson is perhaps that while states are in many ways shaping migration processes, this influence is primarily felt through 'non-migration' policies.

Guide to Further Reading

Some valuable overviews of migration theory are available: Massey *et al.* (1993) is a seminal review of migration theories. Brettell and Hollifield (2014) survey theoretical contributions across a broad range of disciplines. Massey *et al.* (1998) give a comprehensive overview of theories and development impacts of migration. Skeldon's (1997) *Migration and Development* is essential reading for those wishing to understand how development drives migration. Hatton and Williamson's (1998) book illustrates how massive migrations from Europe to North America in the late nineteenth and early twentieth centuries were embedded in demographic and socioeconomic transformation processes on either side of the Atlantic. Important collections on migration theory can be found in special issues of the *Journal of Ethnic and Migration Studies* (Castles *et al.* 2010; Carling and Collins 2018) and *Population, Space and Place* (Smith and King 2012).

Sassen (1988) gives an original perspective on the political economy of migration, while Borjas (1990; 2001) presents the neoclassical view. Stark (1991) offers a comprehensive introduction to the new economics of labour migration, while Stark and Bloom (1985) provide a useful summary. Carling (2002) introduces the concept of *involuntary immobility,* while de Haas (2009; 2014a) provides introductions to the aspirations-capabilities model. Massey (1990) provides an introduction

▶

into cumulative causation theory. De Haas (2009) reviews theories on the internal dynamics of migration processes.

Students are encouraged to read classic theoretical texts which are still surprisingly relevant. Ravenstein's (1885; 1889) pioneering articles which analyse British and West-European migration are still useful to gain a foundational understanding of migration patterns and the history of migration research. Building upon Ravenstein's work, Lee's (1966) *Theory of Migration* provides a concise, well-written and useful introduction into key patterns of migration. Mabogunje (1970) provides a useful introduction into migration systems theory, primarily based on African examples. Zelinsky's (1971) *Hypothesis of the Mobility Transition* provides a comprehensive overview into how demographic transitions and modernization have reshaped migration and mobility patterns. Piore's (1979) *Birds of Passage* is one of the best books ever written about migration, explaining why the demand for migrant labor is structurally built in the economies of industrialized countries.

Extra resources can be found at: **www.age-of-migration.com**

4 Migration, Ethnicity and Identity

Although each migratory movement has its own specific historical patterns, it is possible to generalize on the social dynamics of the migratory process. Most economic migrations are undertaken by young, active people with a significant level of education, skills and ambitions. Particularly in developing countries, labour migrants are often target-earners, who want to save money to improve conditions at home, by buying land, building a house, getting married, setting up a business, or paying for education. After a period in the destination country, some of these migrants return home, but others prolong their stay, or return and then re-emigrate. The latter pattern is known as 'circular migration'. As time goes on, many erstwhile temporary migrants get their spouses to join them, or find partners in the new country. With the birth of children, settlement generally takes on a more permanent character. This also applies to many non-economic forms of migration, such as refugee migration.

It is these powerful internal dynamics of migration processes (see Chapter 3) that often confound the expectations of the migrants and undermine the objectives of policy-makers (Böhning 1984; Castles 2004). In many migrations, there is no initial intention of family reunion and permanent settlement. However, when governments try to stop flows they may find that the movement has become self-sustaining through network dynamics. What started off as a temporary labour migration is often transformed into family reunion or undocumented migration. This is a result of the maturing of the migration process and of the migrants themselves as they pass through their life cycle.

The frequent failure of policy makers and analysts to see migration as a *social process* with its own internal dynamics is at the root of many political and social problems. The source of this failure is often a one-sided reliance on outdated push–pull models or neoclassical models, which see migration as a simple response to market factors (see Chapter 3). This has led to the belief that migration can be turned on and off like a tap, by changing policies which influence the costs and benefits of mobility for migrants. But migration may continue due to social factors, even when the economic factors which initiated the movement have been completely transformed.

Such developments are illustrated by the Western European experience of settlement following the 'guestworker'-type migration from 1945 to 1973. Comparable outcomes arose from movements from former colonies to the UK, France and the Netherlands, and migration from Europe, Latin America and Asia to the US, Canada and Australia (see Chapters 6–9). One lesson of the last half-century is that it is very difficult for countries with democratic rights and strong legal systems to prevent migration turning into settlement. The situation is somewhat different in labour-recruiting countries which lack effective human rights guarantees, such as the Gulf states or some East and Southeast Asian countries, where restrictions may hinder family reunion and settlement. But even these countries find it increasingly difficult to prevent long-term settlement (see Chapters 8 and 9).

Although the initial migration dynamics can be somewhat different in the case of refugees and asylum seekers, outcomes may be similar. Forced migrants primarily leave their homes because persecution, human rights abuse or generalized violence made life there unsustainable. Most remain in the neighbouring countries of first asylum – which are usually poor and often politically unstable themselves. Onward migration to countries which offer better opportunities is only possible for a small minority – mainly those with financial resources, knowledge and social networks in destination countries (Zolberg and Benda 2001). Onward migration is motivated by the imperative of flight from violence, but also by the hope of building a better life elsewhere. Attempts by policy makers to make clear distinctions between economic and forced migrants are often hampered by these 'mixed motivations'.

The formation of ethnic minorities

The long-term effects of migration on destination societies emerge particularly in the later stages of the migratory process when migrants settle more permanently. Outcomes can be very different, depending on migrants' characteristics and the actions of destination states and societies. At one extreme, openness to settlement, granting of (dual) citizenship and acceptance of cultural diversity may allow the formation of *ethnic communities*, which are seen as part of a multicultural society, and which generally blend in or assimilate within a few generations. At the other extreme, denial of the reality of settlement, refusal of citizenship and rights to settlers, and rejection of cultural diversity may coincide with the formation of *ethnic minorities*, whose presence is frequently regarded as undesirable or problematic and can become the target of discrimination and racist violence.

Critics of immigration portray ethnic minorities as a threat to economic well-being, public order and national identity. Yet to a certain extent these ethnic minorities may in fact be the creation of the very people who fear them. Ethnic minorities may be defined as groups which

- Have been assigned a subordinate position in society by dominant groups on the basis of socially constructed markers of phenotype (that is, physical appearance or race), origins or culture;
- Have some degree of collective consciousness (or feeling of being a community) based on a belief in shared language, traditions, religion, history and experiences.

Ethnic minorities do not necessarily have to be migrants. Jews as well as Roma and Sinti ('Gypsies') in Europe, native Americans in the Americas, African–Americans in the US, Imazighen ('Berbers') in North Africa and Aboriginal populations in Australia are examples of population groups who have been around for many centuries, or whose presence even precedes the presence of majority groups. In some cases, particularly on the local and regional level, 'minority groups' may even be larger in numbers than majority groups. What characterizes an ethnic minority is its subordinate position vis-à-vis more powerful groups. So, it is power asymmetry rather than population size *per se* that defines a minority.

An ethnic minority is therefore a product of both 'other-definition' and of 'self-definition'. *Other-definition* means ascription of undesirable characteristics and assignment to inferior social positions by dominant groups. *Self-definition* refers to the consciousness of group members of belonging together on the basis of shared cultural, religious and social characteristics. The relative strength of these processes varies. Some minorities are mainly constructed through processes of marginalization (which may be referred to as *racism* or *xenophobia*) by the majority or dominant group. Others are mainly constituted on the basis of cultural and historical consciousness (or ethnic identity) among their members.

Ethnicity

In popular usage, *ethnicity* is usually seen as an attribute of minority or migrant groups, but most social scientists argue that everybody has ethnicity, defined as a sense of group belonging based on ideas of common origins, history, culture, experience and values (Fishman 1985: 4; Smith 1986: 27). These ideas change only slowly, which gives ethnicity durability over generations and even centuries. But that does not mean that ethnic consciousness and culture within a group are homogeneous and static. Rather, ethnicity is based on the linguistic and cultural practices through which a sense of collective identity is 'produced and transmitted from generation to generation, *and is changed in the process*' (Cohen and Bains 1988: 24–25, emphasis in original).

Scholars differ in their explanations of the origins of ethnicity; one can distinguish primordialist, situational or instrumental approaches. The anthropologist Geertz, for example, saw ethnicity as a *primordial attachment*, which results 'from being born into a particular religious community, speaking a particular language, or even a dialect of a language and following particular social practices. These congruities of blood, speech, custom and so on, are seen to have an ineffable, and at times, overpowering coerciveness in and of themselves' (Geertz 1963 quoted from Rex 1986: 26–27). In this approach, ethnicity is not primarily a matter of choice; it is pre-social, almost instinctual, something one is born into.

Other anthropologists prefer a concept of *situational ethnicity*. Members of a specific group decide to 'invoke' ethnicity as a criterion for self-identification. This explains the variability of ethnic boundaries at different times. The markers chosen for the boundaries are also variable, generally emphasizing cultural characteristics, such as language, shared history, customs and religion, but sometimes including physical characteristics (Wallman 1986: 229). In this view there is no essential difference between the drawing of boundaries on the basis of cultural difference or of phenotypical difference (or race).

Some sociologists reject the concept of ethnicity altogether, seeing it as 'myth' or 'nostalgia', which cannot survive against the rational forces of economic and social integration in large-scale industrial societies (Steinberg 1981). Yet it is hard to ignore the continued significance of ethnic mobilization. Studies of the 'ethnic revival' of the 1960s by the US sociologists Glazer and Moynihan (1975) and Bell (1975) emphasized the instrumental role of ethnic identification: phenotypical and cultural characteristics were used to strengthen group solidarity, in order to struggle for market advantages, or to make claims for increased allocation of resources by the state. This notion of ethnic mobilization as instrumental behaviour has its roots in Max Weber's concept of *social closure*, the process whereby a status group establishes rules and practices to exclude others, in order to gain a competitive advantage (Weber 1968: 342). This does not imply that markers such as skin colour, language, religion, shared history and customs are not real, but rather that the decision to use them to define an ethnic group is a strategic choice.

Whether ethnicity is primarily seen as 'primordial', 'situational' or 'instrumental', the key point is that ethnicity leads to identification with a specific group, but its visible markers – phenotype, language, culture, customs, religion, dress, behaviour – may also be used as criteria for exclusion by other groups. Ethnicity only takes on social and political meaning when it is linked to processes of boundary-drawing with other groups. Becoming an ethnic minority is not an automatic or necessary result of immigration, but rather the consequence of marginalization processes.

Race and racism

The visible markers of a phenotype (skin colour, features, hair colour and so on) correspond to what is popularly understood as *race*. Most scientists agree that classification of humans into races is unsound, since genetic variance within any one population is greater than differences between groups. Race can therefore be seen as a *social construct* produced by racism (Miles 1989). *Racism* may be defined as the process whereby social groups categorize other groups as different or inferior, on the basis of phenotypical or cultural markers. This process involves the use of economic, social or political power, and generally has the purpose of legitimating exploitation, discrimination or exclusion of the group so defined.

The use of the term 'race' varies from country to country and from language to language. In Anglo-American usage, race and racism historically referred to differences of skin colour, but have now been extended to include all types of 'visible minority' (the latter originally a Canadian term), including those distinguished by cultural markers like clothing (for instance headscarves). In Europe, race and racist ideologies around inherent white and Christian superiority have been used to justify colonial conquest, slavery and exploitation, as well as with regard to anti-Semitism. However, particularly as a consequence of the Holocaust (see Chapter 5) – and its association with Nazi eugenics – the term race is controversial and has been largely banned from public debates in many continental European countries, to be partly replaced by terms such as 'discrimination' and 'xenophobia'.

In any case, the debate over terminology can become rather sterile: it is more important to understand the phenomenon and its causes. Racism (or xenophobia) towards certain migrant (and non-migrant) groups can be found in virtually all societies. Racism implies making predictions about people's character, abilities or behaviour on the basis of socially constructed markers of difference. The power of dominant groups is sustained by developing institutions and structures (such as laws, policies, administrative practices and discourses) that discriminate against subordinate minority groups. This aspect of racism is generally known as institutional or structural racism.

Racist attitudes and discriminatory behaviour on the part of members of the dominant group are referred to as informal or everyday racism (Essed 1991). Racist behaviour can often be largely unconscious, as members of majority groups are often barely aware of their racist biases and the underlying superiority thinking. Racist attitudes tend to be deeply ingrained into cultural norms language, and social practices that are taken-for-granted and seem the natural order of things. Some researchers use the term *racialization* to refer to public discourses which imply that a range of socioeconomic or political problems are a 'natural' consequence of certain ascribed physical or cultural characteristics of minority groups (Murji and Solomos 2005). An example is the way in which the perceived lack of integration of lower-skilled migrants is often ascribed to their religion (such as Islam) or 'backward' (for instance, rural or 'traditional') culture.

The historical background for racism in Western Europe and in post-colonial settler societies (like the US, Canada, Australia or New Zealand) lies in traditions, ideologies and cultural practices, which have developed through ethnic conflicts associated with nation-building and colonial expansion (see Miles 1989). Opposition to immigration in East Asian countries such as Japan and South Korea is linked to the idea that it would threaten ethnic 'homogeneity'. Recent experiences of post-colonial nation building in much of Africa and the Middle East and the concomitant need to construct a particular national identity have regularly coincided with racism, discrimination and sometimes violence against migrant and non-migrant groups that were not seen as fitting into the new nation. For instance, there has been a rise of *nativist* discourses in parallel with an increased politicization of migration throughout much of Africa (Geschiere 2009; Mitchell 2012).

An important reason for the occurrence of racism in contemporary societies is that immigration and the presence of ethnic minorities question the nation state ideologies and concomitant ideas and myths about common ancestry, identity and culture. For instance, up to the mid-twentieth century, large-scale Catholic immigration from countries such as Ireland, Poland and Italy as well as immigration of Jews and Japanese provoked nativist reactions in the US by Protestant majority groups who saw this immigration as a threat to American 'WASP' (White Anglo-Saxon Protestant) identity. As the nation's first Catholic president, John F. Kennedy's religious affiliation was still a major issue in the 1960 US election. The fact that this almost seems unimaginable now also shows that notions of national identity shift, and that outsiders often become insiders, and part of the nation, over time. The same applies to the ways in which the Chinese have incorporated themselves in much of Southeast Asia, although remnants of prejudice still exist.

Another reason for the persistence of racism seems to lie in fundamental social transformations: over the past decades, 'neoliberal' economic restructuring, deregulation and privatization have been experienced by many sections of the population as a threat to their livelihoods, jobs and social security. Since ethnic minorities have emerged at the same time, their presence is often linked to these unsettling changes, even though there is no direct causal relation. Large-scale immigration and settlement can therefore lead to reactive reassertions of nativist nationalism (Hage 1998). The tendency has been to perceive the newcomers as the cause of the threatening changes: an interpretation eagerly encouraged by the extreme right, but also taken over by mainstream politicians and parties (Davis 2012).

Nativism is a particularly virulent form of racism because it represents minorities as an outright threat, going beyond more common fears that migrants steal jobs or undercut wages. As Higham (2002: 3) argued, 'no age or society seems wholly free from unfavourable opinions on outsiders'; however, only under some circumstances does it take on the form of nativism. Analysing the emergence of anti-foreign parties in New York and other cities around 1835, Higham linked nativism to a particular kind of nationalism: 'whether he was trembling at the Catholic menace to American liberty [or] fearing an invasion of pauper labor', the nativist believes 'that some influence originating abroad threatened the very life of the nation from within' (Higham 2002: 4). We can therefore define nativism as an intense opposition to an internal minority on the grounds of its foreign (for instance 'un-American') connections (Higham 2002: 4). Nativists typically accuse minorities of a lack of loyalty to the nation and plotting a takeover from within.

In Western societies, Jews have typically been the target of nativist attacks, which was often fuelled by conspiracy theories. From the late twentieth century onwards, racism has increasingly been directed towards Muslims, not only in Western countries, but also in a number of Asian countries such as Myanmar (against the Rohingya) and India. Partly as a reaction to this, Christian and other religious minorities in the Middle East and Asia have been targeted by racist violence, sometimes encouraging their emigration.

Moreover, neoliberal policies and economic transformations have weakened the labour movement and working-class cultures, which might otherwise have provided some measure of protection, community and identity for people who feel they have lost out on globalization and economic liberalization. Some have argued that the decline of working-class parties and trade unions, as well as the decline of the middle class, and the erosion of local community networks may therefore have created the social space for racism to become more virulent (Vasta and Castles 1996; Wieviorka 1995).

Gender and migration

Racial and ethnic divisions are crucial dimensions of social differentiation affecting migrants. Other important dimensions include gender, religion, social class, position in the life cycle, generation, location and legal status. None of these aspects of social differentiation are reducible to any other, yet they constantly cross-cut and interact, affecting life-chances, lifestyles, culture and social consciousness. Immigrant groups and ethnic minorities are just as diverse as the rest of the population.

Since the nineteenth century, a large proportion of migrant workers have been female. As Phizacklea (1983) pointed out, it was particularly easy to ascribe inferiority to women migrant workers, just because their primary roles in patriarchal societies were defined as wife and mother, dependent on a male breadwinner. They could therefore be paid lower wages and controlled more easily than men. Today, migrant women still tend to be overrepresented in the least desirable occupations, such as repetitive factory work and positions in personal care and other, often lower-skilled services sectors, although there has also been increased mobility into white-collar jobs. In line with dual labour market theories discussed in Chapter 3, complex patterns of division of labour along ethnic and gender lines have developed. How gender affects work experiences of migrant women will be further discussed in Chapter 12. Apart from labour migrants, many women migrate for other reasons: as refugees, for education, for marriage or through family reunion. For instance, there has been a growth of marriage migration in Asia, which frequently leads to situations of dependency and exploitation for women from poorer countries (see Chapter 8).

The role of gender in ethnic closure is evident in immigration rules which still often treat men as the principal immigrants while women and children are mere 'dependents'. For instance, Britain used gender-specific measures to limit the growth of the black population. In the 1970s, women from the Indian subcontinent coming to join husbands or fiancés were subjected to 'virginity tests' at Heathrow Airport. The authorities also sought to prevent Afro-Caribbean and Asian women from bringing in husbands, on the grounds that the 'natural place of residence' of the family was the abode of the husband (Klug 1989: 27–29). Today, in many countries, women who enter as dependants or marriage migrants do not have an entitlement to residence in their own right and may face deportation if they get divorced. Female domestic workers, particularly if they have irregular status, are often entirely dependent on the benevolence of their employers in highly unequal patron–client relationships.

Anthias and Yuval-Davis (1989) analyse links between gender relations and the construction of the nation and the ethnic community. Women are not only seen as the biological reproducers of an ethnic group, but also the 'cultural carriers' who have the key role in passing on the language and cultural symbols to the young. Racism, sexism and class domination are three specific forms of 'social normalization and exclusion' which have developed in close relationship to each other (Balibar 1991: 49). Racism and sexism both involve predicting social behaviour on the basis of allegedly fixed biological or cultural characteristics. The interrelationship of gender, class, race, ethnicity and other social divisions is also known as *intersectionality*, particularly with regard to the insight that disadvantages and discrimination along gender and race lines often reinforce each other (Crenshaw 1989; see also Yuval-Davis 2006). Essed argued that racism and sexism 'narrowly intertwine and combine under certain conditions into one hybrid phenomenon. It can therefore be useful to speak of *gendered racism* to refer to the oppression of black women as structured by racist and ethnicist perceptions of gender roles' (Essed 1991: 31, emphasis in original).

However, gender can also become a focus for migrant women's resistance to discrimination (Vasta 1993), as this can also be the case with class through trade union activism. Some studies suggest a more positive interpretation of female migration: it can reinforce exploitation of women, but migration can also help women from patriarchal societies to gain more control over their own lives by earning their own income, establishing their own social networks and acquiring rights in destination countries (Phizacklea 1998; Schewel 2018). This frequently makes women reluctant to return to their countries of origin out of fear for losing new-won rights and freedoms (de Haas and Fokkema 2010; King *et al.* 2006: 250–251).

Culture and identity

Culture, identity and community often serve as a focus of resistance to centralizing and homogenizing forces of globalization and capitalism (Castells 1997; see also Geschiere 2009). As already outlined, cultural difference serves as a marker for ethnic boundaries. Culture plays a central role in community formation: when ethnic groups cluster together, they establish their own neighbourhoods, marked by distinctive use of private and public spaces. In turn, ethnic neighbourhoods are perceived by some members of the majority group as confirmation of their fears of a 'foreign takeover'. They symbolize a perceived threat to the dominant culture and national identity. Dominant groups may see migrant cultures as static and regressive, while clinging onto ideas about the homogeneity and static nature of their own culture. Linguistic and cultural maintenance is taken as proof of backwardness and inability to come to terms with an advanced industrial society and their unwillingness or inability to integrate. According to this position, those who do not assimilate 'have only themselves to blame' for their marginalized position.

For migrant communities and ethnic minorities, on the other hand, culture plays a key role as a source of identity and as a focus for resistance to exclusion and discrimination. Identification with the culture of origin can help people maintain self-esteem in a situation where their capabilities and experience are undervalued. But a static culture cannot fulfil this task, for it does not provide orientation in a new, strange and sometimes rather hostile environment. Migrant and minority group cultures are therefore constantly changing in their interaction with the wider society and its institutions (Schierup and Alund 1987; Vasta *et al.* 1992).

Struggles for socioeconomic inclusion and cultural integration can eventually change national identities and culture, with migrants and their descendants eventually becoming part of the 'mainstream' and, hence, redefining the nation. This almost inevitably involves some level of tension and conflict. An apparent regression, such as religious fundamentalism, may in fact be a counterreaction to a form of modernization that has been experienced as discriminatory, exploitative and destructive of group identity. Conversely, discontent can also morph into trade union activism and social protest movements.

It is therefore necessary to understand the development of ethnic cultures, the formation of ethnic communities and changes in national identity and 'mainstream' culture as different dimensions of a single process. This process is not self-contained: it depends on constant interaction with the societies and states of destination and origin countries. Immigrants and their descendants typically do not have a static, closed and homogeneous ethnic identity, but rather dynamic, multiple, transnational and sometimes diasporic identities (see next section). However, this insight should be extended to culture and identity in general, whether of individuals, groups or nations, and whether of migrants, minority or majority groups. Instead of static or homogeneous, people's identities as individuals, groups and nations are always multiple, situational and subject to change (see Sen 2007).

Particularly when cultural change is caused by outside forces, such as globalization or immigration, this can be received with curiosity, discomfort or hostility, depending on the values and preferences of different majority groups. Culture is therefore often politicized in societies experiencing high levels of immigration and settlement. Particularly in contexts where ideas of racial or ethnic superiority have lost their ideological strength and are no longer seen as acceptable, exclusionary practices against minorities increasingly focus on issues of cultural difference, although some may argue that this is racism in a different guise. At the same time, the politics of minority resistance and majority group outrage crystallize more and more around cultural symbols – as is illustrated by the political significance given to Islamic dress in France, Britain and the Netherlands (see Scott 2005).

Transnationalism and diaspora formation

Globalization has increased the ability of migrants to maintain network ties over long distances. Researchers have argued that globalization has therefore led to a rapid proliferation of transnational communities (Vertovec 1999). *Transnationalism* refers to the 'multiple ties and interactions linking people or institutions across the borders of nation-states' (Vertovec 1999: 447). Although improvements in transport and communication technologies have not increased migration they have boosted non-migratory mobility (see Chapter 3), and made it easier for migrants to foster close links with their societies of origin through (mobile) telephone, (satellite) television and the internet, and to remit money through globalized banking systems or informal channels. This has apparently increased the ability of migrants to foster multiple identities, to travel back and forth, to relate to people and to work, and to do business and politics simultaneously in distant places. This *de facto* transnationalization of migrants' lives and identities seems to challenge traditional models of migrant integration based on assimilation into the culture and society of destination societies (de Haas 2005).

Transnational activities can be defined as 'those that take place on a recurrent basis across national borders and that require a regular and significant commitment of time by participants' (Portes 1999: 464). Portes and his collaborators emphasized the significance of transnational business communities, but also noted the importance of political and cultural communities. They also made a useful distinction between *transnationalism from above* – activities 'conducted by powerful institutional actors, such as multinational corporations and states' – and *transnationalism from below* – activities 'that are the result of grass-roots initiatives by immigrants and their home country counterparts' (Portes *et al.* 1999: 221).

A much older term for transnational communities is *diaspora*. This concept goes back to ancient Greece: it meant 'scattering' and referred to city–state colonization practices. Diaspora is often used for peoples displaced or dispersed by force (for example, the Jews; African slaves in the New World), but it has also been applied to certain trading groups such as Greeks in Western Asia and Africa, the Lebanese, the Chinese or the Arab traders who brought Islam to Southeast Asia, as well as to labour migrants (Indians in the British Empire; Italians in the US; Moroccans and Turks in Europe) (Cohen 1997; Safran 1991; Van Hear 1998).

Although the term diaspora is now popularly – and often too loosely – used to denote almost any migrant community, researchers stress that diaspora communities have particular features that set them apart from other migrant communities. To defines what characterizes a diaspora, Cohen (1997) established a useful list of criteria, which include: dispersal from an original homeland, often traumatically, to two or more foreign regions; the expansion from a homeland in pursuit of work or trade, or to further colonial ambitions; a collective memory and myth about the homeland; a strong ethnic group consciousness sustained over a long time; and a sense of empathy and solidarity and the maintenance of 'transversal links' with co-ethnic members in other countries of settlement.

Source: Getty Images/Bettmann/Lewis Hine

Photo 4.1   Group of Italian arrivals ready to be processed at Ellis Island, New York City, c. 1905

Glick-Schiller (1999: 203) suggests the use of the term 'transmigrant' to identify people who participate in transnational communities based on migration. Levitt and Glick-Schiller (2004: 1003) state that 'the lives of increasing numbers of individuals can no longer be understood by looking only at what goes on within national boundaries'. However, there is a danger of overstating this point. First of all, it would be misleading to think that past migrations could be entirely understood within the context of the state. Second, although modern technology may have arguably increased its scope, transnationalism as such is anything but a new social phenomenon, as the historical cases of the Jewish and Armenian diasporas show. Ironically, public concerns about transnationalism may actually reflect the *increasing* importance of states for people's lives and identities and the increasing homogenization of national culture (Duyvendak and Scholten 2012) rather than a fundamental change in the way migrants behave and identify as such.

Third, many migrants do wish to fit in and do assimilate in fact. Second and third generations often show considerable degrees of cultural and economic assimilation, going along with the weakening of transnational ties (see also Chapter 13). Based on their survey of transnational political engagement among three Latin American immigrant groups (Colombians, Dominicans and Salvadorians) in four US metropolitan areas, Guarnizo *et al.* (2003) concluded that the number of immigrants regularly involved in cross-border political activism is relatively small. They also found that there was no contradiction between transnational activism and participation of immigrants in the political institutions of the US. This challenges the idea that transnational identities and activities necessarily stand in the way of migrant integration in destination societies. Transnational political and business activities are generally not the refuge of the marginalized, but they often include migrants with relatively high socioeconomic status who have the resources to do so.

Inflationary use of such terms as 'diasporas' and 'transnational communities' should therefore be avoided. The majority of migrants probably do not fit the transnational pattern. Temporary labour migrants who sojourn abroad for a few years, send back remittances, communicate with their family at home and visit them occasionally are not necessarily 'transmigrants'. Nor are permanent migrants who leave forever and retain only loose contact with their homeland.

State and nation

Large-scale migration, growing diversity and transnationalism may have important effects on political institutions and national identity. In the contemporary world, the approximately 200 nation states are the predominant form of political organization. They derive their legitimacy from the claims of providing security and order and representing the aspirations of their people (or citizens). The latter implies two further claims: that there is an underlying cultural consensus which allows agreement on the values or interests of the people, and that there is a democratic process for the will of the citizens to be expressed. Such claims are often dubious, for most countries are marked by diversity, based on ethnicity, class and other cleavages, while only a minority of societies consistently use democratic mechanisms to resolve value and interest conflicts. Nonetheless, the democratic nation state has become a global norm, and practices of accountable government and the rule of law are gaining ground (Giddens 2002; Habermas and Pensky 2001; Shaw 2000; Tilly 1990).

Immigration of culturally diverse people presents nation states with a dilemma: incorporation of the newcomers as citizens may undermine myths of cultural homogeneity; but failure to incorporate them may lead to divided societies, marked by severe inequality and conflict. Pre-modern states based their authority on the absolute power of a monarch over a specific territory. There was no concept of a national culture which transcended the gulf between aristocratic rulers and peasants. By contrast, the modern nation state implies a close link between cultural belonging and political identity (Castles and Davidson 2000).

A *state* can be defined as, '... a legal and political organization, with the power to require obedience and loyalty from its citizens' (Seton-Watson 1977: 1). In this traditional view, the state regulates political, economic and social relations in a bounded territory. Most modern nation states are formally defined by a constitution and laws, according to which all power derives from the people (or nation). It is therefore vital to define who belongs to the people. Membership is largely marked by the status of citizenship, which lays down rights and duties. Non-citizens are excluded from at least some of these. Citizenship is the essential link between state and nation, and obtaining citizenship is therefore of central importance for newcomers to a country.

A *nation* by contrast may be defined as 'a community of people, whose members are bound together by a sense of solidarity, a common culture, a national consciousness' (Seton-Watson 1977: 1). Such subjective phenomena are difficult to measure, and it is not clear how a nation differs from an ethnic group, which is defined in a similar way (see above). Anderson provides an answer with his concept of the nation: 'it is an *imagined* political community – and imagined as both inherently limited and sovereign' (Anderson 1983: 15). The implication is that an ethnic group that attains sovereignty over a bounded territory becomes a nation and establishes a nation state. Although the national identity may be an imagined one, and reflects an ideology that overlooks diversity and variation in the way people shape their identities, this does not matter as long as members of society perceive it as real. Anderson (1983) argued that nationalism can be a very strong, socially bonding force, as it creates bonds of solidarity and common identification amongst large groups of people that do not know each other and that can transcend divisions along class, religion and ethnicity. Nationalism should therefore not be judged negatively a *priori* unless it descends into divisive nativism and superiority thinking.

Gellner (1983) argues that nations could not exist in pre-modern societies, owing to the cultural gap between elites and peasants, while modern industrial societies require cultural homogeneity to function, and therefore generate the ideologies needed to create nations. However, both Seton-Watson (1977) and Smith (1986) argue that the nation is of much greater antiquity, going back to the ancient civilizations of East Asia, the Middle East and Europe. However, all these authors seem to agree that the nation is essentially a belief system, based on collective cultural ties and sentiments. These convey a sense of identity and belonging, which may be referred to as national consciousness.

In modern nation states, as they originally evolved in Western Europe and North America and spread around the world since, being a citizen is associated with membership in a certain national community, usually, but not always, based on the dominant ethnic group of the territory concerned. Thus, a *citizen* was always also a member of a nation, a *national*. Nationalist ideologies demand that ethnic group, nation and state should neatly coincide – every ethnic group should constitute itself as a nation and should have its own state, with all the appropriate trappings: flag, army, Olympic team and postage stamps.

In fact, such perfect congruence is rarely achieved: nationalism has always been an ideology trying to achieve such a condition, rather than an actual state of affairs, and has constantly been challenged by the never-stopping forces of social transformation and cultural change, of which migration is an intrinsic part.

The historical formation of nation states has involved the geographical extension of state power, and the territorial incorporation and control over hitherto distinct ethnic groups. These may or may not coalesce into a single nation over time. Attempts to consolidate the nation state have often meant exclusion, assimilation or even genocide for minority groups. It is possible to keep relatively small groups in situations of permanent subjugation and exclusion from the 'imagined community'. This has applied, for instance, to Jews, Catholics and Roma in various European countries, to indigenous peoples in the Americas and Oceania, and to the descendants of slaves and contract workers in the Americas, the Caribbean and East Africa. Political domination and cultural exclusion are much more difficult if the subjugated nation retains a solid territorial base, like the Kurds in Turkey, Iraq and Syria; the Imazighen (Berbers) in North Africa; native Americans in the Andes countries; the Uighurs and Tibetans in China; and the Somali in Kenya; the Scots, Welsh and Irish in the UK; the Basques in Spain and France; and the Frisians in the Netherlands.

As a consequence of rather arbitrary border drawing by colonial powers, some states, particularly in Africa and the Middle East are characterized by such high degrees of ethnic diversity that effective nation-state formation has been severely hampered and that efforts by the dominant group to impose their version of nationhood have in several cases coincided with considerable level of conflict and forced migration. While nation states provide protection to 'nationals' they can be intrinsically violent for those who are not – or no longer – seen as part of the nation, as happened with Jews under Nazi rule (when Jews were stripped of their German citizenship), Uganda's Indian heritage population under Idi Amin's rule, and the Rohingya in contemporary Myanmar.

However, high levels of diversity do not necessarily hamper nation-state construction, as long as national identity accommodates this diversity. In contrast to *ethno-states*, whose nationality concepts have been constructed around the idea of common ancestry, other states have a greater ability to recognize and accommodate considerable regional diversity in culture, language and (hyphenated) identity while retaining national political unity. Examples include states such as Indonesia, India, Canada, Belgium, Switzerland, Nigeria, South Africa, most Caribbean countries, and to some extent also the United States, the United Kingdom, Brazil and New Zealand. However, even in these cases, particular groups often still dominate in politics and economics.

Traditional models of the relationship between state and nation are particularly challenged by high international migration and the concomitant growth of ethnic diversity. Widespread fear of 'ghettoes' highlights the fact that minorities are generally perceived as most threatening when they do not assimilate fast and concentrate in distinct areas. For nativist nationalists, an ethnic group is a potential nation which does not (yet) control any territory, or have its own state. Confronted with the reality of settlement and the formation of migrant communities, many modern states have made conscious efforts to achieve cultural and political integration of minorities. Mechanisms include citizenship itself, the propagation of national languages, universal education systems and creation of national institutions like the army, or an established church (Schnapper 1991, 1994). As migrants become citizens, their political influence increases, as well as the potential of migrants to redefine the nation (see Box 4.1). The problem is similar everywhere: how can a nation be defined, if not in terms of a shared (and single) ethnic identity? How are core values and behavioural norms to be laid down, if there is a plurality of cultures and traditions?

Box 4.1 Ethnic voting bloc and the political salience of migration: The US case

The settlement of migrant groups and the formation of ethnic minorities can fundamentally change the social, cultural and political fabric of societies, particularly in the longer run. This became apparent during the US presidential election in 2012. The burgeoning minority population of the US voted overwhelmingly in favour of Barack Obama whereas the Republican presidential candidate Mitt Romney won most of the white non-Hispanic vote. According to analysis of exit polls, Latinos voted for President Obama over Romney by 71 per cent to 27 per cent. Latinos comprised 10 per cent of the electorate, up from 9 per cent in 2008 and 8 per cent in 2004. Hispanics make up a growing share of voters in key battleground states such Florida, Nevada and Colorado (Lopez and Taylor 2012).

Republican opposition to immigration reform, and estrangement from the daily lives and concerns of many Latino and other minority voters seemed to come at a political price. This particularly relates to the inability of President George W. Bush, Jr to secure immigration reforms and, more generally, it relates to strong Republican opposition with regard to immigration reform allowing the legalization of the approximately 10.7 million undocumented migrants living in the US, who are primarily of Mexican and Central American origin.

The same patterns were repeated during the 2016 elections. About 66 per cent of Latino voters voted for the Democratic candidate Hillary Clinton compared to 28 per cent for Donald Trump. Latinos comprised 11 per cent of the electorate, further up from 10 percent in 2012. By contrast, Donald Trump received 64 per cent of votes of non-college-educated whites, while 28 per cent voted for Clinton. This indicates an estrangement between the Democratic Party and lower-educated white voters.

Attitudes on race and gender were powerful forces in affecting the 2016 presidential vote. The election witnessed a dramatic polarization in vote choices of citizens based on education (Schaffner *et al.* 2018). Very little of this gap can be explained by the economic difficulties faced by less-educated whites. Rather, most of the divide appears to be associated with sexism and denial of racism, especially among whites without college degrees (Schaffner *et al.* 2018).

In the longer run, the political weight of minority groups is bound to increase. A recent study estimated that 40 million Latinos will be eligible to vote in 2030, up from 23.7 million in 2010 (Taylor *et al.* 2012). This has provoked debates amongst Republicans about the importance to increase the party's appeal to non-white voters. It might partly do so by stressing conservative values around family and religion, which are shared by several immigrant groups. As migrants and minorities get progressively incorporated in political systems, politicians can no longer afford to ignore their problems and preferences.

Citizenship

States of immigration countries have devised various policies and institutions to respond to immigration and ethnic diversity (see Aleinikoff and Klusmeyer 2000, 2001). The central issues are defining who is a citizen, how newcomers can become citizens and what citizenship means. In principle, the nation state only permits a single membership, but immigrants and their descendants often have relationships to more than one state. They may be citizens of two states, or they may be a citizen of one state but live in another. These situations may lead to 'transnational consciousness' or 'divided loyalties' and undermine the nationalist ideal of cultural homogeneity. Thus, large-scale migration and settlement inevitably provoke debates on citizenship.

Citizenship designates the equality of rights of all citizens within a political community, as well as a corresponding set of institutions guaranteeing these rights (Bauböck 1991: 28). However, formal equality rarely leads to equality in practice. In many societies, women have been excluded from full citizenship rights in the past and sometimes still are today. Moreover, the citizen has generally been defined in terms of the cultures, values and interests of the majority ethnic group. Finally, the citizen has usually been explicitly or implicitly conceived in class terms, so that gaining real political rights for members of the working class (such as the right to vote or to be a member of parliament) has been one of the central historical tasks of the labour movement.

The first concern for immigrants, however, is not the exact content of citizenship, but how they can obtain it, in order to achieve a legal status formally equal to that of other

Source: Getty Images/AFP Contributor

Photo 4.2   French President Emmanuel Macron (centre) poses with people who received French citizenship during a ceremony in Orléans, central France, in July 2017

residents. Access has varied considerably in different countries, depending on the prevailing concept of the nation. The research literature has distinguished the following ideal types of citizenship:

1. The *imperial model*: definition of belonging to the nation in terms of being a subject of the same power or ruler. This notion pre-dates the French and American revolutions. It allowed the integration of the various peoples of multi-ethnic empires (the British, the Austro-Hungarian, the Ottoman, the Russian, etc.). This model remained formally in operation in the UK until the Nationality Act of 1981. It also had some validity for the former Soviet Union. The concept almost always has an ideological character, as it helps to veil the actual dominance of a particular ethnic group or nationality over the other subject peoples.

2. The *folk or ethnic model*: definition of belonging to the nation in terms of ethnicity (common descent, language and culture), which means the exclusion of minorities from citizenship and from the nation. Germany came close to this model until the introduction of new citizenship rules in 2000; Japan and Turkey are other examples. Many African, Arab and East-Asian countries also belong to this model.

3. The *republican model*: definition of the nation as a political community, based on a constitution, laws and citizenship, with the possibility of admitting newcomers to the community, providing they adhere to the political rules and are willing to adopt the national culture. This assimilationist approach dates back to the French and American revolutions. France and the United States are the most obvious current examples, while the Roman Empire provides an ancient example.

4. The *multicultural model*: the nation is also defined as a political community, based on a constitution, laws and citizenship that can admit newcomers. However, in this model they may maintain their distinctive cultures and form ethnic communities, providing they conform to national laws. This pluralist or multicultural approach became dominant in the 1970s and 1980s in Sweden, the Netherlands, Australia and Canada, and was also influential elsewhere, although there has been a backlash against multiculturalism since the 1990s.

All these ideal types have one factor in common: they are premised on citizens who belong to just one nation state. Migrant settlement is seen as a process of transferring primary loyalty from the state of origin to the new state of residence. This process is symbolically marked by naturalization and acquisition of citizenship of the new state.

Several scholars have questioned the extent to which these models have reflected reality, and the applicability of these models to specific countries will be discussed in Chapter 13. Such models are neither universally accepted nor static even within a single country (Bauböck and Rundell 1998: 1273; Finotelli and Michalowski 2012) and many countries apply a mix of these models. Moreover, the distinction between citizens and non-citizens is becoming less clear-cut. Immigrants who have been legally resident in a country for many years often obtain a status tantamount to 'quasi-citizenship' or *denizenship* (Hammar 1990). This may confer such rights as permanent residence status; rights to work, seek employment and run a business; entitlements to social security benefits and health services; access to education and training; and limited political rights, such as voting rights in local, but not in national, elections. Such arrangements create a new legal status, which is more than that of a foreigner, but less than that of a citizen.

A further element in the emergence of quasi-citizenship is the development of international human rights standards, as laid down by bodies like the UN, the International Labour Organization (ILO) and the World Trade Organization (WTO). A whole range of civil and social rights are legally guaranteed for citizens and non-citizens alike in the states which adopt these international norms (Soysal 1994). However, the legal protection provided by international conventions can be deficient when states do not ratify them or do not incorporate the norms into their national law, which is often the case with international measures to protect migrant rights. This particularly applies to authoritarian states. Many countries, particularly in Africa, the Middle East and Asia, make it very difficult or quasi-impossible for foreigners to acquire rights and citizenship. Such denial of fundamental human rights to migrants is much more difficult for democratic states, which also limits their legal ability to control migration, for instance through family reunion or asylum (see Chapter 11).

The EU provides the furthest-going example for a new type of citizenship. The 1992 Maastricht Treaty established Citizenship of the European Union, which includes the right to freedom of movement and residence in the territory of member states and to vote and to stand for office in local elections and European Parliament elections in the state of residence (Martiniello 1994: 31; see Chapters 6 and 10).

The meaning of citizenship may change, and the exclusive link to one nation state has become more tenuous, particularly when migrants develop transnational identities. Dual or multiple citizenship has become increasingly common. Many countries have changed their citizenship rules over the last 40–50 years – sometimes several times. More and more countries accept encourage dual citizenship (at least to some extent) although such practices remain contested (Faist 2007). Emigration countries like Mexico, India, Morocco and Turkey have also loosened their rules on citizenship and nationality, in order to maintain links with their nationals abroad.

The number of countries allowing most of their citizens to hold dual citizenship has more than doubled between 1992 and 2012 (Agunias and Newland 2012). Some developing countries recognize dual membership selectively. For instance, India allows dual citizenship to non-resident Indians living in wealthy, industrialized countries but denies it to Indians working in less developed countries (Agunias and Newland 2012). Several countries – including China, Iran, Greece and Morocco – have no provision for relinquishing citizenship. For instance, Chinese-born are still considered Chinese citizens if they take another nationality, regardless of whether they desire or claim dual citizenship (Agunias and Newland 2012). In the same vein, the Moroccan state considers and treats second and third generation descendants of Moroccan migrants as Moroccan citizens (de Haas 2007a). This can be seen as an attempt by origin states to maintain political control on their emigrant populations, which can be perceived by destination countries as running counter to their integration policies.

Conclusion

This chapter has examined the long-term processes through which the settlement of migrants leads to the formation of ethnic minorities and communities, and how this often challenges dominant ideas of national identity and citizenship. One central argument has been that migration and settlement are closely related to other economic, political and cultural linkages being formed between different countries as part of larger globalization processes. A second argument is that the migration process has internal dynamics which often lead to outcomes not initially intended either by the migrants themselves or by the states concerned. The most common outcome of a migratory movement, whatever its initial character, is the settlement of a significant proportion of the migrants and the subsequent formation of ethnic communities or minorities. Thus, the emergence of societies which are more ethnically and culturally diverse must be seen as an inevitable result of initial decisions to recruit foreign workers, or to permit more spontaneous forms of migration.

The third argument concerns the processes that lead to the formation of ethnic minorities. Most minorities are formed by a combination of other-definition and self-definition. Other-definition pertains refers to various forms of exclusion and discrimination. Self-definition has a dual character. It includes assertion and recreation of ethnic identity, centred upon pre-migration cultural symbols and practices. It also includes political mobilization against exclusion, racism and discrimination, using cultural symbols and practices in an instrumental way. When settlement and ethnic minority formation take place at times of economic and social crisis, migration can become highly politicized. Issues of culture, identity and community can take on great importance for the receiving society as a whole. The fourth argument focuses on the significance of immigration for the nation state. The settlement of migrants and increasing diversity can contribute to changes in central political institutions, such as citizenship, and may affect the very nature of nation states. Migrants often maintain links in two or more countries, and sometimes form transnational communities which live across borders. This is evident in the growing salience of dual citizenship, which is seen as a threat to national identity by some.

These factors help to explain the political salience of issues connected with migration and ethnic minorities in immigration countries around the world. Continuing migrations and semi-autonomous processes of settlement and minority formation will inevitably cause new transformations, both in societies already affected and in further countries now entering the international migration arena. The more descriptive accounts of historical and contemporary migration trends that follow will provide a basis for further discussion of these ideas. Chapters 5–9 are mainly concerned with the early phases of the migratory process, highlighting the vital role of states and labour recruitment in starting migration, and showing how initial movements give rise to migratory chains, network formation and long-term settlement. Chapters 10 and 11 discuss the role of states and policies in shaping processes of migration and settlement. Chapters 12 and 13 discuss the ways in which settlement and minority formation affect the economies, societies and political systems of destination countries. Chapter 14 looks at the 'other', unfortunately often ignored, side of migration, by examining the significant impact of migration on development and change in countries of origin.

Guide to Further Reading

Two general studies on Europe (Penninx *et al.* 2006) and the US (Portes and Rumbaut 2006) provide valuable background on migration and settlement. Vertovec (2009) is a key introductory text into the concept of transnationalism and Cohen (1997) provides an extremely well-written and accessible account into the concept

▶

of diasporas. Analyses of the relationship between migration and citizenship can be found in Bauböck (1991, 1994a, 1994b), Bauböck *et al.* (2006), Aleinikoff and Klusmeyer (2000, 2001) and Castles and Davidson (2000). Useful works on gender and migration include Phizacklea (1998), Andall (2003), Pessar and Mahler (2003), Brettell (2016) and Amelina and Lutz (2018). Cohen and Bains (1988), Miles and Brown (2003), Balibar and Wallerstein (1991), Essed (1991), Solomos (2003) and Back and Solomos (2013) are good introductions on racism. Anderson's (1983) *Imagined Communities* is essential reading for gaining an understanding of modern nationalism. Gellner (1983) and Ignatieff (1994) also provide stimulating analyses of nationalism, while Smith (1986, 1991) discusses the relationship between ethnicity and nation.

Extra resources can be found at: **www.age-of-migration.com**

5 International Migration before 1945

Recent migrations may be new in scale and scope, but population movements have always been part of human history. Indeed, recent evidence from a range of scientific disciplines, including archaeology, genetics, historical linguistics and anthropology, has shown that all human beings originated in evolutionary processes that started some 7 million years ago in Africa. The spread of *Homo erectus* – a new hominid species with superior brain capacity – out of East Africa started about 2 million years ago. *Homo erectus* was able to colonize large areas of Africa and went on to establish settlements as far afield as the Middle East, the Caucasus, Java and South China. Then, about 200,000 years ago, a new hominid species – *Homo sapiens* – emerged in Africa. *Homo sapiens* – the modern human – had superior adaptive capacities, and was able to displace all other hominids, moving stage by stage to people the entire earth (see Goldin *et al.* 2011; Manning 2005).

Warfare, conquest, formation of nations, the emergence of states and empires, and the search for new economic opportunities have all led to migrations, both voluntary and forced. The enslavement and deportation of conquered people was a common early form of forced labour migration. This applies to all regions of the world (Hoerder 2002; Manning 2005; Wang 1997). From the end of the Middle Ages, the emergence of European states and their colonization of the rest of the world gave a new impetus to migrations – both voluntary and forced – of many different kinds. At the same time, the formation of states with fixed territorial boundaries, standing armies and growing bureaucracies increased the need to control people's movement.

In Europe, 'migration was a long-standing and important facet of social life and the political economy' from about 1650 onwards, playing a vital role in modernization and industrialization (Moch 1995: 126; see also Bade 2003; Moch 1992). Particularly from the mid-nineteenth century, the consolidation of modern national states and the social transformations engendered by the Industrial Revolution in Europe as well as Japan would give an impetus to large-scale rural–urban migration as well as trans-oceanic migrations towards the Americas. European capital and industrialization also provided a context for the worldwide growth of Chinese emigration (McKeown 2010: 104). This would lead to the large-scale movements from Europe, China, Japan and India to the Americas, Southeast Asia and North Asia, each of which received over 50 million migrants in the period between 1846 and 1940 (McKeown 2004, 2010).

As highlighted by theories discussed in Chapter 3, accelerating emigration was the direct consequence of broader transitions from agrarian to industrial societies and colonial pursuits pushing empires to seek resources in frontier lands, as well as the increasing globalization of trade, travel and transport. As part of these transformation processes, a growing number of former peasants, serfs and artisans would seek opportunities for wage labour or business in cities or in foreign lands.

Emigration as a way of seeking new opportunities or an escape from oppressive conditions is an important part of the history of modern societies. Millions of Europeans and Asians were able to build new lives by moving to the Americas or other settler colonies. Many never returned, but a lot did. Although data is sparse, it is believed that between 25 and 40 per cent came back to origin areas, where they improved their economic and social situation using resources gained through a sojourn abroad. Migrants and their descendants helped build new (often democratic) societies in the New World, while the broader perspectives gained through migration often proved a ferment for reform in origin areas. The movement of people was one of the great forces of change in the nineteenth and twentieth centuries. However, such change was not always positive for everyone – certainly not for the native peoples of Africa, the Americas, Asia and Oceania, who experienced displacement, marginalization, enslavement and even genocide through European colonization.

Individual liberty is often seen as one of the great moral achievements of capitalism, in contrast with earlier societies where liberty was restricted by traditional bondage and servitude. Neoclassical theorists portray the capitalist economy as being based on free markets, including the labour market, where employers and workers encounter each other as free legal subjects, with equal rights to make contracts. But this harmonious picture often fails to match reality. As Cohen (1987) has shown, capitalism has made use of both *free* and *unfree* workers in every phase of its development. Labour migrants have frequently been unfree workers, either because they are taken by force to the place where their labour is needed, or because they are denied rights enjoyed by other workers, and cannot therefore compete under equal conditions. Even where migration is formally voluntary and unregulated, institutional and informal discrimination may limit the real freedom and equality of migrant workers.

Since economic power is usually linked to political power, mobilization of labour often has an element of bureaucratic control or outright coercion, sometimes involving violence and military force. Examples include the slave economies of the Americas; serfdom in the Russian empire; indentured colonial labour in Asia, Africa and the Americas; mineworkers in colonial Southern Africa; Japan's use forced labour during its colonial rule of the Korean peninsula from 1910 to 1945; forced labourers in the Nazi war economy; 'guestworkers' in post-1945 Europe; and undocumented migrants denied the protection of law in many countries today. Trafficking of migrants – especially of women and children for sexual exploitation – can be seen as a modern form of highly exploitative bonded labour.

One important theme requires more intensive treatment than is possible in the present work: the devastating effects of international migration on the indigenous peoples of colonized countries. European conquest of Africa, Asia, America and Oceania led either to the domination and exploitation of native peoples or to genocide, both physical and cultural. Nation-building – particularly in the Americas and Oceania and to a certain extent also in the Russian empire – was based on the importation of new populations. Thus, immigration contributed to the exclusion and marginalization of indigenous peoples. One starting point for the construction of new national identities was the idealization of the destruction of indigenous societies: images such as 'how the West was won' in the US or the struggle of Australian pioneers against the Aborigines became powerful myths. The roots of racist stereotypes – today directed against new immigrant groups – often lie in historical treatment of colonized peoples. Nowadays there is increasing realization that appropriate models for intergroup relations have to address the rights of indigenous populations and other minorities, as well as those of immigrant groups.

Migration during European imperialism

Imperialism has always given rise to significant voluntary and forced population movements. For instance, military conquest by the Roman Empire, the Umayyad Caliphate, the Ottoman Empire, the Moghul Empire, as well as imperial China under the Qing dynasty brought various peoples under one single rule. Such imperial expansion coincided with significant migration and cultural interchange, but also assimilation into the dominant culture and institutions of the imperial state. European imperialism would give such processes a global dimension, laying the foundations for the modern world system. This chapter will mainly focus on the evolution of migration from a Western perspective. Later chapters will analyse the evolution of migration from the perspective of other world regions.

Colonialism and slavery

European colonialism gave rise to various types of migration (see Map 5.1). One was the large outward movement from Europe, first to Africa and Asia, then to the Americas, and later to Oceania. Europeans migrated, either permanently or temporarily, as sailors, soldiers, farmers, traders, missionaries and administrators. Some of them had already migrated within Europe. For instance, around half the soldiers and sailors of the Dutch East India Company in the seventeenth and eighteenth centuries were not Dutch but immigrants, mainly from poor areas of Germany (Lucassen 1995). The mortality of these migrant workers through shipwreck, warfare and tropical illnesses was very high, but service in the colonies was often the only chance to escape from poverty. Such overseas migrations helped to bring about major changes in the economic structures and the cultures of both the European countries and the colonies.

An important, and extremely violent, antecedent of modern labour migration is the system of *chattel slavery*, which formed the basis of commodity production in the

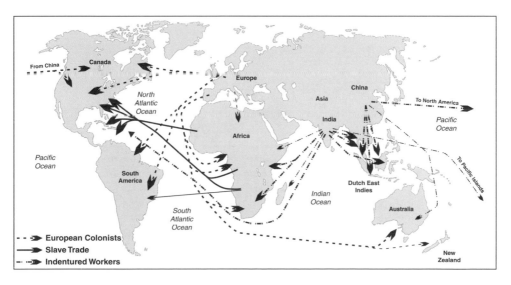

Map 5.1   Colonial migrations from the seventeenth to nineteenth centuries

Note: The arrow dimensions give an approximate indication of the volume of flows in the 1945–1973 period. Exact figures are often unavailable

plantations and mines of the New World from the late seventeenth century to the mid-nineteenth century. The production of sugar, tobacco, coffee, cotton and gold by slave labour was crucial to the economic and political power of Britain and France – the dominant states of the eighteenth century – and played a major role for Spain, Portugal and the Netherlands as well. By 1770 there were nearly 2.5 million slaves in the Americas, producing a third of the total value of European commerce (Blackburn 1988: 5). The slave system was organized in the notorious 'triangular trade': ships laden with manufactured goods, such as guns or household implements, sailed from ports such as Bristol and Liverpool, Bordeaux and Le Havre, to the coasts of West Africa. There, Africans were either forcibly abducted or were purchased from local chiefs or traders in return for the goods. Then the ships sailed to the Caribbean or the coasts of North or South America, where the slaves were sold for cash. This was used to purchase the products of the plantations, which were then brought back for sale in Europe.

An estimated 12 million African slaves were taken to the Americas before 1850 (Lovejoy 1989). Figure 5.1 shows that the numbers of African captives being shipped to the Americas went up steadily until the early nineteenth century and that the risk of dying during the journey was significant, between levels of Seven and 30 per cent. For women, hard labour in the mines, plantations and households was frequently accompanied by sexual exploitation. The children of slaves remained the chattels of the owners. In 1807, slave trafficking was abolished within the British Empire, while most European states followed suit by 1815. A number of slave rebellions broke out – notably in Saint Domingue (later to become Haiti) (Schama 2006). Although slave trafficking from Africa was abolished, slavery itself was not abolished until 1834 in British colonies, 1863 in Dutch colonies, 1865 in the Southern states of the US (Cohen 1991: 9), 1886 in Cuba and 1888 in Brazil, the last American nation to abolish slavery. Despite this abolitionist

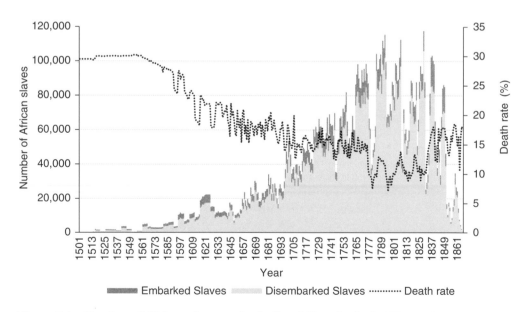

Figure 5.1   Number of African slaves embarked and disembarked, 1501–1866

Source: The Trans-Atlantic Slave Trade Database (www.slavevoyages.org)

movement, slavery therefore grew in significance over the nineteenth century. The number of slaves in the Americas doubled from 3 million in 1800 to 6 million in 1860, with corresponding growth in the area of plantation agriculture in the Southwestern US, Cuba and Brazil (Blackburn 1988: 544).

Slavery and serfdom had existed in many pre-capitalist societies around the world, but the colonial system was new in character. Its driving force was the emergence of global empires, which began to construct a world market, dominated by merchant capital. Slaves were transported great distances – across oceans – by specialized traders, and bought and sold as commodities. Slaves were economic property and were subjected to harsh forms of control to maximize their output. The great majority were exploited in plantations which produced for export, as part of an internationally integrated agricultural and manufacturing system (Blackburn 1988; Fox-Genovese and Genovese 1983). The slave trade led to the permanent settlement of sizeable African origin communities across the Americas, and particularly in Brazil, the Caribbean region and the Southern United States, where in many regions they now form a sizeable part, or the majority, of populations.

Indentured labour

In the latter half of the nineteenth century, slaves were gradually replaced by *indentured workers* as the main source of plantation labour in the Americas and the Caribbean. Indenture (or the *coolie* system) involved recruitment of large groups of workers, and their transportation to another area for work, on the basis of contracts that bound them to a particular employer for a fixed period of time. Contracts often allowed employers to sell the indentured worker's labour to a third party. Although indentured workers usually received limited cash payments, their labour could also serve to pay off debts. Particularly when it involved debt bondage, indenture comes close to what is nowadays often seen as trafficking (see Chapter 2). Although indentured workers signed up voluntarily, working and living conditions were often not what were expected. However, return rates were low as upon completion of their contracts, indentured servants were given their freedom and conditions were often still better than at home.

The practice of indentured labour is rather old, going back at least to thirteenth-century England, but became more common during European colonization of North America. During the seventeenth and eighteenth centuries, over half of European immigrants to British North America served an average period of three years' servitude as farm labourers, domestic servants or craftspeople in order to repay the cost of their transportation (Klepp and Smith 1992). This form of debt bondage was a common way for poor Europeans to migrate to the American colonies: they signed an indenture in return for a costly passage. Young men and women from countries such as Britain and Germany would work out an arrangement with the ship's captain, who would not charge money for the trans-Atlantic passage (see Klepp and Smith 1992), but sold on the indenture to an employer in the colonies who needed workers. So, shipowners were acting as contractors, hiring out their labourers. Although such practices involved various forms of exploitation that are nowadays often associated with 'trafficking', this highlights that migrants involved often have an interest in such arrangements, since they allow them to achieve long-term improvements in their living conditions.

Historians estimate that between one half and two thirds of white migrants to the American colonies between the Puritan migrations of the 1630s and the American

Revolution (1765–1783) had come under indentures (Smith 1947: 336, cited in Galenson 1984: 1; see also Donoghue 2013). In fact, unskilled servants, mostly from Britain and Ireland, made up the majority of the plantation workforce in the West Indies and several North American plantation economies until the late seventeenth century, after which sugar and tobacco planters transitioned to a labour force of perpetually enslaved Africans (Donoghue 2013: 893). So this form of indentured servitude which was initially a 'device used to transport European workers to the New World' (Galenson 1984: 1) would dwindle as black slavery grew in importance in the British colonies.

Following the abolition of slavery, indentured servitude thus reappeared in the Americas in the mid-nineteenth century as a means of transporting workers from Asia and elsewhere to the Caribbean sugar islands and South America (Galenson 1984), but also gained importance in European colonies elsewhere in the world. Particularly Chinese and Indians, but also Javanese, Japanese, Tonkinese, Madeirans, Cape Verdeans, Madagascans, Senegalese, Congolese, Mozambicans and Zanzibaris left their native land to work in the colonies of the Americas, the Caribbean, Indian Ocean (Réunion, Mauritius), the Pacific (Fiji), Asia (Indonesia) and in recently conquered territories in Africa.

British colonial authorities recruited workers from the Indian subcontinent for the sugar plantations of Trinidad, British Guiana and other Caribbean countries. Others were employed in plantations, mines and railway construction in Malaya, East Africa and Fiji. Between 1834 and 1920, an estimated 1,500,000 indentured labourers, of whom 85 per cent from India, were sent to British colonies, about one third to Mauritius, one third to the British West Indies, and the rest to Natal in South Africa (Chaillou-Atrous 2016). The British also recruited Chinese 'coolies' for British Malaya (contemporary Malaysia and Singapore) and Portuguese (from Madeira) for Guyana (see also Vezzoli 2015).

The Dutch recruited indentured workers in the Indian subcontinent and Java to work in Suriname, and used Chinese labour on construction projects in the Dutch East Indies (contemporary Indonesia) such as in Sumatra. In total, an estimated 750,000 Chinese left for the Dutch East Indies as well as Malaysia, Cuba, the British West Indies or La Réunion. Tens of thousands of Indian workers emigrated to the French colonies of La Réunion, Martinique, Guadeloupe and French Guiana (Chaillou-Atrous 2016). Up to 1 million indentured workers were recruited in Japan, mainly for work in Hawaii, the US, Brazil and Peru. Their descendants, the *Nikkeijin*, would often migrate to Japan in the late twentieth century (Shimpo 1995; see Chapters 7 and 8)). Tens of thousands of African workers were recruited to work in the West Indies and French Guiana (mostly from Congo and Senegal) as well as La Réunion, Mauritius and Natal (mostly from Mozambique or Zanzibar) (Chaillou-Atrous 2016).

Indentured workers were deployed in 40 countries by all the major colonial powers. The system involved from 12 to 37 million workers between 1834 and 1941, when indentureship was finally abolished in the Dutch colonies (Potts 1990: 63–103). Partly because of declining demand, and the demise of plantation economies, the immigration of indentured labourers gradually declined after the First World War in favour of more spontaneous migratory movements (Chaillou-Atrous 2016).

Settlement and creolization

Indentured workers were bound by strict labour contracts for a period of several years. Wages and conditions were generally very poor, workers were subject to rigid discipline and breaches

Box 5.1 Creolization and the birth of jazz

Processes of cultural mixing, also known as 'creolization' (see Cohen 2007) that went along with slavery and colonialism, have led to the emergence of new identities and forms of cultural and artistic expression in religion, food and music. The legacy of slavery is still present in widespread endemic racism and painful, and sometimes violent, political struggles for recognition and equal rights.

Some of these groups, such as the Maroons (descendants of runaway slaves) have retained some of their ancestral African traditions, blended in with Amerindian and various European influences. Processes of creolization have been strongest in the Caribbean, where descendants of forced and voluntary migrants from highly diverse backgrounds contributed to the creation of entirely new societies and cultures.

This gave rise to new, vital Creole cultures and new syncretic religions such as Winti in Suriname, Voodoo in Haiti and Candomblé in Brazil, as well as new languages, cuisines, habits and music. Creole forms of cultural expression draw from at least two prior traditions, but, essentially, they 'cannot be reduced to either' (Cohen 1987: 373). Despite, or perhaps *because of*, the fact that they were born out of contexts of slavery and extreme violence, creole cultures have often gained extraordinary power and international influence, against the odds.

The influence of creole cultures has been extremely profound, particularly in the effects of jazz on global music. From the late nineteenth century, African–Americans in the Southern US developed vital new musical forms in blues and jazz, which emerged as a fertile dialogue between black folk music in the US, often derived from the plantations and rural areas, and black music influenced by European music based in urban New Orleans:

> The field hollers met parlour music. Negro spirituals met those who liked the opera. Those who played low-down blues met those who danced the waltz, the mazurka, the polka, and the quadrille. In the parades, the funeral dirges, the popular songs for picnics and parties, jazz developed as a Creole music par excellence. The honky tonks, the brothels, the picnic grounds, parks, and the streets of New Orleans were the testing grounds for a music that first captured the American South, then generated what is probably the world's most powerful music form since the development of European classical music. (Collier 1978: 59–64, citing Cohen 2007, 373)

The creation and worldwide spread of jazz gave an extremely powerful impetus to musical art, which stood at the basis of the development of modern popular music. Most forms of popular music, including soul, rock 'n' roll, rock, country, rap and their various spin-offs, further transformations and subsequent mixing with local styles around the world, are indebted to the legacy of African–American music.

Although born from the cruelty and suffering of slavery and racism, the emergence and spread of blues and jazz testify to the refusal of the oppressed to give up as well as their determination to express their pain, anger and hope.

of contract were severely punished. Indentured workers were often cheaper for their employers than slaves (Cohen 1991: 9–11). On the other hand, work overseas offered an opportunity to escape poverty and oppression, such as the Indian caste system. Upon completion of
their indenture period, many workers were ultimately not repatriated to their origin country
at the end of their indenture period, and opted to stay and bring in dependants (Chaillou-
Atrous 2016). Many workers thus remained as free settlers in East Africa, the Caribbean, Fiji
and elsewhere, where they could obtain land or set up businesses (Cohen 1995: 46). Some
remained labourers on large estates, while others became established as a trading class, medi-
ating between the white and mixed-race ruling class, and the black majority.

The Caribbean experience shows the devastating effects of colonial, profit-driven
labour practices on dominated peoples: the original inhabitants, the Caribs and Arawaks,
were wiped out completely by European diseases and violence. With the development of
the sugar industry in the eighteenth century, Africans were brought in as slaves. After
emancipation in the nineteenth century, former slaves generally became small-scale
subsistence farmers, and were replaced with indentured workers from India. Indenture
epitomized the principle of divide and rule, and a number of postcolonial conflicts (for
example, hostility against Indian heritage populations in East Africa, Guyana, Suriname
and Fiji, or against Chinese in Southeast Asia) have their roots in such divisions. For
instance, freed slave populations often considered that immigration of indentured workers undermined their struggle for better working conditions and pay.

Despite attempts to reduce native American groups and diverse ethnic groups imported
from Africa and Asia to a subhuman and docile workforce of slaves, serfs and bonded
labourers, efforts by colonial powers to suppress the customs and beliefs of slaves and
servants were never completely successful, as they often retained their religion, language,
music, dance and culinary practices (Chaillou-Atrous 2016). The mixing of people and cultural exchange gave rise to new syncretic cultures in a process also known as *creolization*
(Cohen 2007), contributing to the extraordinary ethnic, cultural and religious diversity
and vitality of colonial societies in the Americas, the Caribbean and elsewhere, particularly on island nations such as Mauritius, Réunion and Fiji. This led to the emergence of
new forms of music, such as blues, jazz and reggae, which would transform musical cultures around the world (see Box 5.1). The movement of indentured workers also laid the
roots for the later expansion of worldwide diasporas, particularly of Indians and Chinese.

Industrialization and trans-Atlantic migration

The wealth accumulated in Western Europe through colonial exploitation provided much
of the capital which was to unleash the industrial revolutions of the eighteenth and nineteenth centuries. In Britain, profits from the colonies were invested in new forms of
manufacturing, as well as encouraging commercial farming and speeding up the enclosure
(through purchase or dispossession) of small landholdings and common arable land by large
landowners. The displaced tenant farmers swelled the impoverished urban masses available
as labour for the new factories. This emerging class of wage-labourers was soon joined by
destitute artisans, such as hand-loom weavers, who had lost their livelihood through competition from the new manufacturers. Herein lay the basis of the new class which was crucial
for the British industrial economy: the 'free proletariat', which was free of traditional bonds,
but also of ownership of the means of production, and hence available to sell their labour –
this process is also known as *proletarianization*.

However, *unfree labour* continued to play an important part. Throughout Europe, draconian poor laws were introduced to control the displaced farmers and artisans – the 'hordes of beggars' who were seen as threatening public order in towns and cities. Workhouses and poorhouses were often the first form of manufacturing, where the disciplinary instruments of the future factory system were developed and tested (Marx 1976: Chapter 28). In Britain, 'parish apprentices', orphan children under the care of local authorities, were hired out to factories as cheap unskilled labour. This was a form of forced labour, with severe punishments for insubordination or refusal to work. In Russia, serfdom was officially abolished only in 1861, although practices of bonded labour would endure for decades.

However, for many peasants, farm workers and former serfs, migration did often represent hope for a better future, and a genuine chance to escape from the constraints as well as uncertainties of peasant life. The transition from agrarian-rural to industrial-urban economies in combination with demographic transitions and improved access to information as well as cheaper, faster and safer travel (mainly as a result of the advent of train transport and steamships), as well as exposure to the stories and letters from family and friends who already left, increased the number of young people eager to migrate (see Thomas and Znaniecki 1918; Hatton and Williamson 1998). While this primarily translated into increasing rural-to-urban migration within and between fast urbanizing European societies, it would increasingly spill over in long-distance migration for the growing number of people who could afford the trip (see Map 5.2). In addition, some non-European groups from colonies in the Middle East and North Africa, such as the Lebanese, would join these trans-Atlantic migrations. Between 1846 and 1939, some 59 million people left Europe (Stalker 2000: 9), mainly for areas of settlement in North and South America, but also for Australia and New Zealand, as well as settler colonies in Algeria and parts of East and Southern Africa.

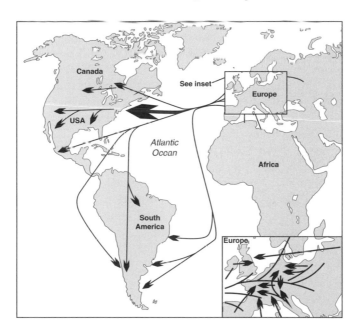

Map 5.2　Labour migrations connected with industrialization, 1850–1920

Note: The arrow dimensions give an approximate indication of the volume of flows. Exact figures are often unavailable

Migration to the Americas

Not coincidentally, the peak of the Industrial Revolution was the main period of British migration to America: between 1800 and 1860, 66 per cent of migrants to the US were from Britain, and a further 22 per cent were from Germany. From 1850 to 1914 most migrants came from Ireland, Italy, Spain and Eastern Europe, areas in which industrialization came later. America offered the dream of becoming an independent farmer or trader in new lands of opportunity. Often this dream led to disappointment: the migrants became wage-labourers, building roads and railways across the vast expanses of the New World; 'cowboys', gauchos or stockmen on large ranches; or factory workers in the emerging industries of the Northeastern US. However, many settlers did eventually realize their dream, becoming farmers, white-collar workers or businesspeople, while others were at least able to see their children achieve education and upward social mobility.

The US is generally seen as the most important of all immigration countries and epitomizes the notion of free migration. The peak period was from 1861 to 1920, during which 30 million people came, more than half of the estimated 54 million people that entered between 1820 and 1987 (Borjas 1990: 3). This large-scale migration is seen by some economic historians as a crucial feature of the 'greater Atlantic economy' (Hatton and Williamson 1998). Until the 1880s, migration was largely unregulated: anyone who could afford the ocean passage could come to seek a new life in America. Because of economic growth and decreasing costs of transportation, an increasing number of Europeans were able to afford the passage, which also contributed to the demise of the indenture system in the early nineteenth century. However, indentured labour would continue to exist for Chinese and other Asian migrants to California, Hawaii and the West Indies, for whom debt servitude was often the only way to fund the trans-oceanic passage (Galenson 1984).

An important US Supreme Court decision of 1849 affirmed the 'plenary power' of the federal government to regulate international migration, thereby thwarting attempts by eastern seaboard municipalities to prevent the arrival of Irish migrants (Daniels 2004). However, American employers continued to organize campaigns to directly recruit potential workers, and a multitude of agencies and shipping companies helped organize movements. Many of the migrants were young single men, hoping to save enough to return home and start a family. But there were also single women, couples and families. Transnational networks facilitated chain migration. For instance, towards the end of the nineteenth century, around a quarter to a half of Danish, Norwegian and Swedish migrants to the US came on steamship tickets prepaid by migrants already in America (Tilly 1976: 17). Through visits and extensive letter-writing, migrants kept in close touch with family back home, facilitating the flow of information about opportunities and life in America (Thomas and Znaniecki 1918). Such 'social remittances' often increased migration aspirations amongst those staying behind.

Slavery had been a major source of capital accumulation in the early US, but the industrial take-off after the Civil War (1861–5) was fuelled by mass immigration from Europe. After the abolition of slavery, the racist *Jim Crow system* (entrenched discrimination against black people) served to keep the now nominally free African–Americans as sharecroppers in the plantations of the Southern states, since cheap cotton and other agricultural products were central to industrialization. The sharecropping system was a form of semi-bonded labour: black families rented small plots of land in return of which they had to give a share of the crop to white landowners.

While the British, Irish, Germans, Scandinavians and other Northwestern Europeans dominated trans-Atlantic migration up to the mid-nineteenth century, the emigration potential of these countries declined as they entered more advanced stages of urban-industrial development and demographic transitions. The largest immigrant groups from 1860 to 1920 became Irish, Italians, Russians and Jews from Eastern Europe, but there were people from just about every other European country, as well as from Mexico. Patterns of settlement were closely linked to the emerging industrial economy. Labour recruitment by canal and railway companies led to settlements of Irish and Italians along the construction routes. Some groups of Irish, Italians and Jews settled in the East coast ports of arrival, where work was available in construction, transport and factories. Chinese immigrants settled initially on the West coast, but moved inland following recruitment by railway construction companies (see Chapter 7).

Similarly, early Mexican migrants were concentrated in the southwest, close to the Mexican border, but many moved northwards in response to recruitment by the railways. Some Central and Eastern European peoples became concentrated in the Midwest, where the development of heavy industry at the turn of the century provided work opportunities (Portes and Rumbaut 2006: 38–40). The American working class thus developed through processes of chain and network migration which led to patterns of ethnic segmentation. Racist campaigns led to exclusionary laws to keep out Chinese and other Asians from the 1880s. For Europeans and Latin Americans, entry remained largely free until 1920 (Borjas 1990: 27). The census of that year showed that there were 13.9 million foreign-born people in the US, making up 13.2 per cent of the total population (Briggs 1984: 77).

Canada received many loyalists of British origin after the American Revolution. From the late eighteenth century there was immigration from Britain, France, Germany and other Northern European countries. Many African–Americans came across the long frontier from the US to escape slavery: by 1860, there were 40,000 black people in Canada. In the nineteenth century, immigration was stimulated by the gold rushes, while rural immigrants were encouraged to settle the vast prairie areas. Between 1871 and 1931, Canada's population increased from 3.6 million to 10.3 million. Immigration from China, Japan and India also began in the late nineteenth century. Chinese came to the West coast, particularly to British Columbia, where they helped build the Canadian Pacific Railway. From 1886 a series of measures was introduced to stop Asian immigration (Kubat 1987: 229–235). Canada received a large influx from Southern and Eastern Europe over the 1895 to 1914 period. In 1931, however, four preferred classes of immigrants were designated: British subjects with adequate financial means from the UK, Ireland and four other domains of the crown; US citizens; dependants of permanent residents of Canada; and agriculturists. Canada discouraged migration from Southern and Eastern Europe, while Asian immigration was prohibited from 1923 to 1947.

South America was another important destination for European migrants, particularly from Spain and Italy, during this age of industrialization. After its independence from Spain in 1816, Argentina started to encourage immigration, and at the end of the 1880s, the government went so far as to subsidize immigrant boat passages. Over seven million people migrated to Argentina between 1870 and 1930 (Jachimowicz 2006). Over the same period, between 2 and 3 million migrants settled in Brazil, mostly from Europe, but also from the Middle East and Asia (Wejsa and Lesser 2018). In 1960, with 2.4 million immigrants Argentina was the second largest destination country in

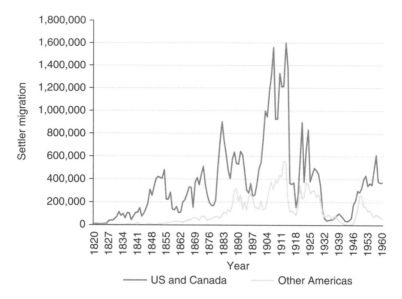

Figure 5.2   Settler migration to the Americas, yearly arrivals, 1820–1960

Source: DEMIG TOTAL database (www.migrationinstitute.org)

the world after the US, while Brazil was home to 1.3 million immigrants (Czaika and de Haas 2014). These numbers reflect the legacy of large-scale immigration over the preceding century.

Figure 5.2 shows how (predominantly European) migration to the Americas showed an increasing trend over the nineteenth century – with short-term fluctuations mainly explained by business cycles and concomitant highs and lows in labour demand – to peak around the turn of the century, and how it would plummet during the interbellum (1918–1939) as a consequence of economic crises and rising rationalism. It also showed that migration to other Latin American countries (mainly Argentina, Brazil, Uruguay and Cuba) gained momentum only from the late nineteenth century, but followed similar trends as migration to North America between the late nineteenth century and 1950.

Migration to Australia and New Zealand

For *Australia*, immigration has been a crucial factor in economic development and nation-building ever since British colonization started in 1788. The Australian colonies were integrated into the British Empire as suppliers of raw materials such as wool, wheat and gold. The imperial state took an active role in providing workers for expansion through convict transportation (another form of unfree labour) and the encouragement of free settlement. Initially there were large male surpluses, especially in the frontier areas, which were often societies of 'men without women'. But many female convicts were transported, and there were special schemes to bring out single women as domestic servants and as wives for settlers. Large-scale European settlement went along with the murder and brutal repression of indigenous Aboriginal populations.

Source: Getty Images/Dean Treml

Photo 5.1   In 2004, an estimated 10,000-plus crowd of Māori demonstrators make their way through the streets of Wellington, New Zealand toward parliament as part of a protest against legislation affecting their land

When the surplus population of Britain became inadequate for labour needs from the mid-nineteenth century, Britain supported Australian employers in their demand for cheap labour from elsewhere in the Empire: China, India and the South Pacific Islands. The economic interests of Britain came into conflict with the demands of the nascent Australian labour movement. The call for decent wages came to be formulated in racist (and sexist) terms, as a demand for wages 'fit for white men'. Hostility towards Chinese and other Asian workers became violent. The exclusionary boundaries of the emerging Australian nation were drawn on racial lines, and one of the first Acts of the new Federal Parliament in 1901 was the introduction of the White Australia Policy (see de Lepervanche 1975). In parallel with developments in North America (Fitzgerald and Cook-Martín 2014), the growth of the labour movement and democratization thus went along with a rise in xenophobia and the introduction of racist immigration rules.

New Zealand was settled by British migrants from the 1830s. The 1840 Treaty of Waitangi between the British Crown and some 540 chiefs of the indigenous Māori people was the prelude to dispossession and marginalization of the Māori. Entry of British settlers (including white British subjects from elsewhere in the Empire) was to remain unrestricted until 1974. The government provided assisted passages virtually only for Britons, while 'non-Britons' required a special permit to enter. When quite small numbers of Chinese workers were recruited as miners and labourers from the 1860s onwards, public agitation led to strict control measures and a 'white New Zealand' policy. The great majority of the population considered themselves British rather than New Zealanders. British migrants were regarded as 'kin', and a sharp distinction was drawn between 'kin' and 'foreigners'. Māori, of course, were not 'foreigners', as the Treaty of Waitangi made them British subjects (McKinnon 1996).

Labour migration within Europe until 1914

In Europe, overseas emigration took place side by side with internal migration and intra-European migration. For instance, of the 15 million Italians who emigrated between 1876 and 1920, nearly half (6.8 million) went to other European countries (mainly France, Switzerland and Germany (see Cinanni 1968: 29), the others went to North and South America. This general mobility increase was an intrinsic part of industrialization and concomitant social transformations, that increased the eagerness to migrate among new generations in search of new opportunities, away from agrarian livelihoods based on sharecropping, subsistence farming or irregular farm labour.

As Western Europeans went overseas, often in the hope of escaping proletarization and finding better opportunities abroad, workers from more peripheral areas, such as Poland, Ireland and Italy, were drawn in as replacement labour for large-scale agriculture and industry. In Russia too, industrialization was associated with increasing rural out-migration. The abolishment of serfdom, in several ways Russia's equivalent of American slavery, by Czar Alexander II in 1861, was partly intended to create a cheap labour force, a mobile proletariat, that could serve industrial development by releasing them from the control of land-owners. In Asia, Japan also went through an industrial revolution since the Meiji Restoration of 1868, generating large-scale internal migration and significant emigration, for instance to Brazil and Peru (see also Taeuber 1951 and Chapter 6).

However, the speed and characters of modernization and capitalist development differed significantly across industrializing countries, and so did the resulting internal and international migration patterns. While in countries such as Britain, Ireland and Italy, these transitions generated mass emigrations, in other countries, such as France and Japan, for various reasons these transitions were less disruptive and abrupt, and rural-to-urban migration would be largely contained within borders. To illustrate this variation in development-migration trajectories, the remainder of this section will focus on a few contrasting cases of social transformation and migration: Britain, Ireland, Germany and France.

British and Irish migration

As the earliest industrial country, *Britain* was the first to experience large-scale labour migration. The new factory towns quickly absorbed labour surpluses from the countryside. Atrocious working and living conditions led to poor health, high infant mortality and short life expectancy. Low wage levels forced both women and children to work, with disastrous results for the family. Natural increase was inadequate to meet labour needs, so Britain's closest colony, *Ireland*, became a labour source. The devastation of Irish peasant agriculture through absentee landlords and enclosures, combined with the ruin of domestic industry through British competition, had led to widespread poverty and unemployment.

The famines of 1822 and 1845–7 contributed to massive Irish migrations to Britain as well as to the US and Australia, partly because they set in motion changes in agrarian inheritance which decreased the access of young rural populations to farmland. Most sections of the Irish population participated in this emigration. Also the poor, illiterate and non-English speaking were often able to migrate because they could rely on assistance from relatives already living abroad (see Hatton and Williamson 1993). This combination of factors contributed to massive depopulation. In 1845 the total Irish population was estimated at about 8.5 million to drop to 6.5 million in 1851. While some 800,000 people had died of disease or starvation, a further million had emigrated during the fateful five years after the failure of the potato crop in 1846. In the following year, the annual volume of emigration generally exceeded 200,000, to surpass a quarter of a million in 1851 (Cousens 1960).

By 1851 there were over 700,000 Irish in Britain, making up three per cent of the population of England and Wales and seven per cent of the population of Scotland (Jackson 1963). They were concentrated in the industrial cities, especially in the textile factories and the building trades. Irish 'navvies' (a slang term derived from 'navigators') dug Britain's canals and built its railways. Engels (1962) described the appalling situation of Irish workers, arguing that Irish immigration was a threat to the wages and living conditions of English workers (see also Castles and Kosack 1973: 16–17; Lucassen 2005). Hostility and discrimination against the Irish in Britain (as in the US) were common far into the twentieth century. This was true of Australia too, where Irish immigration accompanied British settlement from the outset. Both in Britain and Australia, it was the active role played by Irish workers in the labour movement which was finally to overcome this split in the working class, just in time for its replacement by new divisions after 1945, when black and Asian workers came to Britain and Southern Europeans to Australia.

The next major migration to Britain was of 120,000 Jews, who came as refugees from the pogroms of Russia between 1875 and 1914. Most settled initially in the East End of London, where many became workers in the clothing industry. Jewish settlement became the focus of racist campaigns, leading to the first restrictive legislation on immigration: the Aliens Act of 1905 and the Aliens Restriction Act of 1914 (Foot 1965; Garrard 1971). The Jewish experience of social mobility is often given as an example of migrant success. Many of the first generation managed to shift out of wage employment to become small entrepreneurs in the 'rag trade' (clothing manufacturing) or the retail sector. They placed strong emphasis on education for their children. Many of the second generation were able to move into business or white-collar employment, paving the way for professional careers for the third generation. Interestingly, one of Britain's newer immigrant groups – Bengalis from Bangladesh – now live in the same areas of the East End, often working in the same sweatshops and worshipping in the same buildings, synagogues converted to mosques.

Labour recruitment and German industrialization

Irish and Jewish migrant workers may not be categorized as 'unfree workers'. The Irish were British subjects, with the same formal rights as other workers, while the Jews rapidly became British subjects. It is in Germany and France that one finds the first large-scale use of the status of foreigner to restrict workers' rights. In *Germany*, the heavy industries of the Ruhr, which emerged in the mid-nineteenth century, attracted agricultural workers away from the large estates of Eastern Prussia. Conditions in the mines were hard, but preferable to semi-feudal oppression under the Junkers (large landowners). The workers who moved west were of Polish ethnic background, but had Prussian (and later German) citizenship, since Poland was at that time divided up between Prussia, the Austro-Hungarian Empire and Russia.

By 1913, it was estimated that 164,000 of the 410,000 Ruhr miners were of Polish background (Stirn 1964: 27). The Junkers compensated for the resulting labour shortages by recruiting 'foreign Poles' and Ukrainians as agricultural workers – an example of 'replacement migration' (see Chapter 3). Often workers were recruited in pairs – a man as cutter and a woman as binder – to work in Germany, leading to so-called 'harvest marriages'. However, there was fear that settlement of foreign Poles might weaken German control of the eastern provinces. In 1885, the Prussian government deported some 40,000 Poles and closed the border. The landowners protested at the loss of up to two thirds of their labour force (Dohse 1981: 29–32), arguing that it threatened their economic survival (see also Lucassen 2005: 50–73).

By 1890, a compromise between political and economic interests emerged in the shape of a system of rigid control. 'Foreign Poles' were recruited as temporary seasonal workers only, not allowed to bring dependants and forced to leave German territory for several months each year. At first they were restricted to agricultural work, but later were permitted to take industrial jobs in Silesia and Thuringia (but not in western areas such as the Ruhr). Their work contracts provided pay and conditions inferior to those of German workers. Special police sections were established to deal with 'violation of contracts' (that is, workers leaving for better-paid jobs) through forcible return of workers to their employers, imprisonment or deportation. Thus police measures against these bonded migrant labourers were deliberately used as a method to keep wages low and to create a split or 'dual' labour market (Dohse 1981: 33–83).

Foreign labour played a major role in German industrialization, with Italian, Belgian and Dutch workers immigrating alongside the Poles. In 1907, there were 950,000 foreign workers in the German Reich, of whom nearly 300,000 were in agriculture, 500,000 in industry and 86,000 in trade and transport (Dohse 1981: 50). The authorities did their best to prevent family reunion and permanent settlement. Both in fact took place, although the exact extent is unclear. This system developed to control and exploit foreign labour was a precursor both of forced labour in the Nazi war economy and of the 'guestworker system' in the German Federal Republic after 1955.

The French exception: Low emigration and high immigration

In *France*, the number of foreigners increased rapidly from 381,000 in 1851 (1.1 per cent of total population) to 1 million (2.7 per cent) in 1881, and then more slowly to 1.2 million (3 per cent) in 1911 (Weil 1991: appendix, table 4). The majority came from neighbouring countries: Italy, Belgium, Germany and Switzerland, and later from Spain and Portugal. Movements were relatively spontaneous, though some recruitment was carried out by farmers' associations and mines (Cross 1983: Chapter 2). Foreign workers were mainly men who carried out unskilled manual work in agriculture, mines and steelworks – the heavy, unpleasant jobs that French workers were unwilling to take (see also Rosenberg 2006).

The peculiarity of the French case lies in the reasons for the shortage of labour during industrialization. Birth rates fell sharply after 1860. Peasants, shopkeepers and artisans followed birth control practices, which led to small families earlier than anywhere else in Europe (Cross 1983: 5–7). According to Noiriel (1988: 297–312) this *grève des ventres* (belly strike) was motivated by resistance to proletarianization. Keeping the family small meant that land property could be passed on intact from generation to generation, without endless subdivision, and that there would be sufficient resources to permit a decent education for the children.

Unlike Britain and Germany, France therefore saw relatively limited overseas emigration during industrialization. The only important exception was the movement of settlers to Algeria, which France invaded in 1830. Rural–urban migration was also fairly limited and generally a smoother, less disruptive process, certainly compared to Britain, where a combination of large-scale dispossession (through enclosures) and industrialization had fuelled rapid urbanization. In France, the 'peasant worker' developed: the small farmer who supplemented subsistence agriculture through sporadic work in local industries. Where people did leave the countryside, it was often to move straight into the new government jobs that proliferated in the late nineteenth century: so straight from the primary (agricultural) to the tertiary (service) sector. The French case illustrates that although industrialization and urbanization inevitably boost rural out-migration, there is high variation across countries in resulting levels and patterns of migration.

Particularly because of the limited domestic supply of native French industrial workers, the shift from small to large-scale enterprises, increasingly necessary because of growing international competition from about the 1880s, could only be made through the employment of foreign workers. Thus, labour immigration played a particularly important role in the emergence of modern industry and the constitution of the working class in France. Immigration was also seen as important for military reasons. The nationality law of 1889 was designed to turn immigrants and their sons into conscripts for the impending conflict with Germany (Schnapper 1994: 66). From the mid-nineteenth century to the present, the labour market has been regularly fed by foreign immigration, making up, on average, 10–15 per cent of the working class. It has been estimated that, without immigration, the French population in the mid-1980s would have been only 35 million instead of over 50 million (Noiriel 1988: 308–318).

The interwar period: Protectionism, nativism and racism

The outbreak of the First World War in 1914, which lasted until 1918, would mark the end of the first era of globalization which had started around 1870. This would mark an era of economic crises, nationalism, xenophobia and the rise of totalitarian regimes in the form of fascism and Soviet communism. Rising nationalism would go along with growing hostility towards immigrants as well as ethnic and religious minorities in Europe and North America. The period from 1918 to 1945 was one of reduced international labour migration, primarily because of economic stagnation and crises, such as the Great Depression, economic protectionism and immigration restrictions. This was also the era in which the modern passport system was fully established (Torpey 1998). It marked the completion of a longer historical shift, coined by Zolberg as the *exit revolution*, in which from the mid-nineteenth century, the focus of migration policies started to shift from controlling exit to controlling entry (Zolberg 2007: 55) (see Chapter 10).

Warfare, repression of foreign workers and forced migration

Yet the economic crisis and immigration controls did not entirely stop migration for various reasons. At the onset of the First World War, many migrants returned home to participate in military service and munitions production. However, labour shortages soon developed in the combatant countries. The German authorities prevented 'foreign Polish' workers from leaving the country, and recruited labour by force in occupied areas of Russia and Belgium (Dohse 1981: 77–81). The French government set up recruitment systems for workers and soldiers from its North African, West African and Indo-Chinese colonies, as well as from China (about 225,000 in all). They were housed in barracks, paid minimal wages and supervised by former colonial overseers. Workers were also recruited in Portugal, Spain, Italy and Greece for French factories and agriculture (Cross 1983: 34–42).

Britain, too, brought soldiers and workers to Europe from its African and South Asian colonies during the conflict, although in smaller numbers. All the warring countries also made use of the forced labour of prisoners of war. Many Africans were pressed into service as soldiers and 'carriers' within Africa by Germany, Britain and other European countries. Official British figures put the military death toll in East Africa at 11,189, while 95,000 carriers died. The estimates for civilian casualties go far higher – for instance, at least 650,000 in Germany's East African colonies (Paice 2006). The demise of the Ottoman Empire and the subsequent formation of the Turkish nation state in 1923 would coincide with the drive to homogenize populations, resulting in a large forced emigration of Greeks out of Turkey, and ethnic Turks from Greece to Turkey, as well as the mass killings and deportation of up to 1.5 million Armenians (see Muller 2008) (see also Chapter 9).

As part of the economic crisis and immigration restrictions, also migration to Australia, fell to low levels and did not grow substantially until after 1945. An exception was the encouragement of Southern Italian migration to Queensland in the 1920s: Sicilians and Calabrians were seen as capable of back-breaking work in the sugar cane plantations, where they could replace South Pacific Islanders deported under the White Australia Policy. However, Southern Europeans were treated with suspicion. Immigrant ships were refused permission to land and there were 'anti-Dago' riots in the 1930s. Queensland passed special laws, prohibiting foreigners from owning land, and restricting them to certain industries (de Lepervanche 1975).

In *Germany*, the crisis-ridden Weimar Republic (1918–1933) had little need of foreign workers: by 1932 their number was down to about 100,000, compared with nearly a million in 1907 (Dohse 1981: 112). Nonetheless, a new system of regulation of foreign labour developed. Its principles were: strict state control of labour recruitment; employment preference for nationals; sanctions against employers of illegal migrants; and unrestricted police power to deport unwanted foreigners (Dohse 1981: 114–117). This system partly reflected the influence of the strong labour movement, which wanted measures to protect German workers, but it reinforced the weak legal position of migrant workers.

France: The complicity of native workers in exploiting migrant workers

France was the only Western European country to experience substantial immigration in the interwar years. The 'demographic deficit' had been exacerbated by war losses: 1.4 million men had been killed and 1.5 million permanently disabled (Prost 1966: 538). Yet there was no return to the pre-war free movement policy; instead the government and employers refined the foreign labour systems established during the war. Recruitment agreements were concluded with Poland, Italy and Czechoslovakia. Much of the recruitment was organized by the *Société générale d'immigration* (SGI), a private body set up by farm and mining interests. Immigration from France's North African colonies Morocco, Algeria and Tunisia was also increasing. In addition, a 1914 law had removed barriers to movement of Algerian Muslims to Metropolitan France. Although they remained noncitizens, their numbers increased from 600 in 1912 to 60,000–80,000 by 1928 (Rosenberg 2006: 130–131).

Foreign workers were controlled through a system of identity cards and work contracts, and were channelled into jobs in farming, construction and heavy industry. However, most foreign workers probably arrived spontaneously outside the recruiting system as networks facilitated further migration once migrant communities had been established at the destination. The non-communist trade union movement cooperated with immigration, in return for measures designed to protect French workers from displacement and wage cutting (Cross 1983: 51–63; Weil 1991: 24–27).

Nearly 2 million foreign workers entered France from 1920 to 1930, about 567,000 of them recruited by the SGI (Cross 1983: 60). Some 75 per cent of French population growth between 1921 and 1931 is estimated to have resulted from immigration (Decloîtres 1967: 23). In view of the large female surplus in France, mainly men were recruited, and a fair degree of intermarriage took place. By 1931, there were 2.7 million foreigners in France (6.6 per cent of the total population). The largest group were Italians (808,000), followed by Poles (508,000), Spaniards (352,000) and Belgians (254,000) (Weil 1991: appendix, table 4). Large colonies of Italians and Poles sprang up in the mining and heavy industrial towns of the north and east of France: in some towns, foreigners made up a third or more of the total population. There were Spanish and Italian agricultural settlements in the southwest.

During the Depression of the 1930s, hostility towards foreigners increased, motivating a policy of discrimination in favour of French workers. In 1932 maximum quotas for foreign workers in firms were fixed. They were followed by laws permitting dismissal of foreign workers in sectors where there was unemployment. Many migrants were sacked and deported, and the foreign population dropped by half a million by 1936 (Weil 1991: 27–30). Cross therefore concluded that in the 1920s, foreign workers 'provided a cheap and flexible workforce necessary for capital accumulation and economic growth; at the same time, aliens allowed the French worker a degree of economic mobility'. In the 1930s, on the other hand, immigration 'provided a scapegoat for the economic crisis' (Cross 1983: 218). Both the German and the French cases seem to confirm Piore's (1979) argument that the native labour force (workers possessing citizenship) is often complicit in the exploitation of foreign workers.

Nativism and immigration restrictions in the United States

In the US, nativist groups claimed that Southern and Eastern Europeans were 'unassimilable' and that they presented threats to public order and American values. Congress enacted a series of laws in the 1920s designed to limit drastically entries from any area except Northwest Europe (Borjas 1990: 28–29). This national-origins quota system stopped large-scale immigration to the US until the 1960s (see Figure 5.2). But the new mass production industries of the Fordist era had a substitute labour force at hand: black workers from the South. The period from about 1914 to the 1950s was that of the *Great Migration*, in which African–Americans fled segregation and economic exploitation in the Southern states for better wages and – they hoped – equal rights in the Northeast, Midwest and West. Often they encountered new forms of segregation in the ghettoes of New York or Chicago, and new forms of discrimination, such as exclusion from the unions of the American Federation of Labor. As a form of replacement migration, the Great Migration partly filled the void left by the end of large-scale European immigration.

Meanwhile, Americanization campaigns were launched to ensure that immigrants learned English and became loyal US citizens. During the Great Depression, Mexican immigrants were deported by local governments and civic organizations, with some

Source: Getty Images/Chicago History Museum

Photo 5.2   African American men, women and children who participated in the Great Migration to the North, with suitcases and luggage placed in front, Chicago, US, 1918

cooperation from the Mexican and US governments (Kiser and Kiser 1979: 33–66). Many of the nearly 500,000 Mexicans were forced to leave, while others left because there was no work. In these circumstances, little was done to help Jews fleeing the rise of Hitler in Germany. There was no concept of refugees in US law, and it was difficult to build support for admission of Jewish refugees when millions of US citizens were unemployed. Anti-Semitism was also a major factor, and there was never much of a prospect for large numbers of European Jews to find safe haven before World War II.

The Second World War, forced labour and genocide

The Second World War and the Nazi occupation of most of Europe led to unprecedented population displacements and mass killings. Feeding on and further reinforcing endemic European anti-Semitism, Nazi Germany embarked upon a massive genocide of Jews in Germany and occupied territories. The *Holocaust* (known as Shoa in Hebrew) was the systematic persecution and murder of around 6 million Jews by the Nazi regime. It was the first time a state threw its entire power and bureaucracy behind the idea that an entire people should be exterminated. Starting in 1933 with the rise of Hitler to power in Germany, and particularly after the 1938 'Kristallnacht', attacks on Jews became increasingly violent. German Jews as well as Romani ('Gypsies') had already been stripped of their German citizenship in 1935.

After the invasion of Poland in 1939 and other countries in 1940, the Nazi regime established ghettoes to segregate Jews, starting a policy of systematic exclusion, violence and killings. In Germany and through occupied Europe, Jews were fired, their property confiscated, they were employed as slaves and deported to death camps. In 1941 the Nazi leadership started to execute their systematic policy of extermination called 'The Final Solution to the Jewish Question'. While around 1.3 million Jews were killed in mass shootings between 1941 and 1945, between 1942 and 1945 millions were transported to extermination camps. In total an estimated 6 million Jews were killed, or about two thirds of the European Jewry. Besides Jews, also hundreds of thousands of Roma and up to ten million Slavs (ethnic Serbs, Poles, Russians), both considered as subhuman by the Nazis, were subjected to systematic persecution, forced labour, enslavement and genocide, while gay men, disabled people and Jehovah's Witnesses were also persecuted and killed. The mass deportation and murder of European Jews would not have been possible without the large scale collaboration of bureaucrats, police and ordinary citizens, either out of fear, antisemitism or financial gain (see Goldhagen 1996).

In the pre-war years, Jews fleeing from Nazi Germany were not necessarily welcome in other European countries, as part of widespread anti-Semitic sentiment in Europe and related opposition to immigration of Jewish refugees from Germany. In the late 1930s, after Hitler singled out the Jews as the primary cause of Germany's troubles and began his extermination campaign, Jews started to leave the country in growing numbers (Gellman 1971). While the situation of Jews in Nazi Germany became increasingly dangerous, European nations and the United States only accepted limited numbers of Jewish refugees. Before embarking on the systematic, planned killing of Jews, the Nazis initially saw emigration as an important 'solution' to what they called the 'Jewish problem', including emigration to Palestine. However, European and American countries became increasingly reluctant to host significant numbers of Jewish refugees, while the British closed Palestine to Jewish immigration in 1939. When MS St. Louis, a German ocean liner, tried to find a refuge for 930 German Jews, they were denied entry to Cuba, the US and Canada, before returning to Europe, where many were murdered during the Holocaust (see Gellman 1971).

Also, neighbouring countries like the Netherlands and Switzerland closed their borders for Jewish refugees, and many were sent back to Germany – although, with the help of smugglers, thousands were able to escape despite stringent border controls. Immigration restrictions were defended with the argument that the crisis-stricken European countries could not bear the burden of large-scale Jewish immigration, but widespread anti-Semitism played an important role too. For instance, in 1938 the Dutch prime minister Colijn proclaimed that border closure was actually in the interest of Dutch Jews themselves, because allowing in more refugees would further fan the flames of anti-Semitism. In an official statement the Dutch government proclaimed that 'a further intrusion of alien elements will be harmful of the maintenance of the Dutch race. The government finds that, in principle, our limited territory should be reserved for its own people' (MVA 1938).

The Nazi regime recruited enormous numbers of foreign workers – mainly by force – to replace the 11 million German workers conscripted for military service. The occupation of Poland, Germany's traditional labour reserve, was partly motivated by the need for labour. Labour recruitment offices were set up within weeks of the invasion, and the police and army rounded up thousands of young men and women (Dohse 1981: 121). Forcible recruitment took place in all the countries invaded by Germany, while some voluntary labour was obtained from Italy, Croatia, Spain and other 'friendly or neutral countries'. By the end of the war, there were 7.5 million foreign workers in the Reich, of whom 1.8

million were prisoners of war. It is estimated that a quarter of industrial production was carried out by foreign workers in 1944 (Pfahlmann 1968: 232). The Nazi war machine would have collapsed far earlier without foreign labour.

The basic principle for treating foreign workers declared by Sauckel, the Plenipotentiary for Labour, was that: 'All the men must be fed, sheltered and treated in such a way as to exploit them to the highest possible extent at the lowest conceivable degree of expenditure' (Homze 1967: 113). This meant housing workers in barracks under military control, the lowest possible wages (or none at all), appalling social and health conditions, and complete deprivation of rights. Poles and Russians were compelled, like the Jews, to wear special badges showing their origin. Many foreign workers died through harsh treatment and cruel punishments. These were systematic; in a speech to employers, Sauckel emphasized the need for strict discipline: 'I don't care about them [the foreign workers] one bit. If they commit the most minor offence at work, report them to the police at once, hang them, shoot them. I don't care. If they are dangerous, they must be liquidated' (Dohse 1981: 127). The Nazis took exploitation of rightless migrants to an extreme which can only be compared with slavery, yet its legal core – the sharp division between the status of national and foreigner – was to be found in both earlier and later foreign labour systems.

It has been estimated that 18.5 million persons were displaced, not including the six million Jews deported to concentration camps (Kulischer 1948). Outside Europe, an additional 3 to 10 million people were killed as a consequence of Japanese aggression in Asia between 1931 and 1945 (Rummel 1998). All of these developments involved massive suffering and loss of life. This shows the extent to which massive human displacements are a characteristic outcome of warfare. The end of the Second World War witnessed mass population movements in Europe of Holocaust survivors, displaced persons and ethnic groups, such as the approximately 12 million ethnic Germans expelled as part of ethnic cleansing policies in Eastern Europe (see Ther 1996). These displacements would be an important impetus for the establishment of organizations such as the International Organization for Migration (IOM) and the United Nations High Commissioner of Refugees (UNHCR), in an effort to come to find more effective international responses to situations of human displacement.

Conclusion

This chapter has described the key role of labour migration in colonialism and industrialization. It has shown how contemporary migratory movements and policies are profoundly influenced by historical precedents. Labour migration has been a major factor in the rise of industrial societies and the construction of a capitalist world market. Historical migration patterns are rooted in practices of colonization and the desire of employers – as well as states representing their interests – to mobilize and control labour for production. Whether this happened through slavery, indentured labour, forced labour or temporary recruitment of migrant workers, the common dominator of such practices is that the lower status and lack of rights of migrant workers was used as a tool for exploitation.

It is therefore only possible to explain contemporary migration patterns by understanding how European (and Japanese) imperialism shaped the modern world system and forged worldwide connections between nations through language, culture and trade, and how the need to mobilize migrant labour for colonial economies and, from the nineteenth century onward, industrial development, laid the foundations of contemporary migration patterns. The current global migration map is to a large degree a legacy – and an inevitable consequence – of that imperial past. This shows that the role of states and recruitment in migration cannot be reduced to 'intermediate' factors affecting cost-and-benefit calculations of migrants – as neoclassical and push-pull models tend to do – but that they are forces that actively initiate and drive migration processes.

While often rooted in a past of violent conquest, subjugation and unfree labour, this chapter has also showed that, once settled at the destination, migrants started to organize their own journeys through network connections, while their migration gained an increasingly permanent character, leading to the formation of migrant communities, new creole and transnational identities and worldwide diasporas. This shows the dangers of reducing migrants to passive victims of colonial and capitalist forces, and bears testimony migrants' agency and resolve to improve their destiny despite the formidable challenges they often face. Time and again, historical experiences have shown us that migrants, however oppressed they are, have been able to defy attempts to prevent them from staying, getting organized and demanding rights.

Notwithstanding the substantial struggle this often involves, this has often resulted in their partial or full emancipation. In most of the colonial world – in America, Asia and Africa – immigrant groups achieved high levels of socioeconomic and cultural integration within a few generations, although others formed ethnic groups or minorities with distinct cultural traits, either out of free will (internal closure) or as a result of exclusion (external closure), or both. In the plantation economies of the Caribbean, some parts of the Southern US (such as Louisiana), Brazil and islands like Réunion, Mauritius and Fiji, the arrival and settlement of highly diverse groups went along with the development of new, syncretic cultures in a process of creolization.

This chapter has illustrated the usefulness of various theories reviewed in Chapter 3 to in understanding the evolution of migration patterns in the modern era. The experiences of migration under European colonialism and industrialization highlight the fact that states and recruitment have played a central role in shaping contemporary migration patterns, as emphasized by historical structural theories. However, they also show the validity of theories such as the new economics of labour migration, that emphasize that migration is often a deliberate attempt by people to overcome local constraints and improve the long-term well-being of their families rather than a 'desperate' flight from misery. Forced movements of slaves do obviously not fit into theories that emphasize people's agency. However, for most other forms of migration, including indentured work, migrants' agency played some role. Despite substantial exploitation they saw a clear long-term interest in migrating. At the same time, structural factors such as colonialism and recruitment by states and companies were important migration drivers in their own right and put heavy constraints on people's free choice in terms of destinations and work circumstances, and workers often felt deceived when they found out about real circumstances.

In the case of migrations from Europe to America and Oceania in the nineteenth and early twentieth centuries, most migrants went by free choice, which fits better with liberal (or neoclassical) ideas on migration. Many young men and women went with the intention of permanent settlement, but quite a few went in order to work for a few years and then return home. Some did return, but in the long run the majority remained in the New World, often forming new ethnic communities. Many migrants moved under difficult and dangerous conditions. Sometimes their hopes of a better life were dashed.

Yet in general they had good reasons to take the risk, because the situation was usually even worse in the place of origin: poverty and a lack of perspective, domination by landlords, exposure to arbitrary violence – these were all powerful reasons to leave, and in this sense their migration was rational. At the same time, people's aspirations increased, and geographical horizons expanded under the influence of education, access to information and contacts with, and letters from, people who had already migrated. This further boosted the desire to leave in search of greener pastures. And many – indeed most – migrants succeeded in building a better life in the new country – if not for themselves, then for their children. There are important parallels with today's migrations: migrants still experience many hardships, but they often do succeed in escaping a lack of perspective in their place of origin and finding new opportunities elsewhere. Migrants are willing to endure initial hardship and take significant risks to achieve long-term improvements of their families' lives. Being a migrant can be very tough but staying at home is often worse.

The period from about 1850 to 1914 – also known as the first era of globalization – was a period of mass migration in Europe and North America. State formation and industrialization were causes of internal rural-to-urban migration and both emigration and immigration. After 1914, war, nationalism, economic stagnation and increased state control (such as the introduction of passports) caused a considerable decline in migration. The large-scale movements of the preceding period seemed to have been the results of a unique and unrepeatable constellation. When rapid and sustained economic growth got under way after the ravages and genocide of the Second World War, the new age of migration was to take the world by surprise.

Guide to Further Reading

Manning (2005) gives a history of migration from the earliest times to the present, as does Part 1 of Goldin *et al.* (2011). Hoerder (2002) and Wang (1997) also cover the broad sweep of global migration history. Cohen (1987) provides an historical overview of migrant labour in the international division of labour. Blackburn (1988) and Fox-Genovese and Genovese (1983) analyse slavery and its role in capitalist development, while Schama (2006) charts the history of abolition and its meaning for British and US politics. Archdeacon (1983) examines immigration in US history, showing how successive waves of entrants have 'become American'. Hatton and Williamson (1998) present an economic analysis of 'mass migration' from Europe to the US. Portes and Rumbaut (2006) analyse historical patterns of entry into the US. McKeown's overviews of Asian and Chinese migration between 1846 and 1940 are essential reading for those wishing to move beyond Eurocentric accounts of migration (McKeown 2004, 2010). Bade (2003) and Lucassen (2005) analyse the role of migration in European history. Moch (1992) gives an overview of earlier European migration experiences, while many contributions in Cohen (1995) are on the history of migration. Lucassen *et al.* (2006) examine the history of immigrant integration in Western European societies. French readers are referred to the excellent accounts by Noiriel (1988, 2007). Jupp (2001, 2002) provides detailed accounts of the Australian experience.

Extra resources can be found at: **www.age-of-migration.com**

6 Migration in Europe since 1945

Since the sixteenth century, Europeans have been moving *elsewhere* through conquering, colonizing and settling in lands elsewhere on the globe, as well as by transporting slaves from Africa to the Americas and moving indentured workers from South and East Asia, to work in the Caribbean, East Africa, Mauritius and elsewhere. These patterns would largely reverse in the second half of the twentieth century. Under the influence of decolonization, demographic change and rapid economic growth, Europe emerged as a major global migration destination. This would take most European societies by surprise, as large-scale immigration and settlement were neither planned nor generally desired. There have been four main phases in post-WWII European migrations:

- 1945–1973: West European decolonization, rapid economic growth, decreasing emigration to North America and Oceania, post-colonial migration from Africa, Asia and the Caribbean, recruitment of Mediterranean 'guestworkers', and the settlement of Russians in Soviet republics;
- 1973–1989: Economic stagnation and restructuring, settlement of migrant workers in Western Europe alongside family migration, increasing labour migration to Eastern Europe from communist countries, and planned Soviet migration;
- 1989–2008: End of Cold War, demise of Soviet empire, growing East–West migration and intra European circulation following the creation of the Schengen zone and EU enlargements, increasing labour migration to Southern Europe, alongside the increasing politicization of migration across Europe;
- Since 2008: Post-Great Recession slump followed by a resurgence of labour migration to the European economic core areas, along with a further politicization and securitization of asylum and undocumented migration.

While migration of Europeans to the Americas and Oceania rapidly declined in the 1960s and 1970s, Western European economies started to attract increasing numbers of migrants. Initially most migrant workers came from former colonies and countries located on the European periphery, but would start to include an increasingly diverse array of European and non-European origin countries. Although governments have often struggled to come to terms with these new realities, immigration levels in much of Europe have come to match or exceed those of classical immigration countries. In 2017, an estimated 82 million foreign-born people were living in Europe (including former Soviet states), of whom 34 million were born outside Europe, compared to 55 million foreign-born in North America and 41 million foreign-born in the Middle East. The drying up of Europe as a global source of migrants would also change the face of migration to North America, Australia and New Zealand, which would become increasingly Latin American and Asian in character. The structural increase in immigration from non-EU countries as well as intra-EU migration has contributed to a growing diversity of populations, particularly in Western Europe. Figures 6.1 and 6.2 summarize the evolution of European immigrant

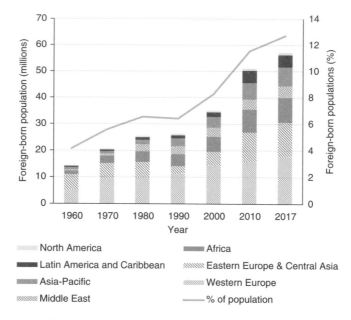

Figure 6.1   Foreign-born populations in Western Europe, by origin region, 1960–2017

Source: UNDESA 2017 and Global Bilateral Migration Database

Note: 'Western Europe' includes all European countries that were not part of the communist East Bloc. Eastern Europe and Central Asia include former communist countries in Central and Eastern Europe and most of the former Soviet Union

populations since 1960. They show the fast increase in immigrant populations in Western Europe, which represented 12.7 per cent of the total population in 2017. Migration in Eastern European and Central Asia (comprising the former Soviet Union) is also significant but lower and more stable, with most people moving within the region.

Migration in the post-war boom (1945–1973)

While global migration rates have remained relatively stable, one of the fundamental changes in post-WWII migration geography has been the emergence of Europe as a global migration destination. Between 1945 and the early 1970s, various types of migration led to the formation of new, ethnically distinct populations in advanced industrial countries in Northwest Europe:

- Mass movements of European refugees and displaced people at the end of the Second World War, particularly in Germany, Poland and the Soviet Union;

- Migration from former colonies in Asia, Africa and the Caribbean to the previous colonial powers, particularly France, the United Kingdom, the Netherlands and Belgium; consisting of repatriation of colonial settlers as well as migrant workers;

- Migration from 'labour frontier' countries in the European periphery including Finland, Ireland, Italy, Spain, Portugal, Greece, Yugoslavia, Turkey and Morocco to the industrial core nations of Western Europe, sometimes through 'guestworker systems';

- State-led settlement of Russian populations in other Soviet Republics, high internal and rural-to-urban migration, and increasing student migration to the Soviet Union.

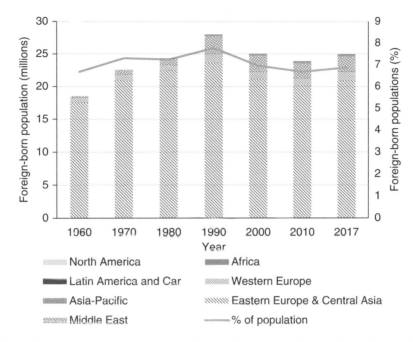

Figure 6.2   **Foreign-born populations in Eastern Europe and Central Asia, by origin region, 1960–2017**

Source: UNDESA 2017 and Global Bilateral Migration Database

Map 6.1 gives an idea of some of the main migratory flows of this period from a European perspective, with declining movements to the New World being gradually replaced by large-scale migration towards Europe from countries in the European periphery and former colonies.

Foreign workers and 'guestworker' systems

In the first phase from 1945 to the early 1970s, the chief economic strategy of West European governments and large-scale enterprises was the concentration of investment and expansion of industrial production. As a result, large numbers of migrant workers were employed or recruited from less developed countries in the Mediterranean region as well as from Ireland and Finland into the fast-growing industrial areas of Western Europe. With the exception of former Yugoslavia, Communist countries in Eastern Europe were largely sealed off from Western Europe, and apart from some refugee movements, migration between 'East' and 'West' was minimal. All the highly industrialized countries of Northwestern Europe used temporary labour recruitment at some stage, although this sometimes played a smaller role than spontaneous entry of foreign workers, which was facilitated by the development of migrant networks.

Immediately after the Second World War, the *British* government brought in 90,000 mainly male workers from refugee camps and from Italy through the European Voluntary Worker (EVW) scheme. The scheme was fairly small and only operated until 1951, because it was easier to make use of colonial workers. A further 100,000 Europeans entered Britain on work permits between 1946 and 1951, and some of these migrations continued subsequently (Kay and Miles 1992). *Belgium* also started recruiting foreign workers immediately

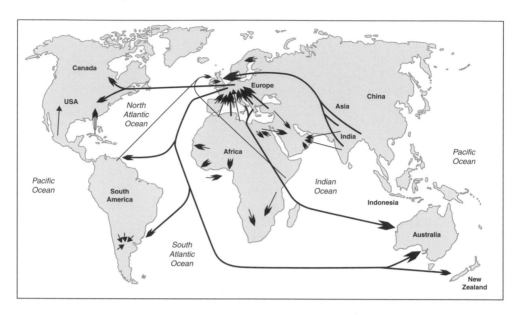

Map 6.1   International migration, 1945–1973

Note: The arrow dimensions give an approximate indication of the volume of flows in the 1945–2015 period. Exact figures are often unavailable

after the war. They were mainly Italian men, and were employed in the coalmines and the iron and steel industry. The system operated until 1963, after which foreign work-seekers were allowed to come of their own accord. Many brought in dependants and settled permanently, changing the ethnic composition of Belgium's industrial areas.

France established an *Office nationale d'immigration* (ONI) in 1945 to organize the recruitment of workers from Southern Europe. Essentially continuing pre-war trends, immigration was seen as a solution to post-war labour shortages. In view of continuing low birth rates and war losses, massive family settlement was envisaged. ONI also coordinated the employment of up to 150,000 seasonal agricultural workers per year, mainly from Spain. By 1970, 2 million foreign workers and 690,000 dependants resided in France. Many found it easier to come as 'tourists', get a job and then regularize their situation. This applied particularly to Portuguese and Spanish workers, escaping their respective dictatorships, who generally lacked passports. By 1968, ONI statistics revealed that 82 per cent of the foreigners admitted by the ONI came as 'clandestines'.

Switzerland pursued a policy of large-scale labour import from 1945 to 1974. Foreign workers were recruited abroad by employers, while admission and residence were controlled by the government. Job changing, permanent settlement and family reunion were forbidden to seasonal workers until the mid-1960s. Considerable use was also made of cross-frontier commuters. Swiss industry became highly dependent on foreign workers, who made up nearly a third of the labour force by the early 1970s. The need to attract and retain workers, coupled with diplomatic pressure from Italy, eventually led to relaxations on family reunion and permanent stay, so that Switzerland, too, experienced settlement and the formation of permanent migrant communities.

The examples could be continued: *the Netherlands* and *Belgium* brought in 'guestworkers', mainly from Spain, Italy, Turkey and Morocco, but also from Portugal, Greece and former Yugoslavia. In the 1960s and early 1970s, *Luxembourg*'s industries were also highly dependent on foreign labour, while Sweden employed workers from Finland and Southern European countries. In *Italy*, migration from the underdeveloped south was crucial to the economic take-off of the northern industrial triangle between Milan, Turin and Genoa in the 1960s. This was internal migration, but rather similar in its economic and social character to foreign worker movements towards other European countries.

The key case for understanding the 'guestworker system' was the Federal Republic of Germany (FRG), which set up a highly organized state recruitment apparatus (Castles and Kosack 1973; Castles *et al.* 1984). The former West-German Government started recruiting foreign workers in the mid-1950s. The Federal Labour Office (Bundesanstalt für Arbeit, or BfA) set up recruitment offices in the Mediterranean countries. Employers requiring foreign labour paid a fee to the BfA, which selected workers, tested occupational skills, provided medical examinations and screened police records. The workers were brought in groups to Germany, where employers had to provide initial accommodation. Recruitment, working conditions and social security were regulated by bilateral agreements between the FRG and the main origin countries: first Italy, then Spain, Greece, Turkey, Morocco, Portugal, Tunisia and Yugoslavia.

The number of foreign workers in the FRG rose from 95,000 in 1956 to 1.3 million in 1966 and 2.6 million in 1973. This massive migration was the result of rapid industrial expansion and the shift to new methods of mass production, which required large numbers of low-skilled workers. Foreign women workers played a major part, especially in the later years: their labour was in high demand in textiles and clothing, electrical goods and other manufacturing sectors.

German policies conceived of migrant workers as temporary labour units, which could be recruited, utilized and sent away again as employers required. To enter and remain in the FRG, a migrant needed a residence permit and a labour permit. These were granted for restricted periods, and were often valid only for specific jobs and areas. Entry of dependants (family members) was discouraged. A worker could be deprived of his or her permit for a variety of reasons, leading to deportation.

However, it was impossible to prevent family reunion and settlement. Often migrant workers were able to get employers to request their wives or husbands as workers. Competition with other labour-importing countries such as Belgium and the Netherlands to attract migrant workers contributed to relaxation of restrictions on entry of dependants in the 1960s (Bonjour 2011). Families became established and children were born. Foreign labour was beginning to lose its mobility, and social costs (for housing, education and healthcare) could no longer be avoided. When the Federal Government stopped labour recruitment in November 1973, the motivation was not only the looming Oil Crisis, but also the belated realization that permanent settlement was taking place (Castles and Kosack 1973; Castles *et al.* 1984).

The FRG represents, in the most developed form, all the principles – but also the contradictions – of temporary foreign labour recruitment systems. These include the belief in temporary sojourn, the restriction of labour market and civil rights, the recruitment of single workers (men at first, but with increasing numbers of women as time went on), the inability to prevent family reunion completely, the gradual move towards longer stay, and the inevitable pressures for settlement and community formation. The FRG took the system furthest, but its central element – the legal distinction between citizen and foreigners as a criterion for determining political and social rights – was found throughout Europe (see Hammar 1985).

Multinational agreements were also used to facilitate labour migration. Free movement of workers within the European Economic Community (EEC) which came into force in 1968, was relevant mainly for Italian workers, while the Nordic Labour Market facilitated migration of Finns to Sweden. However, in the 1960s and early 1970s, the labour movement within the Community was actually declining, owing to gradual equalization of wages and living standards within the EEC, while migration from outside the Community, particularly from Turkey and Morocco, was already increasing.

Colonial workers

Migration from former colonies was important for Britain, France, the Netherlands and later also for Portugal. Since they lacked colonial empires, Germany, Austria and Switzerland would continue to mainly rely on Mediterranean 'guestworkers' from Italy, Spain, Yugoslavia and Turkey. *Britain* experienced a net inflow of about 350,000 migrants from Ireland, its traditional labour reserve, between 1946 and 1959. Irish workers provided manual labour for industry and construction, and many brought in their families and settled permanently. Irish residents in Britain enjoyed all civil rights, including the right to vote. Immigration of workers from the New Commonwealth (former British colonies in the Caribbean, the Indian subcontinent and Africa) started after 1945 and grew during the 1950s. The 1948 British Nationality Act gave all Commonwealth subjects the status of British citizen and, hence, the full right to work and settle in the UK. Some workers came as a result of recruitment by London Transport, but most migrated spontaneously in response to labour demand. By 1951, there were 218,000 people of New Commonwealth origin (including Pakistan, which subsequently left the Commonwealth), a figure which increased to 541,000 in 1961. Entry of workers declined after 1962, partly through the Commonwealth Immigrants Act of 1962 which abolished free entry of Commonwealth citizens, and partly because of the early onset of economic stagnation in Britain. One unexpected result of these restrictions was a 'beat the ban' migration rush from the Caribbean just before the Commonwealth Immigrants Act came into effect (Peach 1968).

Most of the Commonwealth immigrants had come to stay, and family reunion grew in importance, until it in turn was restricted by the 1971 Immigration Act. The population of New Commonwealth origin increased to 1.2 million in 1971 and 1.5 million in 1981. Most Afro-Caribbean and Asian immigrants and their children in Britain enjoyed formal citizenship (although this no longer applies to those admitted since the 1981 Nationality Act). Most black and Asian workers found unskilled manual jobs in industry and the services, and a high degree of residential segregation emerged in the inner cities. Educational and social disadvantage became a further obstacle to mobility out of low-status positions. By the 1970s, the emergence of ethnic minorities was inescapable.

France experienced large-scale immigration from its former North African colonies Algeria, Morocco and Tunisia as well as from Southern Europe through recruitment and spontaneous migration, while many French *colons* were forced to leave Algeria after independence in 1962 (see Chapter 9). Some of these migrants came before independence, while they were still French citizens. Others came later through preferential migration arrangements, or irregularly. Migration from Algeria was regulated by bilateral agreements which gave a unique status to Algerian migrants. Moroccans and Tunisians, by contrast, were admitted through the ONI. Many people also came from the overseas departments and territories such as Guadeloupe, Martinique and Réunion. Non-European immigrants were relegated to the bottom of the labour market, often working in highly exploitative conditions. Housing was frequently segregated, and very poor in quality; indeed, shanty towns (known as *bidonvilles*) appeared in France in the 1960s (see Chapter 13). Extreme-right groups began to subject non-European immigrants to racial violence: 32 North Africans were murdered in 1973.

The Netherlands had two main inflows from former colonies. Between 1945 and the early 1960s about 375,000 'repatriates' from the former Dutch East Indies (now Indonesia) entered the Netherlands. Although most had been born overseas and many were of mixed Dutch and Indonesian parentage, they were Dutch citizens. Although these 'repatriates' awaited a cold reception in the Netherlands, this group of 'Indos' has now largely assimilated into Dutch society. The exception is the roughly 12,500 Moluccan migrants. They served as soldiers in the Dutch colonial army, who wished to return to their homeland to achieve Moluccan independence from Indonesia, a promise on which the Dutch government had failed to deliver. They remained segregated, and rejected integration into Dutch society – an example of *internal closure*. In the late 1970s, a strong sense of injustice led to several violent incidents, including train and school hijackings by second generation Moluccan-Dutch. After 1965, increasing numbers of migrants from the Caribbean territory of Surinam came to the Netherlands A peak was reached in the years leading up to independence in 1975, and 1980, when visas were introduced. This 'now or never' migration surge led to an increase in the number of Surinamese in the Netherlands from 39,000 in 1973 to 145,000 in 1981 (van Amersfoort 2011; Vezzoli 2015) (see Chapter 7).

Soviet migration

Similar to West European colonialism, Russian imperial expansion would create political, cultural and economic ties that would lead to a regional migration system centred around the economic core areas of Russia. The Communist Revolution of 1917 had led to the creation of the Soviet Union in 1922, which consolidated the czarist empire into a modern state. Although nominally a federal union of Soviet Republics, its government was highly centralized and ethnic Russians dominated the Soviet regime (Pipes 1997). Particularly under Stalin, this would often go along with brutal oppression of non-Russian peoples, such as the deliberate starvation of millions of Ukrainians and Kazakhs during the 1932–1933 famine, a case of 'negligent genocide' (Ellman 2007), and the mass deportation and killing of (Muslim) Crimean Tatars in 1944 (Uehling 2004). As part of Soviet imperialism, large numbers of Russian settler colonists had been directed to migrate to Latvia, Estonia, Ukraine, Belarus, Moldova and Kazakhstan (Kuzio 2002).

In the Soviet Union, industrialization, urbanization and agrarian mechanization stimulated large-scale rural-to-urban migration, particularly in Russia, Ukraine and other economically advanced republics. While in 1917 only 17 per cent of the Soviet population lived in urban areas, in 1961 this proportion had risen to half, and by 1987 two thirds of Soviet citizens lived in urban areas (Zickel 1989). Urbanization was most advanced in Russia, Belarus and Ukraine. Although the process was started later in economically underdeveloped, peripheral regions, most of the Soviet republics underwent intensive, rapid industrialization and urbanization (Siegelbaum and Moch 2014). The Soviet Union would reach its peak of military and economic power after the Second World War, in the 1950s and 1960s, when it consolidated its hegemonic sphere of influence over Central and Eastern European satellite states as well as communist and socialist countries in the 'Third World', such as Cuba, Afghanistan, Syria, Egypt, Algeria and Vietnam.

Through draconic exit controls, the state prohibited people from leaving the Soviet Union as well its Central and Eastern European satellite states. At the same time, it tried to regulate internal migration with the aim of redistributing the labour force where workers were most needed. Mobility was partly regulated by the registration (*propiska*) system, although this system did not inhibit migration (Chudinovskikh and Denisenko 2017). Apart from rural-to-urban migration *within* Soviet republics, millions of people moved *between* Soviet republics in attempts by the state to stimulate resettlement to sparsely populated, natural resource-rich regions, such as Kazakhstan and Siberia in the north and the far-east. From the early 1950s to the mid-1970s, the dominant direction of migration was therefore out of Russia, with 2.7 million people moving to other Soviet republics. As part of this 'imperial emigration', Russian migrants moved to Ukraine and Belarus for post-war reconstruction and development, to the Baltic republics, to Kazakhstan for the development of agriculture and to Central Asia to build newly industrialized economies (Chudinovskikh and Denisenko 2017).

In the broader context of Cold War politics, the Soviet Union, and Russia in particular, would also evolve into an important destination for student migrants, particularly from countries with socialist-inclined regimes (White 2009). In 1960, 13,500 foreign students were in the Soviet Union, a number that would further rise to 180,000 by 1990. During the Soviet period, more than 60,000 people from 165 countries studied at the People's Friendship University of Russia, which was established to train students from newly independent countries in Africa and Latin America (Chudinovskikh and Denisenko 2017).

Settlement and the growth of dual labour markets (1973–1989)

The 1973 Oil Shock brought economic fortunes to oil-producing countries, but marked the start of a period of economic stagnation in Western Europe and North America. In Western Europe, rising unemployment led to a recruitment freeze, and the general expectation was that that 'guestworkers' would return and that the post-war era of large-scale immigration had come to an end. However, not only did such large-scale return not happen, but immigration would also continue during this period, against expectations and political rhetoric arguing in favour of border closure. The main trends include:

- A decline of government-organized labour migration to Western Europe and a recruitment freeze;
- Family reunion of foreign workers, and the resulting growth of permanent migrant communities;
- Increasing labour migration to emerging new economic core areas in Spain and Italy, particularly from North Africa and Latin America;
- Recruitment of migrant workers in communist developing countries to work in Eastern Europe and the Soviet Union.

The curbing of organized recruitment of manual workers by West European countries in the early 1970s was part of a larger process consisting of a fundamental restructuring of the world economy and the emergence of a new international division of labour, with industrial production increasingly relocated to low-wage countries and a growth of higher- and lower-skilled service sectors in high-income countries, which was to generate new forms of mobility and migration.

Economic restructuring and the growth of dual labour markets

The period of recessions following the 1973 Oil Shock gave further impetus to a trend of restructuring of the world economy. This restructuring involved capital investment in new industrial areas in developing countries, altered patterns of world trade and involved the introduction of new technologies. Most notably, this economic restructuring led to a decline in employment in the industrial ('secondary') sector in industrialized countries, which accelerated a general trend typical for post-industrial economies of increasing concentration of employment in the service ('tertiary') sector. One the one hand, this shift from industry to services was a consequence of advanced mechanization and automation of production. On the other hand, this reflected relocation or 'outsourcing' of industrial production to low-wage countries, particularly in East and Southeast Asia, a process which again was prompted by increasing wage costs in Europe. This process of economic restructuring led to the mass dismissal of factory and mine workers, many of whom were migrant workers. However, the recruitment freeze did not lead to the large-scale return of workers. Although significant return did happen, many migrant workers preferred to stay and to bring over their families. Family migration largely explains why migration to Northwest Europe continued at relatively high levels even during this period of economic stagnation and unemployment.

The 1970s and 1980s were also marked by a departure from post-WWII economic strategies, which focused on state intervention and income redistribution through progressive taxation, state pensions and welfare as well as an expansion of public services such as health care, housing and education. These policies were replaced by new economic policies, frequently dubbed as 'neoliberal', which pinned their hopes on economic deregulation and flexibilization of labour markets and the privatization of state companies. This went along a partial dismantling of welfare states and tax reforms that increased inequality between poor and rich. These policies of regulation and decreased state interference reinforced the growth of 'precarious' labour market segments, consisting of low-paid, insecure, often informal jobs in the service sector, construction and agriculture, which increasingly contrasted with stable, well-paid jobs in more specialized services.

The resulting growth of dual labour markets (Piore 1979: see Chapters 3 and 12) led to a growing split between a *primary labour market* consisting of stable, formal jobs in the primary sector and precarious jobs in the *secondary labour market*. This labour market segmentation stimulated renewed labour immigration to Europe, as the supply of native workers willing to do precarious jobs had dwindled, partly due to demographic change and increasing education levels, partly due to the low status, pay and prospects such 'dead-end' or '3D' (dirty, difficult and dangerous) jobs offer. From the mid-1980s, when economic growth resumed, the rapid growth of the service sectors would start to draw in more and more migrant workers from increasingly diverse skill levels, while labour demand in sectors such as intensive agriculture and construction was maintained. Economic reforms of the 1973–1989 period thus laid the groundwork for the resumption and rapid growth of labour migration to Europe over the 1989–2008 period.

Settlement of 'guestworkers' and new labour migrations

The immediate post-1973 period was one of consolidation and demographic 'normalization' of immigrant populations in Western Europe through family reunion. Active recruitment of foreign workers and colonial workers largely ceased. For migrants in Britain,

France and the Netherlands, trends to family reunion continued. In fact, the recruitment freeze and other immigration restrictions interrupted circulation by pushing many migrants into permanent settlement, as they feared that they would not to be able to re-emigrate if their return were not successful. Governments initially tried to prevent family reunion, but with little success. In several countries, the law courts played a major role in preventing policies deemed to violate the protection of the family contained in national constitutions and international treaties (see Chapter 11). The settlement process, and the emergence of second and third generations born in Western Europe, led to the development of community structures and consciousness. By the 1980s, migrants and their descendants had become clearly visible social groups, sparking debates about the need to 'integrate' groups whose presence had long been considered as temporary.

In the same period Southern European countries started to experience migration transitions. Economic growth, combined with a sharp fall in birth rates and increasing labour shortages, ended the post-War emigration surge to the Americas and Northwest Europe. Italy, Spain, Greece, and, to a more limited extent, Portugal, all became countries of immigration, using labour from North Africa, Latin America, Asia and – later – Eastern Europe for low-skilled jobs (King *et al.* 2000). Although growing labour demand was often not matched by immigration policies, governments influenced by economic lobbies were often willing to turn a blind eye towards the illegal employment of migrant workers. These policies of lax enforcement would lead to large-scale migration of North African, East European and Latin American male and, increasingly, female workers in the post-1989 period, particularly to Spain and Italy.

Meanwhile, significant migration occurred within and between communist countries. There was substantial migration *within* the Soviet Union. Imperialist Russification policies went along with continued Russian emigration to non-Russian Soviet States. While between the early 1950s and the mid-1970s, the dominant movement had been of Russian 'colonists' moving to other Soviet republics, a reverse migration movement, from other Soviet republics to Russia, had gradually gained momentum. Between 1975 and 1991 these reverse flows dominated and increased Russia's population by 2.5 million. Initially, these migrations were directed at remote, resource-rich areas aided by state investments in the development of oil and gas fields in West Siberia and mineral resources elsewhere in Eastern Russia (Chudinovskikh and Denisenko 2017) instead of cities. These movements can be considered as *frontier migration* into 'resource niches' (see also Skeldon 1997), typical of imperial and colonial states vying to 'develop' peripheral zones such as by exploiting their resources.

The Soviet satellite states in Central and Eastern Europe also recruited migrant workers from abroad. East Germany, one of the most industrialized communist countries, set up its own version of a 'guestworker' system by recruiting workers from Hungary (from 1967), Poland (from 1971), Algeria (from 1974 to 1984), Cuba (from 1978), Mozambique (from 1979), Vietnam (from 1980), Angola (from 1985), and North Korea and China (from 1986). In 1989 the number of foreigners in East Germany reached a peak of 190,000 and represented 1.2 per cent of the total population (see also Schwenkel 2014; Mac Con Uladh 2005); Hungary and Czechoslovakia started to recruit migrant labour from Cuba, Vietnam and elsewhere as planners thought that economic growth was hampered by labour shortages as a result of low fertility (Perez-Lopez and Diaz-Briquets 1990). From the 1970s the Soviet Union also recruited thousands of temporary migrant workers from other communist countries, mostly from Bulgaria, North Korea, and Vietnam to work in major Soviet cities and other areas of Russia (Chudinovskikh and Denisenko 2017). The fact that such recruitment happened despite the stagnation of the Soviet economy from the mid-1970s reveals that Piore's (1979) central argument that industrialized economies have a chronic, built-in demand for migrant workers (see Chapter 3) is not only applicable to 'capitalist' societies, but also to communist and other state-capitalist models of industrialization.

Migration during neoliberal globalization (1989–2008)

West and East European migrations would undergo fundamental changes in the wake of the fall of the Berlin Wall in 1989, the dissolution of the Soviet Union in 1991, the downfall of communist regimes in Central and Eastern Europe, and their gradual incorporation into the European Union (EU). The 1992 *Maastricht Treaty* would lay the basis for the reinforcement and expansion of federalist European institutions. In 1995, the Schengen Agreement came into effect and eliminated internal border controls. This was followed by enlargement rounds in 1995, 2004 and 2007 which increased the number of EU members from 12 to 25 (see Chapter 10 for further details). The end of the Cold War went along with a period of 'market triumphalism' and neoliberal globalization, a fast increase of world trade and accelerating immigration of lower- and higher-skilled migrants from an increasingly diverse array of European and non-European origin countries. In this period of neoliberal globalization, European migration underwent the following structural shifts:

● Post-Soviet repatriation of Russian-speakers from former Soviet Republics and the subsequent consolidation of Russia as a destination country for labour migrants from former Soviet republics;

● Initial refugee movements because of nationalist tensions and violent conflict in the wake of the dissolution of the Soviet Union and Yugoslavia;

● Increased migration and circulation from Central and Eastern Europe to Western Europe following EU enlargement and the abolition of internal border controls among signatories of the Schengen agreement;

● Increasing labour migration from the Maghreb (Morocco, Algeria and Tunisia), West Africa (countries such as Senegal, Mali, Ghana and Nigeria), Latin America (countries such as Ecuador, Bolivia and Brazil) and to Western and Southern Europe;

● Transition of Southern European countries and also Turkey from countries of emigration into countries of immigration and transit.

Migration in the former Soviet space

The fall of the Berlin Wall in November 1989 precipitated the collapse of the closed communist migration system, although deeper regional connections and migration systems created by Soviet imperialism and socialist internationalism would remain intact. The end of the Soviet Union and its extended Eurasian empire would create a series of political crises and economic shocks and transformations. In total, 13 out of 14 non-Russian successor states of the Soviet empire (except for Belarus) underwent post-imperial transitions involving nation state building (Kuzio 2002; Shevel 2009). Increasing nationalism and the concomitant stoking up of anti-Russian feelings (Janmaat 2007) would ignite 'post-imperial' migrations of Russians living in former Soviet Republics to Russia.

Source: Getty Images/Gerard Malie

Photo 6.1   West Berliners crowd in front of the Berlin Wall on 11 November 1989 as they watch East German border guards demolishing a section of the wall in order to open a new crossing point between East and West Berlin, near the Potsdamer Square

In the early 1990s, significant resettlement took place of Russian-speaking people from former Soviet Republics that no longer felt at home after independence (Ivakhnyuk 2009). Anti-Russian nationalist backlashes, including laws elevating languages other than Russian to official language states and armed conflicts, prompted ethnic Russians to move to Russia (Chudinovskikh and Denisenko 2017; Kuzio 2002). This was similar to the post-colonial 'repatriations' experienced by France, the Netherlands and the UK in the 1950s and 1960s.

On the other hand, following the dissolution of the Soviet Union in 1991 and the establishment of the Commonwealth of Independent States (CIS, comprising ten post-Soviet republics) in 1992, and again in parallel with the earlier experiences of West European colonial powers, Russia underwent a post-imperial migration reversal (a structural change in the dominant direction of migration flows): Russian emigration to the former Soviet republics would largely stop while Russia reinforced its role as a destination for non-Russian labour migrants from former Soviet republics. Initially, around 2 million ethnic Russians left or were displaced from the Baltic states, Ukraine and other parts of the former Soviet Union (Münz 1996: 206). Additionally, there were nearly 1 million refugees from various conflicts (such as in the Caucasus, Moldova and Tajikistan) and some 700,000 internally displaced persons (IDPs) mainly from areas affected by the 1986 Chernobyl nuclear disaster (Wallace and Stola 2001: 15).

The dissolution of the Soviet Union and the creation of independent states meant that millions of internal migrants became international migrants overnight. Also non-migrant ethnic groups, such as the millions of Russians in Ukraine and other former Soviet republics, found themselves increasingly marginalized (Pilkington 1998). The reclassification of internal into international migrants caused a (somehow artificial) hike in the global international migrant numbers in the early 1990s. This exemplifies the somewhat blurred distinctions between internal and international migration. On the other hand, the drawing of new borders had real consequences for migrants and ethnic minorities, through a change in their legal position, discrimination or increased pressure to integrate.

The fall of the Berlin Wall also allowed ethnic Germans (*Aussiedler*) to move to Germany and Jews to depart for Israel, Germany or the United States. According to official statistics, from 1987 to 1991, 134,000 people moved from Russia to Israel, 102,000 to Germany and 15,000 to the United States (Chudinovskikh and Denisenko 2017). Between 1989 and 2002, more than 1.5 million Jews and their relatives emigrated from the former Soviet Union. About 940,000 (or 62 per cent) went to Israel, and the rest were mainly divided between the US and Germany (Tolts 2003).

After an initial hike of repatriation of ethnic Russians to Russia, refugee migration and departure of ethnic and religious minorities, labour immigration to Russia became increasingly important. Reflecting the deep cultural, political and economic connection forged by centuries of Russian imperial domination, migration from former Soviet Republics remained mainly oriented on Russia, with the exception of the Baltic republics (Estonia, Lithuania and Latvia), who were to join the EU in 2004, and to some extent also Ukraine and Moldova. Migration patterns can differ across gender, related to the structure of labour demand in destination countries. For instance, most Moldovan men migrate to Russia and other CIS countries to work in construction, while the majority of women seek work in Italy in the care and service sectors (Yanovich 2015). Between 1991 and 2015, citizens from former Soviet republics represented more than 90 per cent of all arrivals and more than two thirds of departures in Russia. Over this period, total estimated immigration to Russia was 11.8 million and total emigration 5.3 million (Chudinovskikh and Denisenko 2017). This caused a fundamental change in the composition of immigrant populations in Russia (Abashin 2017; Schenk 2018).

While immigrant populations from Ukraine, Belarus and the Baltic states have shrunk (partly because migration became increasingly oriented towards Western Europe), inflows from Central Asia and Transcaucasia (Armenia, Azerbaijan and Georgia) have increased – from 15 per cent of total immigration in the 1990s to almost 40 per cent in the early 2010s. This reflects a gradual transition of temporary to permanent immigration. In 2010, main countries of birth of international migrants in Russia were Ukraine (about 3 million people, or 26 per cent of all migrants), Kazakhstan (2.5 million, 22 per cent), Uzbekistan (1.1 million, 10 per cent), and Belarus and Azerbaijan (740,000, or 6.6 per cent each) (Chudinovskikh and Denisenko 2017).

Migration in Central and Eastern Europe until EU accession

The fall of the Berlin Wall and the political and economic instability this caused initially precipitated movements of migrant workers and refugees to Western Europe. The demise of communism led to the disintegration of several multinational federal states – the Soviet Union, Yugoslavia and Czechoslovakia – from which a multitude of new sovereign nations arose, most of which had little or no recent history of independent statehood (Janmaat and Vickers 2007). The political turmoil that followed would lead to significant population mobility in Central and Eastern Europe particularly from former Yugoslavia. The collapse of communism also undermined many of the barriers that had kept population mobility in check, and led to an increase of East–West migration, although at lower structural levels than expected. After an initial hike, East–West migration did not gain the massive proportions as once expected (or feared), and could not substitute for increasing labour migration from outside Europe, as many expected (and hoped) in the 1990s (de Haas *et al.* 2019b).

In more structural terms, this phase saw the opening up of a new 'labour frontier' in Central and Eastern Europe: Poland, Ukraine, Romania, Bulgaria and the Baltic republics emerged as important new source regions of migrants in Western and Southern Europe, but would also become transit and destination countries in their own right. The fall of communist regimes and the sudden demise of Soviet-Russian imperialism heralded the rapid incorporation of most countries in Central and Eastern Europe (CEE) into expanding West European migration systems. This preceded the formal accession of most of these states to the EU in 2004, which did not start but rather consolidated migration patterns that were already established over the 1989–2004 period (see de Haas *et al.* 2019a).

The post-1989 transition from Communist rule to democracy and market economies transformed states and societies in Central and Eastern Europe. The lifting of exit restrictions led to a largely temporary surge in East–West migration. Many Roma ('Gypsies') from Southeastern Europe (especially Romania) participated in such movements. Initially, this fuelled fear of mass migrations in Western Europe. Populist politicians and media spoke of a 'migration crisis' (Baldwin-Edwards and Schain 1994), and warned that 'floods' of desperate migrants would 'swamp' Western European welfare systems and drag down living standards (Thränhardt 1996).

Such fears were reinforced through a hike in refugee migration as a result of the wars in former Yugoslavia, especially in Bosnia and Croatia (1991–3) and Kosovo (1999). This led to a large influx of largely temporary asylum seekers, particularly from Bosnia, to Germany and other Northwestern European countries as well as Hungary, the Czech Republic and Poland (de Haas *et al.* 2019b). Asylum-seeker entries to European OECD countries peaked at 695,000 in 1992 in response to the Yugoslav civil wars and then declined. There was another peak in 2001 (425,000 application in the EU 27) after which the number of applications fell to around 200,000 (Seilonen 2016). While during the Cold War refugees from communist countries were often warmly welcomed, asylum seekers were increasingly received with suspicion, particularly if they came from non-European countries such as Sri Lanka, Iran or Somalia.

East–West migration increased as well. Initially most migrants were members of ethnic minorities moving to so-called ancestral homelands, where they had a right to entry and citizenship: ethnic Germans (Aussiedler) to Germany (Levy 1999; Thränhardt 1996: 273), Russian Jews (officially known as *jüdische Kontingentflüchtlinge*) to Germany, Bulgarian Turks to Turkey, and Pontian Greeks (see below) to Greece. Germany was the most important destination of such 'ethnic migration' from East to West. Since 1989, approximately 2.2 million ethnic Germans and 235,000 Jews have migrated to Germany (Dietz and Roll 2017).

Migration from Central and Eastern European countries was stimulated by often bumpy transitions towards democracy and market economies that initially increased unemployment and inequality, the dismantling of social security, socio-economic hardship and interethnic tensions. This fuelled migration to Western Europe, particularly among new generations of aspiring youth. However, by the mid-1990s it had become clear that an 'invasion' was not going to take place. Movements of Polish, Ukrainians, Russians and other East Europeans to Western Europe in search of work increased, but did not reach the levels originally predicted, while much of their migration turned out to be temporary and circular. There were some exceptions to this rule, such as Moldova and the Baltic republics, where emigration levels reached very high levels.

At the same time, more economically advanced states like Poland, Hungary and the Czech Republic started attracting immigrants in significant numbers, which was a partial continuation of migration patterns established under communist rule. For instance, the emigration of Polish workers after EU accession on 1 May 2004 increased employers' fears of labour shortages. Poland, as other CEE countries, is experiencing low fertility and an ageing population alongside rapid economic growth. It was in this context that Poland lifted restrictions on short-term workers from Belarus, Georgia, Moldova, Russia and Ukraine, which would stimulate 'replacement migration' from these countries.

The EU and Schengen: Expansion and integration

The processes of EU enlargement and the concomitant removal of internal borders stood at the basis of an increasingly integrated European migration system. The current EU results from a long process that started in 1951 when Belgium, Germany, France, Italy, Luxembourg and the Netherlands founded the European Coal and Steel Community, and, subsequently in 1957, the European Economic Community (EEC). The EEC was renamed the European Community (EC) in 1993 and became the European Union (EU) in 2009 by the Treaty of Lisbon. In seven successive accession waves from 1973 to 2013, the current EU enlarged from the original 6 to 28 members in 2013. This enlargement process started with the accession of Denmark, Ireland and the UK in 1973. Greece joined the EEC in

Source: Getty Images/Dan Kitwood

Photo 6.2　A Polish Supermarket in Grimsby, England, in October 2018

1981, to be followed by Spain and Portugal in 1986. Austria, Finland and Sweden became EU members in 1995 (de Haas *et al.* 2019).

On 1 May 2004, 10 new member states gained accession to the EU: the Czech Republic, Cyprus, Estonia, Hungary, Latvia, Lithuania, Malta, Poland, Slovakia and Slovenia (known as the EU10). Most of the existing member states (the EU15) decided to restrict migration from the new Eastern and Central European member states over a seven-year transitional period, but Ireland, the UK and Sweden opted not to. This stimulated major immigration from Poland and the Baltic republics, especially Lithuania, to the UK and Ireland. Finally, Bulgaria and Romania joined the EU in 2007, and finally Croatia in 2013, creating the current EU28.

In parallel, the 1985 signature of the Schengen Agreement led to a borderless European Area in 1995, which included all then EU members, except for the United Kingdom and Ireland, which opted out. Since then, all new EU members as well as non-EU members Iceland, Norway, Switzerland and Liechtenstein joined the Schengen Area, which in 2018 comprised 26 members. This contributed to a growing integration of European migration systems and a drastic expansion of the European free migration space.

These developments stimulated East–West migration, but not to the extent once hoped (by EU bureaucrats) or feared (by populist politicians). The biggest emigration occurred from Poland. One million Poles emigrated between 1 May 2004 and April 2007, principally to the UK, Ireland and Germany. However, part of such increases may be statistical artefacts. The 2004 enlargement process had a significant *legalization effect* for EU10 workers that were already working in the EU employed irregularly in the EU15 states before May 2004: several hundred thousand benefited from de facto legalization (Tomas and Münz 2006).

Over this period, migrant workers of countries like Poland (particularly in the UK and Germany), Romania (particularly in Spain), Bulgaria and Baltic countries have come to occupy an important role in West European economies. While the removal of migration barriers led to migration surges – particularly when economic gaps between origin and destination countries were large – these tended to be temporary, with migration after a few years consolidating at lower levels and becoming increasingly circular (de Haas *et al.* 2019). The major exception on this rule was the UK, where large-scale settlement of Polish and other East European workers occurred.

While the removal of barriers for travel and migration has boosted intra-European mobility, the increase in intra-European migration was not as large as expected. Intra-European migrations remained too modest to replace immigration from non-EU countries, dashing such hopes among some EU bureaucrats. The main effect of the liberalization of border regimes following the fall of the Berlin Wall in 1989, the Maastricht Treaty in 1992 and the successive EU enlargements has been an increase in *circulation* while permanent migration has remained limited with a few notable exceptions, such as the large-scale settlement of Poles in the UK (de Haas *et al.* 2019).

Totally contradicting expectations, immigration from non-EU countries showed a structural increase over this period. In fact, both intra-EU and extra-EU immigration increased at the same time (see Figure 6.1). Origin countries of non-EU migrants no longer mainly comprised North Africa, Turkey or former colonies, but migration from Latin America, Asia and sub-Saharan Africa also increased. This immigration would also become increasingly diverse in terms of education and skills of migrants. Although lower-skilled workers kept on responding to labour demand in service, agriculture and construction sectors, migration from higher-skilled workers and students, particularly from Asia, increased over this period. The effects would be most significant in Southern Europe, and particularly Spain and Italy, which, after having been prominent emigration countries for many generations, transformed to major lands of immigration.

Southern European migration transitions

In the period between 1989 and 2008, Southern Europe – the main labour reserve for Western Europe, North America, South America and Australia for over a century – established its new position as destination region. Until the 1970s, Italy, Spain, Portugal and Greece were primarily emigration countries. Then, at somewhat different junctures, and as a result of economic growth, demographic transitions, democratization and EU membership, they started to experience declining emigration and increasing immigration. As a result of these migration transitions, they started to share many of the characteristics and concerns of Northern EU states, yet remain distinct by the central role played by the informal economy in shaping migration, the importance of undocumented migration as well as a low governmental willingness and capacity to regulate international migration (Reyneri 2001). Many women from Latin America and Asia have migrated to Southern Europe to work as caretakers, which is linked to the lack of state-subsidized facilities for childcare and, particularly, elderly care. Although many have irregular status, their essential role in care provision makes them into socially accepted, 'tolerated' workers (Ambrosini 2015).

In *Italy*, numbers of foreigners with residence permits doubled between 1981 and 1991, from 300,000 to 600,000, and then rose to 1.4 million by 2000 and 4.6 million in 2010, representing 7.6 per cent of the total population living in Italy (OECD 2012; Strozza and Venturini 2002: 265). These steep increases reflect a series of regularizations of undocumented migrant populations between 1986 and 2009. Entry of non-EU citizens for employment into Italy is officially governed by annual quotas. However, the quotas are very low and do not reflect real labour demand, resulting in substantial rates of overstaying and irregular entry. Romanians form the largest immigrant group. In 2017, their number reached 1 million. The leading group of non-EU residents are Albanians (458,000) and Moroccans (435,000) (OECD 2018) (see Box 13.6 for more data on immigrants in Italy).

Spain went through a similar migration transition to Italy. Prior to 1900, Spain was mainly a land of emigration. That status quo began to change with post-Franco democratization after 1975 and EC membership (alongside Portugal) in 1986. The 1986–99 period saw rapidly increasing immigration particularly from North Africa, which was boosted by economic growth, the large-scale incorporation of Spanish women into the labour market which generated a demand for reproductive labour (such as domestic work) and regulation of entry through recruitment and regularizations. After 2000, growth consolidated, migration from Latin America was encouraged and, as a consequence of permanent settlement of workers, family migration grew in significance. The registered foreign population in Spain grew from 279,000 in 1990 to 1.3 million in 2000. By 2005, it had shot up to 4.1 million only to increase further to 5.7 million in 2010, (OECD 2012). As in Italy, a large share of immigrants entered legally and subsequently overstayed visas while a minority entered Spain irregularly. Between 1985 and 2005, Spain carried out 12 legalization campaigns (Plewa 2006: 247).

In 2017, the numbers of foreign-born people had reached 6 million, representing 12.4 per cent of the Spanish population. The biggest groups are Moroccans (700,000), Romanians (607,000), Ecuadorians (409,000), Colombians (362,000). Other significant immigrant groups include British (298,000, often pensioners and 'lifestyle migrants') and Argentinians (256,000) (OECD 2018). In line with dual labour market theory (see Chapter 3), foreign workers have become specialized according to their origins and gender: Moroccans and other Africans tend to work in agriculture. Latin American women often work in domestic service and care work, while Latin American men tend to work in construction and services and Europeans in industry (OECD 2006: 214).

Other Southern European countries also witness increased immigration, although to a lesser extent than Italy and Spain. Migration to *Portugal* has remained predominantly post-colonial. From the mid-nineteenth century to the mid-1970s, Portuguese emigrated in large numbers, leaving a legacy of some 5 million Portuguese and their descendants living abroad (OECD 2004: 254). The revolution of 1974, the end of dictatorship, and decolonization prompted significant postcolonial migration from former Portuguese possessions in Africa. As in Spain and Italy, most recent immigrants overstayed visas or arrived irregularly and there have been recurrent legalizations (Cordeiro 2006; OECD 2007: 276). According to the most recently available official figures, in 2011 the most sizeable immigrant groups were Angolans (163,000), Brazilians (140,000), French (95,000), Mozambicans (73,000) and Cape Verdeans (62,000) (OECD 2018). In contrast to Italy and Spain, however, Portugal has never ceased to be an important emigration country, reflecting the relative weakness of its economy.

Until 1990, international migration to *Greece* mainly involved repatriation of ethnic Greeks from abroad and arrivals of refugees in transit. After 1990, immigration soared and in 2010 the estimated 810,000 foreigners constituted 7.1 per cent of Greece's population. About a third of Greece's migrants are of Greek ethnic descent: Pontian Greeks from the former Soviet Union (150,000) and ethnic Greeks from Albania (Triandafyllidou and Lazarescu 2009). In 2010, the largest groups of registered non-EU immigrants were from Albania (491,000), Ukraine (21,500), Georgia (17,000) and Pakistan (16,000). The largest groups of EU nationals living in Greece came from Bulgaria and Romania (OECD 2012: 234). From 2015, Greece would become the focal point of large-scale sea crossings by Syrian and other refugees from Turkey (see below).

The politicization of migration and the 'global race for talent'

The end of the Cold War also prompted the politicization of migration, with politicians in Northwest Europe stoking up fears of massive East–West and South–North migrations. Politicians started to scapegoat migrants for problems not of their making, such as the closure of factories and mines and defunding of public housing. High unemployment and welfare dependency amongst low-skilled migrant workers led to a perception that immigration was a threat to the welfare state, an idea which gained increasing traction. From the late 1980s and early 1990s, politicians and media in countries such as the UK and the Netherlands started to portray asylum seekers as welfare scroungers in disguise. Although refugees represented a minority of immigrants, the 'bogus asylum seeker' became a favourite subject of political fearmongering. Governments vied with each other in introducing tougher asylum rules, although they struggled in implementing these rules because of constitutional constraints, humanitarian concerns and the practical difficulties associated with – and frequent public protests against – the deportation of rejected asylum seekers and their families.

Confronted with the growth of sizeable communities of former migrant workers from North Africa and Turkey and their descendants, and the growing perception that their integration had 'failed' or at least remained far behind expectations, there were growing political calls to prevent admission of lower-skilled migrant workers. In addition, concerns continued about 'secondary family migration' as a result of new unions between the children of former guestworkers and their spouses living in origin countries, which continued legal immigration even after primarily family reunification had been completed in the 1980s, reinforced attempts to make such migration more difficult by introducing language tests, income requirements and (rather controversial) investigations into partners' romantic relationships to prevent 'sham marriages'.

However, governments encountered legal constraints in the extent to which they could infringe on the fundamental rights (such as to family life) of migrants, particularly if they had become citizens. Family reunion has therefore long remained the largest single immigration category in the many West European countries. The relative importance of family reunion for these migrant groups would decrease over the 2000s, partly because of policy restrictions and partly because integration went together with a decreasing social pressure in migrant communities and decreasing willingness among members of the second and, particularly, third generation to marry partners from origin countries.

This coincided with a growing importance of labour migration including 'free movement' migration from Eastern Europe to Western and Southern Europe, increased migration from outside the EU – lower-skilled workers from Latin America, Africa and former Soviet Republics such as Ukraine and Moldova, as well as skilled workers, particularly from Asia. As part of an increasingly intensive 'global race for talent', many governments introduced preferential entry rules for skilled migrants and students in a bid to attract 'the best and the brightest' (Czaika and Parsons 2017; Kapur and McHale 2005). Yet the continued demand for lower-skilled labour migrants continued to be unmatched by long-term admission schemes, and continued to be met either through limited temporary and seasonal recruitment schemes, or through undocumented migration.

Figure 6.3 shows the evolution migration within and towards the area comprising the EU25, the 25 countries that formed the EU from 2004 to 2006. It excludes Bulgaria, Romania and Croatia, which joined the EU in 2007 and 2013, as no data was available for these countries. The graph distinguished (1) migration *between* countries of the EU25 (2) migration from non-EU countries considered part of Europe (such as Russia, Ukraine and Switzerland) to the EU; and (3) migration from outside Europe (including, somewhat controversially, Turkey) towards the EU. The picture is clear:

- There has been a structural long-term increase in migration to Europe;
- The immigration peak of the 1960–1973 period mainly concerned immigration from Mediterranean countries and Turkey;
- Intra-EU mobility has gradually decreased in relative importance since 1963 with Eastern European migration to Western Europe largely replacing Southern European migration over time;
- Post-1989 immigration to the EU soared to unprecedented levels and has been mainly been driven by a structural increase of non-EU legal immigration;
- Business cycles largely explain short-term fluctuations in immigration, highlighting the importance of destination country labour demand in driving migration.

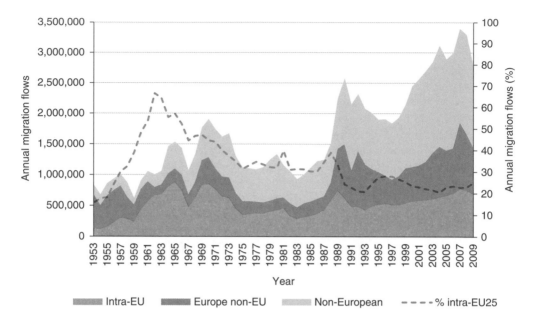

Figure 6.3   Annual migration flows within and towards the EU25, 1953–2009

Source: Adapted from de Haas *et al.* 2019b; DEMIG C2C database, International Migration Institute (www. migrationinstitute.org); data for Bulgaria, Romania and Croatia not available

Annual total inflows into the current EU28 countries were above 2 million between 1998 and 2009. The long-term trend is upward: from around 1 million over most of the 1970s, then increasing fast since the late 1980s stabilizing at levels of around 2 million yearly entries over the 1990s, and then accelerating again to reach immigration levels of up to 3 million over the 2000s. While entries to many Northwestern European countries also increased, the biggest increases in the number of legal entries occurred in Southern Europe, which peaked in 2007 at 921,000 in Spain and 515,000 in Italy. Intra-European migration has remained rather stable over the long term, and has therefore not been a substitute for immigration from outside the EU. Instead, extra-EU immigration reached levels of between 2 and 2.5 million per year over the 2000s. Increasing legal migration from Africa and Latin America are the most important factors explaining this increase (de Haas *et al.* 2019b).

Free intra-EU migration and the establishment of the Schengen zone resulted in the establishment of common visa rules for non-EU citizens and provided the impetus to step up external border controls. For instance, this led to the introduction of Schengen visas for citizens of Morocco and Tunisia who, before 1991, could previously travel freely back and forth to Spain and Italy. More stringent visa rules and increasing border controls stimulated undocumented migration and smuggling and could also not stop the structural increase in legal immigration, by remained far the biggest source of entry to Europe. Immigration restrictions interrupted circulation by pushing migrants into permanent settlement by discouraging their return. While labour demand continued to fuel labour migration through legal and illegal channels, family reunification facilitated the continuation of legal migration, this time particularly to Southern Europe.

Box 6.1 Europe's recurrent migration crises

Since the fall of the Berlin Wall in 1989, Europe has been gone through a series of recurrent 'migration crises'. The usual pattern is one where some politicians exploit hikes in arrivals to create a climate of fear by fabricating doomsday scenarios about an impending migrant invasion. For instance, the post-communist lifting of exit restrictions in Central and Eastern Europe fuelled fear of mass migrations in Western Europe. Populist politicians and media spoke of a 'migration crisis' (Baldwin-Edwards and Schain 1994), and warned that 'floods' of desperate migrants would 'swamp' Western European welfare systems and drag down living standards (Thränhardt 1996).

When it became clear that this invasion would not happen, the fear of mass migration was increasingly projected on to Europe's southern border. The introduction of a visa requirement for North Africans by Spain and Italy in the early 1990s launched the phenomenon of 'boat migration'. Intensified border patrolling prompted a diversification of crossing points further east, such from Libya to Italy, and, later, from West Africa to the Canary Islands, where the arrival of 31,000 Africans in 2006 provoked a highly mediatized 'migration crisis' with politicians claiming that Europe would be swamped (de Haas 2008a).

In 2011, the Arab Spring prompted a temporary hike in undocumented emigration from Tunisia to Italy and, from there on, to elsewhere in Europe (Natter 2014). This provoked a political crisis in the EU, prompting governments of some countries including France and Denmark to temporarily reintroduce controls at Schengen borders. When conflict broke out in Libya in the same year, Italian interior minister Maroni warned of a 'biblical exodus' and 'an invasion … that would bring any country to its knees', although movements remained relatively small.

The politicization of migration in Europe came to a boiling point during the 'refugee crisis' of the mid-2010s. In Syria, violent attacks by the Assad regime on pro-democracy protestors in 2011 provoked a devastating cycle of violence and the outbreak of civil war. Up to 2019, this led to the internal displacement of 6.2 million while 5.6 million Syrian refugees moved to neighbouring countries. A lack of prospects motivated increasing numbers to move to Europe, provoking a peak in refugee migration in 2015, when more than 1 million asylum seekers, mostly Syrians, but also Afghanis and Iraqis, crossed from Turkey into Greece, and from there further into Europe.

This precipitated a major political crisis, highlighting the inability to establish effective international cooperation on EU level. The failure to agree on 'burden sharing' undermined political support for generous refugee policies in countries such as Germany (Pries 2018). This increased pressure to step up border controls to prevent asylum seekers from arriving in the first place, such as through closing borders on the 'Balkan route'. In 2016 the EU concluded a deal with Turkey, with the Turkish government agreeing to take back refugees in exchange for major financial assistance and the promise to abolish EU visas for Turkish citizens. This exemplifies the extent to which irregular migration has become a negotiation chip in relation between European and transit states in the Middle East and Africa (see Box 13.4 and 13.6, (Paoletti 2011; Brachet 2016).

▶

However, the real refugee crisis did not happen in Europe but in the Middle East. In 2019, about 1 million Syrian refugees lived in Lebanon, of a total population of 6 million, while over 3.5 million lived in Turkey, the largest refugee hosting country in the world. By comparison, in 2015 about 0.4 per cent of the total EU population was a refugee (Postel *et al.* 2015). However, the most severe consequences are for those who flee: border controls and three decades of attempts by the EU to 'externalize' border controls through bilateral agreements with transit states have increased the costs and risks of migrating. Between 2000 and 2017 more than 33,000 people have died while crossing the Mediterranean, making it by far the world's more lethal border area (Fargues 2017).

Migration after 2008: Destination Europe

The Great Recession marked the end of – or at least a pause in – the previous period of accelerated globalization, neoliberal policies and EU expansion. The Recession initially led to a slump in, particularly undocumented, immigration as unemployment amongst lower-skilled migrant groups, such as Pakistanis and Bangladeshis in the United Kingdom and North African and South American immigrants in Spain increased fast (BBC/MPI 2010). However, large-scale return of migrant workers did not occur, and the Recession did not fundamentally change the long-term patterns of non-EU immigration. Return migration has generally been highest for EU member state citizens enjoying free mobility (such as Romanians in Italy or the Polish in the UK). Return rates were lower for non-EU migrants, such as Moroccan, Ecuadorian, Peruvian and Bolivian workers in Spain. As return would often be tantamount to giving up residency rights and the possibility to re-emigrate, many preferred to stay put, on the safe side of the border. Similar dynamics had made many 'guestworkers' stay in Northwest Europe after the 1973 Oil Shock.

The migratory consequences have been largest for the countries where the crisis hit hardest, particularly Ireland, Spain, Portugal, Italy and Greece and some Eastern European countries, which would experience plummeting immigration and increasing out-migration. For instance, in Spain the unemployment rate of foreigners climbed to 36 per cent in 2011, explaining the decline in inflows from 921,000 in 2007 to 248,000 in 2013. In parallel, outflows also increased, although more modestly from 199,000 in 2007 to 459,000 in 2011. The crisis also prompted new Spanish emigration, principally towards the UK, France, Germany and the US, but also to Argentina, Brazil, China and Australia (González Ferrer 2012). Since then, immigration has rebounded and returns declined, and in 2015 Spain turned into a net immigration country again.

Although migration to crisis-hit countries in Southern Europe would plummet in the wake of the crisis, it would not stop, but rather stabilize at lower, but still significant, levels. This continuity can be explained by the specific segmentation of dual labour markets, which perpetuates a certain demand for migrant workers in particular sectors even in times of economic crisis and soaring domestic unemployment. For instance, although the crisis most severely hit (male) migrants working in the construction sectors (see González Ferrer 2012), labour demand in other sectors has remained less affected, or even increased, such as the demand for personal care-workers in increasingly ageing societies (Ambrosini 2015).

The Recession neither prompted a massive return, nor did it lead to a substitution of non-European migrant workers by Eastern or Southern European labour migrants. Although migration from particularly crisis-hit Southern European countries such as Spain and Greece to Northwestern European countries such as Germany and the UK would increase, intra-European migration levels have in fact remained surprisingly modest (de Haas *et al.* 2019b), and much below expectations. Moreover, such intra-European migrants are generally higher skilled than previous generations and generally less motivated to pick up jobs typically done by non-European migrant workers in domestic work, cleaning, care, construction, catering and agriculture.

In Central and Eastern Europe, the effects of the Recessions varied a lot. The crisis hit hard in Czech Republic, the Slovak Republic, Hungary, and particularly in the Baltic republics of Estonia, Latvia and Lithuania, leading to decreasing immigration and increasing emigration parallel with steadily worsening labour market conditions (OECD 2012: 248). By contrast, the Polish economy has kept on growing throughout the crisis, and Poland has entered a phase of decreasing emigration and increasing return migration (OECD 2012: 260). Foreign immigration reached 107,000 in 2018, which may signal Poland's future transition into a destination country.

The poorer economies of Romania, Bulgaria and Ukraine remained emigration countries throughout the Recession (see Kubal 2019). Worsening economic conditions continued emigration from Romania (primarily to Italy, Spain and Hungary), Bulgaria (primarily to Turkey and Spain) and Ukraine (primarily to Poland, Russia, Italy and the Czech Republic), although return migration increased as a result of the crisis affecting South European destinations. Russia continued to attract immigrants from former Soviet Union republics (such as Kazakhstan, Uzbekistan, Tajikistan and Ukraine) and China. At the same time, growing numbers of Russian workers and asylum seekers moved to the Czech Republic, Turkey, Poland and Germany.

While total entries into European OECD countries declined to about 3 million in 2009 and 2010, the decline has been much smaller than anticipated. In most Northwestern European countries including Germany, France, the UK and the Netherlands immigration trends were barely affected at all. The recession did not reverse the long-term migration trends, principally because it did not fundamentally affect the economic and demographic transformations that had perpetuated the demand for migrant workers over the post-1989 period. After economic growth resumed in the 2010s, immigration therefore rebounded to and even beyond pre-crisis levels. In European countries such as Germany, the UK, the Netherlands, Belgium, Switzerland, Russia and Scandinavia, immigration reached record high levels. Particularly Germany, the world's second most important migration destination after the US, would consolidate its position as Europe's main migration destination, with annual foreign immigration surpassing the 1 million mark after 2012 (see Figure 6.4).

According to Eurostat data, 2.4 million immigrants entered the EU from non-EU countries in 2017, similar levels as before the Great Recession. Figure 6.5 shows that Germany, Russia, UK, France, Spain and Italy have evolved as Europe's main migration destinations, each home to 6 million or more migrants in 2017. Ukraine, Switzerland, the Netherlands, Sweden, Austria and Belgium also have sizeable immigrant communities. In most European countries, migrants now represent between 10 and 15 per cent of the population, although these percentages are higher in some countries ranging such as 17.6 per cent in Sweden up to 29.6 per cent in Switzerland. The data also show that

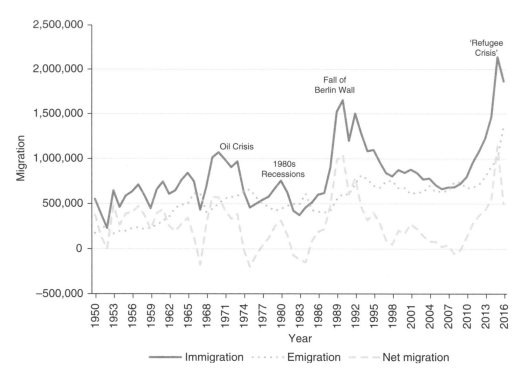

Figure 6.4   Migration to and from Germany, 1950–2016
Source: Statistisches Bundesamt

non-European immigrants form the majority of immigrant populations in former colonial powers such as the UK, France, Spain, Portugal and the Netherlands, while most other countries mainly draw on European sources of migrant labour.

The period after 2008 has also been characterized by a continued politicization and securitization of migration, particularly marked by the rise of populist, anti-immigration and anti-EU parties. The securitization of migration had already increased since the 9/11 terrorist attacks, but after the crisis more and more politicians were drawn towards rhetorics representing immigration, particularly of Muslims, as a fundamental threat to the security and identity of European societies. This resulted in electoral successes of anti-immigrant parties across Europe and also tempted mainstream parties to adopt more restrictive and sometimes xenophobic positions (see Davis 2012). However, the gap between tough migration rhetoric and more liberal policy practice remained high. While most entry policies did *not* become more restrictive, labour market demand for migrant workers continued as the European immigration policies have been increasingly redesigned to serve economic interests (see Chapter 11).

However, the politicization of migration and the staging of a series of 'migration crises' over the 2000s and 2010s (see Box 6.1) would have real political consequences. The politicization of migration came to a boiling point during the 'refugee crisis' of 2015 and 2016, which led to a major political crisis following the influx of about 1 million asylum seekers into Europe, mostly via Greece. This led to a massive increase in European investments in border controls, and increased hostility towards asylum

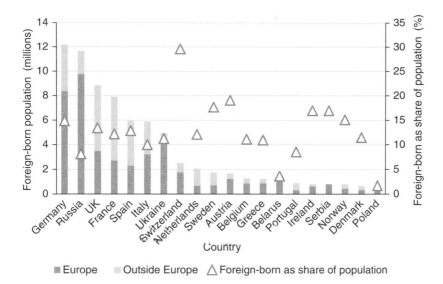

Figure 6.5   Foreign-born populations in European countries, by origin, 2017
Source: UNDESA 2017

seekers, and attempts to prevent humanitarian organizations from assisting or sav-
ing migrants from drowning in the Mediterranean (see also Chapter 9, Box 13.4 and
Box 13.6). In the UK, fears of mass immigration – stoked up by decades of immigrant
scapegoating by media and politicians – were a decisive factor in the 2016 Brexit vote –
alongside a more general trend towards nationalism and protectionism (see Chapter
10). Although the outcome of such processes is uncertain, this shows that public
xenophobia have far-reaching geopolitical and economic consequences. If anything, it
shows the political potential – and risks – of playing the migration card.

Conclusion

The upsurge in migration in the post-1945 period shows the extent to which large-scale
immigration has become an structural feature of European societies and economies. This
marks the transformation of Europe from a region of mass emigration in the nineteenth
and early twentieth centuries, to a region of large-scale inflows from an increasingly
diverse array of origin countries. This shift has been driven by demographic transitions,
economic transformations and the related growth of dual labour markets, that have
increased the demand for migrant labour and subsequent recruitment of 'guestworkers' in
Mediterranean countries.

Besides recruitment of workers, decolonization processes played an important
role in stimulating non-European migration to West Europe's industrial heartlands,
particularly in the UK, France, Netherlands and Russia. The dismantling of West
European and the Soviet empires between 1945 and 1991 led to a post-colonial migration reversal in the form of a turnaround in the dominant direction of migration flows
between centre and periphery. Initially, such reverse flows often consist of returning
colonial settlers, the 'repatriation' of their (mixed) descendants (such as Algerian *colons*

to France, Indonesian *Indos* to the Netherlands, and ethnic Russians to Russia), and other minority populations who left because of rising nationalism and anti-colonial sentiment. This illustrates how nation state building in newly independent countries often led to decreasing tolerance of ethnic and religious minorities. On the other hand, after an initial surge of 'repatriates' reverse flows started to become dominated by labour migrants from former colonies, such as Maghrebis to France, Pakistanis and Bengalis to the UK, or non-Russians from former Soviet Republics to Russia. These labour migrations and family migrations they triggered became dominant over time and contributed to the increasing diversity of European destination countries.

The fall of the Berlin Wall in 1989 and the subsequent dismantling of communist regimes in Central and Eastern Europe facilitated East–West migration and provided an impetus for intra-European migration. After the collapse of the Soviet Union in 1991 – by then the world's last remaining colonial empire (Kuzio 2002) – Russia evolved into the second-most popular European migration destination after Germany, mainly attracting migrants from former Soviet republics. As they went through migration transitions, the former labour reserves of the European periphery (Southern Europe, Ireland and Finland) as well as several Central and Eastern European countries have transformed from net emigration to net immigration countries, coinciding with the outward expansion of the European 'labour frontier' (see also Skeldon 1997).

These developments coincided with European integration and enlargement. The creation of the borderless Schengen Area in 1995 and the EU enlargements of 2004 and 2007 stimulated migration within expanding European migration systems. However, intra-European migration remained lower than expected and largely circular, and did not replace non-European immigration, which only kept on increasing in relative importance. This reveals an striking paradox: the creation of an of open border area has not led to massive internal migration, while external border controls were not able to curb rising legal immigration of non-EU and non-European citizens.

The post-War emergence of Europe as a global migration destination exemplifies a major theme of this book: migration needs to be seen as an intrinsic part of broader processes of development and socioeconomic transformation. Fundamental processes of structural change such as decolonization, demographic ageing, the shift from industrial to service sector jobs, and the growing importance of precarious labour in increasingly segmented labour markets have created the conditions for large-scale immigration to occur. The recent European immigration history highlights the crucial role of active labour recruitment in facilitating migration, either through state-led efforts as with the 'guest-workers', or through recruitment agencies and other migration intermediaries. The latter form part of a burgeoning 'migration industry' (see Chapter 3), whose room for manoeuvre has expanded as a result of labour market deregulation.

This shows the relevance of historical-structural and dual labour market theories highlighted in Chapter 3: states have played a central role in shaping migration patterns, directly – through immigration policies and recruitment practices, as well as indirectly – through 'non-migration' policies affecting economic growth, inequality, labour markets and social security. Dual labour market theory is particularly useful to understand why labour migration to Europe has continued even in times of economic crisis and rising unemployment. Population ageing, increasing education and the massive entry of women in the formal labour forces have depleted traditional domestic supply – school-leavers and women – for low-status and informal sector work. In addition, decades of immigration have contributed to the 'racialization' of particular labour market segments. This particularly applies to low-skilled and often informal jobs in agriculture, construction, catering, cleaning and care. Their subsequent labelling as 'migrant jobs' further decreased the status of such occupations in the eyes of native workers.

Not only West European countries, but also several communist states in Central and Eastern Europe started to recruit migrant workers from the 1970s. Although numbers were comparatively lower, this illustrates that the key factors behind dual labour market formation (particularly shrinking domestic labour supply and job status, (see Piore 1979) and the resulting 'chronic' demand for migrant workers affect all types of industrialized, highly-developed societies with ageing, increasingly educated populations. Although such migrants were widely considered to be temporary, network effects and migration system dynamics (see Chapter 3) explain why, once certain number of migrants had settled at the destination, migration processes tended to gain their own momentum and became increasingly difficult for governments to control, and why immigration restrictions had the unintended effect of stimulating their settlement, prompting large-scale family migration.

The demand for lower- and higher-skilled migrant workers has become structurally embedded in the European economies characterized by increasingly complex and segmented labour markets. This also explains the long-term 'resilience' of immigration to economic downturns such as the Great Recession, because they hardly affected the long-term structural drivers behind immigration. Particularly because of the drying up of traditional Southern European labour reserves, and the transition of South European countries into migration destinations in their own right, migrant workers have increasingly come from outside Europe, particularly from Africa, Latin America (for Western Europe) and Central Asia (for Russia). As a result, many European societies have been confronted with an unprecedented, largely unintended and unplanned increase in non-European migration and a concomitant increase in ethnic and cultural diversity, particularly in urban areas. In 2018, 22.8 million non-EU citizens were estimated to live in the EU (representing 4.4 per cent of the EU28 population) alongside 17.6 million EU citizens living in another EU country.

The politicization and securitization of migration of the 2000s and 2010s seems to have been partly driven by the realization that immigration has gained increasingly permanent features, that migrants were 'there to stay', and considerable political reluctance to accept that new reality. Different from former European settler societies such as the United States and Canada, where immigration has been part of nation building, many European nation states are built on ideas around common ancestry and culture. The new reality of large-scale immigration and settlement did often not resonate with the self-image of most European societies and often clashed with dominant nation state ideologies which struggle to accommodate ethnic diversity. This inevitably generated significant debate and political tensions, particularly as migrants started to settle and started to demand their place and rights in their new homelands.

In times of growing inequality and decreased economic security, some politicians have exploited the unease and fears created by immigration by fanning the flames of xenophobia. They have done so by classic migrant scapegoating and by creating an image of a European continent that is besieged by hordes of fortune seekers or potential welfare scroungers, and bogus asylum seekers and illegal migrants. However, tough migration rhetoric often contrasts with policy practice, which has continued to allow legal immigrants in while often turning a blind eye towards the presence of undocumented migrants that are eager to do jobs native workers are not willing or able to do.

Guide to Further Reading

For a general overview and analysis for migration in Europe, the annual *International Migration Outlook* of the OECD is invaluable. Many European national statistical offices now provide data online (often in English), while reports from interior ministries and other government authorities are also often available. The Migration Policy Centre (www.migrationpolicycentre.eu) at the European University Institute is a good resource for recent analysis on European migration, while the Migration Observatory (migrationobservatory.ox.ac.uk) at Oxford University is a good resource for UK migration issues. The Clandestino project (irregular-migration.net) gives information and estimates on undocumented migration in Europe.

To understand the emergence of Western Europe as a destination for migrant workers, Castles and Kosack (1973) is a comparative study of immigrant workers in France, Germany, Switzerland and the UK from 1945 to 1971, while Miller (1981) provides an early analysis of the political effects of migration. Castles *et al.* (1984) continue the story for the period following the ending of recruitment in 1973–4. Boswell and Geddes (2011) analyse post-Cold War European migration patterns and their relation to political and policy making processes. Schierup *et al.* (2006) examine the 'European dilemma' of migration and increasing diversity. Düvell (2006) provides a good overview of undocumented migration and Düvell (2012) and Collyer *et al.* (2012) give critical analyses of transit migration.

De Haas (2008) reviews myths and facts on of African migration to Europe. Givens and Maxwell (2012) give an overview of ethnic minority and migrant political participation in Western Europe. King *et al.* (2000) gives a useful overview of Southern European migrations. Baganha and Fonseca (2004) gives an overview of the origins of East European migration Southern Europe. Lutz's (2016) gives a European perspective on the relation between undocumented migration and domestic work. Ambrosini (2016) gives an overview of the crucial function of undocumented migration (of women) in providing 'invisible welfare', between official 'political rejection' and 'practical tolerance'. Kubal (2016) gives insights into the socio-legal integration of Polish migrants in the UK. Fargues (2017) gives an overview of four decades of trans-Mediterranean undocumented migration. Schenk (2018), Kubal (2019) and Heusala and Aitamurto (2016) give good overviews of post-Soviet migration and the emergence of Russian migration systems.

Extra resources can be found at: **www.age-of-migration.com**

7 Migration in the Americas

The Americas have been made and re-made by the migrations of the past 500 years. Conquest, slavery, immigration and settlement were crucial components of the process of colonization that began in the late fifteenth century when Spain, Portugal, the Netherlands, England and France fought to gain control over the 'New World'. The fifteenth to nineteenth centuries were marked by violent conquest and resource extraction, at first from gold and silver mines and then from sugar, tobacco and other plantations (Galeano 1973). While indigenous populations decreased because of the diseases, massacres and forced labour brought by colonization, the arrival of European settlers and African slaves, along with the consequent process of *mestizaje* (mixing of ethnic groups) and *creolization* (mixing of cultures), generated deep and lasting changes, which would shape the contemporary face of the Americas and the world at large.

The Americas have been strongly affected by large-scale migration throughout modern history. Immigration is an integral part of the national identity of countries like the US, Canada, Brazil and Argentina. Between about 1850 and 1960, up to 60 million Europeans migrated to the 'New World', not only to the US and Canada, but also to South America, especially to Argentina and Brazil. Although the vast majority of immigrants were Europeans, also people from other, particularly Asian, countries came to the Americas. For instance, Chinese migrants have a long historical presence in the US, with the first Chinese arriving in the mid-1800s during the California Gold Rush years, when workers arrived in the West Coast to do manual jobs in railroad construction (particularly the First Transcontinental Railroad), agriculture, mining and other low-skilled jobs.

The First World War ended the previous period of globalization and marked the beginning of a period of economic depression, protectionism and xenophobia that lasted until the late 1930s. The combination of these factors led to rapidly plummeting migration to the Americas. With xenophobia in the US reaching a high, the Immigration Act of 1917 introduced literacy requirement for immigrants and stopped immigration from most Asian countries. The Immigration Act of 1924 further limited the number of immigrants, and strongly favoured immigration from Northwest Europe. Except for the Philippines, then an American colony, Asians were completely barred from entering. These restrictions led to an increase in undocumented migration across the Mexican and Canadian borders, while from 1942 the *Bracero* programme allowed the temporary migration of Mexican farmworkers to fill labour shortages during World War II.

For much of their modern history, the economic heartlands of Argentina, Uruguay and Brazil in the 'Southern Cone' of Latin America have experienced large-scale immigration, particularly from Spain and Italy. No other Latin American country was transformed by immigration as profoundly as Argentina. In a relatively short period of time – from roughly 1870 to 1930 – the arrival of approximately six million migrants contributed to the creation of a modern, predominantly urban and industrial society (Maier 2015). Similar to trends in North America, large-scale European migration to the Southern Cone declined by the 1930s because of immigration restrictions and the economic depression (Barlán 1988). However, European immigration would partly be replaced by

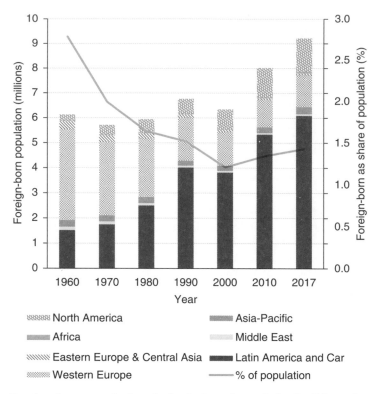

Figure 7.1   Foreign-born populations in Latin America and the Caribbean, by origin region, 1969–2017

Source: UNDESA 2017 and Global Bilateral Migration Database

other migrations, such as the Great Migration, the South-North movement of African-Americans in the US (see Chapter 4). In Brazil, migration from Japan became significant. Japanese migrants arrived in Brazil in two waves: the first, between 1925 and 1936 and the second one between 1955 and 1961 (Amaral and Fusco 2005). These migrants were the ancestors of the *Nikkeijin*, the Japanese heritage population that would migrate to Japan in the late twentieth century.

After the Second World War, the character of American migration would fundamentally change. The main trends were:

- Decreasing European migration to the Americas over the 1950s and 1960s;
- The demise of South America as a global migration destination accompanied by increasing intra-regional migration;
- Increasing migration from Latin America, the Caribbean and Asia to the US and Canada.

These developments led to declining immigration and diversity of Latin American societies, particularly because the Southern Cone region lost its previous position as the second most popular migration destination for European migrants (see Figure 7.1). By contrast, the ethnic and religious diversity of North American societies would increase rapidly through large-scale immigration of populations of non-European origin (see Figure 7.2).

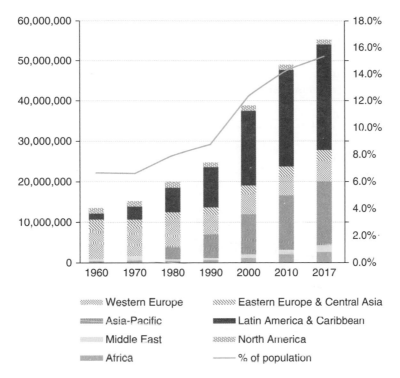

Figure 7.2   Foreign-born populations in North America, by origin region, 1969–2017
Source: UNDESA 2017 and Global Bilateral Migration Database

Regional migration patterns

In 2017, there were an estimated 64.5 million international migrants in the Americas. This is equal to 26 per cent of the world's migrant population, while the region represents only 14 per cent of the world's population. The largely majority of them live in North America (the US and Canada) (see Figures 8.1 and 8.2). Migration in the post-WWII period consisted of the following phases:

- 1945–1973: Decreasing migration from Europe to the Americas; increasing migration of Mexicans, Puerto Ricans and Cubans to the US; colonial and postcolonial migrations from the Caribbean to the UK and the Netherlands; immigration law reforms in US and Canada in 1965 and 1967; increasing intraregional migration in South America, alongside the rise of Venezuela as a migration destination;

- 1973–1989: Increasing diversion of Mexico-US migration towards irregular channels in response to immigration restrictions; growing emigration from Central America to the US resulting from civil wars and US military intervention; intensification of Asian migration particularly from the Philippines, Korea and Vietnam; labour and refugee migration to the US from the Southern Cone to Europe and Japan;

- 1989–2008: Politicization of migration in the US alongside diversification of Latin American immigration and rising inflows of high-skilled migrants particularly from

China and India to the US and Canada; instability and increasing migration from the Andean area to Europe and North America; resumption of migration to Brazil and Argentina;

- Since 2008: Continuation of migration to North America despite the harshening of political climate around immigration culminating in the election of Donald Trump;

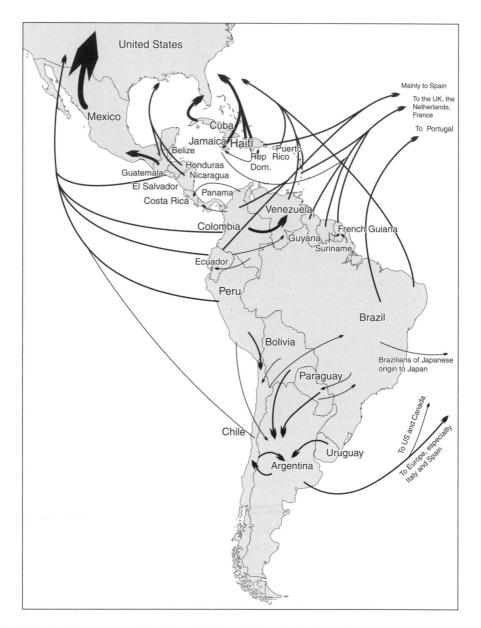

Map 7.1   Contemporary migration within and from Latin America

Note: The arrow dimensions give an approximate indication of the volume of flows in the 1945–2015 period. Exact figures are often unavailable

declining immigration from Mexico alongside increasing migration from crisis-torn Central America; rise of Chile as a regional destination alongside increasing Haitian migration to Mexico and South America, and increasing refugee migration from Venezuela.

Migration movements and migrant numbers vary considerably within the region. Map 8.1 gives an overview of the main migration movements over the post-WWII period. We can distinguish five sub-regions, which have been affected by immigration and emigration in different ways:

* The *US* and *Canada* have traditionally had substantial populations of European origin because of large-scale European immigration. With decreasing European immigration since the 1950s, immigration from Latin America and Asia has become more important. The US also has a large minority population of African–Americans, who are mostly descendants of slaves, forming about 13 per cent of the population;

* *Mexico* and *Central American* societies largely comprise populations of *mestizo* and indigenous background. Although in geographical and political terms Mexico is part of North America, high emigration to the US and its position as a country of transit migration draws it closer to its Central American counterparts. As the world's largest migration corridor, Mexico–US migration dominates public debates on migration in both countries;

* *Caribbean* countries have mixed populations resulting from the impact of colonization, forced migration (slavery) from Africa and indentured migration from Asia (see Chapter 5 and Flahaux and Vezzoli 2016). This region has sizeable populations of African, European, Asian and mixed origin. Decolonization after 1945 generated substantial migration to former colonial powers. In recent decades, migration from many Caribbean countries has become increasingly oriented towards North America, while Haitians have also moved to South America;

* The *Andean area* in the north and west of South America comprises Ecuador, Peru, Colombia, Bolivia and Venezuela. The region has a population of predominantly indigenous and *mestizo* background. With the exception of oil-rich Venezuela, immigration from Europe has been relatively low and most movement is intra-regional. In recent decades, the Andean area has experienced significant migration from Ecuador, Colombia and Peru to Spain, and from Colombia to the United States and Venezuela. Since 2014, the crisis in Venezuela generated significant refugee migration, particularly to neighbouring countries;

* The *Southern Cone* includes Brazil, Argentina, Chile, Uruguay and Paraguay. Particularly the zone located between São Paulo and Buenos Aires is the industrial heartland of South America. These countries have substantial populations of European origin reflecting large-scale immigration and settlement from Europe. There were also substantial inflows from elsewhere: for example, African slaves in the nineteenth century, and Japanese workers. In recent decades, emigration to Europe increased while immigration from other parts of Latin America and the Caribbean has grown.

Box 7.1 US hegemony and migration

Large-scale Caribbean and Asian migration to the US has deep historical roots in US hegemony, military occupation and recruitment of migrant workers and soldiers. In the 1845 the US annexed the Republic of Texas, and after the ensuing Mexican–American War the US annexed large parts of Mexican territory (California, Nevada, Arizona, Colorado, New Mexico west of the Rio Grande, and Southwestern Wyoming). The Spanish–American War of 1898 marked the further rise of US power through the capture of Puerto Rico, the Philippines and Guam from Spain, while Cuba was reduced to a US client state (until the 1959 communist revolution). The rise of the US as a military and economic power, combined with high labour demand, created the conditions for large-scale migration from Mexico, Puerto Rico and the Philippines to the US mainland (see Duany 2012).

Recruitment often played a crucial role in starting such migrations, as it created the patterns and transnational ties that would facilitate large-scale migrations over the twentieth century. After the US occupied the country in 1898, Puerto Ricans started migrating as contract labourers, first to sugarcane plantations in Hawaii (another US colonial possession before its incorporation as a state in 1959), and then to destinations on the US mainland. After Puerto Ricans were extended US citizenship in 1917, they started moving in larger numbers (Thomas 2015).

Starting in 1906, the first Filipino migrant workers were recruited to work on the sugarcane and pineapple plantations of Hawaii and other US Pacific territories such as Guam (Asis 2006). Some Filipinos subsequently left Hawaii to work in agriculture in California, Washington and Oregon or the salmon canneries of Alaska, while others were employed in the merchant marine (Asis 2006). Filipinos were considered 'nationals' but not US citizens. This ambiguous legal status as colonial subjects nevertheless allowed them to move freely to the US until Congress established an immigration quota for Filipinos in 1934 (Asis 2006). Many Filipinos served in the US Navy, which also provided a route towards US citizenship.

Immigration from Korea was also encouraged by US foreign influence and recruitment practices. The Chinese Exclusion Act of 1882 encouraged employers to recruit Korean workers. The first significant Korean immigration occurred between 1903 and 1905 with the recruitment of at least 7,226 Koreans to work plantations in Hawaii. They were followed by about 2,000 'picture brides' of plantation workers working in Hawaii as well as California.

Religion played an important role too: American Presbyterian and Methodists missionaries who converted many Koreans to Christianity, also helped Koreans to migrate to America. Although immigration dropped after Congress passed the Oriental Exclusion Act of 1924, it permitted Korean (and other Asian students) to study at US universities. When Korea was under Japanese rule from 1910 to 1945, this attracted Korean refugees and intellectuals (Zong and Batalova 2017c). After the Second World War, US involvement in the 1950–1953 Korea war boosted immigration from South Korea while US military involvement in Vietnam between 1955 and 1973 also led to large-scale migration.

These examples illustrate how military and economic interference abroad triggers migration in the reverse direction because of the social, political and economic ties this usually creates.

America's migration transition (1945–1973)

After an initial post-war surge, since the 1950s the scale of European migration to the Americas rapidly declined as Latin America entered a period of economic stagnation and political instability. Within Latin America, intra-regional migration took the place of European immigration. Most of this migration was spontaneous or unregulated, and was generally not viewed as a problem until the late 1960s (Lohrmann 1987). In North America, migration from Mexico and the Caribbean took the place of European immigration. In this period, the US Amendment of Immigration and Nationality Act of 1965 and the Canadian introduction of the *points-based system* in 1967 laid the legal basis for growing immigration from an increasingly diverse array of Latin American and Asian countries in later periods.

Mexican immigration and policy reform in North America

Large-scale labour immigration to the US developed later than in Western Europe, partly because of restrictive legislation enacted in the 1920s. Immigration levels averaged 250,000 persons annually in the 1951–60 period, and 330,000 annually during 1961–70: a far cry from the average of 880,000 immigrants per year between 1901 and 1910. In 1970 the number of overseas-born people had declined to 9.6 million (only 4.7 per cent of the population) (Briggs 1984) compared with 13.9 million (13.2 per cent of the population) in 1920.

However, as in Western Europe, fast economic growth, demographic change and restructuring of the US labour market would eventually lead to a major upsurge in immigration. Employers, particularly in agriculture, recruited temporary migrant workers, mainly men, in Mexico and the Caribbean. Already in 1942, severe labour shortages led to the Bracero programme, which lasted until 1964 and allowed for the temporary recruitment of workers from Mexico. The Bracero programme mobilized 4.5 million young men to work in 24 different US states as temporary migrants in agriculture and railway track maintenance (Alba 2010; Durand 2004). Trade unions were highly critical, arguing that immigration would displace domestic workers and hold down wages. Government policies varied: at times, systems of temporary labour recruitment were introduced; in other periods, recruitment was formally prohibited, but tacitly tolerated under pressure of employers' lobbies, leading to the growth of the number of undocumented workers.

The legal basis of US immigration was fundamentally overhauled in the 1960s. The 1965 amendments to the Immigration and Nationality Act were seen as part of the civil rights legislation. This legislation aimed to combat racism and discrimination against African–Americans. Yet it also removed the discriminatory national-origins quota system, which previously favoured (white) European immigration. Although this legislation was neither intended nor expected to boost non-European immigration (Borjas 1990), the removal of racist immigration rules created a system of worldwide immigration, in which job offers or kinship with US citizens or residents became the main admission criteria. The Act boosted the diversification of US immigration and made family migration the largest category of permanent entry. A second factor contributing to this diversification was the decline in European immigration as a result of increased prosperity in Western Europe and emigration prohibitions in Soviet-dominated Central and Eastern Europe.

Canada followed policies of mass immigration after 1945. At first only Europeans were admitted. Most entrants were British, but the number of Eastern and Southern European migrants increased quickly. Germans, Italians and Dutch leaving the ravages of post-WWII Europe were the largest immigrant groups in the 1950s and 1960s. However, In parallel to developments in the US, the abolition of racist immigration rules and the introduction of a non-discriminatory points system in 1967 (a new migration policy tool, which determines non-citizens' eligibility for immigration on factors as education, wealth, age and job offers, and that would be adopted by more and more countries in subsequent decades, see Chapter 11), opened the door for non-European migrants.

Latin America's migration transition

With the deterioration of economic conditions and the rise of dictatorship in the region, large-scale immigration from Europe to the Southern Cone nations would largely come to an end. As immigration from Europe waned, intraregional migrations increased in importance. Social transformations, urbanization and increased infrastructure links increased immigration from previously isolated agrarian areas. For instance, Paraguayan and Chilean labour migrants began to find employment, especially in agriculture, in Northeastern Argentina and Patagonia in the 1950s and 1960s, respectively (Jachimowicz 2006). Single – mostly male – migrants were soon joined by their families, creating neighbourhoods of undocumented immigrants in some cities. Argentina also witnessed its first significant emigration: between 1960 and 1970, an estimated 185,000 mostly high-skilled Argentines relocated to countries such as the US, Spain and Mexico (Jachimowicz 2006).

Over this period, migration from Mexico became almost exclusively directed to the US. In the early twentieth century, the Mexican government had initially discouraged potential emigrants from answering the calls of US recruiters to move north for work. As part of a rise in nationalism, it also aimed at attracting 'back home' the Mexican heritage population who had remained in the territories lost to the US in the mid-nineteenth century (see Box 7.1). However, these policies were given up later and after the outbreak of the Second World War, the US and Mexican governments negotiated the Bracero programme.

While Mexican workers started moving in large numbers to the US, movement in Central American countries remained mainly internal and intra-regional. For instance, Guatemalan agricultural workers travelled every year to the south of Mexico to labour on coffee plantations; while Salvadorians went to work in the cotton-growing areas of Guatemala and Nicaragua and the banana plantations of Honduras (Castillo 2006; Hamilton and Stoltz Chinchilla 1991). Migration to destinations outside the region, in particular to the US, only reached significant levels with the outset of military conflicts in the 1960s (USINS 1999).

The Andean area: Internal and intra-regional migration

Reflecting low development levels and relative isolation, migration from the Andean area remained small-scale up until the 1970s, and was dominated by rural-to-urban migration tied to urbanization and infrastructure development. For instance, in *Peru* the growth of urban centres and the spread of capitalism to the hinterlands broke down the isolation of rural villages, leading to increasing migration from the Andean highlands to urban centres, which took place within a well-defined urban hierarchy dominated by the capital city of Lima (Skeldon 1977). Some internal movement would spill over in cross-border migration. From *Ecuador*, for instance, there was limited migration to Venezuela and the US from the 1940s (Jokisch 2007). Bolivian migration to *Argentina*, which had started in the mid-1930s, lasted for decades until mechanization of Argentinian agriculture reduced labour demand.

Unlike the US, Canada, Argentina and Brazil, and reflecting their weak economic status, most Andean countries never attracted large numbers of European immigrants. *Colombia* attempted to implement immigration programmes after the Second World War, but failed because of political instability (Bérubé 2005). *Venezuela* was a major exception in the region, after the military regime of Perez Jiménez established a policy of 'open doors' to meet labour demand created by the booming oil sector. This attracted many agricultural and skilled workers from Spain and Italy between 1949 and 1958 while migration from Colombia increased as well (Álvarez de Flores 2006–2007). Although many of the approximately one million immigrants who moved (Kritz 1975) would return, about one third of them, mainly of Italian origin, settled in Venezuela (Picquet *et al.* 1986; Grau 1994).

After 1958, Venezuela tightened immigration policies, which curbed immigration from Europe but stimulated large-scale spontaneous immigration from other Latin American countries. Colombian workers continued to move into the rural sector, partly to replace Venezuelan rural–urban migrants who were leaving farms for the cities over the 1960s (Kritz 1975; Sassen-Koob 1979). In 1979, Bolivia, Peru, Venezuela, Columbia and Ecuador signed the Andean Pact to boost regional integration. Although the Pact obliged member states to legalize undocumented migrants (Picquet *et al.* 1986), Venezuela legalized only 280,000 to 350,000 residents (Meissner *et al.* 1987). Some 1.2 to 3.5 million undocumented residents lived in Venezuela in 1980, out of a total population of around 13.5 million (Meissner *et al.* 1987).

Caribbean: Increasing emigration

Intra-regional migration has a long tradition in the Caribbean. For instance, by 1930 an estimated 100,000 Haitians and 60,000 Jamaicans lived in Cuba (Portes and Grosfoguel 1994). Migration often followed colonial patterns. After the abolition of slavery, West Indian islanders went to British Guiana and Trinidad and other countries in the Caribbean basin (Chaney 1989). By the mid-twentieth century, migration had become an essential part of Caribbean life, particularly in the British colonies (Thomas-Hope 1996). From the mid-twentieth century, Caribbean migration diversified increasingly beyond the region (Peach 1991; Vezzoli and Flahaux 2017). Britain recruited thousands of Caribbean men for military service in both World Wars, with many eventually settling in the UK. After 1945, labour shortages in the UK encouraged migration from the Caribbean, in particular Jamaica and Barbados (Glennie and Chappell 2010; Peach 1968). The post-Second World War reconstruction efforts also attracted increasing numbers of West Indians to the US (Levine 1987; Thomas-Hope 2000; Vezzoli 2015).

Source: Getty Images/Haywood Magee/Stringer

Photo 7.1   West Indian immigrants arrive at Victoria Station, London, after their journey from Southampton Docks

Puerto Rico had been the leading source of migration from overseas possessions of the US since Spanish-Cuban-American War of 1898, when the US invaded and took possession of the island. Emigration to the US accelerated between 1945 and 1965. More than half a million Puerto Ricans moved to the US mainland during this period. New York evolved into the main destination, but sizeable Puerto Rican communities also sprouted up in New Jersey, Connecticut, Philadelphia and Chicago (Duany 2012; Thomas 2015). Most were unskilled workers who were incorporated into the lower rungs of the labour market, such as manufacturing, domestic work and seasonal agriculture (Duany 2012).

Cuban migration soared after the 1959 Cuban Revolution (Portes and Grosfoguel 1994), when Fidel Castro led a communist takeover and ousted the US-backed Batista regime. The Cuban Adjustment Act (CAA), which US Congress passed in 1966, allowed Cubans to become lawful permanent residents after living in the US for one year. As a result, the Cuban population in the United States would grow almost six-fold within a decade, from 79,000 in 1960 to 439,000 in 1970, with the majority settling in Florida (Batalova and Zong 2017).

Overseas emigration from French and Dutch Caribbean colonies also increased in this period (Calmont 1981; Milia-Marie-Luce 2007). Movement from the French Antilles to metropolitan France was encouraged by the state-organized recruitment system serving the double goal of filling low-level civil service positions in France while reducing unemployment and preventing social unrest in the French Antilles (Condon and Ogden 1991). Migration from the Dutch Caribbean (Suriname and the Antilles) primarily concerned students pursuing higher education in the Netherlands (Oostindie 2009), although employment opportunities in the Netherlands also encouraged migration of workers.

The globalization of American immigration (1973–1989)

Diversification of migration to North America

The 1973 Oil Shock and the ensuing economic recessions barely influenced the structural trend of increasing migration to the US and Canada that had started after the Second World War. In Canada, the introduction of the points-based system in 1967 had encouraged immigration from outside Northwest Europe. The main origin countries in the 1970s were Jamaica, India, Portugal, the Philippines, Greece, Italy and Trinidad (Breton *et al.* 1990). Throughout the period, family entry was encouraged, and immigrants were generally seen as settlers and future citizens.

Also, immigration to the US grew steadily as the demand for foreign labour increased, with the 1965 Immigration and Nationality Act contributing to accelerating immigration from Latin America and Asia. The total number of foreigners (called 'aliens' in the US) granted legal permanent resident status ('Green Cards') rose from 4.2 million in 1970–9 to 6.2 million in 1980–89. Mexico would remain the predominant source of lower-skilled migrant labour, with border restrictions leading to an increasing reliance on family migration and irregular border crossings (see Box 11.1). Although much smaller then Mexican immigration, US intervention in Central America and Andean countries during the Cold War contributed to a destabilization of the region and increasing migration from countries such as El Salvador, Guatemala and Colombia (Massey and Pren 2012).

Cuban refugee migration to the US would remain significant too. Cubans were generally welcomed as in the Cold War context they were fleeing a hostile political regime that challenged US hegemony in the region. Though Cuban authorities would generally prohibit emigration, on several occasions they allowed people to leave for the United States, partly as an escape valve for dissidence and excess labour (Batalova and Zong 2017), partly to test and expose the limits of US 'hospitality'. The largest migration hike occurred during the 'Mariel boatlift' of 1980. After President Fidel Castro temporarily allowed people to leave, about 124,800 Cubans made the sea crossing to Florida in overcrowded boats (Batalova and Zong 2017). This mass influx embarrassed US president Jimmy Carter, forcing him to negotiate an agreement with the Cuban government to end this migration (Engstrom 1997).

Rising immigration from the Philippines, Korea and Vietnam to the US

Soaring migration from the Philippines, Korea and Vietnam heralded the increasing predominance of Asian migration to the US. Migration from all three countries had their origins in US hegemonic ties, military interference and recruitment going back to the late nineteenth century (see Box 7.1), but would accelerate from the 1970s. In the first half of the twentieth century there had already been significant migration from the *Philippines* to the US. After 1946, when the Philippines became independent from the US, Filipino immigration started to pick up again. Initially, this mostly concerned brides of US servicemen and as recruits into the armed forces, while others came to train as healthcare workers (Zong and Batalova 2018). After the 1965 US Immigration and Nationality Act abolished national origin restrictions, Filipino migration increased sharply, with immigrant numbers reaching 913,000 in 1990 (Zong and Batalova 2018). Filipino emigration was also facilitated by the policies of the 1965–1986 Marcos regime in the Philippines to encourage labour migration for development purposes and as a poltical-economic 'safety valve' (see Chapter 9).

Immigration from *Korea* would also surge in this period. After a first movement of plantation workers of the early twentieth century, a second surge in Korean immigration occurred during and after the Korean War (1950–1953), including Korean wives of American soldiers ('war brides') and war orphans adopted by American families, as well as students, businessmen and professionals. The 1965 Immigration and Naturalization Act facilitated recruitment of skilled workers from South Korea and immigration of their family members. High unemployment and the military dictatorship in South Korea stimulated migration until the early 1980s (O'Connor and Batalova 2019), when rapid economic development and democratization would decrease emigration. Over this period, *Vietnam* also became another important source of migration to the US. This began after the US defeat in the Vietnam War in 1975, leading to the US-led evacuation of about 125,000 Vietnamese refugees. As the humanitarian crisis in the Indochina region (Vietnam, Cambodia and Laos) intensified, more refugees and their families were admitted to the US. Facilitated by social networks, this fuelled increasing migration from Vietnam and other countries in the region.

Latin America: Repression, violence and increasing emigration

While seasonal labour migration from Central America to Mexico has a long history, Mexico's role as a country of transit and destination increased in the 1970s and 1980s following the civil wars and economic insecurity in Nicaragua, El Salvador and Guatemala, with many people fleeing to Mexico or travelling through to the country in order to claim asylum in the US (Ángel Castillo 2006; Smith 2006). Conservative estimates suggest that up to a million Central Americans sought refuge in the US during the 1980s, yet the US administration denied refuge to many (Gzesh 2006).

Gang violence became an increasing problem in Central America. Violent criminal organizations such as the *Mara Salvatrucha*, which were formed in the US, have been 'exported' to Central America as gang members were deported due to criminal convictions (Migration Dialogue 2007; Portes 2010). This phenomenon first emerged in El Salvador and Guatemala, to spread to Honduras later (Cruz 2010). There are also significant levels of intraregional migration, with relatively wealthy Mexico, Costa Rica and Belize attracting migrants from neighbouring countries (see Mahler and Ugrina 2006). The experience of Costa Rica diverges from the rest of Central America. Because of the absence of armed conflicts and relative economic prosperity, it has low emigration and attracts migrants from surrounding countries, particularly Nicaragua.

With the deteriorating of economic and political conditions, this period also saw the consolidation of Latin America's Southern Cone as an emigration region. From the late 1960s until the 1980s, military regimes forced thousands of people to flee Argentina, Chile and Uruguay (Pellegrino 2000). Between 1976 and 1983, many Argentinians fled to Spain, Mexico and the US (Maier 2015). Economic woes, particularly during the 1980s, combined with restrictive immigration policies adopted by military regimes, discouraged immigration. These factors contributed to the Southern Cone's reverse migration transition, from the world's second most important migration destination (after North America) until the Second World War to a region of predominant out-migration. Compared to other parts of Latin America, where migration became increasingly focused on the US, emigration from the Southern Cone has been more diverse, with significant migration to Europe. Many would move to their ancestors' countries of origin in Europe (particularly Italy, Spain and Portugal) and Asia (Japan), often under preferential agreements (IOM 2005).

For instance, by 2000 over 1.8 million Brazilians lived abroad. Of these, 442,000 lived in Paraguay, 225,000 in Japan (the *Nikkeijin*) and 799,000 in the US (Amaral and Fusco 2005). Persecution and economic instability also encouraged migration of skilled workers and their families from Argentina, Chile and Uruguay to Venezuela as a result of the post-1973 Oil Shock petroleum boom. Until the economic downturn of the 1980s, Venezuela also continued to attract high-skilled workers from the US, Italy, Spain and Portugal as well as millions of lower-skilled workers from Colombia, who often lacked regular status.

Geographical reorientation of Caribbean migration

Since the 1970s, Caribbean emigration, particularly from Anglophone countries, became increasingly focused on the US and, to a lesser extent, Canada (Ferrer 2011; Vezzoli 2015; Vezzoli and Flahaux 2017). The number of Caribbean immigrants in the US grew from 194,000 in 1960 to 2,953,000 in 1990 (Batalova and Zong 2016). As US citizens, Puerto Ricans are not included in such figures, as they are considered internal migrants. Half of all Caribbean immigrants in the US came from Cuba and the Dominican Republic (McCabe 2011). Other important origin countries are Jamaica, Haiti and Trinidad and Tobago. In Canada, Jamaicans became the largest Caribbean migrant group (Lindsay 2001).

Increasing Caribbean migration to North America was part of a geographical reorientation of migration from former British colonies. This reorientation had its roots in policy changed in previous decades. Until the 1962 Commonwealth Immigrants Act ended the right of British Commonwealth citizens to settle in Britain, most British Caribbean migrant workers moved to the UK (Peach 1968). The announcement of Commonwealth Immigrants Act in 1961 had initially triggered an immigration surge as Caribbean migrants 'rushed to beat the ban' (Peach 1968). After this effect waned, new migration would increasingly reorient towards the US and Canada. This geographical diversion of Anglo-Caribbean migration was facilitated by the common English language, booming labour demand in North America and the abolition of racist immigration policies (Vezzoli 2015).

Migration from former or current French and Dutch colonies would remain more strongly oriented towards France and the Netherlands because of strong linguistic, cultural and political ties. Uncertainties around Suriname's independence from the Netherlands in 1975 – which was hastened by the Dutch government in a covert bid to stop free migration of Dutch-Surinamese citizens – and the subsequent introduction of travel visas in 1980 triggered a 'now or never' migration surge in which about 40 per cent of Surinamese migrated to the Netherlands (van Amersfoort 2011; Vezzoli 2015). Also in the post-1980 period, migration from Suriname (and the Netherlands Antilles, which did not achieve independence) would remain strongly oriented towards the Netherlands because of strong historic, linguistics and social ties (Vezzoli 2015). Migration from French Guiana, a French *département*, whose inhabitants are French citizens, remained oriented towards France, although emigration has remained strikingly low, partly because the absence of migration restrictions has been associated with more relaxed attitudes towards migration and the concomitant absence of a strong urge to leave (Vezzoli 2015).

An elusive quest for control? American migration since 1989

Latino immigration to the US: Diversification and intensification

In the post-1989 period of globalization and economic liberalization, migration from Mexico and other Latin American countries to the US further intensified and diversified. While economic growth and labour demand continued to attract migrant workers, growing border enforcement had the unintended effect of interrupting circulation and encouraging permanent settlement, which boosted the formation of permanent Latino communities throughout the US (see Box 11.1). During this period, Latino culture, music and cuisine as well as the Spanish language became an integral feature of American (urban) life.

The establishment of the North American Free Trade Agreement (NAFTA) in 1994 accelerated economic integration between Mexico and the US through the growth of *maquiladoras* (mainly US-controlled factories in the Mexico–US border zone), which attracted internal Mexican migrants. Migration to the US would continue though because of strong labour demand particularly in agricultural and service enterprises (Delgado Wise and Covarrubias 2009). The Clinton administration argued that NAFTA trade liberalization would reduce immigration by encouraging economic growth in Mexico. This ignored predictions that NAFTA would at least initially create economic dislocations and rising unemployment, such as by driving Mexican peasants out of business through imports of US agrarian products, thereby boosting migration (Martin 1993; Martin and Taylor 1996). A study by Mahendra (2014b) suggests that such a 'migration hump' (see Chapter 3) indeed occurred in the first two decades following the implementation of NAFTA.

Source: Getty Images/Pacific Press

Photo 7.2　The Puerto Rican community in New York City, protesting against the United States government's response to hurricane Maria in 2017

The post-Cold War period was also marked by a growing politicization of immigration of low-skilled workers (mainly from Mexico) and Cuban refugees. The Immigration Reform and Control Act (IRCA), which was signed into law by the Reagan administration in 1986, intended to curb immigration by criminalizing undocumented hiring and expanding border controls. However, its main initial effect was a massive regularization ('amnesty') of about 2.7 million undocumented migrants, most of them of Mexican origin, around 1991 (visible in the hike in legal migrant admissions in Figure 7.4 below), which further consolidated permanent Latino presence in the US.

In the post-Cold War context, attitudes towards Cuban refugees became less hospitable. By the mid-1990s, rising boat migration to Florida prompted the Cuban and US governments to negotiate new migration terms. The resulting 1995 accords established the 'wet-foot, dry-foot' policy: unless citing fears of persecution, Cubans intercepted at sea would be returned to Cuba or a third country. Those reaching the US, whether by land or sea would be permitted to stay and apply for residency after a year. The unintended effect was a major surge in Cuban arrivals, with some 650,000 admitted to the US between 1995 and 2015 (Batalova and Zong 2018; Migration Dialogue 2007).

Puerto Rican migration would also continue over this period. Almost 8 per cent of the population of Puerto Rico moved to the US mainland during the 1990s alone, and this would further accelerate over the 2000s. In 2007, more people of Puerto Rican origin lived on the US mainland (4.12 million) than in Puerto Rico (3.96 million) (Duany 2012). A particular feature of Puerto Rican migration is the intensity of back-and-forth movements (Duany 2012). Such circularity is facilitated by the fact that Puerto Ricans are US citizens.

The persistence of US immigration

The terror attacks of 11 September 2001 – which were perpetrated by Islamist radicals, mostly from Saudi Arabia – contributed to the securitization of migration. The 2001 US Patriot Act increased funding for surveillance and deportation of foreigners without due process. US refugee admissions decreased in the wake of the attacks – due to more stringent security requirements in asylum processing. Middle Eastern and African refugees were particularly adversely affected by the securitization of migration, although numbers would increase again after 2005.

Notwithstanding the politicization of migration and increased border enforcement, the US would continue to admit large numbers of higher- *and* lower-skilled workers. Influenced by labour shortages and employers' lobbies, the US government expanded temporary work-related visa schemes. Although these do not provide permanent residence, temporary migrants can apply for permanent residence if they are sponsored by their employers. Between 2007 and 2016, a yearly average of 1.1 million temporary immigrants were admitted, besides a similar average of permanent immigrants, mainly family members of US citizens and residents. Temporary workers admissions include highly skilled workers on employer-sponsored temporary (H-1B) visas, to fill jobs usually requiring a university degree. In 2018, 180,000 skilled workers were admitted on H-1B visas, in addition to 153,000 L-1 visa intra-company transferees. The intake of seasonal agricultural workers (H2A visas) increased from 28,000 in 2000 to 139,000 in 2010, and 196,000 in 2018, the vast majority (180,000) of them Mexicans. In 2018, 390,000 people entered as students.

Largely similar to experiences in Western Europe, the Great Recession would not reverse the structural increase of migration to North America since the 1950s. One major exception was the decrease of Mexican migration to the US. This was partly linked to the US recession, which particularly hit sectors that employ many migrants, such as construction and manufacturing (Ruiz and Vargas-Silva 2009). But it also reflected improving economic conditions and a slowing down of population growth in Mexico (Villarreal 2014), as the country seems to be entering the right tail of its migration transition, in a similar pattern to the decreasing emigration of Turkish workers to Europe over the same period.

Decreasing migration from Mexico increased the relative importance of US-bound migration from Guatemala, Nicaragua and Honduras – with migrants often using Mexico as a transit country (Olayo-Méndez 2018). Over the 2010s, the Mexican government has been under constant US pressure to crack down on transit migration. The risks to migrants trying to reach the US via Mexico are often extreme because of state and gang violence: kidnapping, extortion and murder are frequent (Olayo-Méndez 2018). Many of these migrants and refugees lack the connections and money needed to secure visas, pay smugglers and travel safely over large distances. If they move, they often do so by walking or hiking on trains, and often they get stuck while in Mexico, and are vulnerable to exploitation and abuse (Olayo-Méndez 2018).

Rising high-skilled immigration from China and India

While the US government struggled to contain Latin American immigration, legal inflows of skilled workers and students from Asia, China and India in particular, further accelerated after 1989. In a major shift, after 2010 Asian immigration would overtake Hispanic immigration in terms of annual arrivals. While Chinese immigration to the US goes back to the mid-nineteenth century, it would pick up again in the 1970s and the 1980s, to further accelerate over the 1990s and 2000s. Besides the 1965 Immigration and Nationality Act, another important factor behind this increase was China's loosening of exit controls in 1978 and the improvement of US-China relations since 1979 (Zong and Batalova 2017a). The Chinese immigrant population has grown more than sixfold since 1980, from 384,000 in 1980, to 1.2 million in 2000, to over 2.3 million in 2016 (Zong and Batalova 2017a). China has become the main source of foreign students enrolled in US higher education, and Chinese receive the second-largest number of employer-sponsored H-1B temporary visas, after India. Chinese have become the third-largest foreign-born group in the US, after Mexicans and Indians. Half of Chinese immigrants live in California (31 per cent) and New York (20 per cent) (Zong and Batalova 2017a).

The same changes in US immigration policy boosted inflows from India. Most Indian immigrants are highly educated, have strong English skills, with many working in 'STEM' (science, technology, engineering and maths) fields. Between 1980 and 2010, the Indian immigrant population grew more than eleven-fold, roughly doubling every decade (Zong and Batalova 2017b). In 2016, Indians were the top recipients of high-skilled H-1B temporary visas and were the second-largest group of international students in the US, after the Chinese. With 2.4 million Indian immigrants, they are the second-largest foreign-born group. Indian migrants are concentrated in California (20 per cent), New Jersey (11 per cent), Texas (9 per cent), New York and Illinois (7 per cent each) (Zong and Batalova 2017b).

Although immigration from the Philippines would decelerate over this period, immigrant numbers grew from 913,000 in 1990 to 1,942,000 in 2017. The biggest Filipino communities live in California (Los Angeles and San Diego in particular), Hawaii, Texas, New York, Illinois and New Jersey (Zong and Batalova 2018). With the improvement of economic and political conditions in South Korea, immigration would decrease over the 1990s and the 2000s, and the Korean immigrant population stagnated around 1.1 million (Zong and Batalova 2018c). Despite the fact that no more Vietnamese were admitted as refugees after the end of the Cold War, Vietnamese immigration continued, with the large majority of immigrants gaining permanent residence ('green cards') through family reunification. As a consequence, the Vietnamese immigrant population has grown fast, roughly doubling every decade between 1980 and 2000, and then increasing 26 per cent in the 2000s. In 2017, more than 1.35 million Vietnamese-born resided in the US.

Although still a relative minor share of US immigration, migrants from sub-Saharan Africa are one of the fastest-growing segments of the US immigrant population, increasing by about 200 per cent over the 1980s and 1990s and by 100 per cent over the 2000s (Capps *et al.* 2012). In 2017, 119,000 Africans obtained permanent resident status, about one tenth of the total of 1.33 million legal admissions in that year. Like other immigrant groups, most black Africans are admitted through family reunion channels, but they are much more likely than other immigrant groups to be admitted as refugees or through the diversity visa programme. In 2014 the most important origin countries of the African foreign-born were Nigeria (258,000), Ethiopia (207,000), Ghana (148,000), Kenya (111,000), Somalia (93,000) and Liberia (79,000) (Morgan-Trostle *et al.* 2016). An increasingly diverse group, black Africans are among the highest-educated immigrant groups in the US, who generally fare well on integration indicators, with high employment rates and college completion rates that far exceed those for most other immigrant groups as well as US natives (Capps *et al.* 2012).

Canada: A quintessential immigration country

While migration became increasingly politicized in the US (and Europe), Canada would remain one of the few Western countries (alongside Australia and New Zealand) with a proactive permanent immigration policy, which aims to admit the equivalent of 1 per cent of its total population of about 34 million each year. In 2017, there were 7.4 million foreign-born residents in Canada, up from 5.5 million in 2000. Foreign-born residents made up 20 per cent of the Canadian population – one of the highest shares in any Western country. In contrast to earlier European migrants, who settled all over Canada, sometimes as farmers, new arrivals have increasingly concentrated in the largest cities: Toronto, Montréal and Vancouver.

In 2017, Canada recorded 286,000 'landings', to use the Canadian term for permanent arrivals. In parallel to the US, entries from Europe have declined. Entries from South and East Asia and the Middle East have grown, while Latin American immigration has remained comparatively modest compared to the US. In 2017, the top three origin countries of permanent residents were India (18%) the Philippines (14%) and China (11%). Economic immigration is dominated by Asian countries, with India (24%), the Philippines (21%) and China (11%) representing over half of all entries (OECD 2018).

Like the US, Canada steadily increased its admissions of temporary foreign workers and students. In addition to permanent residents, 375,300 temporary residents received a first permit in 2017. Study visa permits accounted for 52 per cent, while work permits were at 47 per cent (OECD 2018: 228). This reflects a trend to emphasize skills, education and language abilities in immigration selection criteria, facilitated by the Canadian points-based system. As part of Canada's Global Skills Strategy, since 2017 short-term researchers and highly skilled workers no longer require a work permit. Furthermore, high-skilled workers have work permits and visas processed in two weeks as part of the Express Entry system – with economic migrants from India, the US, Nigeria, the United Arab Emirates, the UK and Pakistan among the main beneficiaries of the programme.

A significant component of Canada's temporary migration programmes is the Seasonal Agricultural Workers Program (SAWP) which, since its inception in 1966, has recruited workers from Mexico and the Caribbean to work in agriculture, particularly in Ontario's tomato industry. While recruitment used to be limited to married men, in more recent years some women have been recruited as well. SAWP has been mentioned as a 'model' for temporary migration programmes due to its high degree of circularity. The key to its effectiveness is that the same groups of workers are allowed to return year after year to work in Canada, which takes away incentives to overstay visas (Newland *et al.* 2008). Yet critics point to excessive employer control and workers' restricted rights (Basok 2007; Preibisch 2010).

Despite some backlash against multiculturalism after 2006, Canada maintained its open attitude to immigration, helped by the fact that the US acts as a geographical buffer against spontaneous migration from the south (Griffith 2017). Canada has remained an 'immigration country by design' that actively encourages immigration, to boost the economy and as an intrinsic part of the Canadian nation building project. Canada also has an extensive refugee resettlement programme. The relative ease with which Canadian society seems to deal with immigrant diversity may also be related to its own multi-ethnic and multi-linguistic make-up, coming out of a process of struggle and compromise between indigenous peoples and French and British settlers, although there should be no doubt that indigenous peoples lost out most in the process.

Social transformation and migration in Latin America

Resumption of migration to the Southern Cone

With the fall of military regimes and the rise of democratic governments, nations such as Brazil, Chile, Argentina and Venezuela started pursuing new development paths. The general trend towards democratization in Latin America from the 1990s also contributed to a dismantling of border restrictions and an accelerated liberalization of immigration policies (de Haas *et al.* 2019a). Together with economic recovery, this contributed to a partial resumption of migration from Andean and Caribbean countries to the Southern Cone countries. While Southern Cone countries evolved into important destinations for intraregional migrants, citizens of these regional destinations themselves would increasingly migrate out of the region, to Europe and the US, particularly during economic downturns. For instance, economic crises during the late 1990s and early 2000s encouraged emigration from Argentina to Spain, Italy and the US (Courtis 2011). Migration to Europe was facilitated by the easy access to residence and citizenship for the many Southern Cone inhabitants with Spanish or Italian ancestors.

As a whole, however, the Southern Cone experienced a period of increased political and economic stability and democratization, and would attract migrants, mainly from poorer Latin American countries (Jachimowicz 2006). In *Brazil*, the return to civilian rule in 1985 facilitated emigration as the new government lifted exit controls and the passport issuance restrictions imposed by the military regime. This prompted increasing migration from Brazil to Portugal and other European destinations. Importantly, Brazilians enjoy

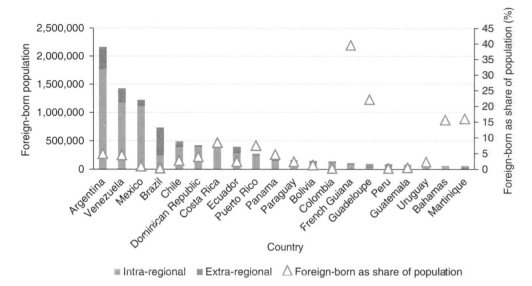

Figure 7.3 **Foreign-born populations in Latin American and Caribbean countries and territories, by origin, 2017**
Source: UNDESA 2017

visa free entry into the Schengen Zone, with significant numbers prospective workers 'overstaying' the duration of their visas (Kubal *et al*. 2011). The election of Lula da Silva as president in 2003 heralded a period of growth and social reform, prompting rising immigration from the Americas, including Bolivians, Haitians and Venezuelans.

Reflecting its prosperity, *Chile* – which in 2010 became Latin America's first OECD member became attractive for immigrants from Bolivia, Ecuador, Peru and even Argentina (Altamirano Rúa 2010; Courtis 2011). An increasing proportion of Peruvian migrants were women, many employed as domestic workers by middle-class Chilean families (Doña and Levinson 2004). Figure 7.3 shows that Argentina, Mexico, Brazil, Chile and, until recently, Venezuela are the most important immigrant immigration countries in the region and that migration is predominantly intraregional.

Rising migration from the Caribbean and Andean area

In the Caribbean and the Andean area, Haiti and Venezuela experienced significant emigration related to economic and political crises, principally to countries within the region. *Haiti*, the poorest country in the western hemisphere, experienced large-scale departures of migrants, particularly in the aftermath of a severe earthquake in 2010. An estimated 500,000 Haitians moved to the neighbouring *Dominican Republic*, where many have been victims of discrimination and deportations (Migration Dialogue 2011), to the point of even removing birthright citizenship to children of undocumented migrants (Aber and Small 2013). Increasing numbers of Haitians move to Mexico, Brazil, Chile, French Guiana and elsewhere in South America in search for work or small-scale business opportunities (Da Silva 2013).

In the Andean region, improving infrastructure and education encouraged migration out of the region. Apart from Southern Cone and the US, Spain and to a more limited extent Italy evolved into a major destination for Andean migrant workers (Vono de

Vilhena 2011), mainly from Ecuador, Colombia, Peru and Bolivia. In 2017, over 1.3 million Andean migrants were registered as living in Spain.

In *Colombia*, sustained political and economic crises generated increasing emigration, mainly to the US and Venezuela (Bérubé 2005). The long-term conflict between government forces, left-wing guerrillas, paramilitaries and drug cartels has led to significant internal displacement of approximately 3.4 million people by the mid-2010s (UNHCR 2011). Economic instability in *Ecuador* in the 1980s and the 1990s generated two emigration movements. The first movement concentrated on the US (Gibson and Lennon 1999). Most of the second movement over the 1990s went to Spain, which had less restrictive entry requirements and offered jobs for low-skilled workers in the informal economy. At the same time, Ecuador became a destination for migrants from Colombia and Peru since 2001 (Jokisch 2007).

Besides increasing migration from Peru to neighbouring countries and Spain, descendants of earlier Japanese immigrants, the *Nikkeijin*, migrated to Japan as workers, following reforms that allowed second- and third-generation persons of Japanese descent easier access to a legal residential status (Kashiwazaki and Akaha 2006). Although most *Nikkeijin* migrated from Brazil, many Peruvian *Nikkeijin* also moved overseas.

Political and economic instability in *Venezuela* encouraged emigration, particularly to the US (IOM 2005), but Spain also became a destination for Venezuelan migrants, mainly from rural areas. However, because of its relative wealth, Venezuela continued to be a destination for migrants from neighbouring countries (O'Neil *et al.* 2005). The deepening of the crisis since 2014 led to large-scale emigration – probably the largest displacement of people in the recent history of Latin America. About 4 million Venezuelans were estimated to live outside of Venezuela in December 2017. Only a small percentage of the Venezuelan displaced have filed asylum claims, although numbers increased fast, from 1157 at the end of 2013 to 279,902 in June 2018. Argentina and Uruguay have been the most welcoming countries in the region, granting Venezuelans legal residence based on the MERCOSUR Residency Agreement in force since 2009 (Freier 2018).

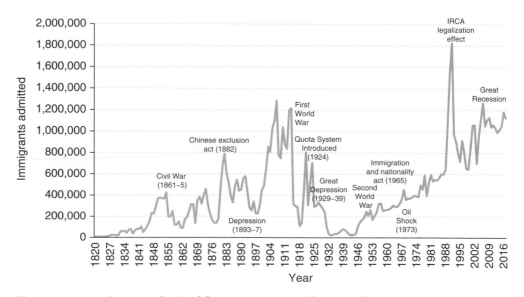

Figure 7.4   Immigrants admitted for permanent residence to the US, 1850–2010

Source: MPI data hub (www.migrationpolicy.org), based on DHS immigration statistics

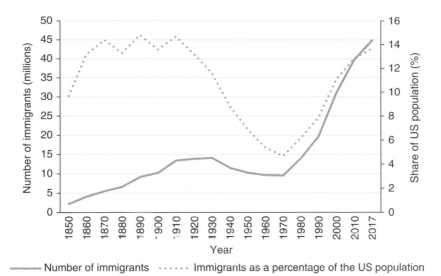

Figure 7.5   Foreign-born population in the US, absolute and as share of population, 1850–2010

Source: MPI data hub (www.migrationpolicy.org), based on US Census Bureau data

Recent developments in the US: A backlash against diversity?

The super-diversification of US immigration

Continuing immigration over the post-War period has reinforced the position of the US as the world's prime migration destination. Figure 7.4 shows the structurally increasing trend in immigration since the 1950s. This data concerns permanent admission. Including temporary admissions, figures become even higher. Over the 2008–2016 period, an average yearly number of 2.16 million migrants entered the US, with permanent and temporary immigration each making up half of these inflows. These are similar levels as immigration to the EU (see Chapter 6) but as the US population (329 million in 2019) is two thirds of the EU population (513 million), this reflects a higher immigration *rate*, with annual inflows representing around 0.65 per cent of the US population. Data on outflows is unfortunately not available.

The officially estimated foreign-born population of the US in 2017 was estimated at 44.5 million, or about 13.7 per cent of the total US population, up from 4.7 per cent in 1970 (see Figure 7.5). That put the proportion of immigrants at the highest level since 1910, when they represented 14.7 per cent of the population. Another change has been the extraordinary diversification of US immigration, as exemplified by a striking drop of the share of Europeans on the total foreign-born population from 84 per cent in 1960 to 26 per cent in 1990 and further down to 13 per cent in 2016 (see Figure 7.6). Although immigration from Mexico is decreasing, migration from other Latin American countries and Asia has been increasing fast. In 2013, India and China supplanted Mexico as the main sources of newly arriving immigrants (Zong and Batalova 2017b). In 2016, the top countries of origin for new immigrants were India, with 126,000 people, followed by Mexico (124,000), China (121,000) and Cuba (41,000) (López *et al.* 2018) (see Box 13.1 for more information about diversity in the US).

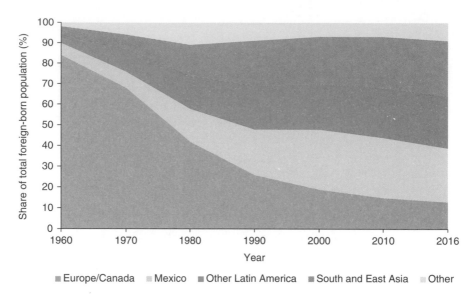

Figure 7.6   Origins of foreign-born population in the US, 1960–2010
Source: Pew Research Center

This extraordinary diversity of immigrant populations – not only in terms of countries of origin and nationality, but also with regards to factors such as religion, language, class and education as well as legal statuses and immigration channels – dubbed 'super-diversity' by Vertovec (1999), has changed life and identity in America, particularly in urban areas. For many this diversity, as well as the civil rights that immigrants and non-white citizens now enjoy, is a cause of celebration vis-à-vis a past of slavery and endemic racism. However, for some Americans increasing diversity has been a source of unease, discontent and sometimes anger. This particularly applies when minorities and immigrants are *perceived* as receiving privileged treatment in the eyes of white working classes, a significant share has felt 'left behind' by a political class perceived as being out of touch with their grievances about job loss, social insecurity and declining living standards. While such problems are primarily the result of 'neoliberal' economic policies and social transformations that are not directly linked to migration, it has been tempting for some politicians to blame migrants and minorities for these problems.

Racism plays a big role in what seems a backlash against diversity, but grievances about a loss of security, identity, community and dignity among low-income groups play an important factor too (see Hochschild 2018). In the representative multiparty systems of continental Europe, a similar 'backlash against multiculturalism' has led to the rise of extreme-right right-wing parties since the 1980s. In majoritarian two-party systems like the US (and the UK), nativist forces could not get a parliamentarian foothold until the 2010s. However, eventually political mobilization of xenophobia found its way to power through the rise of Donald Trump from *within* the Republican Party. Capitalizing on his followers' resentment of politicians, Trump successfully mobilized voters by presenting himself as an anti-establishment outsider (Kellner 2017). Some, such as political commentator Van Jones, have interpreted his election as a 'whitelash', as Trump won the 2018 election running on an anti-immigration platform, particularly directed against Mexicans as well as Muslim migrants and refugees. A minority of people with far-right, white supremacists, sympathies felt emboldened by Trump's election. For instance, Donald Trump's rhetoric on Islam-related topics on social media strongly correlates with anti-Muslim hate crime (Müller and Schwarz 2018). This shows that politicians' rhetoric can further reinforce the latent fears of immigrants and minorities, potentially turning them into overt hate and violence. The perpetrator of the mass shooting in El Paso in August 2019, killing 22 and injuring 24 people, was a white, anti-Hispanic nativist who aimed to kill as many Latinos as possible. His act was inspired by the Christchurch mosque shootings in New Zealand earlier that year (see Chapter 8).

The political salience of migration and diversity

The election of Obama and subsequently of Trump symbolize the fundamental changes America is going through – partly as a result of immigration – as well as the powerful political forces and counterforces this can generate. The election of Barack Obama as US president in 2008 marked the profound transformation that US society has undergone, particularly through the achievements of the civil rights movement, the emancipation struggle of African–Americans and the fast increase in diversity as a result of rising non-European immigration. One generation earlier, it would have been unimaginable for a non-white US citizen to become president.

As such, US immigration policies did not change fundamentally under Obama's presidency (2009–2017), as the political contradictions and paradoxes remained the same. As under previous presidents, the Obama administration remained focused on border enforcement and intensified deportation programs. With more than 2.7 million deportations between 2009 and 2016, Obama expelled more migrants than any other president in US history. These efforts can be partly seen as an effort to appease immigration hardliners in US Congress in the hope to achieve comprehensive immigration reform, particularly with regard to some sort of 'amnesty' for the large undocumented migrant population. At the same time, the US continued to facilitate the immigration of temporary low-skilled labour, high-skilled migrants and students while significant numbers of refugees were admitted.

As previous presidents, Obama failed to forge a political deal that would enable the legalization of undocumented immigrants. According to estimates, in 2016 there were about 10.7 million undocumented migrants living in the US, down from 12.2 million in 2007. Mexicans represent an estimated 5.4 million, or half, of this group, down from 6.9 million in 2007. The remaining 5.2 million are mainly migrants from Central America (Krogstad *et al.* 2018). Critics point out that the current system has created a 'massive underground of persons' who have lived with an undocumented status for decades (Human Rights Watch 2011).

To deal with the most pressing humanitarian issues, in 2012 Obama enacted an executive order, the Deferred Action for Childhood Arrivals (DACA), that allowed hundreds of thousands of undocumented immigrants, mostly Hispanics, who came to the US as children, to remain, work and study without fear of deportation. However, the more comprehensive DREAM (Development, Relief, and Education for Alien Minors) Act, that would include pathways to permanent residence, failed to pass the US Senate. This continues a history of failed attempts to pass similar bills since the first bipartisan DREAM act failed to pass the Senate in 2001.

The 2014 decision by Barack Obama and Cuban President Raúl Castro to normalize relations signalled the end of the 'wet-foot, dry-foot' policy that granted Cuban arrivals automatic residency. Anticipating an end to their special immigration treatment, and as part of a 'now or never' migration surge, Cuban arrivals more than doubled from 24,300 in 2014 to 56,400 in 2016 (Batalova and Zong 2018). During the final days of his presidency in January 2017, Obama announced the immediate cessation of the policy (Batalova and Zong 2018).

The continued relevance of race

While the election of Obama appeared to signal the success of the civil rights movement, the reactions that Obama's rise to the presidency provoked exemplifies the extent to which race is still a hugely sensitive and potentially divisive issue in America. For instance, during his candidacy Obama had to navigate carefully not to associate himself too closely to African–American communities in order to prevent alienation from white electorates. At the same time, some African–Americans opinion makers initially considered Obama, who is of mixed Kenyan and white American heritage, as not 'black enough' because he does not share heritage with the majority of black Americans descending from plantation slaves (Sarmah 2007). Obama had to avoid being seen as 'too black' or 'not black enough'. To solve this dilemma, he had to build an identity as a prototypical representative of the American people, by carefully crafting an 'in-group identity that was oriented to an increasingly socially diverse America – a diversity that he himself exemplified and embodied as a leader' (Augoustinos and De Garis 2012: 564).

The fact that Obama's identity was an issue in the first place exemplifies the continued political relevance of race. This continued during Obama's presidency. One way that Donald Trump launched onto the political stage was through his claims that Obama was not a US citizen (Clinton and Roush 2016). Around 2011, Donald Trump started to push the false idea that Obama was born in Kenya, rather than in Hawaii, and was therefore an illegitimate president. These ideas gained considerable popularity on right-wing media and had considerable effect: a 2016 poll suggested that almost three-quarters of registered Republican voters and almost one-fifth of Democratic voters doubted Obama's citizenship (Clinton and Roush 2016).

Another example was the claim that Obama secretly practises Islam. According to a 2010 poll, 17 per cent of Americans believed that Obama was a Muslim, and only 34 per cent believed he was a Christian (Pew 2010). The fact that Obama felt compelled to publicly deny that he was a Muslim is not only striking because this assertion was an obvious lie, but also because of the underlying assumption that being Muslim would apparently disqualify a US citizen for the presidency. This exemplifies the continuous struggles faced by non-white and non-Christian Americans to be accepted as fully American. Hate crime continues to be significant, and the rise of advocacy for minorities has also coincided with a rise in white nativism (see Box 10.1).

Migration policies under Trump: Between rhetoric and practice

From the beginning of his administration, Donald Trump upended long-standing American traditions and values. The new President immediately moved to restrict admission of Muslim refugees despite several federal judges ruling such measures illegal. According to data from the Department of State, in 2018, refugee admissions dropped to post-Second World War low of 22,491, down from a 2008–2017 average of 67,137. In September 2017 the Trump administration announced plans to phase out Obama's DACA. President Trump threatened to punish 'sanctuary cities' that have pledged to be a safe haven for undocumented migrants, while continuing deportation policies of the Obama administration.

The Trump administration became increasing focused on deterring inflows from the 'Northern Triangle' of Central America (El Salvador, Guatemala and Honduras)

(Chishti and Bolter 2019). Border enforcement also was a priority of the Clinton, Bush and Obama administrations, but Trump stepped this up to another level, particularly in terms of vitriolic rhetoric against immigrants. In November 2018, a presidential decree sought to ban those crossing the border without prior authorization from applying for asylum. Under Trump's 'Remain in Mexico' policy, asylum seekers were sent back to Mexico until their asylum determination hearings would take place (Chisthi and Bolter 2019). In February 2019, Trump shut down the US Federal Government in order to pressure Congress to provide funds for a 'big new beautiful wall' – to deliver on one of his central campaign promises. In the same month, he declared a national emergency at the US–Mexico border in order to fund the wall and also ordered military to the border to curb influxes of Central Americans seeking safe haven. These measures led to the resignation of the Homeland Security Secretary and widespread protests. In June 2019, Trump threatened to impose import duties on all goods from Mexico to pressure the Mexican government to curb the arrival of central-American migrants on the US border.

It is as yet unclear to what extent President Trump will be able to deliver on his promises to step up immigration enforcement at the Mexican border and the US interior, as the administration's goals are 'being consistently stymied by court injunctions, existing laws and settlements, state and local government resistance, congressional pushback, and migration pressures that are beyond the government's ability to swiftly address' (Chishti *et al.* 2019). Besides legal obstacles, resistance by lower governments, practical limits in funding and staffing also thwart efforts to 'seal off' the border'. In spring 2018 the Trump Administration was compelled to cancel its new deterrence policy of separating families arriving at the border because of the public outrage this generated. In November 2018, a US appeals court ruled that the Trump administration must continue the DACA programme (Chishti *et al.* 2019)

Despite the political theatre around border enforcement, Trump's migration policy agenda is split between anti-immigrant rhetoric and border enforcement showmanship on the one hand, and the stated intention of overhauling the American immigration system on the other. This is a continuation of attempts by previous administrations to achieve 'comprehensive immigration reform'. Although there is political consensus that the US immigration system is 'broken', attempts to achieve reform have been unsuccessful. This particularly hinges on the highly controversial issue of amnesty for the millions of undocumented migrants living in the US, which has remained unresolved since the 1990s because of the continuing political stalemate on this issue.

In order to reduce the perceived overreliance of the US immigration system on family migration, the administration has proposed to introduce a points-based immigration system for issuing visas and green cards. Because of high demand for workers in agriculture and services, it also intends to increase legal facilities for the entry of temporary foreign workers (an extended form of H-2B visa). However, this is somehow contradicted by Trump's 'Buy American, Hire American' executive order of April 2017 which has made it more difficult to obtain high skilled-worker (H-1B) visas.

Trump's rhetoric on immigration epitomizes securitization. However, it remains to be seen whether, beyond the rhetoric, this will lead to a real change in policies and immigration trends, as the central challenges around enforcement, legal opposition and political stalemate will remain the same. It also remains to be seen whether the Trump administration will be able to withstand corporate pressures to let more workers in and to continue the practice of turning a blind eye towards illegal employment of migrant workers. Despite the administration's stated intentions to restrict immigration, there is little sign of real reform of immigration law, and strong labour demand has kept immigration at high levels. Given these contradictions, it is likely that issues around immigration, race and identity will continue to dominate political debates in the US in the years to come.

Conclusion

No other region in modern history has been as profoundly affected by international migration as the Americas. Rooted in European conquest and occupation, migration to this region has coincided with the destruction of indigenous states and societies, which have been violently supplanted by states and capitalist systems following European models. It also involved the large-scale forced movement of African slaves across the Atlantic as well as the recruitment of indentured workers from Asia. Only from the mid-nineteenth century were such migrations gradually replaced by more spontaneous forms of migration from Europe to immigration countries located at the northern and southern economic heartlands of the Americas, particularly the United States, Canada, Argentina and Brazil – as well as the immigration of Japanese, Chinese, Filipino, Korean and Puerto Rican workers.

American migration systems were to undergo a profound restructuring in the decades following Second World War. While the massive trans-Atlantic movement of Europeans came to an end, economic crises and authoritarian regimes change meant that Southern Cone countries lost their former appeal for migrants. This heralded a period of increased migration from Latin American and Caribbean countries to North America, particularly through large-scale migration of Mexican workers to the US, while movements from Asian countries such as the Philippines, Korea, China and India to North America also started growing. The disjuncture between economic dynamics and immigration restrictions resulted in large-scale undocumented migration, most notably of Mexican and other Latino migrants in the US.

This chapter highlighted the relevance of several migration theories analysed in Chapter 3. First, it underscored the preponderant role of states and businesses in shaping human mobility by establishing initial patterns of migration through warfare, occupation and recruitment. The best example of this is the 1942–1964 Bracero programme, America's equivalent of the European 'guestworker' programmes. It was the largest temporary migrant workers programme in the US history, which laid the basis for massive migration and settlement of Mexicans in later decades. This shows that states are important migration drivers in their own right and can therefore not be reduced to mere 'factors' affecting migrants' cost-benefit calculations. This is a far cry from neoclassical ideas of a free 'migration market', which is particularly unrealistic for lower-skilled migrants and refugees, who face huge obstacles crossing borders and accessing rights.

Second, the American migration experience exemplifies the insights of migration systems theory that exchanges of people, goods, capital and ideas between places and countries tend to reinforce each other *in either direction*. Massive migration from Mexico, Cuba, Korea and the Philippines to the US highlights how the inextricable military, economic and cultural ties created by hegemonic domination tend to generate migration flows. The initial outward movement of US soldiers, entrepreneurs, specialists and missionaries to the Caribbean, Central America, the Philippines and Korea would trigger reverse movement of workers, students and brides.

Third, the resilience and continuation of Mexican migration to the US despite increased border enforcement confirms the argument of dual labour market theory (Piore 1979) that the demand for migrant labour is structurally built into industrial economies, with migrants doing jobs for which there is insufficient domestic supply. Even in times of recession, destination countries North America and the Southern Cone have continued to attract lower-skilled workers from poorer countries in the region, while business lobbies pressed governments to turn a blind eye to the employment of undocumented workers.

Fourth, the American migration experience highlights the fact that migration tends to take place in complex, multi-layered geographical patterns. Most migration is not about movement from the poorest to the richest countries, but rather to countries one level up in the regional development hierarchy. While Bolivians and Peruvians move to Argentina and Brazil, most Argentinians and Brazilians move to Europe and North America. Migration from the Dominican Republic is mainly towards the US, while the Dominican Republic is a destination for migrants from neighbouring Haiti. Mexicans migrate to the US, but Mexico itself is a destination for migrants from poorer Central American countries. This may suggest a degree of 'replacement migration', with migrants from low-income countries doing jobs shunned by workers in middle-income countries, who themselves may have migrated to jobs that natives in high-income countries are not able or willing to do. This suggests that Piore's (1979) dual labour market theory also applies outside the context of high-income societies.

Fifth, the American experience illustrates that middle-income countries tend to have the highest emigration levels, which confirms migration transition theory. Since migration requires considerable resources, it is not coincidental that Mexico and various Caribbean high-emigration countries, where profound demographic, social and economic transformation have swelled the ranks of aspiring and migration-prone young adults, are *not* among the poorest in the region. It is also not coincidental that, until recently, emigration from poorer countries located in Central America and the Andean area has been comparatively low and predominantly intra-regional. In recent decades, development and increasing global integration of Andean countries has increased the geographical scope of migration, such as is manifested in growing migration to Spain (mainly from Andean countries) and to the US (mainly from Central America). Yet the experiences of Argentina, Brazil and Venezuela underscore that migration transitions can be partly reversed if relative economic and political conditions deteriorate. It is difficult to imagine now that, back in 1960, Argentina was the second most important global migration destination in the world (after the US), with 2.4 million migrants representing 11.9 per cent of its population (Czaika and de Haas 2014: 295).

Sixth, the American experience shows that state policies can have huge, sometimes unforeseen and unintended, consequences for migration patterns, although the 1965 amendments to the Immigration and Nationality Act in the US unintentionally encouraged the unprecedented diversification of US immigration. Beside the experiences with the pre-emptive 'now or never' migration surges from Suriname (to the Netherlands) and the British 'West Indies' (to the UK) in the 1960s and 1970s, the ways in which US border enforcement encouraged permanent settlement of Mexican migrants illustrate that ill-considered migration policies can become counterproductive. Or, perhaps, behind the smokescreen of political rhetoric about cracking down on immigration, it does serve certain (business) interests. For instance, by discouraging Mexican migrants from leaving the US, the militarization of the US–Mexican border has created a huge pool of exploitable workers lacking the legal rights of citizens (Massey 2007).

Seventh, the American experience shows that migration can challenge the ways in which nations define themselves. The social transformations engendered by migration inevitably go along with tensions – particularly when 'native' groups (or, sometimes, previous immigrants) feel that their livelihoods are threatened by newly arriving groups. Such tensions can come to a boiling point when politicians exploit latent worries by stirring up xenophobia and representing immigration as a foreign invasion and threat to jobs, security and identity of the nation. It is perhaps not a coincidence that anti-immigrant sentiment also ran high in the 1910s and 1920s, when immigrant figures reached a previous high.

It remains to be seen whether the election of Trump in the US, and other nationalist leaders such as Jair Bolsonaro in Brazil – who called Haitian, Senegalese, Bolivian immigrants and Syrian refugees arriving in Brazil the 'scum of the earth' – signifies a change towards a more nationalist, protectionist and perhaps more authoritarian era, or whether this is a temporary backlash that will leave the long-term trend towards greater acceptance of diversity largely unaffected. At the same time, race remains a divisive issue and source of deeply entrenched, historically grown inequalities, with Native Americans and descendants of African slaves typically occupying the lower rungs of societies throughout the Americas. Whatever the future holds, however, it likely is that immigration and diversity will continue to remain subject of heated public debate.

Guide to Further Reading

Good studies of migration to the US include Borjas (2001) and Portes and Rumbaut (2006). The works on the political economy of migration edited by Munck *et al.* (2011) and Phillips (2011a) include studies on the Americas. *A Nation by Design*, by Zolberg (2006) is an authoritative study into the historical role of immigration policy in the shaping of US society. *Immigrant America: A Portrait* by Portes and Rumbaut (2006) is a comprehensive overview of how immigration has transformed American society. Massey and Denton (1993) gives essential insights into the causes and consequences of racial residential segregation in the US. The work by Douglas Massey and his colleagues in the context of the *Mexican Migration Project* provides extensive empirical evidence on the unintended consequences of US immigration policies, summarized in Massey *et al.* (2016). García (2006) gives an overview of Central American asylum migration to Mexico, the US and Canada.

Vezzoli (2015) and Vezzoli and Flahaux (2017) provide excellent insights into the role of states, post-colonial ties and independence in shaping Caribbean migration, with a particularly focus on Guyana, Suriname and French Guiana. FitzGerald and Cook-Martín (2014) provide a comparative overview of the history of immigration policies in the US, Canada, Cuba, Mexico, Brazil and Argentina. The website of the Economic Commission for Latin America and the Caribbean (ECLAC) (www.cepal. org) is a valuable source for information on the political economy of Latin America. A detailed overview of migration concerning the Americas is provided by the *First Report of the Continuous Reporting System on International Migration in the Americas* (Organisation of American States 2011). For a recent analysis of emergent patterns at sub-regional levels, see Martínez Pizarro (2011). The website of the Migration Policy Institute (www.migrationpolicy.org) has many data and analyses on US migration trends, while also the Pew Research Center (www.pewresearch.org) provides up-to-date information on a number of migration-related issues in the US.

Extra resources can be found at: **www.age-of-migration.com**

8 Migration in the Asia-Pacific Region

Over the past half century, Asia has massively entered the global migration stage. The expansion of population mobility within and from the region reflects the extraordinarily rapid economic development and social transformations much of Asia has gone through alongside the impact of globalization processes. Asia's regions differ greatly in history, culture, religion, economy and politics. At the same time, within each region and even within each country, there is enormous diversity. It is impossible to cover all the variations in migratory patterns and experiences. This chapter therefore focuses on the key migration trends.

In view of its great diversity and complexity, it makes little sense to speak of a single 'Asia-Pacific migration system'. According to the official UN classification, the Asia-Pacific region also includes the Gulf oil states, Turkey and the rest of the Middle East. That latter area of 'West Asia' will be covered in Chapter 10. While the former Soviet republics of Central Asia were covered in Chapter 8. This chapter will focus on the sub-regions of South Asia (Afghanistan, Bangladesh, Bhutan, India, Nepal, Pakistan, Sri Lanka) as well as Iran, East Asia (China, Japan, Korea and Taiwan), Southeast Asia (Brunei, Indonesia, the Indo-Chinese, Malaysia, peninsula, the Philippines, and Singapore) and Oceania (comprising Australia, many Pacific island states and New Zealand).

A section near the end deals with Oceania, where experiences are rather distinct, most notably because its dominant nations (Australia and New Zealand) have been the result of European conquest, colonization and large-scale immigration. Their experiences are therefore more similar to other European settler colonies in the Americas.

We can summarize the four main phases of Asian migration as follows:

- 1945–1973: Since the Second World War and the dismantling of the British, French, Dutch and Japanese empires, after an initial surge in post-colonial migrations to Europe, labour migration subsided to lower levels and would largely remain internal and intraregional, while warfare in Korea, Vietnam and elsewhere in Indochina prompted significant refugee migration;

- 1973–1989: After the Oil Boom, the Gulf region emerged as a new migration destination for Asian migrant workers while there was increasing migration from countries such as Philippines, China and South Korea to North America and Australia. Large-scale refugee migration ensued after the end of the Vietnam War (1975) and due to Soviet military intervention in Afghanistan (1979–1989);

- 1989–2008: Increase of lower- and higher-skilled migration to the Gulf and an increasing variety of western destinations, alongside acceleration of intraregional migration from lower income countries like Bangladesh, Indonesia, Vietnam and the Philippines to Japan, South Korea, Malaysia, Singapore, Taiwan and Thailand. While Australia and New Zealand became increasingly integrated in Asian migration systems, US-led military interventions in Afghanistan and Iraq generated large-scale refugee migrations;

- Since 2008: Acceleration of extra-regional migration from an increasingly diverse array of origin countries (including Indonesia, Bangladesh, Nepal and Myanmar) to the Gulf, North America and Europe, consolidation of intraregional migration to destinations in East and Southeast Asia following economic and demographic transitions. New refugee migrations ensued, such as from Rohingya fleeing Myanmar.

Figures 8.1 and 8.2 give overviews of the main trends in migration to and from the region between 1960 and 2018. Figure 8.1 shows that the number of Asian migrants living outside the region has increased fast, particularly since 1990. Figure 8.2 shows that intraregional migration has also increased, particularly since the 1990. However, the figures also show that international migration is low in relative terms. In 2017, the Asia-Pacific hosted 37.8 million immigrants, mainly from other countries in the region. This seems a lot, but it only represents 0.9 per cent of the regional population. Immigrants in Asia comprised only 15 per cent of the world total of 247 million migrants. In 2017, about 77.2 million people from the region lived abroad. This is equal to 1.8 per cent of the population of the Asia-Pacific, and 31 per cent of the global migrant population. This is relatively low, since about 56 per cent of world's population lives in the Asia-Pacific.

Although Asian emigration is high in *volume,* with a few exceptions such as the Philippines and a number of smaller countries, emigration *rates* relative to population size are often low. This partly reflects the fact that in large and populous countries such as China, India, Indonesia and Pakistan, most migration is contained with national borders (see Chapter 1 and see Skeldon 1997, 2006).

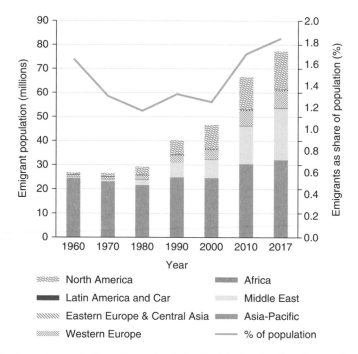

Figure 8.1   Emigrant populations from the Asia-Pacific, by destination region, 1969–2017
Source: UNDESA 2017 and Global Bilateral Migration Database

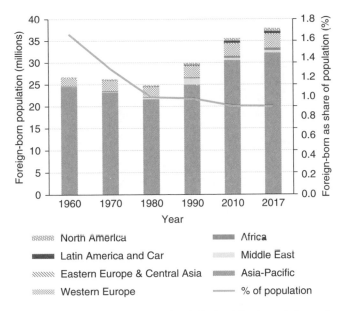

Figure 8.2   Foreign-born populations in the Asia-Pacific, by origin region, 1969–2017

Colonial and post-colonial migration patterns

As other world regions, Asia has always known significant population mobility. The Silk Road connected East- and Southeast Asia to the Middle East, the Mediterranean basin and East Africa over the last two thousand years, bringing about significant trade as well as cultural and religious interchange. Westward movements from Central Asia helped shape European history in the Middle Ages. Arab traders played a central role in the spread of Islam in Southeast Asia (and East Africa), while Chinese migration to South-East Asia goes back centuries (Wang 1997). Throughout the second millennium, Chinese have moved across Southeast Asia to trade, many of them settling permanently in their new homelands, where they would often occupy key commercial positions (Lockard 2013).

In the nineteenth century and early twentieth century there was considerable migration from China and Japan to the US, Canada and Australia. In addition, millions of indentured labourers from contemporary India, China and Java (Indonesia) were recruited to work in East Africa, the Caribbean, Hawaii and la Réunion (see McKeown 2004; Vezzoli 2015), while Japanese were recruited to work in Hawaii, the US, Brazil and Peru (Shimpo 1995) (see Chapter 5). In the colonial period, Chinese settlers in South-East Asian countries (Sinn 1998) and South Asians in Eastern and Southern Africa became trading minorities with an important intermediary role for colonialism (see Chapter 9).

Extra-regional emigration from Asia was low in the early part of the twentieth century owing to racist immigration policies by the US and Canada and European colonial powers. However, movements within Asia continued, often connected with political struggles. Japan recruited 40,000 workers from its then colony, Korea, between 1921 and 1941. Japan also made extensive use of forced labour in World War II. Some 25 million people migrated from densely populated Chinese provinces to Manchuria between the 1890s to the 1930s, with about 8 million staying on 'to reaffirm China's national territory in the face of Japanese expansionism' (Skeldon 2006: 23).

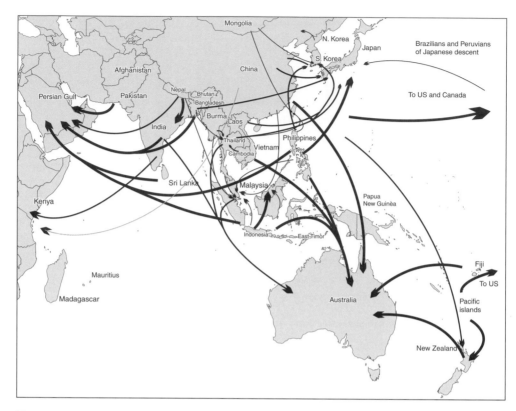

Map 8.1   Contemporary migrations within and from the Asia-Pacific region

Note: The arrow dimensions give an approximate indication of the volume of the flows in the 1945–2015 period. Exact figures are often unavailable

From the late nineteenth century, while the US expanded its influence in the Pacific by annexing Hawaii (1897) and the Philippines (1902), Japan would evolve into a major imperial power, leading to the colonization of Korea and Taiwan (Beasley 1987) and further territorial expansion into Manchuria, China and Mongolia in the 1930s. This culminated into the occupation of much of South-East Asia in the Second World War. It has been estimated that between 3 to over 10 million mostly Chinese, Korean, Malaysian, Indonesian, Filipino and Indochinese people as well as Western colonial prisoners died as a consequence of forced labour, massacre and deliberate starvation during the War (Rummel 1998).

Asian migration patterns would be profoundly altered by the structural changes in the international political economy affecting the world in the post-WWII decades. All in all, these would create the conditions for later increases of Asian migration. A first type of migration was largely forced movement in the wake of conflicts related to post-colonial state formation. For instance, in the violent mass population transfers following Indian Independence in 1947 and the concomitant partition between India and Pakistan, about 5 million Hindus and Sikhs left Pakistan for India and about 6 million Muslims moved into Pakistan from India (Khadria 2008). The Korean War caused significant refugee migration as well as 'bride migration' to the US.

A second type of migration consisted of voluntary and forced migration to the former colonial powers, such as those from Indonesia to the Netherlands, from the Indian sub-continent to the UK and from Hong Kong and Vietnam to France. There were also some smaller movements, like those from Goa, Macau and East Timor to Portugal. In some cases, such moves were partly forced, such as of Indonesian 'repatriates' (see Chapter 6). In other cases, such as with migration from Pakistan, India and Bangladesh to the UK, movements mainly concerned workers.

Post-colonial movements to the UK, France and the Netherlands were often facilitated by preferential immigration rules. Under the 1948 British Nationality Act, citizens of Commonwealth countries such as India and Pakistan could enter without restriction as British subjects. This would change with the introduction of restrictions under the 1962 Commonwealth Immigrants Act. Similar to the 'beat the ban' rush it caused in the 'West Indies' (see Chapter 7), the Act interrupted circulation and spurred permanent settlement as families chose to stay together in the UK rather than risking prolonged separation. In the longer term, however, the Act stimulated a gradual reorientation of South-Asian migration to other countries, mainly the US, Canada and Australia, while Indonesian migration to the Netherlands largely ceased after the West New Guinea Dispute in 1962 deteriorated diplomatic relations.

The entry of Asia on the global migration stage (1973–1989)

Reorientation of Asian migration towards North America and Oceania

In the post-1973 period Asian emigration underwent a fast increase and regional diversi-fication away from the former European colonizers because

- Ex-colonial powers made migration rules for former colonial subjects from Asia more restrictive;

- New migration channels opened up as immigration countries such as the US, Canada, Australia and New Zealand opened their borders for non-white immigrants;

- After the 1973 Oil Shock, the booming oil economies of the Gulf provided unprece-dented migration opportunities for lower-skilled Asian workers.

Asia's large-scale entry onto the world migration stage partly resulted from the opening up of the region to economic and political relationships with industrialized countries. Increasing connections through trade and investment, rapid economic development and improved education endowed more and more people with the resources and aspirations to leave the countryside in search of better conditions and alternative lifestyles in the grow-ing cities or abroad. At the same time, the relocation or 'outsourcing' of labour-intensive industrial production from Western to East Asian countries would contribute to the emer-gence of regional economic hubs, which would start to attract internal and international migrants from *within* the region, accelerating urbanization and intraregional migration.

Since the late 1960s, postcolonial migrations to Europe decreased while the classi-cal immigration countries of the 'New World' abolished their racist immigration rules. Discriminatory rules against Asian entries were repealed in Canada (1962 and 1976), the US (1965) and Australia (1966 and 1973). Increased foreign investment and trade by the US and other destination countries helped create the communicative networks needed for migration. As a result, immigration countries in North America and Oceania developed into important destinations for Asian migrants, initially mainly from countries such as South Korea, Vietnam and the Philippines.

The largest Asian movement was that to the US: most Asians came through the family reunion provisions of the 1965 Immigration and Nationality Act, though refugees or skilled workers were often the first link in the migratory chain. In Canada, it was the introduction of the points-based system in 1967 and the 1976 Immigration Act, with its non-discriminatory selection criteria and its emphasis on family and refugee entry, which opened the door to Asian migration. Asian immigration to Australia also rapidly increased after the repeal of the White Australia Policy, with additional stimulus from the Indo-Chinese refugee movement at the end of the 1970s. Similarly, New Zealand's abandonment of its racially selective entry policies led to considerable inflows from the Pacific Islands and from Asia.

The openness of the US, Canada, Australia and New Zealand to family migration meant that primary movements by workers or refugees were followed by migration of other permanent settlers, usually family members. Through such *migration multipliers* facilitated by chain migration and transnational networks, these migrations gained their own momentum. As a consequence, unexpectedly large movements developed mainly through use of family reunion provisions.

Contract labour migration to the Middle East

Labour migration from Asia to the Middle East developed rapidly after the oil price rises of 1973. The huge construction projects in the Gulf oil countries caused mass recruitment of temporary contract workers. Both unskilled and skilled labour was imported by oil-rich countries, first from India and Pakistan, then from the Philippines, Indonesia, Thailand and Korea, and later from Bangladesh, Sri Lanka and Nepal. As part of 'labour export strategies', governments of countries like India, Pakistan and the Philippines actively marketed their labour abroad, and concluded labour-supply agreements with Gulf countries, while allowing private agencies to organize recruitment (Abella 1995). Because part of the labour demand was in sectors such as domestic work, the rise of the Gulf as a major destination also stimulated the partial feminization of Asian labour migration.

Asian migration to the Gulf would also become increasingly diverse in terms of education and skills. The stereotype of the 'unskilled' Asian worker in the Gulf has always been flawed, as since 1973 skilled personnel – such as engineers, nurses and managers – have also migrated from Asian countries to fill labour shortages and to help Gulf countries in their rapid economic development. While many Asian migrants working in the Gulf remain low-skilled workers in sectors such as construction or domestic work, others would take up semi-skilled or skilled jobs as drivers, mechanics or building tradesmen. Others went with professional or para-professional qualifications (engineers, nurses and medical practitioners).

The Cold War and conflict-related migration

The growing dominance of the US (alongside the Soviet Union) in global geopolitics – at the expense of Western Europe – played a role in stimulating migration to North America. The US military presence in Korea, Vietnam and other Asian countries forged transnational links, as well as directly stimulating movement in the form of 'war brides' of US personnel, refugees and migrant workers.

The Vietnam War caused large-scale refugee movements as did other conflicts in the Indo-Chinese region over the 1970s and 1980s; Over 3 million people fled from Vietnam, Laos and Cambodia following the end of the Vietnam War in 1975. Many left as 'boat people', sailing long distances in overcrowded small boats, at risk of shipwreck and pirate attacks. Over the following 20 years, 2.5 million found new homes elsewhere, while 0.5 million returned. Over a million were resettled in the US, with smaller numbers settling in Australia, Canada and Western Europe. China accepted about 300,000 refugees, mainly of ethnic Chinese origin (UNHCR 2000: 79–103).

The 1979 Islamic revolution in *Iran* would also trigger substantive refugee movements. After the Islamic revolution, many political opponents – such as royalists and communists – and minorities – such as Baha'is, Jews, Christians, Zoroastrians and Sunni Kurds – fled Iran while many young people left to avoid military service during the war with Iraq (1980–1988). In 2000, the number of Iranian refugees and emigrants was estimated at 1 million, although some disputed estimates range from 2 to 4 million. The majority of them live in the US, followed by Germany, the United Arab Emirates (UAE), Canada, Sweden, the UK, Israel and various West European countries (Hakimzadeh 2006).

Up to a third of Afghanistan's 18 million people fled the country during the Soviet military intervention between 1979 and 1989. The overwhelming majority found refuge in the neighbouring countries of Pakistan (3.3 million in 1990) and Iran (3.1 million) (UNHCR 2000: 119). For political, humanitarian, religious and cultural reasons, Pakistan and Iran were willing to provide refuge for extended periods (UNHCR 2000: 118). In this period, Particularly Iran evolved into a major refugee haven for Afghans and to a lesser extent also Iraqis, and became host to one of the largest long-term refugee populations in the world. Refugees were initially welcome in the spirit of Muslim fraternity. At its peak in 1991, the refugee population exceeded 4 million, among whom were 3 million Afghan refugees who fled the Soviet invasion of Afghanistan in 1979, and 1.2 million Iraqis (Hakimzadeh 2006).

The globalization of Asian migration (1989–2008)

Asia's rise to prominence in the 'global race for talent'

In the period of neo-liberal globalization between 1989 and 2008, an increasingly global labour market for skilled personnel as well as fee-paying students emerged, with Asia as the main source – but increasingly the fast-growing and ageing economies of East and South-East Asia started to attract workers themselves. Emigration for employment from countries within the region grew at about 6 per cent a year between 1985 and 2005, with about 2.6 million people leaving their homes in search of work abroad each year (ILO 2006: 37).

Besides North America and Oceania, in this period West European countries also started making efforts to attract skilled Asian migrants, particularly from China, India and the Philippines (OECD 2007; see also OECD 2011b: 46–52). These migrations were generally disconnected from previous 'post-colonial' migrations, and included medical and information technology personnel and increasing numbers of students, as well as female domestic workers and manual workers.

Although most Asian emigration still concerned lower-skilled workers, mobility of higher-skilled workers such as technicians, engineers, academics and managers was also growing. Economic migrants could be found at all skill levels: the lower-skilled would predominantly migrate to the Gulf and other Asian countries while higher-skilled migrants as well as students predominantly moved to North America, Australia and Europe, although the Gulf region also attracts skilled Asian migrants. In the Philippines, for instance, over the 1980s and 1990s about 40 per cent of permanent emigrants had a college education, and 30 per cent of IT workers and 60 per cent of doctors emigrated. For Sri Lanka, academically qualified workers comprised up to one third of emigration (Lowell and Findlay 2002). India, the Philippines and Pakistan were among the top ten countries from which doctors emigrated (World Bank 2011: 10).

High-skilled migrants from Asia would play an increasingly important role in innovation and economic growth in destination countries. For instance, Indian and Chinese IT experts played a key role in the rise of Silicon Valley in the US. Some governments, such as the Philippines, deliberately train surplus medical personnel such as nurses, with the explicit goal of exporting this skilled labour force in order to reap benefits in the form of remittances. Skilled emigrants may come back to their homelands if opportunities present themselves. For instance, the return of Indian IT professionals from the US and other destinations has been a crucial factor in the rise of the Indian IT industry. Sometimes governments seek to encourage such returns through diaspora policies in an effort to stimulate remittances and investments by migrants, and by so doing unlock the development potential of migration (see Chapter 14).

The globalization of domestic work and the feminization of migration

Since the 1990s, South and South-East Asian migration to the Gulf countries continued unabatingly despite frequent calls in the Gulf to 'indigenize' the workforce. The most important origin countries are India, Bangladesh, Pakistan, Indonesia and the Philippines. According to official UN estimates, the number of Asian migrants working in the Gulf increased from an estimated 5.7 million in 1990 to 20.8 million in 2017 (see Figure 9.5. in Chapter 9). Because this migration is often temporary, these figures conceal much higher levels of circulation. Asian migrants from Asia-Pacific occupy three quarters of the migrant work force in Gulf countries. While lower skilled migrant workers have few if any rights and are vulnerable to abuse, the open entry regimes of Gulf countries have created previously unprecedented opportunities for relatively poor Asians to work and gain a much higher income abroad.

Besides construction workers and other manual workers, the demand for work in the expanding service sector of Gulf countries also increased, such as for jobs in catering and driving. There has been a surge in demand for domestic workers, nurses, sales staff and other service personnel, jobs for which women are generally preferred. This would further contribute to the feminization of Asian labour migration. In 2004, 81 per cent of registered new migrant workers leaving Indonesia were women (ILO 2007). The female share among first-time migrant workers from the Philippines rose from 50 per cent in 1992 (Go 2002: 66) to 72 per cent by 2006 (ILO 2007). Female migrant workers came from an increasingly diverse range of origin countries such as the Philippines, Sri Lanka, Indonesia, Nepal and in recent years also Ethiopia (Schewel 2018).

Since 1990, the range of destination countries has expanded and diversified, with other countries in the Middle East, including Lebanon, Jordan and Israel, also becoming labour-importers, especially for Asian domestic workers (Agunias 2011; Asis 2008). In particular, the migration of Filipina domestic workers and nannies spilled over to regions outside the Middle East including Europe and Africa. In Italy, for instance, Filipina workers have played an increasingly important role in informal arrangements for elderly care and childcare around the world (Parreñas 2000; Ambrosini 2013; 2016). Parreñas (2000: 564) argued that as low-wage service workers, Filipinas 'meet the rising demand for cheap labour in the global cities of Asia and Europe and, to a lesser extent, the United States'. On a critical note, she argued that while women in industrialized (Western) countries are often assumed to be more 'liberated' than women in developing countries, many women are able to pursue careers because disadvantaged migrant women 'are stepping into their old shoes and doing the household work for them' (Parreñas 2000: 557–578).

The rise of Asian migration destinations

Besides an increase and geographical diversification of Asian migration to other world regions, migration *within* Asia would increase from 24.6 million in 1990 to 30.5 million in 2010 (to increase further to 32.2 million in 2017). Real figures are likely to be higher because of the significance of irregular migration, particularly in South-East Asia. Rapid economic growth and declining fertility increased labour demand in the new industrial economies of East and South-East Asia, such as Japan, Singapore, Taiwan, Korea, Malaysia and Thailand. Although official political discourses often emphasized the need to discourage immigration, employers' lobbies explain why governments turn a blind eye to irregular migration of lower-skilled workers, while introducing privileged immigration and residence regimes for the higher-skilled. In all the Asian 'tiger economies', migrant workers are doing the 3D jobs (dirty, difficult and dangerous – or just low-skilled and poorly paid) that nationals can increasingly afford to reject (Mendoza 2018). Similar to the role played by internal migrants within China, South-East and South Asian labour migrants in countries such as Thailand, Malaysia and Taiwan provide essential cheap labour input in the textile industry for Western fashion brands (Mendoza 2018).

Many countries in the region experienced both significant emigration and immigration, connected to the proliferation of complex patterns often following multi-layered geographical migration hierarchies. For instance, while Indian emigration has increased fast, at the same time India also attracts numerous migrants from Nepal and Bangladesh. Emigration patterns differ according to class and educational background. For example, industrialization in Malaysia from the 1970s attracted migrant workers from neighbouring Indonesia, whereas better-off Indonesians tried to migrate to destinations further afield like Japan, the Gulf or Europe. While emigration from established labour frontier countries such as Bangladesh, Indonesia and the Philippines to Malaysia continued, migration from new source countries like Vietnam, Cambodia, Laos and Myanmar increased as demographic transitions, economic transformations and increasing education swelled the ranks of migration-prone youth.

Reflecting trends of extra-regional migration from Asia, also intra-Asian migration has been feminizing, following the changing structure of labour demand in destination countries. Most migrant women are concentrated in jobs regarded as 'typically female': domestic work, entertainment (often a euphemism for prostitution), restaurant and hotel staff, as well as assembly-line workers in clothing and electronics. These jobs generally offer poor pay, conditions and status, and are associated with patriarchal stereotypes of female docility, obedience and willingness to give personal service. The increase in domestic service reflects the growth of dual-career professional households in Asia's new industrial countries. Married women have to leave their children in the care of others, with long absences challenging family relationships and traditional gender roles (Lee 2010; Lutz 2018; Parreñas 2000; see Chapter 3 and 12 on *global care chains*).

Despite the increase in intra-Asian labour migration, the contribution to labour forces in destination countries has remained low compared to immigration countries in Western Europe and North America: migrant workers make up only about 4 per cent of labour forces in East and South-East Asia. However, the situation is different in Singapore and Malaysia, where in the late 2000s migrants made up 35 and 19 per cent of the workforces respectively (Department of Statistics Malaysia 2009; Ministry of Manpower Singapore 2011). While early flows were mainly low-skilled, higher-skilled migration increased throughout the region, particularly because demand for health and care workers has been increasing.

Marriage migration

Marriage migration has taken on particular importance in Asia because of the region's particular demographic characteristics. As such the phenomenon is anything but new, and was initially associated with migration to Western countries, such as 'war brides' of US servicemen from Japan, then Korea and then Vietnam and, from the 1980s, 'mail order' brides to Europe and Australia (Cahill 1990). However, since the 1990s, marriage migration became a major driver of *intra-Asian* migration. Foreign brides have been sought by farmers in Japan, Taiwan and South Korea, partly because of the exodus of women from rural areas seeking more attractive lifestyles in urban settings. The Chinese one child policy and the preference for male children in India have led to male surpluses in those countries, and brides started to be recruited for Indians in Bangladesh and for Chinese farmers in Vietnam, Laos and Burma (IOM 2005: 112).

Foreign brides are often recruited through agents. For instance, in 2008 Taiwan received 413,000 marriage migrants, while South Korea received 150,000 (11–13 per cent of all marriages in the 2003–8 period). In Japan some 5–6 per cent of all marriages in the same period involved foreign partners (Lee 2010). Many women migrating for marriage come from South-East Asia (Vietnam, Laos, Thailand, Philippines) and China. Destination countries all have very low total fertility rates (in 2017: in South Korea 1.17 and Japan 1.45). Some regard the high proportion of foreign mothers as a threat to national identity. The Korean and Taiwanese governments have implemented policies to tackle abusive match-making businesses and to support marriage migrants and their children (Bélanger *et al.* 2010; Lee 2010; Kim and Kilkey 2018; Piper and Lee 2016).

Refugee migration in the post-Cold War era

The end of the Cold War emboldened pro-democracy movements and weakened unquestionable support for autocratic regimes by the former West and East blocs. This would encourage the transition to democratic governance in Asian countries such as South Korea, Taiwan, Thailand and Indonesia. In other countries, such movements were met by hard repression, such as in China, where thousands of Chinese sought asylum overseas after the failure of the democracy movement in 1989. In China ethnic and religious minorities such as Tibetans and Uyghurs also faced extremely harsh repression. At least 50,000 North Koreans fled to China (Greenhill 2010). Long-standing regional refugee populations include Tibetans and Bhutanese in India and Nepal, and Burmese in Thailand

and Bangladesh, although migration between these countries is dominated by labour flows. The 26-year civil war in Sri Lanka between 1983 and 2009 led to mass internal displacement as well as Tamil refugee outflows. The war ended with appalling bloodshed in May 2009, and by 2010, there were an estimated 274,000 IDPs in Sri Lanka, while 141,000 persons had left as refugees (UNHCR 2011).

With the retreat of the Soviet forces in 1989, about 1.5 million Afghan refugees returned home. After the Soviet-backed government fell in 1992, the country swiftly descended further into civil war between rival warlords – fuelling further refugee migration (Willner-Reid 2017). The fighting came to a temporary end when the Taliban – an Afghan faction of *mujahedeen* who had resisted Soviet occupation with the undercover backing of the US and Pakistan – seized control in 1996. However, the Taliban's oppressive rule and the war devastation delayed the return of refugees despite the harshening of attitudes towards Afghan refugees in neighbouring Pakistan and Iran. To help fund the costs of rebuilding their villages, increasing numbers of Afghans went to work in the Gulf states, while others sought asylum in Western countries (UNHCR 1995: 182–183).

The events of 11 September 2001 prompted the US-led invasion of Afghanistan of 2001, which was designed to topple the Taliban regime for providing refuge to al-Qaeda and Osama bin Laden (Laub 2014). In the 2001–2011 decade, 4.6 million registered refugees returned home, mainly from Iran and Pakistan; the largest assisted return process in modern history (Willner-Reid 2017). However, the failure to establish security and the continuation of hostilities between the US-led forces and the Taliban hindered further returns. Although the total number of Afghan refugees decreased from 6.3 to 2.5 million between 1980 and 2016, Afghanistan would remain the world's largest refugee origin country, until the Syrian civil war led to a new refugee emergency (see Chapter 9 and Box 6.1).

Migration since 2008: Consolidation and expansion

Most countries in the Asia-Pacific region were only marginally affected by the 2008 Recession, and maintained steady economic growth. Fast increases in income and education levels have continued to boost migratory and non-migratory mobility within and from the region. For instance, the annual number of Chinese travelling abroad for both business and tourism shot up rapidly (Li *et al.* 2010). Particularly Indians and Chinese emigrant communities around the world have grown fast, further consolidating their key importance in low- and high-skilled labour markets in Asia, the Gulf, North America and Western Europe. The increasingly prominent role of Chinese and Indians as global investors, entrepreneurs and scientists is a manifestation of the growing geopolitical clout of Asia, and a gradual eastward shift of the global economic centre in an increasingly multi-polar world.

China in particular has risen fast as a global geopolitical power. For instance, the Chinese Belt and Road Initiative (BRI) involves the construction of a range of transport links throughout Asia, the Middle East, Europe and Africa. Chinese firms are investing heavily in infrastructure and natural resource exploitation around the world, including Africa. Investments and the growing military and diplomatic influence of China also went along with new migrations of investors, engineers and construction workers. The outward movement of Chinese to Africa is increasingly triggering reverse migration movements of African workers and students to China (see Chapter 10).

Economic growth and demographic change have further consolidated intraregional migration mainly directed at Japan, South Korea, Malaysia, Singapore, Brunei, Thailand,

Taiwan and industrialized zones within India and China. New transport and communication technologies and regional integration have opened the way for growing temporary and circular migration. As a result, Malaysia, Singapore, Thailand, Taiwan, Japan and South Korea have reinforced their position as immigration countries, although this has not been matched by policies that facilitate long-term residence and the acquisition of full rights by migrants. Most Asian governments still cling to the official policy that migration is temporary and migrant workers should eventually return home.

Since the 1990s, Asian migration to Europe, North America, Australia and New Zealand was stimulated by selective immigration policies and 'points-based systems' that tried to attract 'global talent'. Perhaps paradoxically, the Great Recession rather reinforced this 'race for talent' as Western governments aimed to boost economic growth through attracting foreign investors, while using foreign students' fees to maintain the financial sustainability of higher education in times of austerity-driven cutbacks in government spending. While extraregional emigration of skilled Asians surged, migration of lower-skilled workers from poorer countries to the Gulf countries also increased. Migration to the Gulf is the main opportunity for a better future for many non-elite groups in South and South-East Asia who cannot afford to migrate further afield. In order to tap new sources of cheap, Gulf employers and agents have increasingly recruited workers in countries such as Nepal, Sri Lanka, Bangladesh and Myanmar (alongside Ethiopia and Eritrea in Africa). This outward geographical shift of the Gulf 'labour frontier' deeper into Asian hinterlands has coincided with increasing emigration from these countries as they became progressively integrated in global migration systems.

The changing geographies of Asia-Pacific migration

Rapid economic and social transformations have led to the emergence of increasingly complex, intra-regional migration systems linking an expanding set of labour frontier countries in South and South-East Asia to already consolidated and emerging industrial core nations in East and South-East Asia. At the same time, the region as a whole has been rapidly incorporated into global migration systems, with lower-skilled citizens from South and South-East Asia increasingly finding their way to the Gulf region, and skilled from across Asia, but particularly China, India and the Philippines, migrating all over the world, but predominantly to North America, Europe, Australia and New Zealand. Growing economic and migratory ties have also increasingly drawn Australia and New Zealand into the Asian sphere. Figure 8.3 highlights the diversity among the main destination countries in the region. In absolute terms, India, Pakistan and Iran have large immigrant populations, but relative to their large populations, these are rather small. In relative terms, Australia, New Zealand, Brunei and particularly Hong Kong, Macao and Singapore come out on top. Notwithstanding the diversity of migration in the region, it is possible to make a distinction between (1) labour export countries in South and South-East Asia, (2) the reluctant immigration countries of East Asia and (3) immigration countries in South-East Asia.

Labour export countries

Just as the migration from its Mediterranean periphery fuelled Western European industrial expansion and just as Mexican, Puerto Rican and Cuban workers facilitated US economic growth, Asia has its own labour frontier: South Asian countries (Bangladesh,

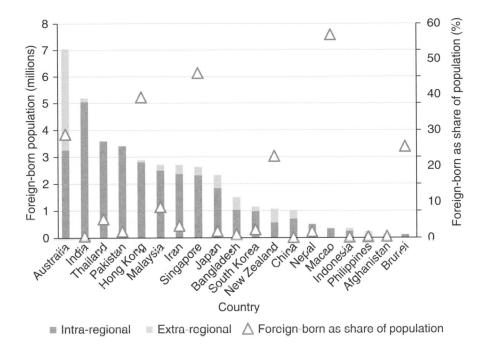

Figure 8.3   Foreign-born populations in countries of the Asia-Pacific, by origin, 2017
Source: UNDESA 2017

Pakistan, India, Nepal and Sri Lanka), and many South-East Asian countries (the Philippines, Indonesia, Vietnam, Cambodia, Laos and Myanmar) have all become major labour providers for the region and increasingly for the rest of the world. Various 'labour exporting' governments set up special departments to manage recruitment and to protect workers, such as Bangladesh's Bureau of Manpower, Employment and Training (BMET) and India's Ministry of Overseas Indian Affairs. Governments stimulate emigration because they see migration as a source of remittances and because they hope it will reduce unemployment and provide training and industrial experience (see Hugo 2005: 28–33).

China is a vast country with huge internal migration. With regard to international migration, China is still mainly an area of emigration, with flows to North America, Europe and – most recently – Africa. The latter is strongly linked to China's emerging trading interests in countries like Mozambique, Zambia, Zimbabwe and Sudan. However, 'the era of cheap labour in China is ending' (Skeldon 2006: 282; see Chan and Selden 2017). The rapid economic expansion and the sharp decline in fertility due to the one-child policy mean that China's rural labour reserves are being depleted. Labour shortages and rising wages have been reported in the industrial cities of the East coast, especially for highly skilled personnel (Pieke 2011), and internal migration to the coast is slowing down with the growth of inland regional cities. High-skilled immigration from Hong Kong, Taiwan and other countries is helping to fill the gap. Another form of high-skilled migration concerns executives and specialists transferred within multinational enterprises, or officials posted abroad by international organizations.

India too has experienced large-scale emigration, and today the 'Indian diaspora' is estimated at over 20 million persons (including non-migrants of Indian origin holding other

Source: Getty Images/Romeo Gacad

Photo 8.1   Returning overseas Filipino workers from the Middle East receive Christmas gifts from government employees at the Manila airport, December 2006, as part of the government's tribute to the labor of the Philippines' large overseas work force

citizenships) across 110 countries (Ministry of Overseas Indian Affairs 2011). Indians still go in large numbers to the Gulf as manual workers, and to Western countries as highly skilled personnel (IT professionals, medical practitioners etc.). High-skilled emigration has been increasingly matched by return flows of skills and capital, which are contributing to the development of modern manufacturing and service industries in some parts of India (Khadria 2008). India is also an immigration country, with inflows of mainly lower-skilled workers from Nepal and Bangladesh, as well as refugees from conflicts countries in the region, including Nepal, Bhutan, Tibet and Sri Lanka.

Over the post-WWII period the *Philippines* has evolved into a major emigration country. Labour export has been an official policy since the 1970s, when it was also seen by the Marcos regime as a political and economic 'safety valve' to reduce domestic discontent. A national 'culture of emigration' has developed: going abroad to work and live has become a normal expectation for many Filipinos. The Philippines has developed strong institutions to manage labour export like the Philippine Overseas Employment Administration (POEA) and to maintain links with the diaspora like the Commission on Filipinos Overseas (CFO).

According to UN data, about 5.7 million people born in the Philippines live abroad. According to Filipino authorities, however, about 10.2 million Filipinos were abroad in 2013, more than 10 per cent of the country's population, but this number includes children of migrants. About half were seen as permanent emigrants and the other half as temporary Overseas Filipino Workers (OFWs), while 1.2 million were registered as irregular migrants. The most popular important countries of the Filipino diaspora population were the US

(3.4 million), Saudi Arabia (1.6 million), Canada (843,00), United Arab Emirates (680,00), Malaysia (570,000) and Australia (385,000), but Filipinos can be found all over the world, particularly in domestic work but also a whole range of other professions such as factory work, nursing, cleaning and catering (Commission on Filipinos Overseas 2013).

The reluctant immigration countries of East Asia

In East Asia a combination of rapid economic growth, fertility decline, ageing and grow-ing undocumented migration has led to serious contradictions, most evident in Japan, but also emerging in South Korea, Hong Kong, Taiwan and China, between official ideologies centred around ethnic homogeneity, and severe labour scarcities in these ageing societies. High labour demand has attracted growing foreign labour forces, a reality which govern-ments have often only accepted grudgingly.

Japan has experienced considerable labour immigration since the mid-1980s. The registered foreign population increased from 817,000 in 1983 to 2.4 million in 2017 (1.9 per cent of the total population). In 2010, 27 per cent of foreigners were considered 'general permanent residents' and another 19 per cent were 'special permanent residents'. The latter are mainly descendants of Koreans who were recruited (sometimes by force) as workers during the Japanese occupation of Korea from 1910 to 1945. According to OECD data, in 2017, 696,000 Chinese, 453,000 Koreans, 244,000 Filipinos (73 per cent of whom women), and 200,000 Vietnamese were registered (OECD 2018). Immigrants include *Nikkeijin*: descendants of past Japanese emigrants in Brazil and Peru now admitted as labour migrants (see also Chapter 7). *Nikkeijin* were recruited because of their Japanese ancestry, but face serious integration challenges. Although they may look Japanese, cul-turally they are often more Latin American.

Official government policies and public attitudes remain opposed to recruitment of foreign labour and to long-term stay, for fear of diluting the perceived ethnic homo-geneity of the Japanese population. Although this ethnic homogeneity is a myth, its tenacity explains political resistance against immigration. In view of the ageing popula-tion and projected future labour demand, especially for elderly care, such policies may be hard to sustain, notwithstanding efforts to use robots as a substitute for human labour. Despite government rhetoric that only skilled migrants are welcome, in practice both lower- and higher-skilled are admitted (Komine 2018). Admission policies have become more open towards low-skilled labour migrants (often entering as 'trainees' and 'interns'), although skilled migrants have easier access to permanent residence and fam-ily sponsorship (Komine 2018). The fastest growing immigrant groups are Vietnamese, Indonesians and Nepalese. Immigration of all kinds is increasing, with inflows reaching levels of 427,000 in 2016, which represents an immigration rate of 0.34 per cent (see Figure 8.4).

South Korea's foreign resident population has risen rapidly from 149,000 in 1996 to 1.2 million (2.3 per cent of its population) in 2017 (see OECD 2018). Annual inflows have gone steadily up from around 200,000 in the early 2000s to around 400,000 in recent years (see Figure 8.4). Similar to Japan, there is considerable public discussion of multi-culturalism and the challenges of immigration for a society very concerned about ethnic homogeneity. South Korea experienced sustained emigration to the US and Australia after the Korean War, and exported labour to the Gulf in the 1970s and 1980s. However, by the mid-1990s rapid economic growth had led to a sharp fall in emigration and rising

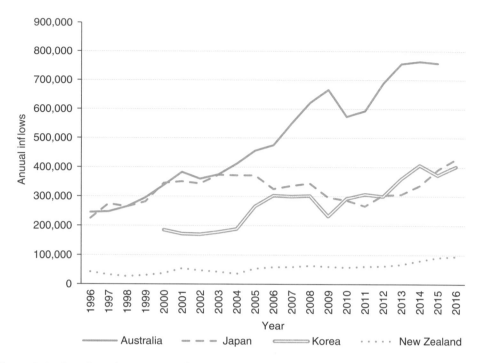

Figure 8.4   Immigration to Australia, Japan, Korea and New Zealand, 1996–2016
Source: OECD 2018

immigration (see Figure 8.4). In 2016, according to OECD data, 549,000 Chinese nationals lived in Korea. The number of Vietnamese has increased rapidly, from 68,000 in 2007 to 138,000 in 2016, while 46,000 Filipinos lived in Korea. Smaller, but fast-increasing groups are male labour migrants from Nepal, Sri Lanka and Uzbekistan.

In 1994, the transition of South Korea into a destination country prompted the government introduced the 'Industrial Trainee System' – a disguised framework for import of low-skilled labour (Hur and Lee 2008; Seol 2001). This scheme failed to reduce irregular migration and was replaced by the 'Employment Permit System' (EPS) in 2004, which gave temporary migrants the same labour market rights as Koreans, but did not permit family reunion. The EPS restricted the sectors for foreign employment to manufacturing, construction, agriculture and fishery (Ministry of Employment and Labor 2010), where employers often prefer male employees. Women wishing to go to Korea therefore often have to seek other options, such as marriage migration (Kim *et al.* 2007; Lee 2008). Unmatched expectations, unequal relationships and communication problems frequently cause conflicts between couples or with in-laws. In recent times, one in four new marriages has been an international marriage, and the number of marriage migrants increased from 25,000 in 2001 to 142,000 in 2011 (Korean Immigration Service 2002; 2012). Many brides are ethnic Koreans from China, but others come from Vietnam, Cambodia and the Philippines.

Between the 1950s and reunification with China in 1997, the former British Crown colony of *Hong Kong* was transformed from a labour-intensive industrial economy to a post-industrial economy based on trade, services and investment. High-skilled temporary immigrant ('expat') workers from North America, Western Europe and India were recruited for finance, management and education. Unskilled workers from China entered illegally in large numbers. After reunification with China in 1997, Hong Kong became a Special Administrative Region (SAR) with its own laws and institutions. Low-skilled workers from Mainland China are officially not admitted to Hong Kong, but also here practices may differ to accommodate labour demand. For instance, some 380,000 mainlanders were allowed in from 1997 to 2004 through family reunion provisions – nearly all women and children. Most of the women were employed as cleaners and restaurant workers (Sze 2007). About two million people born in China lived in Hong Kong in 2017, more than two thirds of its 2.9 million immigrant population. The at least 121,00 Indonesians, 106,000 Filipinos and 17,000 Thais living in Hong Kong are the biggest non-Chinese immigrant groups, mostly women working as domestic helpers in Hong Kong (HKSARG 2011).

Taiwan introduced a foreign labour policy in 1992, permitting recruitment of migrant workers for occupations with severe labour shortages. Duration of employment was limited to two years. Workers came mainly from Thailand, the Philippines, Malaysia and Indonesia. Most recruitment was carried out by labour brokers. Many workers stayed on illegally after two years, or changed jobs to get higher wages and to escape repayments to brokers (Lee and Wang 1996). In 2007, the Taiwanese government set up a 'Direct Hiring Joint Service Center' to assist employers in hiring foreign workers without paying high fees to recruitment agencies. The government signed a Memorandum of Understanding (MOU) with the Philippines government for direct hiring in 2011 and launched an online 'Direct Cross-border Hiring Management Service' system in 2012 (Council of Labor Affairs Taiwan 2012; Wei and Chang 2011). In 2011 there were 461,000 registered foreign residents of whom 365,000 were workers (National Immigration Agency Taiwan 2012).

South-East Asian immigration countries

Reflecting centuries of migrations, colonization and cultural exchanges, South East Asian societies have degrees of ethnic and religious diversity that far outstrip levels of diversity found in East Asian or Western nations. However, governments of South-East Asian immigration countries are therefore often concerned about maintaining ethnic balances, which makes immigration into a politically sensitive issue.

Singapore is a small country lacking in natural resources, which has successfully built a first-world economy through specialization in modern service industries. It relies heavily on the import of labour at all skill levels. Between 1990 and 2010, the non-resident workforce grew from 248,000 to 1,089,000, now more than a third of the total workforce (Ministry of Manpower Singapore 2011). In 2017, about 2.6 million immigrants were estimated to live in Singapore, 46 per cent of its total population of 5.5 million. The vast majority, 2.1 million, of these immigrants are from Asia, particularly Malaysia (1.2 million), but also China (463,000), Indonesia (168,000), India (155,000), Pakistan (133,000) and Bangladesh (83,000), but migrants from all over Asia can be found in Singapore. Singapore has developed as the quintessential Asian 'global city', where people, cultures and cuisines from all over the region meet.

The majority are low-skilled workers (Abella and Ducanes 2009). The government imposes a foreign worker levy for employing unskilled workers and a 25 per cent 'dependency ceiling' (quota) for employing skilled workers to encourage employers to invest in new technology (Ministry of Manpower Singapore 2011; Yue 2011). This has led to downward pressure on migrants' wages, rather than reductions in foreign employment.

Lower-skilled workers are generally not permitted to settle or to bring in their families. Labour migrants tend to work in construction, shipbuilding, transport and services; women mainly in domestic and other services. Due to the rule that ties the unskilled workers to a specific job, which is reminiscent of the Arab *kafala* system (see Box 9.1), employers frequently use the threat of cancelling the visa to force migrants to accept unlawful work conditions (Human Rights Watch 2011).

Malaysia is another industrializing economy in South-East Asia that has become heavily dependent on immigration, particularly from Indonesia. Due to the complex ethnic composition of Malaysian society, immigration has been controversial, and successive governments have struggled to find appropriate approaches. Nevertheless, rapid economic growth since the 1980s has created labour shortages, such as in the plantation and manufacturing sectors. According to UN data, 2017 about 2.7 million foreign born were living in Malaysia, 9 per cent of its population, of whom 1.1 million are from Indonesia, 308,000 from Myanmar, 103,000 from the Philippines and 89,000 from Vietnam. The government imposes levies and quotas to limit the employment of migrant workers to specific sectors (manufacturing and plantations) and from specific countries of origin (Immigration Department of Malaysia 2012). This has boosted irregular migration. Official estimates of undocumented migrants vary from 500,000 to 1.8 million, although employers' associations and trade unions put the figure much higher (Kassim and Zin 2011: 2). Both documented and undocumented migrant workers in Malaysia often face exploitative labour conditions or abusive policing (Amnesty International 2011; Mendoza 2018).

Thailand became an exporter of workers to the Gulf in the 1980s and then to Taiwan, Malaysia, Japan and Singapore in the early 1990s. However, emigration remained relatively low, which seems related to the fact that Thailand has never been colonized and hence lacked extensive historical links with a metropole. Fast economic growth in the 1990s initiated a migration transition, transforming Thailand into a net immigration country. With falling fertility and fast economic growth, many Thais are no longer willing to do '3D jobs' (Skeldon 2006: 285). Construction, agricultural and manufacturing jobs have attracted workers from Myanmar, Cambodia, Laos and Bangladesh. Many people from Myanmar such as the Rohingya, a stateless Muslim minority, have also fled violence in their homeland, mostly to Bangladesh, but also to Thailand, where they often find jobs.

In 2017, the 3.6 million foreign born estimated to live in Thailand represented 5 per cent of the Thai population, forming the third-largest immigrant population in the Asia-Pacific region, after Australia and India. However, many migrants are undocumented. The Thai government has carried out registrations of migrant workers several times – such as in 2009, when about 900,000 migrants (80 per cent from Myanmar) obtained new work permits, and 380,000 renewed their work permits. In 2014, the Thai government started to issue temporary identity cards to undocumented migrants. These cards, commonly known as 'pink cards', give migrants the temporary right to live and work in Thailand and access public health insurance in the provinces in which the cards are issued (Mendoza 2018).

Migration in Australia and New Zealand

Oceania is the world-region with the highest proportion of immigrants: 21 per cent of the population. The great majority of migrants in this region are in Australia (7.0 million in 2017) and New Zealand (1.1 million). The many nations of the Pacific sub-regions of Melanesia, Polynesia and Micronesia such as Papua New Guinea, Fiji, Tonga and Vanuatu are affected by complex patterns of migration. As is the case for most small island states, levels of migration are often very high. In 2017, 72 per cent of migrants from Pacific Islands remained within the region, with New Zealand being the leading country of destination, particularly for migrants from Fiji, Samoa and Tonga, followed by Australia and other Pacific Islands. Canada and the US are home to 25 per cent of the Pacific Islands' migrants. The remainder of this section will focus on Australia and New Zealand. While both countries used to be quintessential European colonial settler societies, the analysis will show that over recent decades these countries have become increasingly integrated into Asia-Pacific migration systems.

Australia

Throughout the post-WWII period, *Australia* has reinforced its position as the region's major destination country. Similar to Canada, Australia is an immigration country by design, originating in British colonial settlement. After the loss of its American colonies, Britain established a new penal colony in New South Wales in 1788, thus starting the colonization process with a form of forced migration. With the end of convict transportation in the mid-nineteenth century, free settlement took over as the main source of immigration. Colonial occupation and migration came at the expense of indigenous 'Aboriginal' populations. Following the British conquest, the indigenous population was dispossessed and mar-

ginalized, with many people dying as a result of massacres, poor living conditions and imported illnesses. Government policies of 'assimila tion' led to the forced removal of Aboriginal and mixed-race children from their families by government agencies and church missions, a prac tice which continued until the 1970s (see Barta 2008). Traditional indigenous ownership of land was legally recognized only in 1992.

Australia initiated a mass immigration programme after 1945, with the official aim of adding the equivalent of 1 per cent of population each year (Collins 1991). The policy, summed up in the popular slogan 'populate or perish', was focused on permanent family immigration in a huge country whose population was only 7.5 million in 1945. The aim was to bring immigrants mainly from Britain. But as this traditional immigration source dried up following rapid post-war economic recovery, Australia began

Photo 8.2   Captain Phillips inspecting convict settlers, Botany Bay. This penal colony was the foundation of Sydney

Source: Getty Images/Time Life Pictures

recruiting post-WWII refugees from the Baltic and Slavic countries, and subsequently labour migrants from Southern Europe. Non-Europeans were not admitted at all, as the White Australia Policy was still in force. From 1947 to 1973 immigration provided 50 per cent of labour force growth, but by the late 1960s, it was becoming hard to attract Southern European migrants. The response was further liberalization of family reunions and recruitment in Yugoslavia and Latin America.

The abolition of the White Australia Policy in 1973 and the introduction of a Canadian-style points-based system opened the door for rapidly increasing Asian immigration, mainly from China, India, the Philippines and Vietnam. Fast-growing Asian immigration coincided with an expansion of temporary worker programmes, in contrast to earlier policies that focused on permanent immigration of entire families from Europe. Many temporary migrants work in the fast-growing mining sector, while others are found in Australia's manufacturing and service industries. The number of overseas students entering Australia has grown even faster. In order to pay high university fees and living costs, many students work up to 20 hours a week, providing a ready labour source for retail and catering businesses.

The difference between temporary and permanent migration is blurred in practice, as legal frameworks for temporary work visa holders and students include pathways to permanent residency. While the Family category has fluctuated between levels of 30,000 and 80,000 over recent decades the Skill category has grown steadily from just 10,100 in 1984–5 to levels well over 110,000 in the 2010s (DIAC 2012; Phillips *et al.* 2010).

Throughout the post-war period and until the present day, successive Australian governments have consistently stimulated immigration to increase the population and to boost economic growth. From the mid-1970s to the early 1990s, there was a political consensus on a non-discriminatory immigration policy and multicultural policies towards ethnic communities. In the post-Cold War era anti-immigration sentiments began to grow, mirroring the growing politicization of migration in Europe. When John Howard became Prime Minister in 1996, leading a coalition of the Liberal and National parties, Australian immigration policy entered a new era, with strong emphasis on recruitment of skilled personnel, cuts in family reunion, draconian measures against asylum seekers and a shift away from multiculturalism. However, despite anti-immigrant rhetoric, both permanent and temporary immigration soared over the 1990–2009 period to levels reaching over 667,000 in 2009. Although the Great Recession led to a temporary decrease in inflows, immigration started to grow after 2010 to reach record high levels of 759,000 immigrants in 2016 (see Figure 8.4).

Australia has a voluntary resettlement programme for refugees selected in overseas camps, yet its response to asylum seekers has become increasingly hostile. Asylum seekers who arrive from countries like Iraq, Afghanistan and Sri Lanka by boat – usually via Indonesia – are detained for long periods in camps, and are labelled as 'queue-jumpers' and 'security threats' by politicians and the media. Although asylum seeker numbers in Australia are relatively low, the recent growth is politically controversial because it is seen as undermining the tradition of strict government control of entries. In 2012, the government decided to send asylum seekers who arrived by boat to await processing in camps on the islands of Nauru and Papua New Guinea (Castles *et al.* 2013). These practices of 'extra-territorial asylum processing' have been criticized by human rights organizations as violating international humanitarian law.

Largely as a result of immigration the Australian population has more than trebled from just 7.6 million in 1947 to an estimated 25.3 million in 2019. The 2011 Census showed an

Aboriginal and Torres Strait Islander population of 550,000 – just 2.4 per cent of the total population. By contrast almost 29 per cent of Australian residents were immigrants in 2017, while many others have immigrant parents. Australia has become one of the world's most diverse countries, with people from close to 200 origin countries. Asian immigration has become increasingly dominant. In 2017, about 3.2 million immigrants were from the Asia-Pacific region (46 per cent of all immigrants), most of whom are from New Zealand (607,000), China (607,000), India (469,000), Vietnam (237,000), the Philippines (246,000), Malaysia (166,000) and Sri Lanka (118,000). Other significant immigrant groups are the 1.2 million migrants from UK as well as migrants born in Italy (210,000), the US (104,000), Germany (124,000) and Greece (117,000) (OECD 2018) (see Box 13.5).

Highlighting increasing global connectedness, emigration from Australia has also increased steadily in recent years, rising from around 88,000 in 1981–2 to 542,000 in 2017 (Productivity Commission 2010: 36; UNPD 2017). It has become an important part of professional or personal experience to live and work abroad. Many Australians go to the UK, the US and New Zealand. Such migrations reflect the linguistic, cultural and economic ties forged by British imperialism. Smaller but increasing numbers of Australians live in Asian business centres like Hong Kong, Singapore, Shanghai and Mumbai. This shows that Australian society is becoming increasingly integrated into Asian economic and migratory systems, and its partial disconnection from events in Britain and (the rest of) Europe.

New Zealand

In the late thirteenth century, Polynesians settled in the uninhabited islands that later were named New Zealand. These settlers would develop a distinctive Māori culture. In 1841, New Zealand became a British colony, which also marked the start of British settlement. Migration has had important consequences for culture, identity and politics in New Zealand. The country has gone from a white settler colony with an indigenous Māori minority to a multi-ethnic society in which about three-quarters are white people of European origin, while around a quarter are Māori, Asians and Pacific islanders. Compared to Australia, New Zealand is more closely integrated in regional migration systems, with a higher degree of migration from Pacific islanders.

In the post-WWII period, New Zealand encouraged 'kin immigration' from the UK, with between 9,000 and 16,000 arriving each year through the 1950s and 1960s. Some white foreigners from other countries were admitted too, mainly from the Netherlands as part of a bilateral recruitment agreement, as well as displaced persons from Eastern Europe. As in Australia, the growing demand for cheap lower-skilled labour in manufacturing stimulated ethnic diversification of immigration, although most new immigrants did not come from Southern Europe as in Australia, but rather from South Pacific islands. Many of these came from New Zealand territories and were not considered foreigners. Temporary foreign workers overstayed their visas, joining the growing Pacific Island population in New Zealand. The economic boom of the early 1970s and governmental attempts to increase the population and labour force led to a record influx of 70,000 persons in 1973–4 (McKinnon 1996).

The full accession of the UK to the European Economic Community (EEC) in 1973 cut free trade ties with Britain and therefore reduced New Zealand's agrarian exports to the UK. The ensuing economic recession and reduced labour demand decreased immigration while unemployment disproportionately affected the Pacific Islander immigrant community. From 1974 to 1979 'Dawn' Raids were carried out by police to remove 'overstayers', mostly Pacific Islanders. From the late-1980s, immigration rebounded following renewed economic growth and new immigration legislation. The New Zealand Parliament passed a new Immigration Act into law in 1987, which introduced a Canadian-style points-based system, which came into effect in 1991. This ended the preference for migrants from Europe or Northern America, instead selecting migrants on their potential contribution to the economy and society. The new points-based system contributed to the further diversification of origin countries and a higher prevalence of temporary migration.

Many recent settlers have come from Asia, especially China, India and the Philippines. Migration from the Pacific Island nations such as Fiji, Samoa, Tonga and the Cook Islands have also been significant, while the UK and South Africa have also remained important source countries. Official government policies still emphasize biculturalism based on the historical relatively peaceful relationship – at least compared to the situation of 'Aboriginals' in Australia – between indigenous Māori and British settlers.

Yet ethnic diversity has become a contentious political issue. For instance, the political party New Zealand First has argued that Asian immigration into New Zealand is too high. Already back in 2004, party leader Peters and former Minister and Minister of Foreign Affairs stated that 'We are being dragged into the status of an Asian colony and it is time that New Zealanders were placed first in their own country', also warning that Māori will be outnumbered by Asian migrants. However, rhetoric would not become as harsh as in Australia, and the politicization of migration would not significantly affect immigration policies on the ground, which have increasingly welcomed economic immigrants.

However, in March 2019, New Zealand was shocked by two terrorist attacks at mosques in Christchurch during Friday prayer, which killed 51 people and injured 49 others. The 28-year-old gunman was a white supremacist and part of the alt-right movement. The attacks led to a nationwide reaction of solidarity.

Since the late 1990s, migration to New Zealand has shown a consistently increasing trend, with annual inflows surging from 27,000 in 1998 to 96,000 in 2016 (OECD data). The foreign-born population had increased to 1.1 million by 2017 (23 per cent of New Zealand's total population). In 2013, the main origin countries of foreign-born people are the United Kingdom (255,000), China (89,000), India (67,000), Australia (63,000), South Africa (54,000), Fiji (53,000), Samoa (51,000) and the Philippines (37,000). Most new immigrants come from China and India, followed by the UK, Australia and the Philippines.

Conclusion

In Australia and New Zealand, there is widespread political recognition that migration has been a major factor shaping society over the last half-century. This is not the case elsewhere in the Asia-Pacific region. Most Asian governments still see migration primarily in economic terms. Governments of destination countries emphasize the temporary nature of labour migration, while origin countries focus on the potential economic benefits of remittances and diaspora investments. Yet despite this utilitarian perspective, Asian migration does not only have economic impacts: it is becoming a major element of demographic, social and political change that is gaining structural and permanent features. However, governments in the region are often slow or reluctant to acknowledge these new realities out of fear of stimulating permanent settlement.

The early twenty-first century has been a period of growing diversity in Asian migration. Virtually all Asian countries experience simultaneous in- and outflows of varying types. Economic migrants can be found at all skill levels: the lower-skilled still migrate out of the region but increasingly also within it. Many high skilled workers and students move to Western countries, but increasing numbers move within Asia, while immigrants from other world regions are attracted to areas of economic growth within Asia. Asian women are in increasing demand in many occupations all over the world, while migration for the purpose of marriage has been growing fast.

New transport and communication technologies have opened the way for growing temporary and circular migration. Often official migration categories do not correspond with social realities. People may visit a country as tourists before deciding to stay or return as migrants. Those who move as permanent settlers may decide to return home, or move back and forth between the country of origin and the destination. Temporary migrants may stay permanently, move repeatedly in both directions or go elsewhere in search of opportunities. New media and ways of communicating facilitate transnational consciousness, with many migrants having affiliations and a sense of belonging in more than one country.

Again it is useful to link the regional experiences to the migration theories reviewed in Chapter 3. Neoclassical ideas on individual economic motives seem to fit well for highly skilled and student migration and possibly also for lower-skilled labour migration. But even here, family and community decision-making processes are often relevant, and these are better explained by the new economics of labour migration approach: families often pool resources and sell assets as land or livestock to finance the migration of one member as a joint investment in a better future for the entire family. The Asian migration experience also highlights the crucial role of labour recruitment in shaping migration patterns. State-sanctioned recruitment has played a central role in enabling the relatively poor to migrate by connecting them to employers overseas.

In a broader perspective, it is clear that the rapid growth of Asian migration can only be understood in the context of the rapid social transformations taking place in the region, along with the uneven nature of these shifts. Many areas have gone through turbulent processes of decolonization, violent conflict and nation-building. Some countries have moved in a short time from low-productivity agricultural economies with traditional social and cultural values, to modern industrial societies. Employment growth has often been linked to rapid demographic shifts, with East Asian countries now exhibiting fertility rates as low as, or lower than, Southern European countries. Others countries such as the Philippines, Myanmar, Nepal and Laos, have changed much more slowly, and still have low average income levels, relatively high fertility and relatively high levels of unemployment or underemployment.

Such transformations give rise both to strong demand for migrant labour in economic growth areas and availability of labour reserves in areas of slower economic growth. Improvements in transport and communications provide the means for migration but also for the growth of transnational communities. The experiences of several Asian countries therefore seem to confirm migration transition theory. While development in poorer Asian countries tends to boost emigration, some of the labour-surplus countries of a generation ago – like South Korea, Thailand and Malaysia – are now poles of attraction. Some former source countries of highly skilled migrants – notably Taiwan, but also South Korea and incipiently China – have successfully reversed the brain drain and are benefiting from the skills of their returnees. In poorer areas of South and Southeast Asia economic development and rising aspirations have rather boosted emigration, a pattern typical for early transitional countries.

As argued in Chapter 4, migration tends to challenge dominant ideas and ideologies on ethnicity, national identity and culture. Asian governments and public opinion often perceive immigration as a threat to their models of the nation-state. The weakness of migration control in some countries contrasts with the dominant official Asian models of migration, based on the principles of strict control of foreign workers, prohibition of settlement and family reunion, and denial of worker rights. East Asian authorities emphasize the importance of maintaining ethnic homogeneity, while most South-East Asian governments wish to safeguard existing ethnic balances. Yet the globalization of migration is bringing about rapid changes and it is far from clear that Asian governments will be able to prevent permanent settlement.

When Western European governments tried to reduce immigration and settlement of foreign populations in the 1970s, they found it difficult: their economies had become structurally dependent on foreign labour, employers wanted stable labour forces, immigrants were protected by strong legal systems and liberal democracies therefore tended to give rights to non-citizens (see also Chapter 11). Such pressures are beginning to make themselves felt in Asia too. There is increasing dependence on foreign workers for the '3D jobs' as ageing kicks in and labour force growth slows, and local workers reject menial tasks. In these circumstances employers seek to retain 'good workers', migrant workers prolong their stays through regular or irregular means, and family reunion or formation of new families in the destination country take place. This applies particularly when migrants have scarce skills – the privileged entry and residence rules for the higher-skilled may well become a factor encouraging permanent settlement and greater ethnic and cultural diversity.

Despite the rapid growth, international movements are still quite small in comparison with Asia's vast population. Migrant workers make up a far smaller proportion of the labour force in countries like Japan and South Korea than in European countries (although the proportion is large in Singapore and Malaysia). However, the potential for growth is obvious. Fast-growing Asian economies such as China, South Korea and Japan may attract increasingly large numbers of migrant workers in the future, while the emigration potential of several Asian countries may decrease in the future as the result of economic growth and ageing. Demand for caregivers is likely to be a major factor in the future, because of population ageing in many destination countries. This will surely affect global migration patterns, with far-reaching social and political consequences for world regions that have been the centre of global migration over the past half century.

For instance, Asian student migration to North America, Europe and Australia might decrease in the future because of the increasing quality of tertiary education in Asia, with Japan, China and Korea all competing for foreign students. Although China is still the number one origin country for international students, increasing numbers of foreign students are coming to China. Japan and South Korea have also experienced substantial growth in foreign students (OECD 2011a). Other Asian countries such as Singapore are investing in their education infrastructure with the explicit purpose of attracting foreign students. At the same time, China's huge investments in higher education and its support for Chinese language education around the world may also increase its role as a destination for student migrants from Asia and beyond. In the longer term, this may challenge the dominant position of North America, Australia and Western European within the international education industry.

The rise of Asia, and particularly China and India, as a global economic powerhouse is likely to contribute to further shifts in migration patterns. For instance, provided that political stability is maintained, China is likely to evolve into an increasingly important regional and global migration destination. Rapid ageing, increasing living standards and rising wages will increase the demand for migrant workers to do jobs for which domestic supply is already getting tight. This will have fundamental repercussion for migration patterns on the regional and indeed global level. While the emigration potential of several Asian countries is likely to decrease in the future, this may lead to increasing global migration towards Asia and the rise of new sources of global migrant labour outside Asia.

Guide to Further Reading

Literature on Asian migration has grown exponentially, yet are few single works that provides a comprehensive treatment. Amrith (2011) provide an overview of historical and contemporary migration trends in Asia. Hugo (2016) and Skeldon (2006) give overviews of the linkages between internal and international migration in East and Southeast Asia, Ananta and Arifin (2004) contains many useful chapters about migration in Southeast Asia. Xiang, Yeoh and Toyota (2013) is an in-depth inquiry into the role of return migration in shaping Asian models of nation, state and development. The chapters on Japan and Korea in Hollifield *et al.* (2013) present useful summaries. Mallee and Pieke (2014) contains various useful chapters on internal and international migration in China. The chapters in Douglas and Roberts (2000) give a detailed insight into how migration challenges Japan's identity as a homogenous society. Other good studies in English on Japan are Komai (1995) and Mori (1997), and Weiner and Hanami (1998) provide other good studies in English. For most other countries, academic journal articles are still the best sources, while national statistical offices provide important data. The website of the Colombo Process (http://www.colomboprocess.org) gives summaries on migration in Asian countries. The IOM *World Migration Reports* provide regional overviews. The Migration Policy Institute (migrationpolicy.org) has many country resources.

Extra resources can be found at: **www.age-of-migration.com**

9 Migration in Africa and the Middle East

Africa and the Middle East have gone through profound political and economic transformations since the end of the Second World War. Many countries have experienced the tumultuous formation of new nation states in the wake of decolonization, and conflicts related to access to oil reserves and other natural resources, alongside more general processes of social transformation and globalization. Some countries, particularly oil-rich countries in the Gulf region, have become extraordinarily wealthy, while other countries, such as in the Horn of Africa and the Great Lakes regions have experienced frequent violence and high poverty. Other countries, such as Turkey, Morocco and South Africa occupy more intermediate positions, with modest levels of development and relatively less violence.

These divergent developments have also shaped divergent migration trends. While Morocco and Turkey have evolved into prime Western European sources of migrant labour, the oil-rich countries in the Gulf and Libya have become an important destination for migrants from poorer countries in the Middle East, Egypt, sub-Saharan Africa and Asia (see Fargues 2011a). However, and with the exception of the *Maghreb* countries (Morocco, Algeria and Tunisia) and Turkey, most migration takes place *within* Africa and the Middle East. This particularly applies to sub-Saharan Africa. Although images of crossings by African migrants in boats across the Mediterranean give the impression of massive irregular migration, the large majority of Africans migrate legally, and only a fraction of migration originating in Africa results in journeys to Europe, the Gulf, the US and beyond (Bakewell and de Haas 2007). African countries as diverse as Côte d'Ivoire, Nigeria, Ghana, Gabon, Libya, Kenya and South Africa have been important migration destinations in their own right.

In the post-1945 period, countries in Africa and the Middle East have been major sources *and destinations* of asylum seekers and refugees as a consequence of political oppression and violent conflict. Upheavals leading to forced migration have included the Israeli–Palestinian conflict, the various wars involving Iraq (and nearby Afghanistan), the conflict around the Western Sahara since 1975, the Somali civil war (since 1980), the wars between North and South Sudan, the West African civil wars in Sierra Leone and Liberia in the 1990s and early 2000s, the recurrent conflicts in the Great Lakes regions, and violent conflicts in Libya, Syria and Yemen in the wake of the Arab Spring. Contrary to popular beliefs, most refugees stay in the region, and only a minority end up in Europe, North America or other overseas destinations.

We can distinguish the following phases with regard to migration trends in Africa and the Middle East:

- 1945–1973: Return and expulsion of colonial settlers, decreasing intraregional migration in the wake of post-colonial state formation and protectionism, large-scale labour migration from the Maghreb and Turkey to Western Europe;

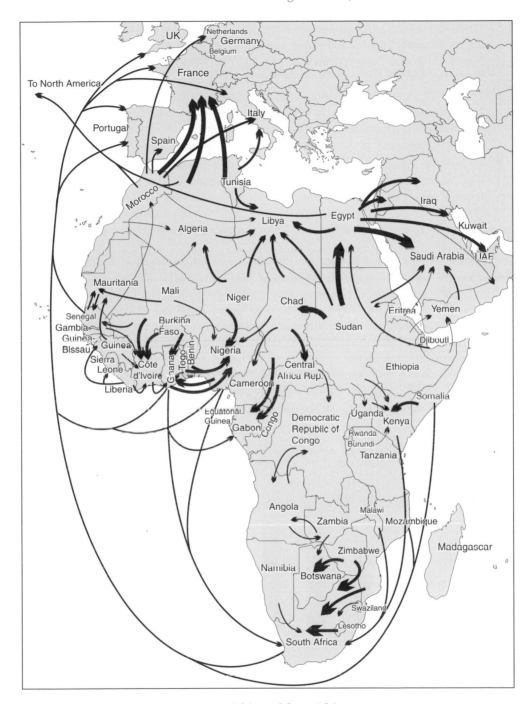

Map 9.1   Contemporary migrations within and from Africa

Note: The arrow dimensions give an approximate indication of the volume of flows in the 1945–2015 period. Exact figures are often unavailable

- 1973–1989: Increasing intraregional migration to the fast-growing oil economies in the Gulf, Libya, Nigeria and Gabon; continuation of Europe-bound migration from the Maghreb and Turkey, particularly through family migration; increasingly refugee migration particularly in the Horn of Africa;

- 1989–2008: Increasing long-distance migration within Africa such as to South Africa and Libya; Asian migration to the Gulf; increasing West African migration to the Maghreb and Southern Europe; increasing reliance on irregular migration from North Africa to Europe in reaction to border restrictions; refugee migration in the Great Lakes district and from Iraq;

- Since 2008: Increasing extra-regional migration from sub-Saharan Africa as a result of social transformations, globalization and economic liberalization; evolution of middle-income countries such as Turkey, Lebanon, Jordan and Morocco into migration destinations; increasing refugee migration and internal displacement following the 2011 Arab Spring, particularly from Syria; increasing Chinese-African migration.

As Map 9.1 illustrates, the bulk of African migrants move within the continent. In fact, Africa has the lowest levels of intercontinental migration of all world regions (Flahaux and de Haas 2016) and this particularly applies to the most isolated and poorest countries in sub-Saharan Africa. Figure 9.1 shows that in the post-WWII period African immigration has been largely stagnant and predominantly intraregional. Migration levels have actually declined in relative terms, and in 2017 only 1.7 per cent of the African population was estimated to be an immigrant.

Figure 9.2 shows that also emigration has been relatively low, although higher than immigration. While absolute numbers have increased, emigrants represent between 2.5

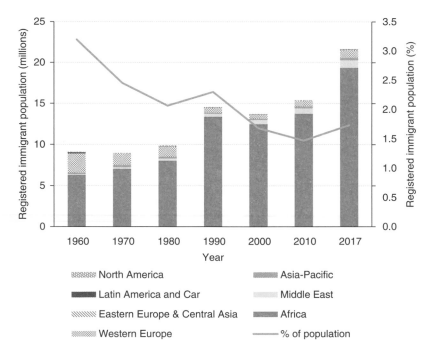

Figure 9.1   Registered immigrant population in Africa, by origin region
Sources: UNDESA 2017 and Global Bilateral Migration Database

and 3 per cent of the African population. The bulk of African migrants living in Europe and the Gulf are from the Maghreb and Egypt, respectively. These figures highlight the fact that sub-Saharan Africa is the least migratory region in the world, although extra-regional migration has increased in recent years. Figure 9.4 shows the fast increase in migrant populations in the Middle Eastern regions, which mainly reflects the rise of the Gulf as a global migration destination. Figure 9.5 shows that emigration from the Middle East to Europe stagnated over the 1990s and 2000s, which partly echoes declining EU-bound emigration from Turkey. However, in the 2010s there has been a fast increase in intraregional migration, particularly as a result of conflict-related movements from Syria.

The historical roots of contemporary African migrations

The Middle East and large parts of Africa have historically been regions of high non-migratory mobility. The existence of large arid and semi-arid areas encouraged the per-sistence of nomadic and semi-nomadic ways of life, while the presence of numerous holy places led to pilgrimages (Chiffoleau 2003). A long history of large empires and ill-defined borders fostered the exchange of goods and knowledge (Laurens 2005: 25–27), but also led to displacement. Such migrations were often temporary or circular in nature. More permanent migrations were often driven by warfare, population growth and economic factors. One of the greatest migrations in human history was that of the Bantu people, who left the area now encompassing Nigeria and Cameroon and formed settlements throughout the entire southern half of the continent, bringing their languages and joining with indigenous groups along the way.

Beginning in the sixteenth century, over three centuries of the Atlantic slave trade resulted in the forced migration of 12 million slaves from the continent (Lovejoy, 1989 see Chapter 5). The legacies of European colonialism would lay the groundwork for many of

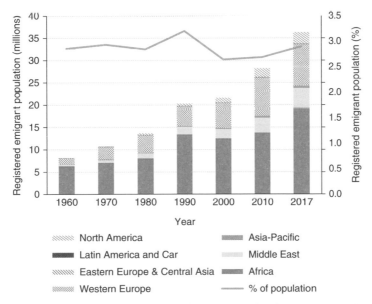

Figure 9.2 Registered emigrant population from Africa, by destination region
Sources: UNDESA 2017 and Global Bilateral Migration Database

the migration patterns that followed. From the nineteenth century, previously free mobility decreased because of colonization, the drawing of formal state borders and increased state regulation. Nomadic and semi-nomadic populations became the primary victims as states implemented sedentarization policies in order to control and tax such groups. Likewise, states started to regulate the movement of pilgrims and traders.

Within Africa, North Africa and sub-Saharan Africa have historically been linked through political and economic relationships (de Haas 2006a; McDougall and Scheele 2012; Scheele 2012a). Throughout known history, there has been intensive mobility between both sides of the Sahara through the trans-Saharan caravan trade in goods and slaves, conquest, pilgrimage and religious education (see Berriane 2012). The Maghreb countries have deep historical ties with West African countries such as Mali and Senegal. The Sahara itself is a huge transition zone, and the diverse ethnic composition of Saharan oases – with their blend of sub-Saharan, Berber, Arab and Jewish influences – testifies to this long history of population mobility (Bakewell and de Haas 2007: 96; Lightfoot and Miller 1996: 78; McDougall and Scheele 2012; Scheele 2010). Today, ancient caravan routes are once again migration routes for Africans crossing the Sahara (Brachet 2012).

For centuries, the various empires of the Middle East and North Africa used migration and population displacement as strategic tools to stabilize and control newly conquered lands. For instance, as the Ottoman Empire expanded, the government ordered Muslim subjects to settle in recently acquired lands, a process known as *surgun* (Tekeli 1994: 204–206). During the last century of the Ottoman Empire from the 1820s to the 1920s, approximately 5 million people sought refuge in the Empire while several million people fled from it (McCarthy 1995). With the contraction of the Ottoman Empire and the creation of new nation-states in its wake, policies of national preference developed. The concurrent expulsion of 'non-nationals' and welcoming of 'nationals' resulted in large-scale, often extremely violent, population transfers (Mutluer 2003: 88–94).

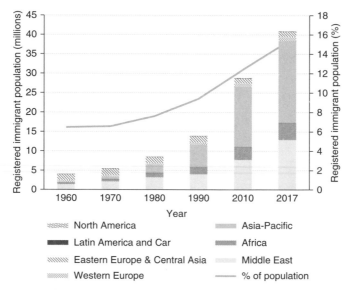

Figure 9.3   Registered immigrant population in the Middle East, by origin region
Sources: UNDESA 2017 and Global Bilateral Migration Database

The drive to create 'homogeneous' nations reflecting official nationalist myths became institutionalized in the immigration policies. The creation of the Turkish Republic out of the ashes of the Ottoman Empire in the early 1920s involved the use of migration as a tool for building national identity: ethnic groups accepted as part of the multi-ethnic Ottoman Empire, like the Greeks and the Armenians, were expelled or persecuted, while people of Turkish ethnicity and Sunni Muslim religion were welcomed as settlers (Kirişçi 2007). This led to the 1923–4 'population exchange' between Greece and Turkey that implied the forced migration of hundreds of thousands of people (Mutluer 2003: 88–94). In 1934, Turkey's 'ethno-state' promulgated the Law of Resettlement, which authorized ethnic Turks from areas formerly comprising the Ottoman Empire to emigrate to and settle in the Turkish Republic (Tekeli 1994: 217). This policy continues to the present: as in the 1980s for instance, 310,000 ethnic Turks from Bulgaria fled to Turkey to avoid persecution, though many of them later returned to Bulgaria.

With the exception of Turkey and a few other countries such as Ethiopia that were never formally colonized, colonial practices would deeply impact African and Middle-Eastern mobility. The nineteenth- and early twentieth-century division of Africa and the Middle East into politico-administrative entities often imposed arbitrary borders, sometimes dividing established nations (Davidson 1992). As a result, members of a single ethnic group could become citizens of two or more states and many new countries have a high degree of ethnic diversity, which often complicated post-colonial nation state building, while people of co-ethnic groups continue to cross national boundaries which often have little real meaning for them.

The colonial period brought European administrators, farmers and entrepreneurs throughout Africa as well as indentured workers from the Indian subcontinent to East

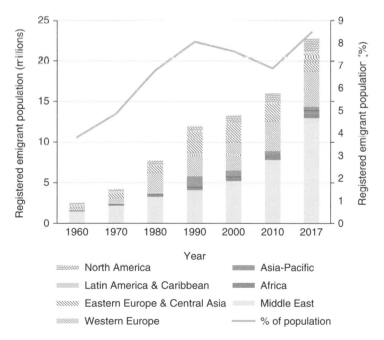

Figure 9.4   Registered emigrant population from the Middle East, by destination region
Sources: UNDESA 2017 and Global Bilateral Migration Database

and Southern Africa, alongside other settlers, such as Syro-Lebanese merchants to West Africa (Leichtman 2005). These Asian and Lebanese populations generally became privileged but vulnerable minorities, often with a key role in trade. Particularly in Africa, colonialism was always concerned with control of mobility, in order to provide labour for European-owned plantations, industry, infrastructure works and mines (Bakewell 2008; Cohen 1987). Although colonial enterprise authorities welcomed low-cost labour, recruitment was legally framed as temporary migration, since permanent concentrations of migrant workers were seen as a potential threat to order. Yet colonial administrators could not prevent permanent settlement.

In the Gulf region, colonial efforts to prevent settlement largely failed as well. In the British dominated city-states on the Gulf, the United Arab Emirate (UAE), Bahrain and Qatar, and in the new state of Saudi Arabia (created in 1932), the influence of British and American oil companies was felt through the management of labour recruitment (Thiollet 2016). From the 1930s, the nascent oil industry created an increasing demand for foreign labour. Technical and managerial staff were recruited from Britain or America while lower-skilled workers were recruited as indentured workers, primarily from India but also Eritrea, Italy and Arab countries. Although indentured labour, which entailed housing segregation and exploitation, was perceived as temporary, in practice many indentured workers settled in their host countries (Thiollet 2016).

The fundamental transformations that affected African and Middle Eastern societies and economies in the colonial era would engender fundamental changes in migrations that continue until the present day. For instance, independent states would continue policies aiming at the sedentarization of nomadic and semi-nomadic populations, and their partial conversion into peasants (Bencherifa 1996; De Bruijn and Van Dijk 2003; Fratkin and Roth 2006; Waithanji 2008). These transformations also set in motion a gradual but irreversible shift from agrarian to urban livelihoods, coinciding with increasing rural-to-urban migration and the concomitant growth of towns and cities.

Postcolonial and guestworker migration (1945–1973)

Most African and Middle Eastern nations achieved independence in the period between 1950 and 1970. During the period of colonial liberation, millions of Africans fled brutal conflicts with colonial powers reluctant to relinquish control (for example, Algeria, Kenya, Congo) or with white settler groups determined to cling to their privileges (for example, Zimbabwe and South Africa). Independence often led to a departure of European settlers, colonial administrators and military, such as from 'British Africa' to the UK and elsewhere in the English-speaking worlds, from French West Africa and the Maghreb back to the 'metropole', and from Portuguese colonies back to Portugal. In some cases, significant white populations would remain, such as in Kenya and Zimbabwe.

Such processes were often violent and sometimes involved the uprooting of entire population groups. In 1962, for instance, over one million *colons* (descendants of European settlers) and *harkis* (Algerians who served with the French army in the war of independence) left Algeria after the National Liberation Front (FLN) succeeded in pushing French colonizers out after an eight year-long independence struggle (Collyer 2003). Resettled in France, this population of *pieds-noirs* lived through experiences of double alienation, longing for their native land as well as feeling removed from mainstream French culture. Nurturing resentment over the perceived French abandonment of its colony, *pieds-noirs* would later form one of the key constituencies of the extreme right-wing *Front National* party (see Comtat 2009).

The significant political turmoil around independence and the drive by new independent governments to assert sovereignty and create a 'nation' generally encouraged the emigration of European settlers and minorities, with declining population diversity as a consequence. With the rise of post-colonial nationalism and economic protectionism, immigration and intraregional mobility decreased (Flahaux and de Haas 2016). In several countries, post-colonial state formation was associated with rising nationalism, xenophobia and decreasing tolerance towards minority ethnic or religious groups. This encouraged the emigration, expulsion or flight of several minority groups, such as Indians from Uganda to the UK and Kenya in 1972, and the large-scale emigration of hundreds of thousands of North African and Middle Eastern *Sefardi* Jews to Israel, France and Québec (Canada) (see Kenbib 1999).

Since the 1960s, Turkey and the Maghreb have experienced significant recruitment-based and spontaneous migration of workers to Western Europe. France, Germany, the Netherlands, Austria, Sweden, Denmark, Switzerland and Belgium started to actively recruit workers in Turkey, Morocco and Tunisia. Migration of workers from Algeria to France was generally more spontaneous, while French industries and mines also recruited workers in Mali and Senegal. From the 1980s and 1990s Spain and Italy also started to attract increasing numbers of migrant workers from the Maghreb and also West Africa. Large-scale labour migration led to the formation of extensive trans-Mediterranean migration systems. Network connections facilitated family reunification which explains the continuation of migration despite the recruitment stop and the imposition of immigration restrictions after 1973 (Berriane *et al.* 2018; de Haas 2014b; Natter 2014).

Except for Turkey, the Maghreb and a few other countries such as South Africa, most migration from Africa and the Middle East would remain intraregional. Extra-continental migration from sub Saharan Africa and the Middle East would remain largely limited to high-skilled workers, students and refugees who could afford the costs of moving. Patterns of colonization strongly shaped migration patterns from Africa to Europe, with Congolese migrating to Belgium, Senegalese and Malians to France, Nigerians, Ghanaians and South Africans to the UK, and Cape Verdeans and Angolans to Portugal.

Intraregional migration: Conflict, urbanization and economic transformation

The defeat of old-style colonialism and the establishment of independent states often did not mean a return to peaceful conditions. During the Cold War, East and West fought proxy wars in Africa and the Middle East. Political and economic pressures, arms supplies, mercenaries and direct military intervention were factors contributing to new conflicts or the continuation of old ones (Zolberg *et al.* 1989). Struggles for domination in Angola, Mozambique, Ethiopia, Eritrea and Yemen involved massive external involvement, with great human costs for local populations. The great majority of refugees would stay within their regions. Many African countries received refugees, often in the middle of their own conflicts: Uganda admitted Rwandans, Burundians and Sudanese; Eritreans and Ugandans went to Sudan; and Burundians, Rwandans, Congolese and Somalis to Tanzania. There were also significant conflict-related long-distance movements of refugees and migrant workers, such as from the Horn of Africa (Somalia, Eritrea and Ethiopia) to Yemen (Thiollet 2014) while significant numbers of Sudanese, Somalis and other Africans moved to Egypt, Syria, Jordan and Israel.

In the Middle East, refugee issues have long remained centred on the *Palestinians* who had been displaced during the creation of the state of *Israel* in 1948. During the

Palestine War, 711,000 out of around 900,000 Muslim and Christian Palestine Arabs fled or were expelled from the territories that became the State of Israel and fled to the West Bank, the Gaza Strip, Lebanon, Syria and Jordan, and countries further afield. The creation of a Jewish state after the Holocaust (during which around 6 million Jews were murdered, around two thirds of the Jewish population of Europe, see Chapter 5) thus led to mass displacement of Palestinians. During the Six-Day War in 1967, hundreds of thousands of Palestinians fled from the territories occupied by Israel in the West Bank, the Gaza strip and the Golan Heights into Jordan, Egypt and Syria. The tensions and nationalism engendered by the Arab–Israeli conflict encouraged the massive emigration of Jews from Arab countries and the virtual disappearance of the sizeable Jewish communities that had been part and parcel of North African and Middle Eastern societies for centuries (Shulewitz 2000).

However, despite the relatively high incidence of conflict-related migration, economic migration predominated. In Africa, important regional migration systems evolved, centring on areas of economic growth such as Libya in the North, Côte d'Ivoire, Ghana and Gabon in the West, Kenya and Mozambique in the East, and Angola, Botswana and South Africa in the South (Bakewell and de Haas 2007: 96). This partly continued migration patterns that were already established in the colonial period. The tendency towards urbanization continued, which led to increasing migration to the cities. This would lead to a fundamental reconfiguration of pre-modern migration patterns, and the emergence of new migration systems centred around fast-growing urban clusters, often located in or close to urban areas, capital cities or mining regions.

Intraregional mobility in West Africa has been dominated by a movement from landlocked countries of Sahel West Africa (Mali, Burkina Faso, Niger and Chad) to the relatively more prosperous plantations, mines and cities of coastal West Africa (predominantly Côte d'Ivoire, Liberia, Ghana, Nigeria, Senegal and The Gambia) (Arthur 1991; Findley 2004; Kress 2006). There was also considerable transversal international migration *within* the coastal zone of mostly seasonal workers to the relatively wealthy economies of Côte d'Ivoire, Ghana (before the 1970s) and oil-rich Nigeria (since the 1970s).

Such coast-bound international migration patterns have often been reproduced inside African countries such as Nigeria and Ghana, with internal migrants moving from (semi-) arid and underdeveloped inland zones to the more humid and prosperous agricultural and urbanized zones, generally located in coastal areas. This highlights the partially artificial nature of distinctions between internal and international migration. Some inland cities such as Kano in Nigeria or the new, centrally located capitals of Yamoussoukro (Côte d'Ivoire) and Abuja (Nigeria) also became internal migration destinations. Also mining areas such as the goldmines of Witwatersrand near Johannesburg in South Africa, the Copperbelt in Zambia, the Southern Katanga province of the Democratic Republic of the Congo and the diamond mines of Sierra Leone attracted settlers and migrants from nearby and more distant places.

International migration in sub-Saharan Africa has largely been voluntary and unregulated. Following patterns elsewhere in the world, in periods of rapid growth, governments have often welcomed labour migrants, while in times of economic crisis migrants have often been expelled in large numbers. One scholar has enumerated 23 mass expulsions of migrants conducted by 16 different African states between 1958 and 1996 (Adepoju 2001). For instance, in the 1950s and 1960s many West African migrants moved to Ghana. After the 1966 coup in Ghana and the subsequent economic decline, the immigrant community became a scapegoat. In 1969 the Ghanaian government enacted the Aliens Compliance Order, leading to a mass expulsion of some 200,000 migrants, mainly from Nigeria, Togo, Burkina Faso and Niger (Van Hear 1998: 73–74).

In the Middle East and North Africa, intraregional migration has been dominated by rural-to-urban migration, fuelled by processes of economic transition. This reinforced the rapid growth of major cities such as Istanbul, Ankara, Aleppo, Damascus, Amman, Cairo and Casablanca. Migration *between* countries in North Africa and the Middle East has often remained limited because of the severe migration restrictions introduced by countries. Such border closure contributed to declining intraregional migration and an increasing orientation of migration out of the region, particularly to Europe and the Middle East.

Box 9.1 The system of sponsorship (*kafala*) in the Gulf

The sponsorship system has been a central feature of immigration policy in Gulf countries. Stemming from local legalization and British colonial practices (AlShehabi 2019), the sponsorship system was originally based on an agreement between the local *emir* (prince) and foreign oil companies in which a *kafil* (sponsor) would find trustworthy men (usually Bedouins) to work on the oil sites. With the oil industry taking off and the national workforce insufficient to fulfil the needs for manpower, the sponsors (*kufala* in Arabic) had to recruit men from abroad. With time, recruiting and 'sponsoring' foreign workers became the main activities of the *kafala*. Today, in order to enter a Gulf country, migrants must find a sponsor. This requirement applies to various forms of migration including construction workers, domestic servants, foreign tradesmen and businessmen (Rycs 2005). The *kafil* and employer are often the same entity, for instance in the case of domestic workers, but it can also be a larger firm or a recruitment company, or an individual who extracts a rent from his 'citizenship' as a migration broker (Thiollet 2016). Princes and princesses enjoy more 'sponsorship' prerogatives and can sponsor many more migrant workers than lay people can (Thiollet 2016).

The sponsorship system externalizes state control over migrants' entry and residence in the Gulf countries (Thiollet 2016). Sponsors regularly exploit migrants by denying them proper wages and conditions, and retaining their passports, or threatening to report them to the police. Employment contracts are often illegally sold on to other employers (Rycs 2005). Yet the sponsorship system also provides a margin for manoeuvre for migrants and can protect them from state control and sometimes offer informal social inclusion (Beaugé 1986; Vora and Koch 2015).

Migrant workers in the Gulf, Jordan, Lebanon and Israel face precarious conditions, often lack rights and are regularly deported (Agunias 2012; Baldwin-Edwards 2005; Jureidini 2003). Many migrants have low education, particularly in the domestic sector, which increases their vulnerability to abuse (Agunias 2012). Since the 1990s, Gulf states including Saudi Arabia, Bahrain and UAE have tried to reform the *kafala* system. For instance, in 2009 Bahrain introduced legislation allowing workers to switch from one employer to another. However, reforms are difficult to implement, mainly because of employers' and other actors' vested interests in maintaining a captive labour force and extracting rents from citizenship (Diop *et al.* 2015).

The rise of the Gulf as a global migration magnet (1973–1989)

The 1973 Oil Shock triggered a reconfiguration of regional and indeed global migration patterns through the rise of the oil-rich Gulf region as a migration destination. The sudden rise in the price of oil generated financial resources to undertake major construction and infrastructure projects, requiring the hiring of thousands of foreign workers. This facilitated the rapid development of the Gulf into a new global migration destination (Sell 1988). Initially, the booming Gulf economies attracted migrants from Arab countries, particularly Egypt, Sudan, Palestine, Syria, Jordan and Yemen. The Gulf economies also started to attract migrants from nearby sub-Saharan African countries such as Somalia, Eritrea and Ethiopia. In parallel, a sub-regional migration system evolved in North Africa around oil-rich Libya, which particularly attracted migrant workers from Egypt, but also other countries such as Sudan and Algeria.

Initially, the common language, religion and culture of Arab migrants were seen as benefits by the states of the Gulf Cooperation Council (GCC), and this also fitted within the pan-Arab ideology prevailing in the 1950s and 1960s (Thiollet 2011). However, over the 1970s, the Gulf monarchies grew worried about the possible political repercussions of their growing Arab migrant populations, who increasingly protested against discrimination by their (fellow Arab) employers or who saw the autocratic regimes of Gulf countries as illegitimate. Palestinians, in particular, were viewed as subversive and politically overactive (Chalcraft 2011). For instance, Palestinians were involved in efforts to organize strikes in Saudi oil fields and in civil strife in Jordan and Lebanon. Yemenis were implicated in various anti-regime activities in Saudi Arabia (Halliday 1985: 674). Non-Saudi Arabs were involved in the 1979 attack on Mecca. In 1987, Saudi Arabia started to limit the number of pilgrims travelling to Mecca and Medina by means of country of origin quotas (Chiffoleau 2003).

One result was the increased recruitment of workers from South and South-East Asia, who were seen as less likely to get involved in politics and easier to control. In the Gulf countries, labour rights were quasi non-existent until the 2000s and the use of the *kafala* system (see Box 9.1), that ties workers to a sponsor, usually their employer, reinforces the vulnerability of migrants (Lavergne 2003; Longva 1999). Partly because Arab workers were seen as a political liability, the proportion of Arab workers in the immigrant workforce started to decline. The share of women among immigrants also increased. This feminization of migration primarily reflected increasing inflows of Asian women working as care workers and domestic servants (see also Chapter 8).

In 1975 there were around two million migrant workers in the region, 68 per cent of whom were from Arab countries and the rest predominantly from Asia (Thiollet 2011). In 1983 the number of foreign workers had increased to five million, of whom 55 per cent were Arabs. Between 1975 and 1985, the relative share of the Arab foreign workforce in Saudi Arabia dropped from 90 to 32 per cent. Asian migration would comprise an increasing diversity of nationalities, besides India and Pakistan, also including South Korea, Taiwan, Indonesia, the Philippines and Thailand (Thiollet 2016).

Several non-oil-producing states in North Africa and the Middle East, such as Lebanon, Jordan and Israel, would also become increasingly dependent on immigrant labour. In Jordan during the mid-1970s, approximately 40 per cent of the domestic workforce were employed abroad, primarily in the Gulf (Seccombe 1986: 378). This outflow prompted a replacement migration of foreign workers into Jordan, including inflows of

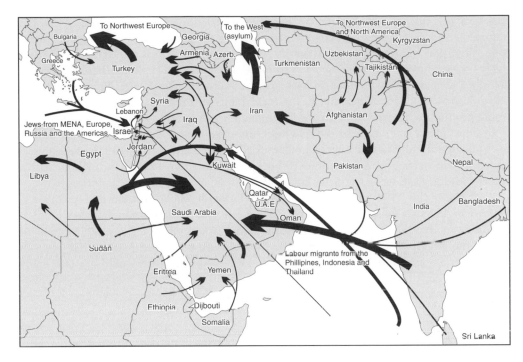

Map 9.2 Migrations within, from and to the Middle East

Note: The arrow dimensions give an approximate indication of the volume of flows in the 1945–2015 period. Exact figures are often unavailable

unskilled Egyptians and Syrians. On the other side of the Jordan River, the Israeli labour market was opened up to workers from Gaza and the West Bank after the 1967 war, as part of a strategy to integrate the occupied territories into the Israeli economy (Aronson 1990). Most of the workers had to commute daily to work in Israel and were required to leave each evening. Palestinians found jobs primarily in construction, agriculture, hotels, restaurants and domestic services (Semyonov and Lewin-Epstein 1987).

Diversifying migration in Africa

The Oil Shock did not only boost migration the Gulf, but would also stimulate fast economic growth and trigger migration to oil-rich countries in Africa, such as Libya, Nigeria and Gabon. With Nigeria's new oil wealth after 1973, millions of Ghanaians and other West Africans sought work there. But corruption and misguided economic policies precipitated a crisis, and in 1983–5 an estimated 2 million low-skilled West Africans were deported from Nigeria, including over one million Ghanaians (Bakewell and de Haas 2007: 104; Van Hear 1998: 73–74). Meanwhile, migration continued to traditional African migration hubs such as Côte d'Ivoire and South Africa.

In Europe, the Oil Shock led to a series of recessions and rising unemployment, and the expectation was that most Maghrebi and Turkish 'guestworkers' would return. However, migration continued, mainly as a consequence of family migration (see Chapter 6).

Paradoxically, this was encouraged by migration restrictions, which interrupted circular migration and encouraged permanent settlement of workers as well as irregular migration. Migration from Turkey and Maghreb countries to the Gulf would remain comparatively limited, mainly because migration from these countries had already become focused on more attractive destinations in Western Europe since the 1960s. From the 1980s onward, increasing labour demand in the formal and informal sectors of Spain and Italy would prompt a surge in Moroccan and Tunisian labour migration to Southern Europe.

Besides labour migration to the new oil economies in the Gulf and Africa as well as continued family and labour migration from the Maghreb and Turkey to Europe, political upheaval and warfare would continue to produce refugee movements throughout Africa and the Middle East, while the situations of Palestinian refugees further deteriorated.

In *Ethiopia* the 1974 revolution overthrew Emperor Haile Selassie and installed a Marxist regime. Forced resettlement, ethnic violence and humanitarian disasters induced hundreds of thousands of Ethiopians to flee. This would prompt the establishment of one of the largest global refugee diasporas. In addition, proxy conflict erupted between Ethiopia and Somalia in 1977, fuelled by Cold War rivalries, which would displace many more. In 1985, the Ethiopian DERG regime began to implement a *villagization* policy involving the forced relocation of an estimated 12 million Ethiopians to villages by 1988 (Terrazas 2007). Although villagization officially served a developmental agenda; in reality they served political purposes, such as the desire to move people away from areas where they could have lent support to anti-government guerrillas into concentrated settlements where they could be more effectively controlled (McDowell and De Haan 1997:12).

In 1975, a violent conflict broke out between *Morocco* and the Polisario independence movement over control of the *Western Sahara*. This led to the forced migration of thousands of Sahrawis to Algerian refugee camps. Since a United Nations-sponsored ceasefire agreement in 1991, the Western Sahara has been under Moroccan control, and the conflict has remained unresolved so far.

African migration in the post-Cold War era

After 1989, a number of political transformations, most notably the end of the Cold War, the 1990–1 Gulf War and the dismantling of South Africa's *apartheid* regime in 1994 coincided with significant changes in migration patterns in Africa and the Middle East:

- Increasing intra-African labour migration to destinations such as South Africa, Gabon, Libya and Equatorial Guinea;
- Further intensification of labour migration from Asia and Africa to the Gulf;
- Refugee migration following the US-led invasion of Iraq in 2003.

The increasing volume and scope of intra-African migration

Although refugee migration receives most attention, the vast majority of people in sub-Saharan Africa migrate for reasons of work, family and study. This also applies to regions known for high levels of conflict such as the Horn of Africa, East Africa and the Great Lakes Region. After a long period of conflict, economic stagnation and border closure, after 1989 intra-African migration would increase, going along with a diversification and increasing geographical scope of destinations and origins (see also Flahaux and de Haas

Source: Oliver Bakewell

Photo 9.1   A Nigerian dress-maker speaks to a Zambian customer in Lusaka, Zambia

2016). While traditional destination countries such as South Africa, Côte d'Ivoire and Nigeria as well as the oil-rich countries of Gabon, Angola, Libya as well as Equatorial Guinea have remained important destinations, other countries – or rather regions within countries – have also attracted migrants. For instance, the Southern Katanga province of the Democratic Republic of the Congo has attracted West African migrants. In East Africa, Djibouti, Uganda, Kenya and Sudan have significant immigrant populations. An increasing number of Ethiopians have found their way to Kenya, South Africa and the Gulf in search of work. Figure 9.5 suggests that South Africa and Côte d'Ivoire have the continent's largest immigrant populations, while sizeable populations are also found elsewhere. In relative terms, levels are highest in oil-rich Libya. However, data is limited particularly with regards to undocumented migration, and real migrant numbers may therefore be higher.

Since the end of apartheid in 1994, *South Africa*, the economic powerhouse of sub-Saharan Africa, started drawing in increasing numbers of migrants. The roots of migration go back to the mine labour system developed between 1890 and 1920 to provide workers for the gold and diamond mines (Cohen 1987). During the apartheid period workers were recruited from Mozambique, Botswana, Lesotho, Swaziland and Malawi, but were expected to go back. In the post-apartheid era, South Africa has attracted migrants from an increasingly diverse array of African countries including Somalia, Democratic Republic of Congo, Kenya, Ghana and Nigeria. Many brought qualifications and experience in medicine, education, administration and business. Others joined the informal economy as hawkers, street food-sellers or petty traders. In Zimbabwe, hundreds of

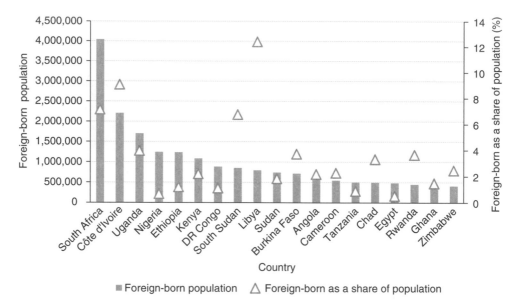

Figure 9.5   Foreign-born populations in African countries, 2017
Source: UNDESA 2017

thousands of people fled deteriorating economic conditions and political repression after 2000, seeking refuge in South Africa (Crush 2008).

As part of its transition from apartheid, South Africa adopted a new constitution which gives strong guarantees on migrant rights. However, in practice governments have often failed to provide migrants with access to residency rights, healthcare and education. Corruption and abuse complicate the enforcement of immigration laws (Vigneswaran 2012). Xenophobia and racist violence have become a major problem (see Gordon 2016; Landau and Freemantle 2009; Vigneswaran 2016). The government introduced heavy penalties for irregular immigration and, since 1994 millions of undocumented migrants have been deported to neighbouring states such as Mozambique, Zimbabwe and Lesotho (Crush 2008).

Over this period, West African migration patterns changed as a consequence of economic stagnation and civil wars in Sierra Leone (1991–2001), Liberia (1989–1996 and 1999–2003), Guinea (1999–2000) and Côte d'Ivoire (2002–2004) (Bakewell and de Haas 2007). This prompted large-scale movements of refugees and internally displaced people (IDPs), and labour migration patterns were temporarily disrupted. In Côte d'Ivoire, West Africa's most prominent migration destination, instability and the launch of an anti-foreigner campaign resulted in over 365,000 persons returning from Côte d'Ivoire to Burkina Faso in 2006 and 2007 (Kress 2006). However, many migrants stayed in Côte d'Ivoire, either because they had integrated into local society or had no real options for return. Migration to Côte d'Ivoire resumed following the end of violent conflict in 2011.

Although most remain within the region, West Africans have increasingly migrated further afield in search of economic opportunities such as in Southern Africa. Improved infrastructure and transport also encouraged West Africans to migrate across the Sahara to North Africa, oil-rich Libya in particular. From the early 1990s, the Libyan leader Gaddafi started to encourage sub-Saharan guestworker migration as part of his pan-African policies (Boubakri 2004; Hamood 2006; Pliez 2005). In the early 1990s, most migrants came from Libya's neighbours Sudan, Chad and Niger, which subsequently developed into transit countries for migrants from further afield (Bredeloup and Pliez 2005). Since 2000, anti-immigrant riots, expulsions and increased xenophobia prompted African workers in Libya to move to Maghreb states and, from there on, to Europe, although Libya remained an important destination for migrant workers in itself.

Migration and settlement along the migration routes revitalized ancient trans-Saharan (caravan) trade routes and desert oases in Mali, Niger, Chad, Libya, Algeria and Mauritania. Desert towns and oases now house significant resident sub-Saharan populations. Similarly, major North African cities, such as Rabat, Algiers, Tunis and Tripoli became home to sizeable communities of sub-Saharan migrants (Bensaad 2003; Boubakri 2004; Bredeloup and Pliez 2005; McDougall and Scheele 2012).

Refugee migration in Africa

Decreasing conflict levels and the strengthening of democracy reduced refugee movements and increased the relative importance of labour migration in various parts of Africa. The number of African refugees recorded by UNHCR declined from 6.8 million in 1995 (UNHCR 1995) to 2.4 million in 2010 (UNHCR 2011).

From 1994, the end of the apartheid regime in South Africa removed a major cause of conflict. In Mozambique, for instance, South Africa had funded and armed the Mozambican National Resistance (RENAMO) rebel movement, and by the early 1990s there were an estimated 1.7 million Mozambican refugees and 4 million internally displaced persons. By 1996, most had returned home (USCR 2001). The fall of the Ethiopian DERG regime in 1991 led to the return of refugees, although by then thriving Ethiopian communities in the US and other destinations had already formed transnational networks, which would facilitate continuing migration (Abye 2004; Schewel 2019).

The early twenty first century also saw the end of conflicts in Angola, Liberia, Sierra Leone and the Great Lakes Region. Where peace agreements have been successfully implemented, large-scale repatriations of refugees and resettlements of internally displaced persons (IDPs) have occurred. However, most returns occurred spontaneously, outside official repatriation schemes. Other refugees opted not to return because they had successfully integrated in destination societies. For instance, in 2010 the Tanzanian government decided to naturalize more than 162,000 Burundian refugees who had been living in the country since 1972 (UNHCR 2012b). In post-conflict situations, the line between economic and forced migration can be blurred. Bakewell (2000) observed that many Angolans living in Zambia who were seen as 'repatriating refugees' by the government and the UN saw themselves as villagers moving in search of better livelihoods. They therefore had little inclination to return, particularly if they lived in communities belonging to the same ethnic groups.

However, in some regions, particularly in the Horn of Africa and East Africa, new conflicts broke out, while other conflicts, such as in *Somalia*, persisted. While earlier Somali emigration in the 1970s and 1980s mainly concerned workers to the Gulf, the outbreak of civil war in 1988 and the fall of the Barre regime in 1991, the effective collapse of the state and the civil war in Somalia created one of the largest refugee diasporas in the world (Al-Sharmani 2007; Horst 2006). The Rwandan Civil War, which had started in 1990, led to the mass slaughter of between 10 and 50,000 Hutus and between 500,000 to 1 million

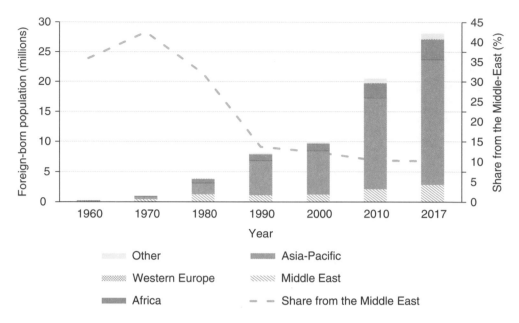

Figure 9.6   Immigrants in GCC countries, by origin region, 1960–2017
Sources: UNDESA 2017 and Global Bilateral Migration Database

Tutsis and involved massive displacement (Mamdani 2014). From 2003, violence in the Western Sudanese province of Darfur, where an estimated 400,000 Sudanese died, led to massive displacement (UNHCR 2012a).

Diversification of migration in the Middle East

In the Gulf, the politicization of migration came to a head during the First Gulf War (1990–1) in response to Iraq's invasion of Kuwait in 1990. Large numbers of Arab workers were deported, particularly Palestinians and other migrants from countries who were seen as sympathetic to Iraq and whose loyalty was questioned. This further increased the impetus to recruit non-Arab workers, particularly from Asian but also from sub-Saharan Africa. Particularly in smaller Gulf states, migrants started to outnumber native populations. Figure 9.6 displays the spectacular growth of immigrant populations in the GCC countries as well as the declining share of Arab migrants alongside fast-rising immigration from Asia. The share of Middle Easterners of all migrants living in GCC countries has dropped from 43 per cent 1970 to only 14 per cent in 1990, mainly because of a spectacular increase in Asian immigration. Official numbers of African immigrants are still relatively small, but increasing fast, from an estimated 230,000 in 1990 to 3.4 million in 2017. By 2017, migrants would make up 86 per cent of the total population of Qatar, 88 per cent in the UAE (up from 70 per cent in 2010), 76 per cent in Kuwait (up from 69 per cent in 2010), 48 per cent in Bahrain (up from 39 per cent), 45 per cent in Oman (up from 28 per cent) and 37 per cent in Saudi Arabia (up from 28 per cent) (UNDESA 2009, 2017). This clear shows the failure of yearlong official efforts to 'indigenize' the GCC workforce.

In the 1973–1989 period, foreign powers had largely abstained from *direct* military intervention in the Middle East, although the US and USSR and their allies would support

Source: Getty Images/Sean Gallup

Photo 9.2   Construction workers from Bangladesh take a break near new highrise office buildings and hotels still under construction in the new City Center and West Bay district in October 2010 in Doha, Qatar

allied regimes, thereby condoning and contributing to violence and repression. This would change through the US-led military interventions in *Iraq*: the first Gulf War in 1990–1 and the second Gulf War (also known as the Iraq War) from 2003 to 2011. Between 1990 and 2002, some 1.5 million Iraqis left their country because of warfare and repression by Saddam Hussein's regime. The US-led invasion of Iraq in 2003 and its aftermath triggered a second wave of Iraqi displacement and was the prelude to a protracted period of violence, civil war and economic decline.

Over this period, the situation of Palestinians did not show any sign of improvement. Although nominally supporting the cause of Palestinians, Arab and other regional states have generally put geopolitical interests above pressing for solutions. On multiple occasions, Palestinian refugees have been victimized in intra-Arab divisions. For instance, after the First Gulf War, Kuwait deported about 200,000 Palestinian migrant workers in response to the support of Palestinian leader Yasser Arafat for Saddam Hussein. During the second *Intifada* between 2000 and 2005, roughly 100,000 Palestinians fled from the West Bank and Gaza Strip. Additionally, most Palestinian refugees living in Iraq had to flee in 2006 after many refugees were killed (Fargues 2007).

In Israel, attacks by Palestinians during the first *Intifada* (the uprising of Palestinians in the West Bank and Gaza that began in 1987) and the First Gulf War, heightened tension. Israeli authorities introduced restrictive regulations which led to a sharp decline in Palestinian employment. By 1991, immigration of Jews from the former Soviet Union prompted the Israeli government to replace Palestinian labour in construction and agriculture, yet these efforts largely failed since most Soviet migrants found the pay and working conditions unsatisfactory (Bartram 1999: 157–161). This prompted increasing recruitment of workers from Romania, the Philippines and Thailand.

Over this period, Turkey and Egypt evolved into crossroads for refugee movements. An origin country for large-scale labour migration to the Gulf and Libya, *Egypt* would become host to many Sudanese and also Palestinian and Somali refugees (Al-Sharmani 2007; Zohry and Harrell-Bond 2003). Turkey received substantial numbers of refugees from the Balkans during the wars in former Yugoslavia (Danış and Perouse 2005: 97; İçduygu 2000: 362–363) as well as Iraqi, Iranian and Central Asian populations, while Turkey was also a country of origin for the many Kurdish refugees who fled to Greece, Germany, Sweden and other countries to escape ethnic conflicts.

Social transformation, the Arab Spring and migration

The Great Recession would only have a limited effect on migration in Africa and the Middle East, as economies in the region remained relatively unaffected by the crisis, although it did cause a temporary slump in migration to Europe. Otherwise, trends of the previous period were largely continued, which consisted of:

- Increasing labour migration of sub-Saharan Africans within the continent and from the region towards Europe, the Gulf, North America and beyond;
- Deceasing migration of Turks and, to some extent, North Africans to Europe;
- Increasing refugee migration within and from the region as a consequence of warfare and repression in the wake of the Arab Spring since 2011;
- A modest but significant increase of migration towards Africa from Europe and, particularly, China.

Migration transitions in Turkey and the Maghreb

Parallel to increasing migration from sub-Saharan Africa, the emigration potential of Turkey and Maghreb, countries that dominated extra-regional migration to Europe for decades, has been declining as they seem to be entering the last stage of their migration transitions. This process is most advanced in *Turkey*, as demonstrated by increasing immigration and decreasing emigration. After having been the main source of non-European migration to Europe between 1960 and 1990, Turkey transformed from a country of net emigration into a country of net immigration (İçduygu 2000; Kirisçi 2007). Since 2000, fast economic growth has attracted increasing numbers of migrants from the Middle East and Africa. More recently, many Syrian refugees are settling on a semi-permanent basis in large cities (Baban *et al.* 2017). Because of these migrations, particularly Istanbul is more and more regaining its old character of a cosmopolitan Euro-Asian hub.

Over the second half of the twentieth century, *Morocco* has evolved into a major emigration country. In many ways, Morocco is 'Europe's Mexico' (de Haas and Vezzoli 2013). Since the mid-1990s, Morocco has taken over Turkey's position as Europe's prime source of non-EU migrant labour. Because Morocco has lower levels of economic development than Turkey, emigration continued to expand and diversify over the 2000s and 2010s. While family networks facilitated continued migration to Northwest European 'guestworker' countries, new Moroccan migration destinations emerged in Southern Europe.

In the previous period, from the mid-1980s, Spain and Italy had already started to attract Moroccans as a consequence of rising demand for migrant labour in agriculture, construction and low-skilled services. Initially, this migration was largely circular as Moroccans could travel freely back and forth. After Italy and Spain introduced visa requirements in 1990 and 1991, respectively, circulation was interrupted as migrants were pushed into permanent settlement. This triggered large-scale family migration, while encouraging prospective migrant workers to migrate illegally across the Strait of Gibraltar using smuggler services. Despite the toughening of border controls – and largely parallel to migration dynamics between Mexico and the US – irregular migration and undocumented stay continued primarily because of ongoing labour demand in Europe (de Haas 2007b).

Moroccans form one of the largest and most dispersed migrant communities in Europe. Of Morocco's current population of 33 million; more than 3 million people of Moroccan descent currently live in Western and Southern Europe. Around 2015, 935,000 Morocco-born migrants were estimated to live in France, 700,000 in Spain, 435,000 in Italy, 215,000 in Belgium and 168,000 in the Netherlands. Recently, a smaller but growing number of Moroccan migrants have settled in Canada and the United States. In addition, 140,000 Morocco-born Jews live in Israel, and with 700,000 the Moroccan Jewish heritage population is the second biggest origin group in Israel after Russian Jews. The Moroccan government estimates the total Moroccan heritage population living abroad at levels of at least 4 million (de Haas 2014b).

Since 2000, changing migration patterns have set the stage for potentially far-reaching changes to the economy, demographics and legal systems. Although Morocco remains primarily a country of emigration it is also becoming a destination for migrants and refugees from sub-Saharan Africa as well as Europe. The growing presence of immigrants confronts Moroccan society with an entirely new set of social and legal issues typical for immigration countries, which do not yet resonate with Morocco's self-image as an emigration country (de Haas 2014b). The presence of numbers of undocumented migrants and refugees compelled the Moroccan king to enact a major legalization campaign in 2014 (Berriane *et al.* 2018; Cherti and Collyer 2015).

Similar dynamics can be detected in Algeria and, particularly, Tunisia (Natter 2014). While some go to North Africa with the intention to move on to Europe, other West African and other sub-Saharan workers and students consider the Maghreb as a destination in its own right. Those failing or not venturing to enter Europe often prefer to stay in North Africa as a second-best option rather than to return to their often more unstable, unsafe and substantially poorer home countries.

The Arab Spring and forced migration

A major new development was conflict-related migration in North Africa and the Middle East in the wake of the political destabilization and popular demands for reform caused by the Arab Spring, and the violent state repression this prompted. In *Tunisia*, the fall of president Ben Ali in January 2011 marked the start of a series of pro-democratic uprisings throughout the Arab world also known as the 'Arab Spring'. Despite a harsh crackdown, mass protests following the self-immolation of street vendor Mohamed Bouazizi on December 17, 2010, launched the revolution that culminated in the toppling of President Ben Ali on 14 January 2011, after 23 years in power (Natter 2014). The security void after the Revolution and the effective absence of Tunisian border controls in early 2011 prompted a temporary hike in irregular boat migration to Europe (see also Box 6.1).

Events in Tunisia triggered pro-democracy protest movements all across the Arab world that led to the installation of democracy in Tunisia and the fall of autocratic rulers in Egypt and Libya in 2011, but also to counter-revolutions which reinforced repression or violent conflicts in countries such as Egypt, Syria, Yemen, Bahrain and indirectly also Mali. This generated substantial voluntary and forced migration, mostly within the region, although some movements eventually spilled over into migration to Europe. In *Libya*, violent conflict in 2011 between pro-Gaddafi militia and rebel groups initially caused the mass return of migrant workers to sub-Saharan African countries such as Niger, Mali and Chad (de Haas and Sigona 2012). Several hundred thousand people – Libyan citizens and migrant workers – crossed the border from Libya into Tunisia in early 2011 following the overthrow of the Gaddafi regime, where they were generally welcomed despite the logistical challenges this created (see Natter 2014). While relatively well-to-do migrant workers were quickly repatriated from Libya, many vulnerable migrant workers from sub-Saharan Africa often did not have the means to return. Migrants who got 'stuck' were vulnerable to racist violence and extreme forms of exploitation bordering on slavery.

From 2012, the violent conflict between opposition and government forces in *Syria* led to the mass flight of citizens, mainly towards Turkey, Jordan and Lebanon. According to UNHCR, over 5.6 million people fled Syria between 2011 and 2016, while about 6.6 million became displaced within Syria. The majority of Syrian refugees live in neighbouring countries. In 2019, Turkey hosted about 3.6 million, Lebanon 949,000, Jordan 672,000, Iraq 253,000 and Egypt about 133,000 Syrian refugees. According to UNHCR data, most Syrian refugees live in urban areas, with only about 8 per cent living in refugee camps. Since 2015, increasing numbers of Syrian refugees started to cross from Turkey into the EU alongside refugees and migrants from Iraq, Afghanistan and Eritrea, prompting a major political crisis in Europe (see Chapter 6 and Box 6.1). Approximately 1 million Syrians applied for asylum in Europe. European countries, Lebanon and Turkey have turned more hostile towards the influx of more Syrians, prompting Turkey to construct a border wall with the partial help of EU funding (see Popp 2018).

In Yemen, following a peaceful political uprising in 2011, failed regime transition and internal fragmentation led to civil war and a military intervention as well as a blockade by Saudi Arabia and the UAE in 2015, causing famine, destruction and large-scale internal displacement. International refugee numbers, however, have remained low, partly because of appalling poverty depriving people of the means to flee and partly because of the closure of neighbouring countries to refugees and because of mass deportations of Yemenis by Saudi Arabia since 2013 (Thiollet 2015; Bruni 2018).

Protracted refugee situations have endured elsewhere in the Middle East and Africa. For Palestinian refugees, issues concerning repatriation, compensation, reparations and access to their homeland have remained unresolved. With the Palestinian population of the West Bank and Gaza in dire economic straits, prospects appear bleak for the repatriation of Palestinian refugees from Jordan, Lebanon and Syria. According to official UNRWA data, Palestinian refugees numbered 5.4 million in 2018, the majority of them living in Jordan (2.2 million), the Gaza Strip (1.4 million), the West Bank (828,000), Syria (551,000) and Lebanon (470,000). While their presence has become permanent – with second and third generations outnumbering the first – Palestinians often lack full rights because of the ideological stance of host regimes, insisting on their right to return but therefore denying them integration and citizenship rights.

The peace agreement between Northern and Southern Sudan in 2005, which ended 22 years of civil war, led to the independence of South Sudan in July 2011. However, the civil war that broke out in South Sudan in 2013 led to mass killings, internal displacement and the departure of about 2.4 million refugees, mainly to Ethiopia, Sudan and Uganda. *Eritrea* had already been a source of regional refugees during and after its war for independence from Ethiopia (1971–1991). However, refugee migration would gain another dimension when Eritrea became an independent state in 2001 and transitioned to dictatorship (Debessay 2003). Harsh oppression and compulsory national (including military) service, involving torture, detention and forced labour, prompted large-scale outflows of refugees. Although most refugees went to regional destinations (including Saudi Arabia), Eritreans also started to join trans-Mediterranean boat migration (Horwood and Hooper 2016).

The entry of Africa on the global migration stage

Until the early 2000s, middle-income countries in North Africa and the Middle East dominated extra-regional migration to Europe and the Gulf (Fargues 2005; de Haas 2008a; Natter 2014). In recent years, sub-Saharan Africans have been increasingly present in such extra-continental migrations. The entering of sub-Saharan Africa on the global migration stage is a consequence of integration into global economic circuits as well as profound social and economic transformations in the region. Improvements in education, infrastructure and transport links, and the rapid diffusion of telecommunication techniques such as mobile phones (see de Bruijn *et al.* 2009; Schaub 2012) have increased the desire for different lifestyles among a new generation of young Africans who also increasingly can afford to migrate as a result of modest, but real, economic growth.

Until recently, much migration, particularly from sub-Saharan Africa, to Europe and North America used to mainly be an elite affair. For instance, in the US, African immigrants possess the highest average level of education of any immigrant group (Capps *et al.* 2012; Slater 2000). However, in recent years African extra-continental emigration has seen increased participation of people from less privileged backgrounds who can now afford to migrate. Increasing numbers of women are migrating independently, for instance as domestic workers from Ethiopia to the Gulf and Yemen (De Regt 2010; Fernandez 2010; Schewel 2018). But although migration of higher- and lower-skilled Africans migrating to Europe, the Gulf and the Americas is still relatively low, it is increasing fast.

While regular and irregular movements from Africa to Europe used to be dominated by the migrants from the Maghreb, the share of sub-Saharan Africans has been steadily increasing since the 2000s (de Haas 2008; Flahaux and de Haas 2016). Over the 2000s, tighter control measures and naval patrols by EU countries have compelled migrants to take longer sea routes, increasing the risks and the death rate (Carling 2007). The result has been a continuous 'cat and mouse game' in which migration routes have constantly shifted in reaction to border controls, but overall migration trends have remained relatively unaffected.

Since border controls have compelled migrants and smugglers to use longer and more dangerous routes, this migration involves significant human suffering and an increasing death toll. According to a UNCHR report, an estimated 2,275 people died or went missing crossing the Mediterranean in 2018 (UNHCR 2018). The report also estimated that the death toll rose from one death per 38 arrivals in 2017 to one death per 14 arrivals in 2018. This increasing death toll is related to policies by European and North African authorities to limit or prevent search and rescue operations by organizations such as *Médecins sans Frontières*. As part of deterrence policies, more and more asylum seekers and migrants are deported to Libya with support of the EU, where they face appalling conditions inside detention centres.

Notwithstanding the extensive media and political attention to 'boat migration', the vast majority of African emigrants leave the continent legally. It has been estimated that about 700,000 Africans migrated legally to OECD countries every year in the mid-2010s while an unknown but significant number (but that must run into hundreds of thousands) of Africans move to Gulf countries on temporary labour contracts (de Haas 2019). By comparison, the estimated numbers of African migrants and asylum seekers crossing the Mediterranean irregularly have generally varied between several tens of thousands to about 100,000 per year (de Haas 2019). In addition, the majority of undocumented migrants living in Europe have entered legally and subsequently overstayed their visas and Africans seem to follow that pattern (Flahaux and de Haas 2016; Hearing and Erf 2001). This highlights the dangers of confunding undocumented *entry* with undocumented *stay* (see Chapter 2).

Chinese–African migrations

Sub-Saharan African migrants are increasingly attracted to fast-growing economies beyond the traditional destinations in Europe, the Gulf and North America. Some Africans have gone as far afield as Russia, Turkey, Japan, India and China, and even Brazil and Argentina (Andres Henao 2009). In conjunction with growing commercial and political ties, migration from Africa to China has been growing fast. China's emergence as a global economic superpower and the relative ease of getting temporary visas has also made it a destination for African migrants (Ghosh 2010; Haugen 2012). Initially, most African immigrants were students (see Hashim 2003; Sullivan 1994). China recruits African students to increase Chinese *soft power* and generate income from fee-paying students (Haugen 2013). While African students are sometimes met with hostility (Hashim 2003) and disappointed with the equality of education, they often stay for prolonged periods and become active in trade (Haugen 2013). Since 2000 growing numbers of West African (particularly Nigerian and Ghanaian) traders have been reported in China (Bakewell and Jónsson 2011; Bodomo 2010).

Although most intend to stay temporarily, some are settling and significant African immigrant communities are emerging in China. A number of Africans who initially came as students are engaging in trade and contributing to the fast-growing export of Chinese products to Africa (Cissé 2013; Haugen 2013). Although there are no official statistics, about 200,000 Africans were believed to be living in China already in the late 2000s, particularly in Guangzhou, a prosperous province in Southern China (Bodomo 2010; Ghosh 2010). Many overstay their visas: about three quarters of undocumented foreigners in the Guangdong province in 2009 were reported to be Africans (Ghosh 2010). Most immigrants are from Nigeria. Other important origin countries include Senegal, the Congo, the Gambia, Somalia, South Africa, Cameroon, Niger and Liberia.

Increasing migration to Africa from China (alongside immigration from Europe and North America) is linked to a liberalization of economic policies in many African countries which has facilitated foreign direct investment and trade as well as the relaxing of immigration restrictions. Also migration in the reverse direction, from China to Africa, is still relatively small, but growing fast. This migration can partly be seen as part of China's more general bid to secure raw materials and export markets, although many Chinese migrate independently to create their own economic opportunities (BBC News 2007).

Many Chinese migrants initially came to Africa as entrepreneurs or as employees of Chinese companies involved in road, hydropower and urban construction (Mohan and Tan-Mullins 2016; Tan-Mullins *et al.* 2017). Although most came with the intention to stay temporarily, a proportion of them stay, often starting independent economic activities as petty traders or starting other small businesses. Others migrate to Africa on their own initiative with the goal of starting their own businesses (Jung Park 2009). Hundreds of thousands of Chinese migrants have been settling in rural and urban areas in Africa, and are mainly involved in agriculture, construction and trade. For many Chinese farmers, Africa presents an inviting opportunity and Chinese authorities encourage them to migrate by giving support to investment, project development and the sale of products (BBC News 2007).

Total estimates range from 580,000 to over 800,000 Chinese on the African continent in the late 2000s. By far the largest number of Chinese (200,000–400,000) live in South Africa, followed by Nigeria. Sudan, Angola, Algeria and Mauritius that also host significant Chinese populations, and their numbers are growing in many other countries (Jung Park 2009; Mohan and Tan-Mullins 2016; Tan-Mullins *et al.* 2017).

Conclusion

This chapter has dealt with vast and diverse regions that are undergoing rapid change. This makes generalization difficult. Yet for all the differences, migration trends in Africa and the Middle East do reflect some general trends in global migration mentioned in Chapter 1. Many countries in Africa and the Middle East show trends towards globalization of migration – that is, more countries are affected more profoundly by significant flows of migrants, to and from an increasing variety of destinations and origins. Growing sub-Saharan migration to North Africa, the Middle East and Europe and the recent proliferation of migration to and from China are the most salient examples of this trend. African countries have been increasingly incorporated in international and global migration systems.

Since the 1970s, migration has partly shifted away from postcolonial patterns, with the rise of the Gulf countries and Libya as new destinations for migrants from poorer countries in the Middle East and North Africa and, increasingly, from Asia and sub-Saharan Africa. While sub-Saharan migration has remained predominantly intraregional, since the 1990s, migration from Africa to Europe, North America and Asia has been growing. This is the result of processes of economic development and social transformations that motivate and enable increasing shares of educated and aspiring young populations to migrate.

While the oil-rich Gulf countries and Libya host large immigrant populations, other countries in the Middle East and Africa receive signficant numbers of migrants, and several countries are going through migration transitions. While countries such as Nigeria, Côte d'Ivoire, Gabon and South Africa have historically already attracted large numbers of migrants, several new migration destinations such as Kenya, Ghana and Angola have emerged more recently. North African countries too have witnessed increasing transit and settlement migration. In the Middle East, Turkey has gone through a full migration transition, attracting migrants and refugees from an increasingly diverse array of European,

Asian and African countries. Depending on economic growth and political stability, countries such as Morocco and Tunisia might follow similar paths in the future.

The analysis highlights the fact that most African migration is economically motivated and particularly those moving out of the continent may perhaps be poor in Western eyes but tend to be relatively well-off by origin country standards. This also seems to confirm the theoretical idea that increasing education, better access to media as well as modest increases in income and better infrastructure have the tendency to increase people's aspirations and capabilities to move.

Migration patterns in Africa and the Middle East also question the usefulness of official migration categories, which often do not correspond to the realities. A pilgrimage to Mecca can also be the opportunity to foster contact with trading partners or to overstay and become a labour migrant; Palestinian migrants in the Gulf can be both refugees and labour migrants; Angolan refugees in Zambia may primarily see themselves as villagers moving in search of better livelihoods (Bakewell 2000); and hundreds of thousands of sub-Saharan migrant workers in Libya became 'involuntarily immobile' when violent conflict broke out in 2011, because these stranded migrants and *de facto* refugees did not have the resources to flee and therefore became vulnerable to abuse and violence (de Haas and Sigona 2012).

Although it is common to distinguish 'receiving states' from 'sending states', such dichotomous distinctions are often difficult to maintain in practice, as many 'sending countries' are destination or transit countries from migrants from other, often poorer and more unstable, countries. While Morocco is an important emigration country, it is a destination country for Senegalese migrants, while Senegal is a destination for Malian migrants. As in other world regions, this reveals the proliferation of replacement migration as part of the emergence of complex, multi-layered geographical migration hierarchies.

This chapter has also shown the usefulness of dual or segmented labour market theory (see Chapter 3) for understanding the continuation of migration despite the persistence of poverty and unemployment in many African and Middle-Eastern destination countries. As elsewhere in the world, migrant workers tend to concentrate in particular sectors, where natives increasingly refuse to work, or where employers prefer immigrant labour. Although governments in Africa and the Middle East have attempted to prevent permanent settlement, and despite restrictive immigration integration legislation, migration is gaining an increasingly permanent character in many countries. This shows the dangers of the usual casting of Africa and the Middle East (outside the Gulf region) as regions of out-migration: many countries host significant immigrant communities and deal with similar issues of settlement, integration and diversity as most other immigration countries in the world.

This echoes the general observation that temporary migration often leads to permanent settlement. Immigration has become a key issue in both popular mobilization and political discourses everywhere in the region. This seems to coincide with an increasing emphasis on national identity, the politicization of migration and significant hostility towards immigrants in many African and Middle-Eastern countries (see Mitchell 2012).

In a long historical perspective, it is possible to see that most migratory movements in Africa and the Middle East have common roots. Occupation by Western powers triggered profound changes, first through colonization and ambiguous border drawing, then through military involvement, political links, the Cold War, and trade and investment. This exemplifies the relevance of historical-structural and world systems theories in explaining migration. While colonial intrusion and concomitant capitalist economic development and urbanization were entangled with massive rural-to-urban migration within and across national boundaries, colonial and postcolonial conflict and nation building created large refugee movements. The influence of colonialism is also visible in extra-continental migrations, which often reflect ties built in the colonial era, with migration from the Maghreb and francophone West Africa predominantly oriented on France, and migration from Southern and East Africa still reflecting connections to the former British Empire. Labour recruitment during and after colonization has played a key role in establishing contemporary migration patterns.

This confirms the predictions of migration systems theory discussed in Chapter 2, that increasing flows of capital, goods and ideas between countries are also likely to stimulate flows of people *in both directions*. Thus, the entry of African countries on the global migration stage is an inevitable consequence of the increasing integration of these areas into the world economy, and into global systems of international relations and cultural interchange. The growing trade, investment and migratory links between Africa and China are a case in point.

It is impossible to predict the future of migration, but it is likely that current patterns will undergo fundamental transformations. Irrespective of its long-term outcomes, the Arab Spring indicates the coming of age of a better educated, more aware and more *aspiring* generation. This generation no longer accepts conditions of economic stagnation, stark inequality, corruption and political oppression, thus increasing pressures for political and economic reform. If reform is successful, many youngsters may prefer to build a future in their own land, although others will certainly emigrate to pursue an education and career abroad. If reform fails, and situations of autocracy and economic stagnation persist, even higher levels of emigration are to be expected.

An important factor of change is the demographic transitions taking place in much of the Middle East and North Africa. These have led to plummeting birth rates and declining rates of population growth. In the coming decades, the slowing down of the growth of young adult populations may decrease unemployment, particularly if political reform creates conditions for equitable economic growth. In that case, other middle-income countries in North Africa and the Middle East such as Morocco and Tunisia may follow the Turkish (and Southern European) scenario and experience declining emigration and increasing immigration. However, if reform fails, this aspiring generation is also more likely to migrate away to seek better economic, cultural and political opportunities, in which case, emigration may continue to plateau at high levels.

At the same time, we may expect increasing migration from and to sub-Saharan Africa as a consequence of globalization and the associated further incorporation of the continent in international migration systems. Although declining fast, the population growth rate is still at high levels in most sub-Saharan countries, and economic and other opportunity gaps with industrialized countries remain huge. In this context, increasing levels of education, income and connectivity are likely to fuel emigration to increasingly distant lands. More generally, African development is unlikely to curb migration as it will probably rather enable and inspire *more* people to migrate. The paradox is that high levels of poverty, high illiteracy and weak infrastructure have often prevented previous generations of Africans from migrating. This situation will surely change *as a result of African development*, which will initially lead to accelerating emigration. While the 1973–2008 period was characterized by the massive entry of Asia on the global migration stage, the next period may well become the age of African migration.

Guide to Further Reading

Overviews of African migration include Manuh (2005), Adepoju (2006), Cross *et al.* (2006), Mafukidze (2006), and Bakewell and de Haas (2007). De Bruijn *et al.* (2001) provide a compelling analysis of the complexity and variety of contemporary mobility in sub-Saharan Africa. Schielke and Graw (2012) is a collection of anthropological studies from across Africa on migration aspirations. Vigneswaran and Quirk (2015) is a rich collection of papers on the impacts of migration power and state building in Africa. Berriane and de Haas (2012) is a (free to download) collection of studies of African migration. Berriane *et al.* (2018) gives an overview of Moroccan migrations. Scheele (2012) highlights the role of oases, religion and smugglers in connecting Saharan societies, while McDougall and Scheele (2012) give in-depth insights into Saharan mobility. Brachet (2016) gives a critical analysis of the role of IOM into Europe's role in outsourcing migration controls to Libya; and de Haas (2008) and Brachet (2018) show how clandestine migration and the phenomenon of smuggling have been 'manufactured' by border control policies. On the political economy of migration in the Gulf, see Fargues (2011a) and Thiollet (2011, 2016). The Gulf Labour Markets, Migration, and Population Programme (GLMM) of the European University Institute in Florence (www.gulfmigration.org) gives detailed information and data on migration to the Gulf. On Turkey, see Kirişci (2006) and İçduygu and Yükseker (2012). On Israel, see Bartram (2005). Collyer *et al.* (2012) contains articles which critically discuss the concept of transit migration in the context of migration from the Middle East and Africa towards Europe.

Extra resources can be found at: **www.age-of-migration.com**

10 The State, Politics and Migration

Migration is an inherently political phenomenon. As Abdelmalek Sayad (1999) famously argued, 'To think of immigration is to think of the state' (*Penser l'immigration, c'est penser l'État*). The very concept of migration implies residency, the idea that people have a fixed place where they belong and where they can be located and controlled. This is why states and politicians often become nervous once large numbers of people start moving on their own, because it is an implicit threat to state sovereignty and elite privileges. In early industrializing societies, such fears have been typically concentrated on large-scale migration from rural to urban areas. Anxieties about the 'rural exodus' of peasants and perceived problems of crowding, poverty, crime, disease and cultural change in urban areas, and the concomitant desire to curb rural-to-urban migration and keep people down on the farm, was prevalent in nineteenth-century Europe, and is now still is a major concern in many developing countries. As De Soto argued,

> Migrants to the cities encountered a hostile world. They soon realized that, while urban people had a romantic, even tender image of the farmers and were quick to acknowledge that all citizens had a right to happiness, they preferred that the good farmers pursue their happiness at home. Peasants were not supposed to come looking for modernity. To that end, virtually every country in the developing and ex-communist world maintained development programmes to bring modernity to the countryside. (De Soto 2000: 83)

Such concerns have given rise to policies to control internal migration, such as the Chinese *hukou* system. Whatever the form and particular purpose, modern states have feared 'free floating' populations, and have invariably aimed to 'settle' those populations. Modern states harbour an inherent hostility towards nomadic or itinerant peoples that resist giving up their lifestyles, such as the Romani ('gypsies'), Travelling people (in the UK and Ireland) or the Tuareg (in the Sahara), as they are difficult to control and tax, and defy modern notions of citizenship. But perhaps most importantly, such groups and their lifestyles are deeply threatening to bourgeois norms about the 'good life'. In other cases, the desire by states to control scattered, potentially rebellious, rural populations has given rise to 'villagization' policies. This amounts to the compulsory resettlement of people in designated areas or villages, such as the large-scale villagization under Ethiopia's communist DERG regime in the 1980s (see Chapter 9).

An extreme case is Cambodia, where between 1975 and 1978 the Khmer Rouge regime evacuated cities and sent entire populations on forced marches to rural work projects, in a (failed) attempt to re-establish traditional agriculture and to destroy forms of cultural expression and economic systems seen as 'Western'.

Although governmental policies may affect patterns and to some extent the pace of urbanization and concomitant rural-to-urban migration, efforts to significantly curb it, let alone reverse it and to make populations stay in rural areas, have invariably failed (Rhoda 1983; Skeldon 1997). This is because they ignore the fact that the massive transfer of economic activities and population from rural to urban areas is an intrinsic and therefore inevitable part of industrialization and modernization. In industrialized societies, where most people already live in cities, fears of hordes of poor rural people bringing in bad habits, illnesses, crime and overcrowding tend to subside and are usually redirected toward immigrants from abroad. In the case of international migration, the perceived need to control immigration is tied up with the privileges associated with modern citizenship, dividing working classes into citizens and non-citizens (see Piore 1979).

The preceding regional chapters have shown that processes of nation state formation and the associated drive towards cultural and ethnic 'homogenization' tend to go along with significant voluntary and forced population movements. Such migrations can become particularly massive and violent when new states are formed, such as around the establishment of the modern Turkish state in 1923 out of the ashes of the Ottoman empire, the 'partition' of India and Pakistan in 1947 and the establishment of the state of Israel in 1948, or after the dismantling of the Soviet empire in 1991. The violence and forced migration following independence in much of Africa and the Middle East, as well as in ex-East Bloc states such as Yugoslavia and various former Soviet republics are part of efforts to forge new nation states, where there was often no place for ethnic and religious groups that were no longer seen as fitting into the ideology of the new nation.

The use of nationalism and anti-colonial fervour by new political elites partly explains why African and Asian states tend to have highly restrictive immigration policies and often deny citizenship to foreigners and minorities. The development of more inclusive notions of national belonging in Western countries is a fairly recent phenomenon, which has been sparked by the ethnic diversification of immigration since the 1960s.

The desire to control migration is thus inherent to modern statehood. To prevent a common misunderstanding, 'control' does not necessarily mean that states want to prevent or restrict movement. In many cases, states aim to encourage immigration of particular groups, such as co-ethnics, refugees from hostile regimes or workers for whom there is a strong labour market demand. Control rather implies that states are focused on surveillance, which relates to the ability of states to monitor population movement, through administrative tools such as population registers, residence permits and censuses. For instance, nowadays owners of the 'right' passports can often breeze through electronic passport gates without any hindrance, but their movement is still closely monitored. Movement can thus be free yet highly controlled. The remainder of this chapter will look from various angles at the ways in which migration is entangled with politics and the inherent drive of states to control population mobility.

The exit revolution

While states have always been concerned about controlling populations, the emphasis of their preoccupations has shifted over time, from controlling emigration to controlling immigration (Zolberg 2007). In the pre-industrial era states were generally more worried about people leaving (emigration) than about people coming (immigration) because population was seen as an essential source of taxation and military power, while employers were usually keen to prevent 'their' slaves, serfs and other bonded labourers from escaping. For instance, mercantilist European states in the seventeenth and eighteenth centuries sought to discourage or bar emigration, as the loss of subjects was thought to detract from state economic and military power (Green and Weil 2007). Before the Meiji restoration in 1867, which heralded a fast transition from a feudal society to an industrialized market economy, Japanese law prescribed the death penalty for Japanese leaving the country without permission.

Torpey (1998) argued that as centralized national governments consolidated in the nineteenth century, they also monopolized the 'legitimate means of movement' of their citizens. The ability of states to maintain written records of the population gave them the power to grant citizens permission to leave, thus, to control movement, based on the issuance of passports and other identity documents. At the same time, industrialization and the spread of modern capitalism, the foundation of modern labour relations on wage labour instead of slavery, servitude and various other patron–client relations, decreased the need for employers to prevent workers from leaving. The growing importance of technology in warfare, combined with rapid population growth, decreased the military significance of population size.

In a process coined by Zolberg (2007) as the 'exit revolution', this led to a long-term process of states gradually losing their motivation to control emigration. The exit revolution started in Europe, where, after the French Revolution and its proclamation of a human right to emigrate, the ability and willingness of European states to deter emigration began to erode. Together with other factors such as economic transformations and rural–urban migration, the falling away of exit restrictions and (in Russia) the abolishment of serfdom, in some instances, state assistance to emigration, these processes contributed to large-scale trans-Atlantic emigration of Europeans between 1820 and 1920 (see Chapter 5).

Yet this *exit revolution* has been anything but smooth (de Haas and Vezzoli 2011). Particularly authoritarian states have persisted in trying to control or prevent emigration. This was the case Germany under the Nazis and Communist states like the Soviet Union or North Korea, which employed draconian measures to prevent emigration. Such efforts were never entirely successful. For instance, tens of thousands of North Koreans succeeded in escaping northward to China and, from there to Thailand and Korea (Greenhill 2010: 227–261). Such cou Even ntries often also tried to tightly control internal migration, such as the Soviet *propiska* and Chinese *hukou* systems (Torpey 2007: 25–28). Upon decolonization, governments of newly independent countries looked at emigration with considerable levels of suspicion or ambiguity, as part of anti-colonial sentiment. After a long and traumatic independence war, which lasted between 1954 and 1962 and left up to 700,000 people dead, Algeria, for instance, tried to discourage emigration to France. Even governments which signed bilateral accords for recruitment of citizens for employment abroad such as Morocco or Turkey had ambiguous attitudes towards emigration, as they saw migration as a 'safety valve' to reduce discontent and generate remittances, but also feared migrants' political activism from abroad.

Other governments have taken active steps to support their citizens' welfare and rights in destination countries. Indeed, protection of migrants and prevention of exploitation of Italian emigrants figured importantly in Italian politics and fostered Italian nationalism (Douki 2007). In recent decades, governments of origin societies have increasingly nurtured a relationship with emigrants through so-called diaspora engagement policies, partly to encourage remittances and investments, but also further political causes (see Gamlen 2006; 2008; Thibos 2014). The idea of 'diaspora' generally goes beyond emigrants, as origin states try to reach out to descendants of migrants (the 'second' and 'third' generations) and other populations considered as 'co-ethnics'. Examples include India's policies giving special status to the Indian heritage population around the world and Turkey's policies towards ethnic Turks in Europe and Central Asia.

Many states try to cater for their citizens or subjects living abroad, for instance through consular services or financing origin country language education abroad. Such 'diaspora engagement policies' are partly driven by economic concerns such as facilitating remittances, but take on broader political significance (Adamson and Demetriou 2007; Gamlen 2008). For example, the emigration of Jews from Israel has long been viewed as posing an existential threat to the Israeli state although the traditional hostility of Israel towards the estimated 500,000 Israelis who reside abroad has given way to a friendlier stance, motivated in part by the hope of facilitating returns to Israel (Lustik 2011). Likewise, since 1989 the Moroccan state has shifted from a focus on repressing and controlling Moroccan subjects living abroad (partly out of fear of political activism), to an approach focused on courting the Moroccan diaspora to prevent alienation and to stimulate remittances (de Haas 2007a). Yet such changes are often mainly rhetorical, with particularly authoritarian states continuing their efforts to control political activism and secure the political loyalties of expatriate populations (Smith 2003).

Migration, diplomacy and state power

The ambiguous relation between origin states and 'their' emigrants

Because migration involves the movement of a citizen or subject of one state to the territory and jurisdiction of another state, this frequently creates political tensions, between migrants and states, but also between states. In their essence, such tensions come down to conflicting claims of sovereignty. The world's states vary widely in their political institutions. Migration taking place between two authoritarian or non-democratic states differs from that taking place between two states that possess democratic institutions. Migrants may arrive in democratic settings from states with authoritarian governments, or the other way around. Migrants to non-democratic settings are unlikely to participate much in political life in receiving states. The millions of mainly Asian- and Arab-origin migrants in the Gulf-area monarchies are largely politically quiescent, although even in those cases migrants can gain some political power by the virtue of their numbers and vital economic importance.

For instance, strikes and protests by mainly South Asian migrants in Gulf countries did achieve some modest, but real reforms (DeParle 2007; Surk and Abbot 2008). Put under pressure by allegations of migrant worker abuse in its preparations for the 2022 Football World Cup, the Qatari government pushed through a number of labour reforms, including a minimum wage and lifting the requirement for migrant workers to get permission from their employers to leave the country (Kanso 2018). Such examples show the potential power of organized labour, and why governments and employers have often been keen on preventing migrants from organizing themselves. One of the motives for European industries to recruit low-skilled and often illiterate workers in rural areas of Morocco was desire for a docile and hardworking workforce that was not liable to join trade unions and communist parties (Lacroix 2005).

In such cases, diplomatic representation in support of migrants' interests by origin country governments takes on particular significance, to reach out to migrants, or to control them, or both. Many origin countries have developed extensive consular services to this end (Delano 2011). For instance, the Mexican state has played an active role in defending the rights of Mexican migrants in the US, and actively helps them to obtain ID documents such as drivers' licences. However, the track record of origin country governments defending the interests of expatriate compatriots is, at best, uneven. For instance, both the Philippines and India tried to mandate minimum wages for their expatriates working in the Gulf states in 2007 and 2008 respectively. Bahraini companies resisted paying higher wages, thereby sparking strikes by Indian workers. The Bahraini Minister of Labour held that India had no authority to enforce the measure. The Filipino effort to secure a minimum wage for expatriate domestic workers resulted in declining employer demand for Filipina migrant workers. The efforts by India and the Philippines to better the lives of emigrants were undercut by the ability of employers in the Gulf states to find labour elsewhere (Surk and Abbot 2008).

The Philippine Overseas Employment Administration (POEA), the Philippines' system of managing the overseas employment of millions of temporary Filipino workers has long served as a model for developing countries hoping to access the benefits of global labour mobility. Yet Agunias (2008) pointed out the limitations of the POEA: it has limited ability to monitor workers' welfare, especially when destination governments do not honour employer–employee contracts, and it is not able to hold foreign employers and private recruitment agencies accountable for complying with mandated standards for overseas deployment (Agunias 2008).

These examples show that international migration often takes place in bilateral relationships characterized by domination and subordination, negatively affecting the ability of origin country governments to protect migrant interests. Moreover, origin country governments sometimes collude with the governments of receiving states in maintenance of the status quo unfavourable to, if not oppressive of, migrants, with both governments agreeing that activism should be avoided at any price. While employers and destination country governments have an interest in avoiding strikes and trade union activism, origin country governments have often been keen to avoid migrants and exiles from forming a rebellious political force from abroad or from integrating 'too much' so that they will stop remitting money.

Also, governments of origin countries often try to control and to spy on emigrants out of fear of political activism from abroad. The preoccupation with 'remote controlling' diasporas is particularly significant in contexts involving emigration from authoritarian countries to more democratic settings, such as the movement of Algerians, Tunisians and Moroccans to France, or Turks to Germany (de Haas 2007a; Thibos 2014). Origin country governments can be suspicious towards migrants because of their potential to foment violent conflict in origin countries (Adamson 2006: 190–191). Migrant communities sometimes provide financial aid and recruits to groups engaged in conflicts in origin states or elsewhere. Kosovar Albanian communities in Western Europe and North America, for instance, provided financing and recruits for the Kosovo Liberation Army which, in the late 1990s, engaged in heavy fighting with Serbian forces in the former Serbian republic. Similarly, Tamil Sri Lankans in Europe, Canada, India and elsewhere have aided the Tamil Tigers' insurrection in Sri Lanka, and migrants have provided support for warring parties in Somalia (Nyberg-Sorensen et al. 2002; Van Hear 2004). Citizens of Muslim background in Europe (often children of North African immigrants) have been recruited for the so-called 'Islamic State' (IS) (see García-Calvo and Reinares 2016).

The potential clash between expatriate voting and integration policies

Another policy to maintain ties with emigrants is through their participation in origin country elections. The modalities for expatriate voting vary considerably. While some countries still require emigrants to return home to vote, many states now permit consular voting. Still others permit absentee voting, as in the US. Although voting participation rates amongst emigrants are typically rather low, the extension of voting rights has an important symbolic function and, in some cases, the emigrant vote can make an important difference in closely contested elections. For instance, absentee balloting by Floridians abroad played a key role in the contested outcome of the 2000 US presidential election. Mexicans living abroad became eligible to cast absentee ballots from abroad in the 2006 Mexican presidential election. By 2007, 115 countries and territories allowed citizens to vote from abroad (Gutierrez and Terrazas 2012: 2) and this number seems to have gone further up since.

Electoral campaigning increasingly reflects the significant weight of voters abroad. Ecuadorian and Dominican Republic presidential candidates have campaigned for votes in New York City, just as Italian and Portuguese parties have in the past campaigned for votes in Paris (Miller 1978, 1981; Snel *et al.* 2006). Some origin countries have created legislative districts to solely represent citizens living abroad (Earnest 2008: 2). Turkish politicians regularly campaign amongst the large Turkish diaspora living in Germany and the Netherlands (Thibos 2014). This can create tensions with destination country governments, who may see this as an infringement on sovereignty, particularly when migrants have acquired citizenship. It may also fuel xenophobic discourses according to which migrants would foster double loyalties and be unwilling to integrate. Destination states may see such policies as running counter to their integration policies, particularly against

Source: Getty Images/Adam Berry/Stringer

Photo 10.1   Supporters of Turkish Prime Minister Recep Tayyip Erdoğan attend a rally at Tempodrom hall in February 2014 in Berlin, Germany

the background of a backlash against multiculturalism, where double citizenship has been under attack (see Chapter 13).

Migration diplomacy and the weapons of weak states

The interests of origin states, however, are different. Many have given up their former resistance against emigrants taking up citizenship of destination states, as they have increasingly realized this need not to endanger emigrants' loyalty, and that foreign citizenship can increase circulation and return by migrants, as they no longer have to fear being unable to re-emigrate. Origin states are often very keen on preserving the allegiance of migrant populations, and therefore actively counteract destination country policies to ask for the undivided loyalty to the destination country. Some states, such as Morocco, have no mechanisms for relinquishing citizenship, making all diaspora members acquiring foreign citizenship double citizens by default (de Haas 2007a). Although many migrants may see double citizenship as an asset facilitating their free circulation between origin and destination, other, particularly politically active, migrants would like to rid themselves of origin country citizenship, as it increases diplomatic protection by the destination state.

This exemplifies that the power imbalance characterizing relations between migrant-sending and migrant-receiving states does not imply that less powerful states have no power at all. On the contrary, on multiple occasions origin and destination states have used the 'migration weapon'. The reversal of Cuba's longstanding policy of discouraging emigration of Cubans in 1994 led to a significant movement of Cubans to the US, forcing the US government to negotiate the 'wet-foot, dry-foot' deal (see Chapter 7). Over the 2000s, Colonel Gaddafi skilfully exploited Libya's role as a destination and transit state for African migrants as a diplomatic resource to achieve broader geopolitical goals, such as the lifting of the UN embargo against the country, in exchange for collaboration with the EU's and Italy's migration policies (Paoletti 2011). In 2016, Turkey used the large-scale refugee migration from its territory into Europe as a lever to negotiate major financial concessions and the promise of future visa-free travel for Turks into the EU as part of the EU-Turkey 'migration deal' (see Box 6.1).

In her book 'Weapons of mass migration', Greenhill (2010: 75–130) sees such cases as examples of coercive diplomacy by a weaker state against a much more powerful neighbour. She argues that liberal democracies are particularly vulnerable to such coercive diplomacy involving international migrants. Democracies embrace humanitarian values which make them more vulnerable to shaming and allegations of hypocrisy than authoritarian states.

Immigration policies as a tool of state power

Contrasting popular perceptions of immigration as a challenge or a threat to the sovereign state, migration can also be seen as *increasing* state power (Adamson 2006). Immigration can boost economic growth and has frequently been viewed as indispensable to a state's well-being and power. This was obvious in Australia's post-WWII policy of 'populate or perish' (see Chapter 8) or Canada's policy to boost immigration. Additionally, immigrants are often recruited into the army, and recruitment for the army is an important source of immigration in itself, as has been the case for North Africans and Senegalese in the French colonial army and for Latino and Filipino migrants in the US army. In addition, intelligence services can tap immigrant expertise and knowledge of languages. As the previous regional chapters have shown, both voluntary and forced migration have been central to nation building experiences of European settler colonies in most of the Americas and Oceania as well as of post-colonial and post-imperial nation state formation in Asia, Africa and the Middle-East, such as Turkey and Greece after the demise of the Ottoman Empire and the partition between India and Pakistan.

Immigration policies can also contribute to 'soft power', which refers to the ability of states to achieve foreign policy and security objectives through political and cultural relations without recourse to military or economic coercion. For instance, the large bodies of foreign students studying in the former USSR and the US can be seen as an important source of soft power, because they help build positive long-term linkages (Nye 2004). Welcoming students can enhance a state's reputation and diplomatic clout. Cuba has massively invested in the training and education of doctors from communist countries in Africa and Latin America (see Feinsilver 2010). In Africa, the Moroccan state has welcomed and given scholarships to sub-Saharan students partly as a way to create a new generation of high-skilled 'friends of Morocco' across the continent (Berriane 2015).

Reputational concerns can also push authoritarian states to improve treatment of immigrants, as was the case in Qatar and other Gulf states after negative exposure in the international media. Likewise, Morocco's desire to maintain its image of a beacon of 'enlightened authoritarianism' in the Middle East and North Africa, and to enhance its influence in Africa, played an important role in the regularization of sub-Saharan migrants and asylum seekers from 2014 (Cherti and Collyer 2015). This was in part a reaction to the bad press Morocco received for the racism and harassment suffered by sub-Saharan migrants (Ustubici 2016).

The securitization of migration

Casting migrants as the enemy 'from within'

One of the most important developments with regard to migration politics in the post-Cold War period has been the linking of migration to security, a process of social construction termed *securitization*. While throughout modern history migrants have often been accused of stealing jobs or depressing wages, securitization takes it to an altogether different level by representing migrants as a fundamental threat to security and the cultural integrity of destination societies. Such securitization is anything but a new phenomenon. For instance, in the US in the late nineteenth and early twentieth centuries, Catholic immigrants from countries such as Italy, Ireland and Poland were often looked upon with considerable suspicion, and Jews, immigrant or not, have been looked upon with suspicion and hostility throughout European history.

Securitization has a mass psychological dimension. In the absence of a real threat, politicians are tempted to manufacture an imaginary threat. As long as the public believes this, migration fearmongering can be an effective political strategy. Demonizing the migrant as a potential 'terrorist' 'creates fear and a perception of threat to ontological security far exceeding actual developments' (Faist 2006: 630). The perceived threats associated with migrant and minorities may be divided into three categories: cultural, socio-economic and political (Lucassen 2005):

- The perception of migrant and migrant-background populations as a threat to *culture and identity* of destination countries have been commonplace in Europe with regard to Muslim immigrants and in the US towards 'Hispanic' migrants;

- Examples of the perceived *socioeconomic threat* that minority and migrant populations would form include Jews in Nazi Germany, Italians in Third Republic France, Chinese diasporas in South-East Asia, Indians in East Africa, Surinam and Guyana, and Syro-Lebanese in West Africa;

- Examples of migrants and minorities as potentially *politically disloyal or subversive* include the Japanese in the US during the Second World War, Palestinians residing in Kuwait and Yemenis in Saudi Arabia prior to the first Gulf War; ethnic Chinese in Indonesia suspected of political subversion on behalf of Communist China in the 1960s; and ethnic Russian populations stranded in Baltic Republics after the collapse of the Soviet Union.

These examples show that xenophobia and the securitization of migration is neither new nor a typically Western phenomenon. The political impetus to portray migrants, 'aliens' and minorities as an imminent threat to security can be very powerful, as it serves to (1) create a climate of fear, which allows politicians to (2) position themselves as strong leaders, thereby (3) drumming up political support to unite behind a common enemy. Such demagoguery highlights the mass-psychological dimensions of securitization, which needs to be distinguished from its additional political functions of (4) distracting the attention away from pressing socioeconomic problems such as unemployment, inequality and a lack of housing and good health care, by (5) blaming migrants for these problems not of their own making.

The securitization of migration and the growth of nativist sentiment is not necessarily a political fringe problem but can dominate the general public mood particularly when advocated by politicians. In the US, Chinese and later Japanese migrants have often been constructed as a 'yellow peril' from the late nineteenth century. In 1882, in response to virulent anti Chinese sentiment drummed up by labour leaders and politicians, the US adopted the Chinese Exclusion Act, prohibiting immigration of Chinese workers and barring them from citizenship (Zong and Batalova 2017a). During the Second World War, the US government interned about 110,000 people of Japanese descent – most of them US citizens – officially out of concern about espionage activities (Chin 2005).

In the US there was also considerable hostility towards the Irish and Southern Europeans. Between 1890 and 1920, about 50 public lynchings of Italians were documented in the US, part of broad anti-Italian sentiment. An 1891 editorial of the New York Times called Sicilians 'the descendants of bandits and assassins' and 'a pest without mitigation' (New York Times, 16 March 1891). Such sentiments were also common in elite circles. In 1914, Edward Ross, a prominent US sociologist, argued that white Americans were committing 'race suicide' by admitting Southern Europeans and those of 'African Saracen, and Mongolian blood' (Fitzgerald 2014: 115). A decade later, Edwin Grant, another sociologist, in the *American Journal of Sociology* called for 'a systematic deportation' that 'eugenically cleanses America' of the 'Scum from the Melting Pot' (Grant 1925, quoted in Fitzgerald (2014: 115)).

In nineteenth century Britain (as well as in the US), there was frequent hostility against (Catholic) immigrants from Ireland (Kenny 2006), while Jews fleeing pogroms in Russia and settling in London were also targeted by racist campaigns. In late-nineteenth Germany, there was widespread anti-Polish sentiment. The prominent sociologist Max Weber warned in 1895 that Polish agricultural migrants of a 'lower race' (*tieferstehende Rasse*) were displacing German farmers (Smith 2011; quoted in FitzGerald 2014: 115). However, in the interbellum, nativist sentiments in Germany and elsewhere in Europe were increasingly directed towards Jews. Drawing on age-old European anti-Semitism and a long history of discrimination and violent pogroms, this culminated in the Holocaust (see Chapter 5).

In the period between 1945 and 1989, migration was generally not viewed as a threat to security, at least not in mainstream Western political discourses. After the Second World War, the revulsion against Nazi war crimes served to delegitimize extreme right parties and other anti-immigrant movements. In addition, the Cold War and the threat of nuclear warfare in particular dominated feelings of insecurity. Mainstream study of security largely reflected *realism*, a school of thought about international relations that traditionally assumed that only sovereign states were relevant to analysis of questions of war and peace. In this perspective, migration seemed of marginal significance for security. During the post-war economic boom, immigration was generally seen as a positive force contributing to economic recovery and growth. While countries of the 'New World' encouraged permanent immigration, in Western Europe, the prevalent assumption characterizing that post-war migrations would be mainly temporary also contributed to this perception.

The status quo that prevailed after 1945 endured until roughly 1973, during which time immigration policies were barely an issue of open political debate, with most migration policies being subject to back-room decision-making. A first harbinger of change came with the *politicization* of migration policies in Western Europe after the 1973 Oil Shock. A series of economic recessions and rising unemployment put governments of countries like Germany, France, Belgium and the Netherlands under pressure to stop the recruitment of migrant workers. Mass unemployment and marginalization of former guestworkers during the 1980s and 1990s led to intense debates in Western European societies about the integration of immigrants, and the gradual, reluctant and often painful realization that immigrants were 'there to stay' (see Castles 1985). Such politicization brought migration issues into the public arena, racist violence by far-right groups has never ceased to exist (see Box 10.1), but did not yet engender securitization in mainstream political discourses.

Box 10.1 Racist violence

The change brought about by migration, or the mere presence of ethnic minorities, can lead to racist violence. Key targets of European racism include Europe's 8 million Roma, as well as Muslims and Jews (Björgo and Witte 1993). Besides violence, racism also involves discrimination in employment, housing and education (EUMC 2006). German reunification in 1990 was followed by outbursts of racist violence. Neo-Nazi groups attacked refugee hostels and foreigners on the streets, sometimes to the applause of bystanders.

Racist violence persists today: in 2012 a neo-Nazi gang murdered nine people of Turkish and Greek descent as well as a German policewoman, and attacked Turkish fast-food stalls at random. The police failed to investigate properly, publicly blaming Turkish criminals for the attacks. A study found that Turkish immigrants in Germany had lost much of their confidence in the German state, with many afraid there would be further killings (Witte 2012).

The US has a long history of white violence against African–Americans. Despite the anti-racist laws secured by the Civil Rights Movement of the 1960s, the Ku Klux Klan, neo-Nazi and white supremacy groups remain a threat, and regular attacks have occurred. However, conflicts between minority groups also arise, as shown by the violence of African–Americans towards Koreans in Los Angeles during the 'Rodney King riots' of 1991.

Since the terrorist attacks of 9/11 in 2001, racism has often targeted people of Muslim background. Anti-Muslim sentiment expressed by some politicians and media create a climate where far-right and racist attacks have flared up (Müller and Schwarz 2018). The sentiment that Muslim immigrants form a 'fifth column' aiming to undermine Western society from within – which is reminiscent of anti-Semitic conspiracy theories – has become a central rallying point of the alt-Right movement in the US, extreme right-wing parties in Europe and nativist groups elsewhere. A minority of people with far-right sympathies felt emboldened by Brexit and the 2016 US elections.

Parallel to attacks by radicalized Islamists, extremist nativist groups have also perpetrated violent attacks. In 2017, the German government recorded at least 950 *Islamophobic* incidents, including the vandalizing of mosques, online abuse and physical attacks on veil-wearing women, leaving 33 people injured (Neue Osnabrücker Zeitung, 3 March 2018). In 2019, a white supremacist killed 51 people and injured 49 others in attacks at mosques in Christchurch, New Zealand.

In the US, the rise of advocacy for minorities has coincided with a rise in white nativism, including extremist attacks responsible for 49 fatalities between 2000 and 2016 (Statista 2019). Although anti-Muslim sentiment has been on the rise, racist violence in the US is still predominantly directed at African–Americans and Latinos. In 2016, the FBI recorded 6,121 official instances of hate crime in the United States (Statista 2019), with 2,220 victims among African–Americans. Hate crime also targeted Jewish and Latino groups, as well as persons of South Asian origin, sometimes in the belief that they were Muslims or Middle Eastern (Human Rights First 2008).

Racist violence has not gone unchallenged. Anti-racist movements have developed, often based on coalitions between minority organizations, trade unions, churches and welfare organizations. However, as long as politicians are eager to make electoral capital out of anti-immigrant or anti-Muslim sentiments, racism will continue to be a problem. Racist campaigns and violence are important factors in the process of ethnic minority formation. By isolating minorities and forcing them into defensive strategies, racism may lead to self-organization and separatism, and even encourage religious extremism. These are stark reminders of the political risks of playing the 'race card'.

The construction of an 'Islamic threat'

This started to change after the end of the Cold War, a period in which the fear of mass immigration has partly replaced the function of the 'Communist threat' in Western politics. A major theme of this securitization of migration – particularly in Europe – has been the idea that Islam would be a threat to secular Western societies and that Muslim migrants would form a 'fifth column' plotting to import an Islamic Revolution. Initially sparked by extremist propaganda, such idea has gained traction in Western European populist parties. In 1989, Ayatollah Khomeini, leader of the 1979 Islamic revolution and head of the Iranian state, issued a *fatwa* ordering Muslims to kill Salman Rushdie, a British-Indian novelist, for alleged blasphemy in his novel *The Satanic Verses*. The ensuing street protests and book-burnings by Muslim migrants in Britain and some other European countries were grist to the mill of extreme-right wing propaganda about the 'Muslim Peril'. At the same time, in Northwest Europe refugees were increasingly framed as 'bogus asylum seekers' and economic migrants as 'welfare tourists'. This fuelled a political climate of suspicion towards the non-European 'other', particularly if they came from Muslim countries.

The 9/11 attacks in 2001 intensified the framing of immigration as a fundamental threat to the security of Western destination societies. In Western countries, the *securitization* of migration after 9/11 emanated from processes of increasing politicization of migration that had already evolved in the 1980s and 1990s (Chebel d'Appollonia 2012: 49–76). A key dynamic involved a blurring of counter-terrorism measures with immigration policy measures. An example is the re-introduction of the legal obligation to carry identity documents in the Netherlands, which had been abolished in post-war years because of its association with Nazi terror and mass-deportation of Jews. Also, stringent border security checks seemed to serve a twin goal of countering undocumented migration *and* terrorism. In this process of 'securitization', the growing presence and visibility of Muslims, most of whom are of immigrant background, have fuelled extremist discourses about an alleged 'Muslim takeover'.

Over the past decades, the rise of Islamist insurgencies in Afghanistan, Iraq, Libya and Syria, the murder of film-maker Theo van Gogh by an Islamist extremist in the Netherlands in 2004, and a wave of terror attacks in various European cities including Madrid, London, Brussels and Paris, are seen by extremist politicians as additional evidence that Islam forms a fundamental threat to modern Western societies. Such attacks are perpetrated by a small minority of home-bred terrorists of Muslim background. The phenomenon of religious radicalization and violent attack by Islamist extremists are issues of serious public concern. However, the vast majority of Muslims do not support such radical movements (see Esposito and Mogahed 2007; Kepel 2002; Pargenter 2008; Roy 2003). Although there is broad support and respect amongst European Muslims for European democracies (Boswell and Geddes 2011) as well as evidence that Muslim migrants gradually absorb much of the host culture (Norris and Inglehart 2012), 'Islamic threat' and 'Islamization' narrative have gained considerable political traction in nativist circles, with various politicians and media fuelling fears that potential terrorists may infiltrate Western countries disguised as asylum seekers.

In some countries, Islamophobia seems to have taken the place of – or has rather been added to – age-old European anti-Semitism which was also shrouded in similar conspiracy thinking. In other countries, such as the UK, the threat narrative around migration has remained focused on the idea that (East European) migrants take away jobs, put pressure on public housing and health care, and undermine the welfare state. This process culminated in the 2010s with major electoral gains for parties and politicians running on an explicitly anti-immigration and anti-Islam platform, resulting in the Brexit vote in the UK and the election of Trump as US president in 2016.

Is xenophobia really on the rise?

It is important not to confound politicization and securitization. In itself, the politicization of migration is not a worrying phenomenon. One could argue that an issue that is so central to sovereignty, national identity and citizenship should be subject to democratic debate and critical scrutiny, and should not be the subject of backroom dealings between political elites and corporate lobby groups. It becomes more problematic if entire groups of migrants and minorities are collectively positioned as an essential threat to security and national identity. A second nuance is that the rise of extreme right-wing parties should not be equated with a general process of securitization of politics and societies as a whole. A more optimistic interpretation of the current situation is that, overall, societies and politicians have reacted with remarkable calm to terrorist attacks by Islamist extremists of immigrant backgrounds.

In fact, Boswell (2007) questioned the idea whether 9/11 and other terrorist attacks encouraged a 'securitization' of migration controls, and argued that both political discourse and policy practice in Europe have remained surprisingly unaffected by the terrorism threat. She argued that attempts to construct a causal linkage between immigration and terrorism proved impossible to sustain. Also, at the policy level, Boswell argued that 'we see little indication that migration control practices have been colonized by security professionals', with the exception of 'the appropriation of migration control instruments for the purposes of enhancing surveillance by security agencies' (Boswell 2007: 606). The breakthrough of various anti-immigration parties and politicians over the 2010s appears to indicate that securitization has gained ground, but it seems to have barely translated into more restrictive entry policies.

While the extent to which 'securitization' has been significant is subject to controversy (see also Squire 2015), this debates also shows the importance of making a distinction between discourse and policy. Although the discourses of political parties on immigration may have moved towards more restrictive stances (see Davis 2012), contrary to what many believe, immigration policies have generally not become more restrictive (see Chapter 11).

There are also reasons to question the idea that xenophobia is generally on the rise. Public opinion research in Europe shows mixed results, and generally suggests that attitudes towards immigration vary considerably across countries but are also rather stable through time, and that racial prejudice may rather be declining over time and seems to be lower amongst younger cohorts (see Ford 2008), who increasingly co-identify as 'European' (see Lutz *et al.* 2006). Interestingly, inhabitants of countries and places with high immigration rates seem to have generally more positive attitudes towards immigration and diversity than people with limited direct experience of migrants. As with the Brexit vote, what various studies rather seem to show is a high degree of polarization amongst populations on issues of migration, globalization and identity along age, education and rural–urban divides. Yet it seems difficult to make a case that public xenophobia is *generally* on the rise.

Multilateral migration governance: A cul-de-sac?

Given the central importance of migration to state sovereignty, international migration politics is an extremely complex issue, particularly because migration is but one element in broader geopolitical strategies of governments and diplomacy. Efforts to achieve multilateral migration governance have therefore often been fraught with setbacks. States have sought to regulate international migration either through bilateral treaties or through regional agreements. While bilateral negotiations on migration often do result in some level of compromise, the resistance to do so is much higher on a multilateral level. Various initiatives to establish such global governance, such as the High Level Dialogue on Migration and Development at the UN in 2006, its follow-up through an annual Global Forum on Migration and Development (GFMD) since 2007, and the UN Global Compact for Safe, Orderly and Regular Migration (GCM) in 2018 have been useful in terms of establishing dialogues but have been non-binding and unable to bridge fundamental divides between origin and destination states.

The fundamental problem is that any effective form of multilateral migration management requires the transfer of national sovereignty over what is seen as one of the core tasks of modern states – the control over the immigration and settlement of non-citizens – to supra-national governmental bodies and legal (adjudicating) institutions. This explains the opposition of many destination states to engage in binding international agreements on free immigration rights – and opposition against the loss of sovereignty in migration issues implied in EU membership also partly explains the 2016 Brexit vote. Given these complexities and political concerns, pleas for 'open borders' seem unrealistic. First of all, there is often considerable confusion about what is meant by 'open borders'. It is important to distinguish three levels of mobility freedoms:

- Free entry (visa-free travel);
- Right of residence (stay); and
- Right to establishment (including the right to work and do business).

In negotiations among member states of regional unions, it is generally is much easier to agree on and implement freedom of travel than to introduce freedom of stay and establishment. The latter would come down to granting free migration rights to nationals of other members states. There are currently at least 20 regional economic unions and communities in the world, such as the European Union, the Association of Southeast Asian Nations (ASEAN) free trade area, the North America Free Trade Agreement (NAFTA), Mercado Común del Sur (MERCOSUR) in South America and ECOWAS in West Africa.

Because of the controversial status of the migration issue, regional unions find it easier to agree on liberalizing trade than freeing up migration. A good example is the North African Trade Agreement (NAFTA). Unlike the EU, NAFTA only created a free-trade area. Restrictions on mobility remained largely unaltered by the treaty as it allows only certain categories of higher-skilled workers to move freely across borders. However, besides the EU, several regional unions including ECOWAS, ASEAN and MERCOSUR have granted free entry rights through mutual lifting of visa barriers.

The mobility rights accorded by the various regional unions around the world to migrants differ substantially, and, with the exception of ECOWAS and the EU, other regional unions have generally not succeeded in ratifying clauses on the right of residence and establishment. This is partly because of concerns about the social, cultural and economic consequences of overnight establishment of free migration regimes, partly because of the extreme complexity of the regularity frameworks and harmonization of legislation needed to establish functioning free migration regimes, including effective regulations on diploma recognition and portability of benefits and pensions.

Freeing up intraregional mobility through visa-free travel can be a first step on the much longer, bumpy and uncertain road towards creating regional free migration spaces. In the longer term, visa-free travel can contribute to the removal of xenophobia, increasing knowledge of other societies and cultures, and the fostering of social and cultural ties, and thus create the conditions for the further long-term liberalization of migration regimes. Although visas are nominally not about migration, they can significantly affect migration patterns. For instance, the lifting of visa requirements allows prospective migrants to reduce costs of migrating, to avoid dependency on smugglers, to assess opportunities and to get objective information about circumstances at the destination. The lifting of visa requirements can be a significant boost to tourism, trade and foreign direct investment (see Czaika and Neumayer 2017). In the Association of Southeast Asian Nations (ASEAN) region, the lifting of visa requirements in 2006 boosted regional mobility and travel, and has drastically brought down the costs of travel, particularly through the emergence of low-cost airlines (see Hirsh 2017).

Establishing visa-free travel is an essential first step towards regional economic, social and political integration. The latter can include provisions for rights of residence and establishment, but has proven to be politically much more controversial, and can only be a viable option provided that that member states of regional organizations fully commit themselves to the long-term goal of evening out national economies (Castles 2006: 749), and are willing to give transfers of sovereign legislative, executive and adjudicating rights on the 'control of the legitimate means of movement' (Torpey 1998) to supranational bodies and courts such as the European Commission and the European Court of Justice.

The European Union's migration regime

European unification and the fear of mass migration

The European Union is the regional union which has achieved the highest level of effective integration in terms of migration governance. Stretching back to the European Coal and Steel Community (ECSC) of the early 1950s and the European Community (EC) up to 1992, the EU and its predecessors have comprised a federalist project with an explicit commitment to eventually supersede member-state sovereignty through the creation of European institutions and governance. The project has been security-driven, as economic and, eventually, political integration was above all a strategy to prevent the recurrence of war between member states.

The Single European Act (SEA) of 1986 aimed to achieve a genuine common market and paved the way for signature of the 1992 Treaty on European Union (TEU, also known as the Maastricht Treaty), which resulted in the reinforcement and expansion of federalist European institutions within the then 15-member-state area. The TEU created three pillars related to the single market, Justice and Home Affairs, and the Common Foreign and Security Policy respectively. Governance procedures in the pillars varied, with the first pillar being the most supranational – that is, controlled by decisions at the EU level rather than by member states. The TEU left immigration and asylum matters in the third pillar, that is, primarily in the hands of the member states.

Photo 10.2   (L-R) Czech Republic Prime minister Mirek Topolanek, European parliament president Hans-Gert Poettering, Poland's Prime Minister Donald Tusk and German Chancelor Angela Merkel attend a ceremony as the border gate is lifted up at the German-Polish border checkpoint Porajow-Zittau in December 2007, in the Eastern German town of Zittau, Germany

Source: Getty Images/Marcel Mettelsiefen/Stringer

Aiming to secure an 'area of freedom, security, and justice', the 1997 Treaty of Amsterdam integrated into the EU body of law all decisions made by the member states of the Schengen Agreement (see below). Issues concerning visas, asylum, immigration and other policies related to free movement of persons were brought under the first pillar of the Union. This establishment of a common immigration and refugee policy introduced a progressive transfer of decisions pertaining to free movement (external border control, asylum, immigration and rights of *third-country nationals*) from national and intergovernmental levels to supranational authority. The Lisbon Treaty, which was signed in 2007 and ratified in 2009, marked the complete inclusion of migration and asylum within the framework of European treaties. Hence, migration and asylum were to become central issues in EU governance.

Migration has always played a central role in the history of European integration. The 1957 Treaty of Rome envisaged the creation of a common market between the six signatory states. Under Article 48, workers from member states were to enjoy freedom of movement if they found employment in another member state. In the 1950s, Italy pushed for regional integration in order to foster employment opportunities for its many unemployed citizens (Romero 1993). By 1968, when Article 48 came into effect, Italy's unemployment problem had eased, due in part to economic development spurred by the infusion of EC structural funds. As a consequence, relatively little intra-EC labour migration occurred (Werner 1973).

The accession of Spain and Portugal in the mid-1980s sparked fears that the rest of the enlarged EC would be flooded with Portuguese and Spanish workers. However, the predicted massive inflow did not occur. Instead, after their accession to the EC in 1986, Spain and Portugal became significant lands of immigration in their own right. Meanwhile, intra-European labour mobility remained behind expectations, partly because intra-European capital mobility substituted for labour mobility (Koslowski 2000: 117).

The establishment of the Schengen zone and EU enlargement

France, Germany, Belgium, Luxembourg and the Netherlands signed the Schengen Agreement in 1985. They committed themselves to hasten the creation of a border-free Europe in which EC citizens could circulate freely internally without passport controls. The establishment of the Schengen free mobility zone also created the need to harmonize external border controls and visa rules towards third-country nationals. The SEA (Single European Act) of 1986 defined the single market as 'an area without internal borders in which the free movement of goods, persons, services, and capital is ensured within the provision of this treaty' (Geddes 2000: 70). Many Europeans, including the governments of several EU member states, balked at the idea of eliminating internal borders, fearing that it would lead to further irregular migration and loss of governmental control over entry and stay foreigners.

In March 1995, the Schengen Agreement finally came into force for those signatory states which had established the necessary procedures: Germany, Belgium, Spain, France, Portugal, Luxembourg and the Netherlands. Border elimination was, however, compensated for by the creation of the Schengen Information System (SIS), a network of information designed to enhance cooperation between states on judicial matters such as transnational crime and terrorism. Effectively, the Agreement created a new class of 'Schengen citizens' to be added to the existing categories of EU citizens and 'third-country' (non-EU) citizens. Austria joined the Schengen Agreement in 1995, followed by Denmark, Finland and Sweden in 1996 (Denmark was able to opt out of certain sections). The UK and Ireland refused to join the Schengen Agreement, insisting on maintaining border controls of people coming from the continent.

As this was the case when Greece, Spain and Portugal joined the union, temporary labour-mobility restrictions were placed on the workers of Central and Eastern European states (the 'A8'), which joined the EU in 2004, by most of the 15 states already comprising the EU, with the exception of the UK, Ireland and Sweden. By 2011, 669,000 A8 citizens were working in the UK. The magnitude of the migration surprised the British government – and sparked considerable xenophobia – but also reflected a legalization effect as a substantial number of those who were registered had been living in the UK prior to 2004 (Boswell and Geddes 2011: 181–187).

The Brexit conundrum: The inextricable links between trade and mobility freedoms

On paper, EU member states largely retained their prerogatives over entry, stay and removal of non-EU citizens. Nevertheless, under the Schengen Agreement, third-country nationals are permitted short-term stays of up to three months in the Schengen area. A 2003 directive defined 'long-term' residents as resident foreigners who had resided in a signatory state for five years or more. It also stipulated that such resident foreigners could live in another member state for more than three months if employed or self-employed but also for purposes of education and vocational training (Boswell and Geddes 2011: 197).

This expanding notion of freedom of movement has remained the most controversial element of European integration. Since Southern European countries introduced visas for North Africans around 1991 as part of the establishment of the Schengen zone, regular surges of trans-Mediterranean boat migration, such as the Canary Island crisis of 2006, the boat migration from Tunisia in the wake of the 2011 Arab Spring, or the large-scale movement of Syrians into Greece and further into Europe in 2015 frequently led to calls to reintroduce intra-European border controls, or to question free European migration rights. Although such crises show an inability amongst European states to come to coordinated responses to such migration surges, proposals to infringe on intra-EU mobility freedoms are generally short-lived. Effective political support for dismantling the European free migration and mobility zone seems limited, mainly because of the huge economic and social benefits derived from free circulation.

Particularly Central and Eastern European states oppose any limitation of free mobility rights, while Western European states are the prime beneficiaries of free trade provisions of the EU as it enlarges their export markets. The linking of the free movement of goods and persons is the cement keeping the union together. This is also why during the Brexit negotiations, the UK government has found it impossible to gain free access to the free trade block without giving EU citizens the free right to enter and settle in the UK.

The international refugee regime

Cold War politics and the emergence of refugee regimes

Unlike voluntary forms of migration, where efforts to achieve international cooperation have mainly been successful at the bilateral and sometimes regional level, the movement of refugees has been subject to more sustained efforts to develop a global policy regime. Forced migration has become a major factor in global politics (Loescher 2001). This is reflected in the changing nature of the international refugee regime. This term designates a set of legal norms based on humanitarian and human rights law, as well as a number of institutions designed to protect and assist refugees. See chapter 2 for definitions of key terms such as refugees, asylum seekers and teh principle of non-refoulement. The core of the regime is the 1951 Convention, and the key institution is the United Nations High Commissioner for Refugees (UNHCR), but many other organizations also play a part: such as the International Organization for Migration (IOM, which in 2016 became a Related Organization of the UN), the International Committee of the Red Cross (ICRC), the World Food Programme (WFP) and the United Nations Children's Fund (UNICEF); as well as hundreds of non-governmental organizations (NGOs) such as Médecins sans Frontières (MSF) and the International Rescue Committee (IRC).

The refugee regime was shaped by two major international conflicts: the Second World War and the Cold War (Keely 2001). Many of the 40 million displaced persons who left Europe in 1945 were resettled in Australia, Canada and other countries, where they made an important contribution to post-war economic growth. During the Cold War, offering asylum to those who 'voted with their feet' against communism was a powerful source of propaganda for the West. Since the 'non-departure regime' of the Iron Curtain kept the overall asylum levels low, the West could afford to offer a warm welcome to those few who made it. Asylum levels remained relatively low with occasional spikes following events such as the 1956 Hungarian Revolution and the 1968 Prague Spring.

Different refugee situations were developing elsewhere. The war leading up to the creation of the state of Israel 1948 went along with massive displacement of Palestinians (see Chapter 9), which prompted the establishment of the United Nations Relief and Works Agency for Palestine Refugees in the Near East (UNRWA). The colonial legacy led to weak undemocratic states, underdeveloped economies and widespread poverty in Asia, Africa and Latin America. Western countries sought to maintain their dominance by influencing new elites, while the Soviet Bloc encouraged revolutionary movements. The escalation of struggles against white colonial or settler regimes in Africa from the 1960s, resistance against US-supported military regimes in Latin America in the 1970s and 1980s, and long-drawn-out political and ethnic struggles in the Middle East and Asia – all led to significant refugee flows (Zolberg *et al.* 1989).

From the 1980s profound social transformations and increased inequalities, as well as the end of the Cold War (which led to the decline and implosion of 'strong states', often backed by the US or USSR), fuelled another round of conflicts, particularly in Europe, Afghanistan, the Middle East and the Horn of Africa, and led to renewed refugee flows. Western states and international agencies responded by claiming that such situations were qualitatively different from the individual persecution for which the 1951 Convention was designed (Chimni 1998). The solution of permanent resettlement in developed countries was not seen as appropriate, except for Indo-Chinese (Vietnamese) and Cuban refugees who fitted the Cold War mould.

In 1969, the Organization of African Unity (OAU, now the African Union) introduced its own Refugee Convention, which broadened the refugee definition to include people forced to flee their countries by war, human rights violations or generalized violence. A similar definition for Latin America was contained in the Cartagena Declaration of 1984. UNHCR followed this broader approach and has taken on new functions as a humanitarian relief organization. It would increasingly help to run camps and provide food and medical care around the world (Loescher 2001). This expanding role has made it one of the most powerful UN agencies.

The end of the Cold War and anti-refugee politics

From the late 1980s, asylum seekers would increasingly find their way to Western Europe and North America from conflict zones in the Balkans, Latin America, Africa and Asia. Numbers increased sharply with the collapse of the Soviet Bloc and the political and economic turmoil this caused in combination with a falling-away of strict border controls characteristic of communist states. The most dramatic flows were from Albania to Italy in 1991 and again in 1997, and from former Yugoslavia during the wars in Croatia, Bosnia and Kosovo. Many of the 1.3 million asylum applicants arriving in Germany, France and Italy between 1991 and 1995 were members of ethnic minorities (such as Roma) from Romania, Bulgaria and elsewhere in Eastern Europe. The situation was further complicated by ethnic minorities returning to ancestral homelands (such as the German *Aussiedler*) as well as undocumented workers from Poland, Ukraine and other post-Soviet states (see Chapter 6).

Global refugee numbers peaked in 1992 at the levels of 17.8 million. This amplified the politicization of refugee migration in Europe, which in the early 1990s particularly focused on asylum seekers. Extreme-right mobilization, arson attacks on asylum-seeker hostels and assaults on foreigners were threatening public order. European states reacted with a series of restrictions in attempts to construct a 'Fortress Europe' (Keely 2001; UNHCR 2000).

- Changes in national legislation to restrict access to refugee status, particularly through the extension of temporary protection regimes instead of permanent refugee status, such as for people fleeing the wars in former Yugoslavia;

- 'Non-arrival policies' to prevent people without adequate documentation from entering Western Europe to stop them from accessing asylum seeker procedures and refugee status. Citizens of an increasing number of states were required to obtain visas before departure. *Carrier sanctions* compelled airline personnel to check documents before allowing people to embark (see Chapter 11).

- Diversion policies: by declaring countries bordering the EU such as Turkey, Morocco and Libya to be 'safe third countries'. This allowed Western European countries to return asylum seekers to these states, if they had used them as transit routes;

- Restrictive interpretations of the 1951 UN Refugee Convention, for instance, excluding persecution through 'non-state actors' (such as the Taliban in Afghanistan);

- European cooperation on asylum and immigration rules, through the Schengen Convention, the Dublin Convention of 1990 and its replacement, the Dublin Regulation of 2003, and EU agreements.

Parallel to developments in Europe, the refugee regime of many other countries has been fundamentally transformed over the last 30 years. It has changed from a system designed to welcome Cold War refugees from the East and to resettle them as permanent exiles in new homes, to an exclusionary regime, designed to keep out asylum seekers from the 'Global South'. As humanitarian and international refugee law imposes clear limits on the extent to which states can deny rights to asylum seekers who have already arrived in destination countries, the main effort has focused on increasing border controls and externalization migration controls to 'third countries', so that the arrival of asylum seekers is prevented in the first place

Trends in refugee migration

Global refugee numbers decreased rapidly after 1992 from its peak of 17.8 million to stabilize at lower levels between 1997 and 2013 between 9.5 and 12.1 million. Although politicians have been keen to attribute this decline to tougher asylum policies, the most important factor behind the post-1992 decline in refugee numbers was declining levels

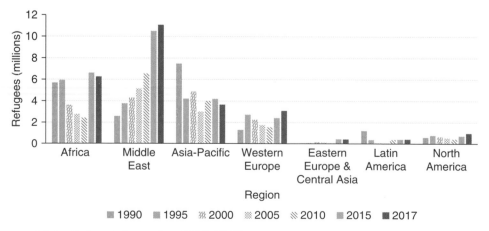

Figure 10.1   Refugees by region, 1990–2017

Source: United Nations Population Division

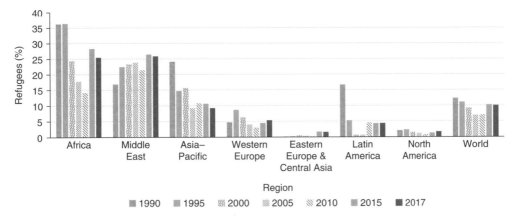

Figure 10.2   Refugees as a share of international migrant populations, by region, 1990–2017

Source: United Nations Population Division

of conflict, such as in former Yugoslavia and Africa, the Soviet retreat from Afghanistan, the end of the Iran-Iraq war in 1988, the collapse of communist regimes and the spread of more democratic modes of governance. The number of international refugees decreased from 17.8 million to 8.7 million between 1992 and 2005. One study estimated that the decline in violence and terrorism explained most of the reduction in asylum seeking in developed countries, with tougher policies only accounting for only about a third of the decline in applications (Hatton 2009).

Since 2013 these numbers have rebounded again, to reach 17.2 million in 2016, mainly as a result of the war in Syria. Figure 10.1 shows long term refugee data in major world regions. These data also include the approximately 5 million Palestinian refugees, who are not covered by the United Nations High Commissioner for Refugees (UNHCR) but by the United Nations Relief and Works Agency for Palestine Refugees in the Near East (UNRWA). Figure 10.2 shows that, since 1990, refugees have counted for between 7 and 10 per cent of the global migrant population, with fluctuations largely depending on conflict levels. Refugees represented about 10 per cent of the global migrant population, thus about 0.3 per cent of the total world population. The highest percentages can be found in the Africa and the Middle East, where about one quarter of all international migrants are refugees. These percentages are lower and generally declining in other world regions, although the refugee crisis in Venezuela is not yet accounted for in these data.

A recent analysis on refugee data between 1951 and 2016 indicated that the number of refugee-generating countries has shown a decreasing trend. This reflects a global decline in conflicts and a concentration of recurrent conflict cycles in a few particular states, such as Afghanistan, Sudan and Somalia (Fransen and de Haas 2019). According to UNHCR data, about 86 per cent of all refugees stay in developing countries, and there has not been a major increase in 'South–North' refugee migration. Countries such as Turkey, Pakistan, Lebanon, Iran, Ethiopia and Jordan currently host the largest refugee populations. Western societies, by contrast, receive a comparatively low number of refugees, and current numbers are not unprecedented (Postel *et al.* 2015).

Although formally the UN refugee convention and the UNHRC provide a global governance structure for refugee migration, the system is partly dysfunctional because of the unwillingness of many states to give permanent residence rights to asylum seekers and the limited willingness to resettle refugees from developing countries. The large majority of refugees thus remain in poor countries, which may lack the capacity to protect them and the resources to provide adequate material assistance. Refugees may spend many years living on subsistence rations in communities with few resources or in isolated camps. UNHCR applies the term 'protracted refugee situation' to refugee populations of 25,000 persons or more in exile for five or more years. Such refugees have few opportunities for work and education, and no prospect of return home to countries still torn by war or ethnic violence. The hopelessness of long years in exile can take a heavy toll: depression, other mental illnesses and interpersonal violence are frequent consequences.

Conclusion

This chapter has shown the centrality of migration to state sovereignty, and how this explains why politicians are keen to be seen 'in control' of population mobility. While nascent empires and bustling economies have often seen immigrants as an economic and political asset that would strengthen the nation, declining empires and stagnating economies have often feared the arrival of strangers. Under such circumstances, politicians get tempted to blame migrants for problems not of their making, such as unemployment, job insecurity, stagnating wages, deteriorating public health care, a lack of social housing and overcrowding generic defunding of public services. Particularly in the post-Cold War context, the politicization of migration has often evolved into the securitization of migration.

Although migration may come with problems, the idea that migrants would be a security threat is not sustained by any evidence. In fact, the threat of 'foreigner hordes' is manufactured by politicians and reproduced by sensationlist media. Such crisis narratives are often shrouded in pseudo-scientific reasoning according to which massive poverty, inequality, climate change and population growth would fuel a migration wave that would threaten welfare, security and social cohesion in wealthy nations. Although such doomsday scenarios have no basis in fact, fear tactics are frequently used by the right and the left to rally support and to call for immediate action, whether through deporting 'illegal aliens', building walls, leaving regional unions such as the EU, or increasing development aid.

As this chapter argues, this 'securitization' of migration, particularly of asylum seekers and potentially undocumented workers, fulfils an important yet dangerous political logic, as migrants – and particularly the most vulnerable amongst them – are easy scapegoats. This largely repeats nineteenth-century fears of peasants threatening the lifestyles of European urban elites and fears of Irish, Jewish, Japanese and Chinse being a threat to the American way of life a century ago. In the long term, such fears never materialize, but the political impacts of immigration fearmongering and public xenophobia can be profound and real, as exemplified by the Brexit vote and the election of Trump as US president.

This chapter also showed the importance of making a distinction between the rhetoric and practices of migration policy making, as there is often a large gap between both. The following chapter will therefore examine how migration policies have evolved *in practice* since the end of the Second World War and will also assess the extent to which these policies have been effective in reaching their stated goals.

Guide to Further Reading

Zolberg (2007) provides an essential introduction to the 'exit revolution'. Torpey (1998) gives insights into the state monopolization of the legitimate 'Means of Movement', while Torpey (2000) provides an overview of the historical establishment of modern passport regime. Boswell and Geddes (2011) provide a comprehensive overview of migration and mobility in the European Union, while Boswell (2007) is a critical discussion of the securitization of migration. Fitzgerald (2019) is a comprehensive overview of ways in which rich democracies repel asylum seekers. The Refugee Studies Centre (RSC) at the University of Oxford (www.rsc.ox.ac.uk) is an invaluable resource on refugees, particularly its key publications *Forced Migration Review* (www.fmreview.org) and the *Journal of Refugee Studies*. The website of the UNHCR (www.unhcr.org) is useful for recent data and information on refugees.

Extra resources can be found at: **www.age-of-migration.com**

11 The Evolution and Effectiveness of Migration Policies

Because the arrival and settlement of migrants often provides a (real or perceived) challenge to state sovereignty and 'national identity', the ability of politicians to control migration and thus exert state sovereignty over the movement of people has become a major political preoccupation. The previous chapter has shown that while early nation states used to be mainly preoccupied with controlling the *departure* (emigration) of their citizens, over the course of the twentieth century the emphasis of migration control policies has shifted to controlling and regulating the *arrival* of foreigners. In order to analyse how states and groups of states have attempted to exert this sovereignty in practice, this chapter gives an overview of the ways in which governments have tried to regulate migration in the modern era, and how effective these policies have been.

In wealthy countries, immigration, in particular of low-skilled workers and asylum seekers from poorer countries, has been increasingly viewed as a problem in need of control. The effectiveness of migration policies has been widely contested in the face of frequent apparent failure to curb immigration and its various unintended effects, opposing those claiming that borders are 'beyond control' (Bhagwati 2003) against those arguing that there is no major migration control crisis (Brochmann and Hammar 1999).

A common perception is that migration has accelerated while migration policies have become more restrictive. As such, the observation that international migration has continued or increased despite policy restrictions is no proof that policies have not been effective, let alone that they have failed. After all, One could argue that immigration would have been even higher *without* migration restrictions. Furthermore, the overview of global migration trends in Chapter 1 challenged the idea that migration has accelerated. Finally, as this chapter will show, it is important not to take politicians' tough migration rhetoric at face value, as there is often a wide gap between what politicians say and the actual policies they adopt and implement on the ground.

Migration policy categories

It is important to distinguish the general role of *states* in migration processes from the more specific role of *migration policies*. We can define migration policies as laws, regulations and measures that states enact and implement with the explicit objective of affecting the volume, origin, direction and internal composition (or selection) of migration (de Haas and Vezzoli 2011). Many '*non-migration*' policies, such as labour market regulations, taxation, social welfare, military and foreign policies affect migration in important and powerful ways, but are not primarily designed to regulate migration, which is what distinguishes migration policies. Migration policies can be categorized along four main criteria: (1) policy areas; (2) citizenship; (3) migrant categories; and (4) policy tools (de Haas *et al.* 2015). First of all, we can distinguish four main *policy areas* which apply to the different phases of the migratory process:

- *Border controls* refer to policies aimed at securing the national territory, particularly with the aim of preventing the entry of 'unwanted migrants' such as asylum seekers;
- *Legal entry and stay* policies regulate legal entry and stay permits, be they for travel or immigration purposes, as well as regularization of migrant groups;
- *Integration* policies regulate the post-entry rights and other dimensions of the integration of migrant groups into destination societies;
- *Exit* policies regulate the forced or voluntary return or exit of migrants and citizens.

These policy areas need to be distinguished because policies may vary substantially across these areas. For instance, more generous post-entry rights can coincide with more stringent entry rules (as is the case in many European countries), or relatively easy entry rules can coincide with limited rights attributed to migrant workers (as is the case in Gulf countries), while high-immigration countries such as Canada seem to combine relative generous admission policies with relatively generous post-entry rights, and many sub-Saharan and Asian states combine restrictive entry and restrictive post-entry migration policies.

The second way to categorize immigration policies is based on the citizenship or national origin of migrants. This has always been an important selection criterion. A classic example is the 'whites only' immigration policies of countries of European settlement such as the US, Canada, Australia and New Zealand. Although such racist immigration policies have been largely abolished since the 1960s and 1970s, national origin still matters a lot in determining people's legal migration opportunities. For instance, regional blocs such as the EU or ECOWAS facilitate free travel and migration between member states, while maintaining restrictions against the settlement of 'third country nationals'. Such regional and national distinctions based on citizenship also play an important role in visa policies.

Third, we can distinguish migration policies in relation to the specific *migrant category* they address. While some policies target particular categories of higher- or lower-skilled workers, others target family migrants, refugees, international students, business people and investors. Often states have developed highly specialized migration policy regimes with rules for particular professions based on assessments of labour market needs – such as in the form of 'points-based systems'. Skills-selective, labour market-driven migration policies have become more important over recent decades. For instance, in 2014 the Dutch government and associations of Asian restaurant owners concluded the *Convenant Aziatische Horeca* (Agreement on Asian Catering). This so-called 'Wok-agreement' enables the legal immigration of Asian cooks who remain tied to their employer (Berghege *et al.* 2018). Besides work, states have also elaborated policies to regulate the immigration of family members of migrants (and citizens), students and asylum seekers.

Political rhetoric often suggests that regulating immigration is about influencing the numbers of migrants coming in. However, in practice modern immigration policies thus rarely resemble the opening and closing of a tap, but rather function as filters by using key criteria such as age, education, skills, nationality and gender to determine the rights of prospective migrants, particularly in terms of their stay and right to work (de Haas *et al.* 2019a). The essence of modern migration policies is therefore not so much their growing restriction, but their increasing focus on migrant selection. This exemplifies the importance of the fourth dimension of migration policies, which are the legal and practical *tools* states deploy to regulate migration.

The migration policy toolbox

Over the past decades states have developed an increasingly diverse and sophisticated array of policies which together aim to regulate the movement and stay of people. These range from work and stay permit systems, quotas, points-based systems to recruitment, carrier and employer sanctions, deportation and regularization campaigns. A key feature of modern migration policies is that they generally aim at affecting the 'selection' of migration in terms of the types of people who are allowed to enter, stay, work and gain access to diverse sets of rights. Rather than a wholesale opening or closing of borders, migration policies thus typically aim to influence the composition rather than the volume of migration. Across the four migration policy areas distinguished above, states use the following main tools to regulate migration:

- *Border control* policies include border surveillance technology including walls and fences, travel visa/permits, rules around passports and identification documents, entry bans, carrier sanctions, detention of undocumented migrants and rejected asylum seekers as well as sanctions against smuggling or illegal employment of migrants;

- *Entry and stay policies* include recruitment and assisted migration programmes, entry or stay permits, work visa and work permits, immigrant quotas (as used by

Source: Getty Images/Alex Grimm

Photo 11.1   A woman puts her passport into a reader during the presentation of the new automated border control system easyPass at Frankfurt International Airport in October 2009

the US), points-based entry systems (as pioneered by Canada in 1967), regularization campaigns (also known as 'amnesties'), refugee status determination policies, refugee resettlement programmes and bilateral or multilateral regional free mobility agreements (such as in the EU, CIS or ECOWAS);

- *Integration* policies are typically associated with policies regulating permanent residency and citizenship, but can also include measures to facilitate language acquisition, citizenship courses, education or (preferential) access to public housing and healthcare. This also includes 'diaspora engagement policies' that regulate states' relations with expatriates and 'heritage populations', such as the extension of voting rights or the rights to 'return', in the hope to secure the loyalty of diaspora members and extract benefits such as remittances (Gamlen 2006);

- *Exit policies* include voluntary and coercive measures. Deportations and readmission agreements with origin or transit states (such as the 'Turkey deal', see Box 6.1) are the most typical example, but exit policies can also include measures to stimulate voluntary return of migrants (such as through 'departure bonuses') or states' efforts to regulate (either stimulate or prevent) the departure of their own citizens, such as through recruitment agreements, selective passport issuance policies or exit visa systems, requiring citizens to ask for permission to leave their country (see de Haas, Natter and Vezzoli 2015).

The table at the end of this chapter gives a full overview of control instruments that form part of the migration policy toolbox (see also de Haas *et al.* 2015). States and national governments are the most important actors in enacting and implementing migration policies. However, they are not the only ones. First of all, sub-national government levels have an important influence on migration policies. In some countries such as Australia and Canada, regional governments have a say in regulating the entry and stay of foreigners. While national regulations determine processes such as refugee status determination and citizenship acquisition, local governments often have an important stake in establishing policies towards migrant integration such as through language, education and housing programmes.

Local governments often have to deal with concrete situations, and tend to be much more acutely aware of the importance of migration for local economies but also the practical challenges and problems immigration can bring. They are more likely to be influenced by local pressure groups and faith organizations defending the rights of migrants to stay. Mayors and councils of big cities with large immigrant populations often lobby governments for more lenient immigration rules. They may also obstruct collaboration with repressive policies such as deportation, such as in the case of 'sanctuary cities' in the US and Canada (Bauder 2017). They are more inclined to provide shelter and basic services to undocumented migrant populations, out of a mix of humanitarian and public order and safety concerns. In 2018, for instance, the Association of Dutch Municipalities (VNG) reached an agreement with the Dutch government, allowing and giving financial support to a limited number of big cities to offer accommodation and food to a group of about 2000 rejected asylum seekers.

Besides regional and local governments, private actors also play an important role in executing migration policies. This particularly applies to the preponderant role of employers, formal and informal recruiters, and temporary employment agencies (such as Randstad, Manpower and Adecco) in the recruitment of – both documented and undocumented – foreign labour, but can also include other private migration intermediaries. Examples include the rising involvement and influence of the private prison industry in US immigration enforcement (Luan 2018), as well as the vested interests of the military-industrial complex involved in border controls, which can all be seen as part of the 'migration industry' (see Chapter 3). Humanitarian organizations can also have an important influence on the execution and effectiveness of migration and refugee policies, either by providing support to complement liberal government policies or by counteracting repressive policies, such as the involvement of *Médecins sans Frontières* in rescuing migrants and refugees in the Mediterranean Sea.

Last but not least, national migration policies are affected by the policies of other states, highlighting the importance of taking account of the wider international context in which national migration policies are made. For instance, large refugee crises can only be managed through effective inter-state cooperation – which is the raison d'être for the UNHCR and refugee conventions. Destination states are dependent on origin or transit states for effective deportation policies, while origin states are dependent on destination countries in their pursuit to enhance the position of their citizens living abroad. Regional free travel and migration agreements such as in the EU, CIS and ECOWAS also show the importance of inter-state collaboration on migration (see Chapter 10).

Between rhetoric and practice

As part of the politicization and securitization of migration, politicians have often adopted discourses emphasizing the need to 'crack down on illegal migration', to limit the entry of low-skilled workers and their families, and to curb the inflows of asylum seekers through toughening entry rules, strengthening border controls, deporting migrants and preventing the unlawful employment of immigrants. To prevent unwanted migration, many Western countries seem to have embarked on a 'quest for control' over cross-border movements. This quest entails efforts to prevent the abuse or circumvention of immigration regulations and policies often shrouded in belligerent terminology such as 'combating illegal migration'. This has given rise to a perception that immigration policies have become more restrictive.

However, a closer look at actual policy developments gives reason to question this perception. It is important not to equate political rhetoric with policy practice. There is often a considerable gap between what politicians say and the actual implementation of policies on the ground. This gap can be explained by either (1) the limited willingness of politicians to implement policies or (2) the practical, financial and moral challenges in implementing such policies, or a combination of both. For instance, politicians often say they want to limit immigration while turning a blind eye towards irregular immigration and employment of migrant workers in vital economic sectors such as personal care, agriculture or catering.

It is possible to analyse migration policies at four levels: (1) official policy discourses; (2) actual migration policies on paper; (3) the implementation of policies on the ground; and (4) policy (migration) outcomes. On this basis, we can identify three 'policy gaps' that can explain perceived or real policy ineffectiveness:

- First, the *discursive gap* between the stated objectives of politicians' (often 'tough') rhetoric or 'discourse', and the (often more watered-down) policies on the ground;
- Second, the *implementation gap* between policies on paper and their actual implementation;
- Third, the *efficacy gap* reflects the degree to which implemented policies have the intended effect on the volume, timing, direction and 'selection' of migration (Czaika and de Haas 2013; de Haas and Vezzoli 2011).

Migration policies are typically a compromise among competing interests, which explains why their objectives are generally multiple and sometimes inherently contradictory. While businesses often lobby for more liberal immigration policies, trade unions have historically seen immigration as a threat to the wages and employment of native workers. This also shows that the migration issue does not neatly cut across the left–right spectrum. Empirical studies suggest that, on average, right-wing governments do not adopt significantly more restrictive policies than left-wing governments (de Haas and Natter 2014). In fact, migration issues tend to divide political parties internally (see Table 11.1). This divide typically pits pro-immigration supporters of economic market liberalism of the right and cosmopolitan-humanitarian streams of the left against anti-immigration cultural conservatives of the right and left-wing economic protectionists (see Massey 1999; Odmalm 2011; Schain 2008). The perception that right-wing parties are 'tougher' on immigration therefore mainly reflects the large gap between rhetoric and practice.

Migration policy making typically involves bargaining and compromising about divergent interests of political parties, businesses, trade unions, or human rights organizations, who favour or oppose the migration of particular groups (Freeman 1995; Boswell 2007). For instance, in the 1960s in Western Europe right-wing political parties and business lobbies favoured the migration of 'guestworkers' and their family members from Mediterranean countries, whereas left-wing parties and trade unions initially saw recruitment as a threat to the interests of native workers (see Bonjour 2011). At the same time, left-wing parties often favour the granting of rights to asylum seekers, undocumented migrants and other vulnerable immigrant groups, while right-wing parties are more likely to oppose this (de Haas and Natter 2014). Such divisions are also reproduced within government bureaucracies. For instance, ministries of justice and interior often take a more restrictive stance toward migration than ministries of economic affairs or labour (Bonjour 2011).

In a similar way, emigration or exit policies tend to be the outcome of various (often countervailing) lobbies as well as power struggles within governments and bureaucracies (Natter 2018, Gamlen 2008), rendering state approaches towards emigration often rather ambivalent. Particularly authoritarian states are often confronted with a trade-off between the perceived benefits of emigration, such as remittances and a political-economic 'safety valve', and the fear that exiles may form a political opposition from abroad. Political leaders in destination and origin countries often pay lip service to 'combating

Table 11.1   Internal divisions on migration issues within political parties

		Left	Right
Restrictive migration policies	Dimension	Economic tradition	Sociocultural tradition
	Ideology	Market protectionism	Value conservatism
	Actors	Labour unions	Cultural conservatives
Liberal migration policies	Dimension	Sociocultural tradition	Economic tradition
	Ideology	International solidarity	Market liberalism
	Actors	Liberal and ethnic groups	Employer lobbies

Source: de Haas and Natter (2014)

illegal migration', while in practice doing little to introduce or enforce migration restrictions, either because they lack the capacity to do so or because they see clear economic and political benefits in migration.

The liberalization of migration policies

Recent studies have challenged the common assumption that migration policies have become more restrictive. Based on an analysis of 6,500 migration policies implemented across 45 countries, de Haas, Natter and Vezzoli (2018) concluded that, since 1945, migration policies have generally become more *liberal*. Drawing on this study, Figure 11.1 shows the average direction of change of immigration policies over the past century. Values below zero indicate that liberal policy measures outnumber restrictive policy measures. It shows that migration policies became more restrictive over the first half of the twentieth century, reflecting the turn towards protectionism and nationalism during and after the Great Depression (see also Timmer and Williamson 1998). This coincided with the introduction of modern passport systems (Torpey 2000) and an increasing focus on controlling immigration.

The period between 1945 and the 1980s saw an accelerated liberalization of entry and post-entry rights for most migrant categories as part of major overhauls of migration regimes. This reflects the increasing commitment of states to international human rights principles, such as the right to family life (giving migrants the right to family reunion) and the right to asylum as well as a gradual extension of facilities for various forms of

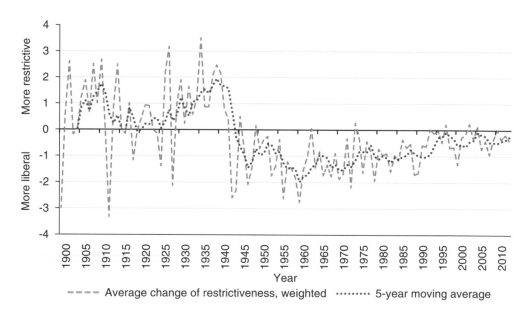

Figure 11.1   Yearly average of weighted changes in migration policy restrictiveness, 45 countries,[1] 1900–2014

Source: de Haas, Natter and Vezzoli 2018, based on DEMIG POLICY data (www.migrationinstitute.org)

[1] The 45 countries included are Argentina, Australia, Austria, Belgium, Brazil, Canada, Chile, China, Czech Republic, Czechoslovakia, Denmark, Finland, France, Germany, German Democratic Republic, Greece, Hungary, Iceland, India, Indonesia, Ireland, Israel, Italy, Japan, Korea, Luxembourg, Mexico, Morocco, Netherlands, New Zealand, Norway, Poland, Portugal, Russia, Slovakia, Slovenia, South Africa, Spain, Sweden, Switzerland, Turkey, Ukraine, United Kingdom, United States of America and Yugoslavia

labour immigration. This also challenges the common perception that after the 1973 Oil Shock, European nations slammed the doors shut to immigration.

Thus post-WWII migration policies have generally been liberalized despite political rhetoric suggesting the contrary. Particularly in Europe, it shows that legal systems gradually came to terms with their new *de facto* status of immigration countries. After 1989, the proportion of restrictive policy changes has increased, heralding a slowing down of the rapid post-WWII liberalization of migration policies. This partly reflects efforts by some governments to restrict access to citizenship as well as immigration of family members of lower-skilled workers, and to require family migrants to take language and integration tests prior to their migration, as well as to tighten asylum criteria.

However, the bulk of restrictive measures focused on stepping up border controls in combination with visa restrictions and *carrier sanctions*, requiring airlines to check the immigration status of passengers prior to boarding. Such measures were meant to prevent asylum seekers and vulnerable groups of undocumented migrants (such as minors and pregnant women) from reaching destination countries and, in doing so, prevent them from claiming various rights. This seems to reflect the increasing politicization and securitization of migration over this period as well as a certain backlash against multiculturalism (see Chapter 13).

Yet the data shows that despite efforts to prevent the arrival of 'undesirable' migrant groups, liberal policy changes have continued to outnumber restrictive policy changes. Rather than a reversal towards more restrictive policies, there has rather been a *deceleration* of liberalization (De Haas et al. 2010). This confirms the considerable gap between what politicians say in public and do in practice. Furthermore, policies differ substantially across policy types and migrant categories: Since 1989, entry and integration policies have become less restrictive, while border control and exit policies have become more restrictive. While policies towards undocumented migrants and family migrants have been tightened in recent years, entry rules for high- and low-skilled workers, students and refugees have generally become more liberal (de Haas *et al.* 2018).

While this trend has occurred in most liberal democracies in Western Europe, North America, Australia and New Zealand, several countries in Asia and Latin America portray an opposite trend, characterized by high levels of restrictiveness up to the 1970s, and an opening-up of immigration regimes since then. In some Asian countries, the liberalization of migration policies is closely tied to broader economic trends, such as the partial dismantling of protectionist economic policies in the 1970s and 1980s. This led to more liberal immigration policies in Japan, China and South Korea, while India and Indonesia abandoned exit restrictions and embarked upon more pro-active 'labour export policies' (see also Kim 1996). Developments in Latin America suggest a link between democratization and the liberalization of migration policies: policy restrictiveness in Latin America peaked in the 1970s and 1980s, a period in which the region was dominated by autocratic regimes. Since then, Latin American states have adopted more liberal and human-rights oriented migration policies (Acosta Arcarazo and Freier 2015; Cantor *et al.* 2013).

The liberal paradox

How can we explain the gap between statements by politicians publicly vowing to control or limit immigration and the much more nuanced practices of immigration policy making? Political scientists have developed several hypotheses to explain why many immigration policies have become more liberal notwithstanding official discourses suggesting the contrary.

Neo-Marxist approaches see immigration policies as part of the class struggle, with the conflicting interests between capital and labour taking centre stage. From this perspective, businesses and capital owners have an interest in easy entry of foreign workers and turn a blind eye towards undocumented migration, as this keeps wages down and boosts productivity and corporate profits. Upper-middle classes and elites in particular benefit from the availability of cheap labour to clean their houses, take care of their children, tend their gardens, pick their vegetables, pack their meat and cook their food. From this perspective, immigration weakens the bargaining power of unions (see Castles and Kosack 1973). Racism and xenophobia facilitate the exploitation of labour by dividing the working class. From this perspective, by making native workers believe that foreign workers take away their jobs and depress wages, anti-immigrant rhetoric by politicians serves the interests of businesses. This can be seen as a divide-and-rule tactic to prevent native and foreign workers from realizing that, in reality, they share the same class interest of obtaining higher wages, more job security and better working conditions.

The *client politics perspective* was pioneered by Freeman (1995) who argued that there is a gap in immigration policy preferences between the political elite and the general public. Client politics refers to the process by which small interest groups gain disproportionate influence on political processes benefits at the expense of the larger public. From this perspective, political elites favoured 'expansive' (more liberal) immigration policies generally opposed by the mass of the public. Freeman argued that immigration generates *concentrated benefits*, especially to employers and investors, and *diffuse costs* borne by the general public. From this perspective, immigration policy has become more liberal because of the effective lobbying of economic interests. The insulation of pro-immigration political elites from electorates generally less supportive of immigration led to a liberalization of immigration policies.

The *embedded legal constraints perspective* focuses on the processes through which liberal democracies put legal constraints on the power of the executive to control immigration. Hollifield (1992) argued that such embedded constraints limit governments' prerogatives in formulating immigration policies. International migrants are human beings with rights, and immigration policies are thereby constrained, particularly in democratic societies. The classic illustration of this came in France in 1977, when a Council of State ruling invalidated the government's effort to prevent family reunification. The French government declared a zero-immigration policy, but could not translate that declaration into policy because France had a bilateral treaty with Portugal that granted legally admitted Portuguese workers the right to family reunification.

Hollifield argued that liberal democracies are trapped in a *liberal paradox*: in order to maintain a competitive advantage, states must remain open to trade, investment and migration. But unlike the movement of goods and capital, the movement of people involves greater political risks, as states are keen to maintain sovereign control of their borders. The central challenge then becomes how to maintain openness and at the same time protect the rights of citizens (Hollifield 1992). In the same way, international human rights law such as the United Nations Refugee Convention curtail the extent to which signatories can expel asylum seekers. This is one of the reasons why states try to avoid their arrival on their national territory, because treaties protect the right of asylum seekers arriving at the border, such as through the *non-refoulement* principle (see Chapter 2). This reveals an important paradox: the expansion of rights for lower-skilled migrants and refugees that have reach their territory has strengthened the incentive for states to prevent their arrival.

The *external legal constraints* perspective argues that, in the post-WWII period, embedded constraints have increasingly gained an international dimension through countries' adherence to international agreements and conventions protecting human rights and the rights of migrants and refugees. Joppke (1998) highlighted the importance of courts in enshrining migrants' rights against the attempt of other state institutions to curtail them. Soysal (1994) viewed the emergence of an embryonic international regime on migrant rights as constraining immigration policymaking of European democracies. Multilateral and bilateral treaties and the influence of international organizations such as the ILO and the Council of Europe empowered international migrants and shaped immigration policymaking. For instance, European governments have not been able to drastically curtail the general right to family reunification introduced in the 1950s and 1960s, as attempts to enact more restrictive policies have been regularly overturned by national and European courts (see also Joppke 2001).

The importance of domestic and international human rights, diplomacy and international norms for migration policy highlight the fact that immigration politics cannot be reduced to the outcome of a class struggle or the influence of powerful business lobbies, but that *ideology* is also an important determinant of migration policies. The liberalization of immigration policies partly reflects an underlying *diffusion of more liberal values*, such as notions of elementary human rights. Bonjour (2011) has therefore argued that this should not be construed as an erosion of the sovereign state, but a conscious political choice informed by changing value systems.

In the same vein, Joppke (1998, 1999) disagreed with the view that any loss of national sovereignty is externally imposed. He argued that such constraints are largely self-imposed by national legal systems. States need to regulate immigration, but by making commitments, such as signing the Geneva Convention on refugees or joining regional unions, they voluntarily impose limitations on what they can do. For instance, EU member states have given up and pooled sovereign prerogatives in order to better achieve immigration policy goals that can be achieved more readily at the regional level (Bonwell and Geddes 2011; Geddes 2003; Lahav 2004). The case of Brexit shows that states have the legal power to reclaim such sovereign rights on migration controls, if they genuinely wish to do so.

Although these theories have been developed to explain policy liberalization in liberal democracies, some of their central arguments may also apply to more authoritarian settings (Natter 2018). There is no automatic relation between democratic governance and liberal migration policies. For instance, democracies such as the US and Canada were among the first and most eager countries to select immigrants by race, and Latin American autocracies the first to outlaw such discrimination (FitzGerald and Cook-Martín 2014). Autocracies in the Gulf have relatively open entry policies although they strictly curtail post-entry rights to labour migrants. Authoritarian governments are also likely to be pressured by business lobbies to let more migrants in or turn a blind eye towards undocumented migration.

Perhaps liberal immigration systems seem a feature of liberal economic systems rather than a characteristic of democratic governance *per se*. After all, in democratic systems elected leaders should also be responsive to popular demands to limit immigration, while autocratic leaders can more easily ignore anti-immigration sentiments (see also Natter 2018). For Africa, for instance, democratization since the 1990s seems to have gone along with a rise of anti-immigration discourses and an increased politicization of immigration (Dodson and Crush 2015; Geschiere 2005; Mitchell 2012).

It is difficult to imagine that citizens of European or North American countries would have supported immigration levels similar to those in the Gulf States. Although Gulf states allow extremely high numbers of migrants in, they give them few if any rights. Ruhs (2013) therefore argued that governments face a trade-off between 'numbers and rights', between openness to admitting migrant workers on the one hand, and the rights granted to migrants after admission on the other. However, beside the normative acceptability of trading off worker rights for economic gains for migrant workers, Castles (2006) questioned whether democratic states will in practice be able to enforce return and prevent the settlement of temporary workers (see also Doomernik 2013). Temporary worker programmes can only succeed if governments take an active regulatory role, which would mean a reversal of neoliberal economic policies leading to decreased labour market regulation and workers' protection (Castles 2006). The danger is to recycle 'guestworker illusions', with politics of denial generating similar negative social outcomes for origin and destination countries as in the past. This exemplifies the complicated dilemmas governments face in designing immigration policies.

Travel visas and the prevention of unwanted migration

While post-WWII migration policies have generally become more liberal, trends differ significantly across policy areas and migrant categories. Entry and integration policies have become more liberal, while border control policies have become more restrictive, particularly through increased funding for walls and fences, border patrolling, border detention, and deportation to origin and transit countries (de Haas *et al.* 2018). Restrictions have mostly targeted 'less-desired' migrant categories (see Bonjour and Duyvendak 2018), including undocumented migrants, prospective asylum seekers and family members of low-skilled migrants. This has been done through a combination of border surveillance, visa policies, carrier sanctions and deportation. These policies seek to prevent migrants from crossing borders because, once on the national territory, they have access to a certain number of rights (which have generally expanded in the post-WWII era).

Although formally not part of immigration policies, states have increasingly used travel visas as a means to block the entry of potential asylum seekers and presumed visa 'overstayers'. As visas can be generally imposed through executive decrees or other administrative measures, and therefore do not require cumbersome legal changes, governments see visas as quick, discrete and effective instruments to curb migration (Czaika *et al.* 2018; Czaika and Neumayer 2017; de Haas *et al.* 2019a). Since the 1980s, destination countries have progressively introduced carrier sanctions to prevent refugees and other migrants without a visa from boarding airplanes and ships. This contributed to the privatization of migration controls through the outsourcing of passport and visa checks to airlines, shipping companies and travel agents (see also Neumayer 2006).

Visas are the rule rather than the exception. In the mid-2010s, around 73 per cent of all country-by-country combinations worldwide required a visa, and this percentage has been rather stable over the past decades (Czaika *et al.* 2018). Visa-free travel is mostly realized between geographically-contiguous countries of integrated regional blocs such as ASEAN, CIS, ECOWAS, the EU, GCC and MERCOSUR. Regional blocs have formed clusters of visa openness and external closure to satisfy requirements that privilege citizens of the regional group for internal mobility. European and North American countries have maintained high levels of entry-visa restrictiveness for citizens from regions such as Africa and Asia. Map 11.1 displays 'inbound visa restrictiveness', which is a measure of the share of nationalities needing a visa to enter a country. Map 11.2 displays 'outbound visa restrictiveness',

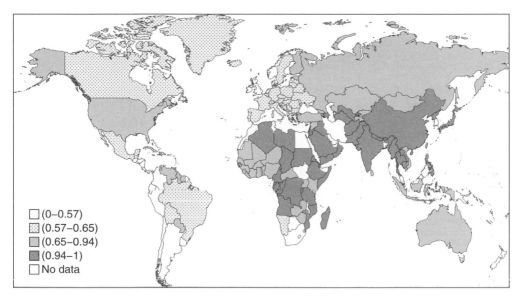

Map 11.1   Inbound visa restrictiveness, world

Source: Czaika, de Haas and Villares-Varela 2018

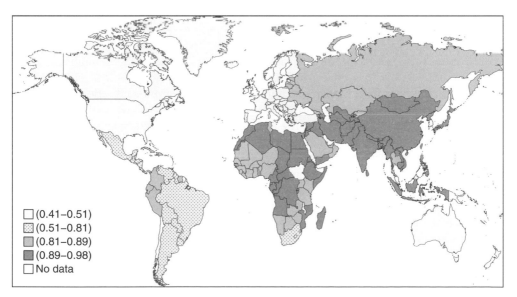

Map 11.2   Outbound visa restrictiveness, world

Source: Czaika, de Haas and Villares-Varela 2018

which measures the share of countries that require a visa for nationals from a country. It shows that some of the most restrictive entry visa regimes of the world are found in sub-Saharan Africa and South and South-East Asia. Citizens of countries of these regions also tend to face high obstacles in terms of visa requirements imposed upon them.

The enforcement challenge

Besides the frequent *discursive gap* between immigration policy discourse and immigration policy practice, the *implementation gap* refers to the difficulty of enforcing policies on the ground. These obstacles can be practical, financial or moral in nature. For instance, it would be considered unacceptable in most countries for the police to raid private homes to check the immigration status of cleaners, nannies and other caregivers. In Italy, for instance, it is a 'public secret' that many private caregivers for elderly people are migrants, often without immigration status, but most Italians understand the crucial social function of these migrant workers, and their large-scale detention and deportation would therefore certainly meet mass resistance (see Ambrosini 2016).

This exemplifies the moral and humanitarian limits of what states can do. Some policies are not or only partly implemented because of practical, planning or budgetary constraints or because of corruption, ignorance or subversion. For instance, the deportation of foreigners is very difficult if origin or transit states refuse to collaborate, if migrants destroy identity documents or if they belong to vulnerable groups (such as minors and pregnant women) that can claim special humanitarian protection. Besides that, deportations are expensive, and emotionally and psychologically taxing for officials carrying out these policies.

Implementation is not only dependent on resources and public support, but is also a matter of discretion of people and organizations involved in the day-to-day carrying out of policies (Eule 2016). For instance, airlines implementing carrier sanctions, asylum case-workers, border agents, or public or private institutions processing visa requests often have considerable leeway in their decision-making (Ellermann 2006; Infantino 2010; Wunderlich 2010). There is often considerable scope for interpretation and political pressure, for instance in refugee status determination and work permit applications (Ellermann 2006). This means that policies can become more or less restrictive without any legal change (Infantino 2014). Challenging the view that the public would favour more restrictive policies than governments, Ellermann (2006) argued that 'street-level bureaucrats' regularly encounter local publics who are opposed to the strict enforcement of migration rules.

The effective enforcement of migration policies has been a particular challenge for policy makers. These challenges testify to the power and agency of crucial actors in the migration process, such as employers, migrants, agents, smugglers and other intermediaries, who often counteract the practical execution of policies. This regularly creates situations of tolerance of irregular entry, employment and long-term stay, which often compel governments to eventually grant legal status to such immigrant groups that were not supposed to be there or only temporarily so. Other obstacles for the effective enforcement can be the moral and psychological burdens the implementation of certain migration policies, such as deportations (particularly of families), can put on police, border guards and government workers.

Governments' limited *willingness* to enforce policies is another important factor behind implementation gaps. This is most obvious in the case of employer sanctions. Most Western countries have adopted laws punishing employers for the illegal hiring of undocumented foreigners. In practice, these programmes have met resistance as employers often have the political clout to prevent effective enforcement of sanctions. Effective enforcement can also be frustrated by insufficient personnel, poor coordination between various agencies, inadequate judicial follow-up, and the adaptation of employers and workers to enforcement measures.

In particular, enforcement levels with regard to irregular employment are low in many countries. For instance, for a century before 1986, US federal law permitted employers to hire undocumented workers (Wishnie 2007). In 1986, the Immigration Reform and Control Act (IRCA) made the hiring of undocumented foreigners a punishable offence. However, the IRCA permitted an assortment of documents to prove employment eligibility, many of which could be easily forged or fraudulently obtained. The Commission for Immigration Reform concluded by 1994 that the employer sanctions system had failed because undocumented foreign workers could simply present false documents to employers (Martin and Miller 2000: 46).

The enforcement of employer sanctions was further hampered by political opposition within the US. Several Hispanic advocacy groups alleged that employer sanctions would increase employment discrimination of minorities, while business interests viewed the restrictions as another government-imposed burden. Others feared that enforcement would disrupt entire industries, such as labour-intensive agriculture, which would result in crops rotting in the fields and higher food costs.

Even the American Federation of Labor and Congress of Industrial Organizations (AFL-CIO), the major confederation of unions, which had been supporting employer sanctions since the early 1980s, announced in 2000 that it no longer supported enforcement of employer sanctions. The AFL-CIO was under new leadership, which emerged from unions with large numbers of immigrant members including many undocumented workers. This faction had close ties to the US Conference of Catholic Bishops, which supported the legalization of the millions of undocumented resident foreigners. This diverse opposition ensured that there was no political consensus in support of the enforcement of employer sanctions.

The overall record of employer sanctions enforcement has not been strong. In 1999, the US government announced that it was suspending enforcement, which resulted in a reduction of the number of Immigration and Naturalization Service (INS) employer investigations from 7,537 cases completed and 17,552 arrests in 1997 to 3,898 cases completed and 2,849 arrests in 1999 (INS 2002: 214). In 2007, the Bush Administration announced a major enforcement initiative centred on so-called 'no-match letters'. Such letters are sent by the Social Security Administration to employers advising them that an employee's name or social security number do not match the agency's records. In response to a no-match letter, employers have 14 or 90 days to resolve the discrepancy, to fire the undocumented worker or risk fines up to US$10,000. Prior to 2007, such no-match letters generally had not been followed up with enforcement; The initiative appeared to confirm that the US never truly had a credible employer sanctions regime.

Enforcement of employer sanctions continued to oscillate under the Obama and Trump administrations. The overall record of workplace enforcement has remained weak. In 2018, 6,484 worksite investigations were opened, and 779 criminal and 1,525 administrative arrests were made (DHS 2019). Although the Trump administration boasted that this was a threefold increase from the year before, these are similar levels as two decades ago. If anything, such figures indicate that only a tiny proportion of the almost 12 million undocumented migrants in the US were affected. Its significance paled in comparison to other enforcement measures, most notably deportations which soared from less than 50,000 in the mid-1999s to 165,000 in 2003 to record levels of 435,000 in 2013.

Legalization: Policy failure or policy tool?

Legalization programmes have been part and parcel of immigration policy regimes in most Western countries. Legalizations can be interpreted as evidence of governmental inability to prevent irregular entry of stay or alternatively as evidence that sovereign states are able to adapt to the realities of international population movements. Opponents of legalization typically contend that legalization undermines the rule of law. Proponents point out that legalized foreigners generally experience improvements in their overall socioeconomic and employment prospects, and that legalization programmes should be considered as part and parcel of migration policies. According to this line of reasoning, as long as states try to regulate migration, there will be people 'falling between the cracks', so some degree of 'illegality' seems almost inevitable in any immigration regime. Since states generally cannot tolerate the long-term presence of large unregistered populations, there will always be a long-term drive towards regularization.

Prior to the 1970s, many national migration policies used 'back door' rather than 'front door' immigration to address labour shortages. During the *Bracero* policy era, nearly 87,000 Mexicans working without authorization were legalized in the US. In France, foreigners who took up employment irregularly after 1947 were routinely legalized and, between 1945 and 1970, legalization comprised the major mode of legal entry into France (Miller 1999: 40–41). Thereafter, the French government declared that legalization would be exceptional, but there were recurrent legalizations throughout the 1970s and 1980s (Miller 2002).

In 1981, the election of a Socialist President and a leftist majority in the National Assembly set the stage for a new French approach to legalization to counter irregular migration and employment. Unlike previous legalizations, trade unions and immigrant associations participated in the legalization effort and additional governmental personnel were mobilized to facilitate the processing of applications. As a result, approximately 120,000 of 150,000 applicants were legalized (Miller 1999: 40–41). Although the legalization benefited its participants, it did not alter the underlying labour-market dynamics attracting undocumented migrant workers.

In 2006, the then Minister of the Interior Nicolas Sarkozy introduced a procedure to process the undocumented resident parents of children in French schools on an individual basis. This measure, along with an increase in deportation orders and the cancelling of a provision granting undocumented migrants legal status after 10 years of residency, marked a tightening of French immigration policy. Nevertheless, France recorded a 55 per cent increase in the number of migrants legalized from 2005 to 2006 (*Le Monde*, 6 April 2007).

The legalization provisions in the US IRCA of 1986 differed in several key respects from its French counterpart. First, the IRCA had a five-year period between the cut-off date for eligibility (1 January 1982) and the effective starting date of the general programme (4 May 1987) known as I-687. The general programme was open to all foreigners who could prove residency prior to 1 January 1982. Almost 1.7 million foreigners applied for legal status under I-687 and 97 per cent were approved. The IRCA also had a programme for Special Agricultural Workers (SAW), which targeted foreigners who could demonstrate evidence of seasonal employment for 90 days between 1 May 1985 and 1 May 1986.

A total of 1.3 million foreigners applied, but widespread fraud contributed to a much lower approval rate than the I-687 programme. In all, approximately 2.7 million undocumented migrants underwent legalization under IRCA provisions (GAO 2006; Kramer 1999; OECD 2006).

Large-scale irregular migration to Southern European countries during the 1980s and the 1990s led to mass legalizations from which more than 3.2 million foreigners benefited. Since 1986, Italy has had six legalization programmes resulting in the regularization of about 1.4 million migrants. Between 1985 and 2005, Spain authorized 12 legalizations (Plewa 2006: 247), granting legal status to over one million immigrants (Sabater and Domingo 2012). Political opponent of such policies argued that it would stimulate new migrants to cross borders illegally through the *efecto llamada* (call-to-effect, known in English as 'honeypot effect') (Sabater and Domingo 2012). Despite the heated debate about irregular migration, numbers should not be exaggerated, certainly not in Europe. It was estimated that in 2008 there were between 1.9 and 3.8 million undocumented migrants in the EU27, representing between 0.4 and 0.8 per cent of the total population and between 7 and 13 per cent of the foreign population (Vogel 2009). The majority of irregular immigrants in the EU have entered legally, on some sort of a visa, and then overstayed (Düvell 2006).

The heated debate concerning legalization continued to resonate across the Atlantic. In many ways, increasingly restrictive immigration policies had encouraged the permanent settlement Latino immigrants in the US, many of whom have undocumented status (see Box 11.1). Over the past decades, several proposals emerged to meet the challenges posed by the large undocumented migrant population in the US estimated at about 10.7 million in the late 2010s (Krogstad *et al.* 2018) Several of these proposals advocated 'earned legalization' which would permit undocumented migrants to benefit from a legal pathway to citizenship upon demonstration of English proficiency, uninterrupted employment, payment of taxes and a clean criminal record. However, such legislation faced opposition across the political spectrum and repeated efforts by successive US presidents to achieve comprehensive immigration reform have been defeated on several occasions (see also Chapter 7).

Box 11.1 How US border closure encouraged Latino migration

The acceleration of Latin American migration to the US in the context of rising border enforcement points to the limited and potentially counterproductive effects of migration restrictions. Massey and Pren (2012) argued that the boom in Latin American migration to the US occurred 'in spite of rather than because' of changes in US immigration law. Although the Immigration and Nationality Act of 1965 opened the door to large-scale immigration from Asia by eliminating past discriminatory policies, Latin Americans faced *more* migration restrictions after 1965 through the introduction of quotas for immigrants from the Western Hemisphere. This was complemented by later measures such as IRCA in 1986 (which criminalized undocumented hiring and expanded border controls), and the US Patriot Act in 2001 (which increased funding for surveillance and deportation of foreigners without due process).

However, instead of curbing inflows, increasing immigration restrictions 'set off a chain of events that in the ensuing decades had the paradoxical effect of producing

▶

more rather than fewer Latino immigrants' (Massey and Pren 2012: 1). Increased border enforcement encouraged overstaying and stimulated irregular border crossings through the use of coyotes (smugglers). Border enforcement increased the risks and costs of smuggling through the use of longer and more dangerous routes (Cornelius 2001; Jimenez 2009). One consequence has been an increase in the number of deaths. Estimates from the US Border Patrol suggest that over the period 1998–2018 there have been 7505 recorded deaths among migrants trying to cross the US–Mexico border.

However, this would not stop people from crossing the border. Border enforcement did not significantly decrease new entries (Massey and Pren 2012), particularly because of the absence of serious efforts to curtail employment of undocumented migrants through worksite enforcement (Cornelius 2001, 2005). The massive increase in border enforcement had the unintended result of reducing returns rather than deterring departures from Mexico (Massey and Pren 2012). Angelucci (2012) observed that stricter border controls increased the size of undocumented migrant populations between 1972 and 1986 (and only had a significantly negative effect after 1997), suggesting that the return-reducing effects of border controls superseded their inflow-reducing effect in the short to medium term.

The imposition of immigration barriers under conditions of continued labour demand thus interrupted what used to be largely circular movements of workers by pushing them into permanent settlement (Massey *et al.* 2015). This triggered a wave of family migration and encouraged naturalization as a pre-emptive strategy to ensure residence rights (Massey and Pren 2012). While immigration continued, return migration from the US plummeted, while the size of undocumented immigrant populations would continue to increase over the 1990s and early 2000s. From 1970 to 2010 the total Latino population in the US grew by a factor of five – from 9.6 million people and 5 per cent of the population to 51 million people and 16 per cent of the population – and the percentage of the population born in Latin America residing in the US more than tripled (Massey and Pren 2012: 24).

Unintended consequence of migration policies

The serious problems around enforcement of particular migration policies do not mean that migration policies are generally ineffective. The political and media attention given to irregular migration can unduly create the impression that borders are indeed 'beyond control'. After all, the majority of migrants travel in possession of required paperwork and therefore through legal channels. For instance, recent estimates of African migration to Europe suggest that about 9 in 10 Africans move to Europe within the law (de Haas 2019). The extensive media attention for Mediterranean 'boat migration' and apocalyptic political rhetoric surrounding it tends to blow out of proportion the true scale of this migration, besides creating an inaccurate impression that African migration to Europe is essentially about 'illegal boat crossings'.

However, migration policies can have unintended side effects that limit their effectiveness, even when migration is legal. These unintended effects occur because the migration flow reduced by a policy restriction is replaced or 'substituted' by another type of migration. Four types of such *substitution effects* can be identified:

- *Spatial substitution* through the diversion of migration via other routes or to destinations;
- *Categorical substitution* through a reorientation towards other legal or illegal channels;
- *Inter-temporal substitution* affecting the timing of migration in the expectation or fear of future tightening of policies; and
- *Reverse flow substitution* if immigration restrictions interrupt circulation by discouraging return and encouraging settlement (de Haas 2011).

Geographical diversion: Spatial substitution

Spatial substitution effects occur when policies divert migrants to countries with more liberal regulations or encourage migrants to use other geographical itineraries without changing their destination. For instance, immigration restrictions by France, Belgium and the Netherlands over the 1970s and 1980s contributed to a diversification of destinations for Moroccan emigrants, particularly to Spain and Italy (Berriane *et al.* 2018; de Haas 2007b). When Spain started to patrol its borders more intensively, this led to a diversification of terrestrial routes and maritime crossing points besides an increasing reliance on smuggling (de Haas 2008).

In the Caribbean, countries whose borders with the (former) colonizing state were 'closed' experienced a higher diversification of migration destinations than countries that retained free mobility with their (former) colonizing state (Flahaux and Vezzoli 2017). Migration restrictions *after* independence encouraged the concentration of migration and the formation of migration-facilitating networks in the former colonizing state, as was the case for Surinamese migration to the Netherlands. Migration restrictions implemented *before* independence tended to divert migration to alternative destinations. For instance, Guyanese migration largely shifted from the UK, its former colonizer, to North America (Vezzoli 2015) (see Chapter 7). Such effects seem to be stronger when destination societies are more similar in terms of culture, language and opportunities, such as the UK and the US.

Category jumping: Categorical substitution

Categorical substitution effects occur when entry through one particular migration channel becomes more difficult and migrants reorient toward other channels. On the one hand, the lack of legal immigration opportunities for low-skilled labour migrants has compelled people who primarily migrated for work to use family, asylum or student channels (Castles 2004; Harris 2002; Massey 2004). For instance, Moroccan and Tunisian migration to Northwest Europe continued after the recruitment freeze in 1973 largely because of a switch to family migration (de Haas 2007b; Natter 2014). In a similar vein, migration from Guyana to the US continued also after the 1976 US Immigration Act made immigration more difficult, through increasing reliance on family reunification, marriage and visa overstaying (Vezzoli 2014).

Restrictions can also divert migration into irregular channels. For instance, the introduction of visa requirements by Italy and Spain in 1990 and 1991 kick-started irregular

Source: Getty Images/Angelos Tzortzinis

Photo 11.2   Migrants wait to be rescued by the Aquarius rescue ship run by non-governmental organizations (NGOs) "SOS Méditerranée" and "Médecins Sans Frontières" (Doctors Without Borders) in the Mediterranean Sea, 30 nautic miles from the Libyan coast, in August 2017

'boat migration' by Moroccans, Algerians, Tunisians and, since the 2000s, increasingly by sub-Saharan Africans (see Chapter 9 and Box 6.1). A statistical analysis of asylum migration to 29 European countries over the 2001–2011 period found that, although restrictive asylum policies reduced the number of persons claiming protection, a ten per cent increase in asylum rejections across Europe raised the number of (apprehended) undocumented migrants by on average about three per cent (Czaika and Hobolth 2016). The deterrence effect of restrictive asylum and visa policies was thus partly counteracted by a reorientation of asylum seekers into irregularity.

'Now or never' migration: Inter-temporal substitution

Inter-temporal substitution or 'now or never migration' may occur if migration surges in the expectation of a future tightening of migration regulations. Restrictions can even become counterproductive when future decreases in immigration are outperformed by the pre-measure surge in inflows. Chapter 7 gave several examples of such inter-temporal substitution effects: Caribbean migration to the UK surged to 'beat the ban' before restrictions were introduced in 1962 (see Peach 1968). In a similar fashion, the Dutch government pushed for Surinamese independence in 1975 primarily because it sought to prevent migration. However, this prompted about 40 per cent of the Surinamese population to emigrate to the Netherlands before visas were introduced in 1980 (Vezzoli 2015).

Conversely, policy *liberalizations* can also generate temporary migration surges. For instance, the removal of migration restrictions tied to EU enlargement led to emigration hikes from Poland and the Baltic republics, although migration consolidated on lower levels after a few years (de Haas *et al.* 2019b). The EU enlargement experiences confirm evidence from large-scale statistical studies that migration flows respond almost immediately to the removal of travel visas by generating a temporary migration hike for a few years, before tapering off and stabilizing at lower levels. Migration surges in the wake of border openings are generally temporary, with migration decreasing and becoming more circular once migrants gain trust that borders will remain open. Yet such surges can make immigration liberalizations self-defeating through pressure on politicians to close the border. In Ecuador for instance, the implementation of universal visa freedom in June 2008 was partially reversed in December 2008 and September 2010 in reaction to public discontent with the almost 30 per cent increase of immigration from newly visa-exempted countries, particularly China (Acosta Arcarazo and Freier 2015; Freier 2013).

Interrupting circulation: Reverse flow substitution

Reverse flow substitution effects occur when immigration restrictions discourage return migration, push migrants into permanent settlement and therefore interrupt circulation. If the return-reducing effect of restrictions is higher than their inflow-reducing effects, their effect on net immigration, and therefore the growth of migrant communities, may even be positive. Several studies have shown that restrictive immigration policies discourage return, such as in the case of Turkish and Moroccan 'guestworkers' who settled in Northwestern Europe after the post-1973 recruitment ban (De Mas 1990; Entzinger 1985), and Mexican migration to the United States (Massey *et al.* 2016; see Box 11.1).

This show the importance to look not only at the effects of policies on *arrivals* but also how they affect returns and overall circulation. In other words, it is important to understand how policies affect *the entire migratory process*. A large-scale statistical analyses indicated that, on average, the immigration-reducing effect of visa restrictions is largely counterbalanced by their emigration- (return) reducing effect (Czaika and de Haas 2017). One study showed that, visa requirements significantly decrease inflows (67 per cent on average), but also outflows (88 per cent on average) of the same migrant groups, yielding a circulation-interrupting effect of 75 per cent (Czaika and de Haas 2017). In addition, the effects of lifting and introducing migration restrictions tend to be asymmetrical: while liberalizing measures often have immediate effects, restrictions tend to have smaller effects and take more time to materialize. The migration-facilitating working of migrant networks seems to largely explain such delayed effects.

Free migration strongly correlates with business cycles in destination societies, with immigration surging during high economic growth, and entries decreasing and returns increasing during economic downturns. By contrast, migration is much less responsive to economic cycles if migrants face immigration barriers such as visa restrictions. Thus, the circulation-interrupting effect of migration restrictions largely undermines the natural responsiveness of (unconstrained) migration to economic fluctuations and job opportunities in destination countries (Czaika and de Haas 2017). One study showed that, over the past decades, Senegalese migrants in France, Italy and Spain have become less likely to return over time due to increasing entry restrictions (Flahaux 2017).

This effect applies to the higher and lower skilled. A study among Indian academics, for instance, found that migration policies do not significantly determine the attraction of destination countries, but they do play a significant role in migrants' retention and subsequent moves. Indian students and researchers with aspirations to move elsewhere or to return to India tend to stay put in destination countries until they obtain permanent residency or citizenship rights as a means of 'insurance' for onward mobility (Toma and Villares-Varela 2019). The acquisition of permanent residency or citizenship sets migrants free to either return or move on without fear of losing their right to re-migrate.

Conclusion

Assessing the capacity of nation-states to regulate international migration is both important and difficult, because the aims of the policies have typically been mixed. Increased openness towards high-skilled workers has coincided with attempts to restrict the rights of low-skilled workers and to prevent the arrival of asylum seekers and prospective undocumented workers. Governments use a complex mix of policy tools to regulate immigration, characteristically including admission regimes for workers, family members, students and asylum seekers, visa policies, border controls, carrier and employer sanctions, deportation and legalization.

Although 'tough' political rhetoric may suggest otherwise, immigration policies have generally become more liberal since the 1950s, showing the considerable gap between policy discourses and more lenient policy practices. More in general, this chapter has highlighted that the popular framing of migration debates in terms of 'open' or 'closed' borders is naïve and overlooks the complexities of migration policymaking. Migration policies increasingly involve complex sets of policy tools that simultaneously encourage and discourage the migration of particular groups according to criteria such as citizenship, age, gender, skills, job offers and income.

Contemporary migration regimes are about selection rather than numbers. Policies have increasingly followed an economically utilitarian logic in determining which migrants are granted preferential access to legal opportunities for migration and settlement. While rules around legal entry, stay and exit of most migrant categories have generally been liberalized, a combination of visa and border control policies have served to prevent the entry of asylum seekers and other 'unwanted migrants', such as prospective undocumented workers.

Rather than limiting or regulating the numbers of migrants coming in, contemporary migration policies aim to increase the ability of states to control the *selection* of immigration. New layers of selection, based on criteria such as skill, wealth or family background of migrants, has been superimposed on national or 'racial' origin criteria that dominated earlier policymaking in the Americas (see FitzGerald and Cook-Martín 2014) and elsewhere. Modern migration regimes thus tend to work as filters rather than taps (de Haas *et al.* 2019a).

Evidence on the effectiveness of immigration policies is mixed. Persisting levels of irregular immigration throughout the world should by no means be interpreted as the general failure of states to control their borders. First, the magnitude of irregular migration is often exaggerated, and in the midst of media coverage and heated debates on 'illegal migration' it is easy to forget that the large majority of migrants move in accordance with the law. Second, most unlawfully residing migrants have entered destination countries legally, and have only become undocumented after their visas expired.

Third, the presence of undocumented migrants is not often as unwanted as 'tough' discourses seem to suggest. Particularly if migrants provide crucial services in agriculture, catering, cleaning, care and other services, the presence of undocumented workers is often tacitly tolerated, reflected in low levels of enforcement. As the regional chapters have shown, countries as diverse as the US, UK, Spain, Italy, Côte d'Ivoire and Thailand have tolerated the presence of large numbers of undocumented migrant workers, with Spain and Thailand even holding public registers for such groups.

While media images and political discourses often suggest otherwise, migration policies are generally effective, with some important exceptions. The large majority of migrants abide by the law and migrate through legal channels, in the possession of visas and other necessary paperwork. In fact, the increasingly sophisticated instruments of migration regimes seem to generally achieve their objectives of influencing the selection (rather than volumes) of migrants.

Although it would be exaggerated to say that borders are generally 'beyond control' (Bhagwati 2003), the capacity of migration policies to 'steer' migration is limited by powerful structural migration determinants that can have unintended consequences. In particular, 'substitution effects' can undermine the effectiveness of migration controls by (1) diverting migration through other geographical routes and destinations (spatial substitution), (2) other legal and irregular channels (categorical substitution), (3) 'now or never' migration surges in anticipation of restrictions (intertemporal substitution) and (4) by discouraging return and interrupting circulation (reverse flow substitution). Substitution effects tend to be particularly strong if strong migration-facilitating networks have already formed combined with a discrepancy between migration policies and more fundamental migration determinants, particularly labour demand in destination countries, and in some cases violent conflict in origin countries.

These unintended effects illustrate the difficulties of reconciling different migration policy objectives, such as the wish to limit immigration on the one hand, and to encourage return and circulation on the other. Immigration restrictions simultaneously reduce immigration and return, which renders the effect on net migration and the growth of migrant communities theoretically ambiguous. The circulation-interrupting effects of immigration restrictions also reduce the much-desired responsiveness of migration to economic fluctuations and job opportunities in destination countries (de Haas et al. 2019a).

While some policy measures are rather effective in regulating migration, others seem less effective, or primarily serve an important symbolic political function as parts of politicians' efforts to show that they are prepared to take a tough stance on migration. For instance, the construction of border walls fulfils an important symbolic function for politicians who want to be seen as taking concrete and visible action, even though their effectiveness as a migration control tool can be questioned. It is the symbolic function of such measures which partly explains the popularity of walls and fences in this Age of Migration.

This evidence does not mean that governments cannot or should not control migration. Rather, it shows that liberal immigration policies do not necessarily lead to mass migration and that ill-conceived migration restrictions can be counterproductive. Free migration is often strongly circulatory, as we see with migration within the EU or within countries. The more restrictive entry policies are, the more migrants want to stay. Restrictive migration regimes limit inflows but may push migrants into permanent settlement. This exemplifies the intense dilemmas and complex trade-offs that migration policymaking usually involves and the importance of having a thorough understanding of *migratory processes* in all their complexity, in order to design effective policies that do not have major unintended and undesired effects.

This also highlights the importance of looking beyond migration policies. A fundamental mismatch between structural migration determinants – such as low-skilled labour demand in the absence of legal migration channels, combined with weak workplace enforcement, or violence and conflict in origin countries in the absence of asylum channels – is likely to translate into increasing irregular border crossings as well as an increasing incidence of migrants 'overstaying' their visas. The more migration policies go against structural migration determinants, the more likely they will have unintended consequences. Migration policy failure is often explained by a failure to understand the complex ways in which structural social, economic and political factors such as destination country labour demand affect migration in mostly indirect, but powerful ways that lie largely beyond the reach of migration policies.

Guide to Further Reading

See de Haas *et al.* (2019a) for an overview of evidence on the effectiveness of migration processes across a large set of countries. De Haas *et al.* (2018) analyses the evolution of migration policies over the past century, and Czaika *et al.* (2018) gives a global overview of the visa regimes between 1973 and 2015. The various publications of the DEMIG project and the International Migration Institute (www.migrationinstitute.org) give up-to-date insights on migration determinants and policy effects in various regions and for various migrant groups, including the higher skilled. FitzGerald and Cook-Martín (2014) give an excellent overview of the history of immigration policies in the Americas. de Haas *et al.* (2019b) track the effects of free mobility regimes in an expanding European Union (EU) on migration within and towards the EU. Miller and Stefanova (2006) and Boswell and Geddes (2011) provide useful comparisons international migration and regionalization processes. The yearly *International Migration Outlook* published by the OECD since 1973, gives an excellent overview of migration policy developments including most recent data on immigration and asylum. The Migration Policy Centre of the European University Institute (www.migrationpolicycentre.eu) and The Migration Policy Institute (www.migrationpolicy.org) are useful resources, particularly for American and European issues around migration governance.

Extra resources can be found at: **www.age-of-migration.com**

Appendix: The migration policy toolbox

Migration policy tool	Policy area: Border controls
Surveillance technology and control powers	Surveillance or registration systems to control the movement and migration status of people, citizens and foreigners alike. This can include the use of information technology, the construction of fences and walls, radars, cameras, fingerprinting, regulations around the number of border guards, the powers of immigration staff and funding for equipment such as vehicles, scanners, cameras, radars, computers and software.
Travel visa/permit	Procedures and eligibility criteria, including fees, applied to foreigners to obtain a travel visa to enter or leave a particular country. This includes measures regulating entry or exit for any purpose (business, family, holidays, religion), but which do not grant residence or working rights to visa holders. Although strictly speaking not applying to migration, travel visa requirements are a crucial instrument of immigration regimes, in order to exert state sovereignty in the selection of immigrants and to prevent the arrival and the access to rights of asylum seekers and undocumented workers. Often used in combination with carrier sanctions.
Identification documents	Regulations on identification documents, such as (biometric) passports, rules on the validity of identity cards, driver licences and other documents.
Entry ban	Policy measures aimed at the categorical exclusion of a specific group (based on citizenship, ethnicity or religion) from the right to enter the country. These groups have no access to a legal channel of entry into the country and are not eligible to apply for an entry visa or permit. An example is the travel bans for Israelis in several Arab and predominantly Muslim countries.
Carrier sanctions	Policy measures that regulate the responsibilities of and requirements for airlines and other transportation companies, including fines and other sanctions for the transportation of people lacking the required visas and other required immigration or emigration permits. This is one of the major and most effective ways in which governments have 'externalized' migration controls to private companies.
Employer sanctions	Measures that regulate the responsibilities of employers related to the employment of foreign workers, such as registration and control requirements or employment permits. This particularly relates to fines and sanctions for the unlawful employment of migrants. This also includes provisions of universities to check the immigration status of students. The effectiveness of such measures strongly depends on the degree of government enforcement.
Other sanctions	Other measures to prevent irregular migration and stay, such as sanctions for document fraud, overstaying, smuggling or human trafficking.

(continued)

Migration policy tool	Policy area: Border controls
Detention	Procedures and eligibility criteria for the detention of foreigners, often in preparation of their return or deportation.
	Policy area: Legal entry and stay
Recruitment/assisted migration programmes	Policies that facilitate and regulate assisted migration schemes or bilateral agreements between governments or between companies and government to organise the recruitment of workers. Assisted migration programmes were particularly associated to the large-scale 'permanent' migration of families from Europe to settler colonies in the Americas and Oceania. Recruitment (by companies in coordination with states acting in their interest) is often organized by private parties or semi-independent agencies, states are generally involved in shaping the legal frameworks and selection rules for such recruitment. The key feature of such programmes is that agencies and states are actively involved in the recruitment and selection of workers.
Entry visa/stay permit	Procedures or eligibility criteria (age, language knowledge, education level, family relations, protection need, fees) to obtain different types of entry visa and stay/residence permits for a specific purpose, such as student visas, investor visas or family visas, but can also refer to the introduction of a compulsory language test or integration contract for entry.
Work visa/permit	Procedures or eligibility criteria (job offer, salary requirements, age, language knowledge, education level, labour market test, fees) to obtain a work visa or permit before or after arrival. This includes working holidaymaker schemes, *au pair* and other youth mobility programmes, but can also refer to the introduction of a compulsory language test or integration contract for entry.
Quota/targets	Policy measures that establish a target or maximum number of persons that are eligible to immigrate for particular migration categories. Quotas were once popular in several destination countries (including still in the US), but are increasingly seen as ineffective in regulating family and labour migration, and have partly given way to points-based systems and other labour-market driven immigration policies that allocate work visas based on labour market demand and concrete job offers.
Points-based systems	Codes policy measures that establish, change or abolish the criteria of a points-based system that gives access to either a work or another visa/permit. Points systems give scores to prospective migrants usually based on age, education and skills criteria. First introduced in Canada and Australia, they have become popular in many destination countries in what is perceived as a 'global race for talent'.

(continued)

Migration policy tool	Policy area: Legal entry and stay
Regularization/ Amnesty	Programmes and legislation that specify eligibility criteria to grant legal status to migrants who lack residency rights. This can be either in the form of special, one-off measures or the continuous provision for the legalization of undocumented migrants. As regulation of migration inevitably 'produces' situations of irregularity, regularization is often seen as an integral and indispensable part of immigration regimes.
Refugee status determination policies	Procedures aimed at verifying whether asylum seekers are entitled to refugee protection. Most signatories to the UN refugee convention have some refugee status determination system in place, although refugee recognition rates show high degrees of variation across countries. In countries with weak institutional frameworks, UNHCR is often involved in refugee status determination as well as in decisions on eligibility for refugee resettlement.
Refugee resettlement programmes	Policy measures that establish, change or abolish programmes that resettle refugees from countries of origin or first asylum already recognized by organizations like UNHCR and grant them residency rights. While spontaneous asylum applicants dominate refugee flows to countries that are in geographical proximity of conflict zones, resettlement programmes are the main source of refugee migration to geographically more distant or isolated countries such as the Canada, Japan or Iceland.
Free mobility rights/ agreements	Bilateral or multilateral agreements in which governments grant reciprocal free right to enter and reside for citizens of each signatory country. This can, but does not necessarily, include the right of 'establishment', that is, to work and do business. The Trans Tasman Travel Arrangement between Australia and New Zealand is an example of a bilateral free mobility agreement, the EU, CIS and ECOWAS are examples of multilateral unions including a free mobility component.
	Policy area: Integration
Access to social benefits and socioeconomic rights	Policies and procedures that give immigrants access to the existing state systems of social benefits and socioeconomic rights. This includes access to social security, health system, education system and unemployment benefits. Such access is often difficult to deny in democracies and liberal political systems wishing to uphold a certain level of human rights.
Access to justice and political rights	Procedures or eligibility criteria giving migrants access to the existing state system of justice and political rights. This includes access to legal aid, the right to vote, the right of appeal, the right to create associations, as well as antidiscrimination legislation and multiculturalism policies.

(continued)

Migration policy tool	Policy area: Integration
Language, housing and cultural integration programmes	Policies and procedures that give migrants access to language programmes, financial assistance or housing programmes, as well as religious and cultural integration programmes especially established for migrants. These have been particularly prevalent in countries following 'multicultural' approaches towards immigrant integration.
Access to permanent residency	Procedures or eligibility criteria that give migrants access to permanent residency, including requirements such as language and integration tests.
Access to citizenship	Procedures or eligibility criteria that give access to citizenship or naturalization, including citizenship and language tests and ceremonies.
Diaspora engagement policies	Policies by origin countries to extend political, civil and social rights to citizens living abroad and descendants of emigrants. Such policies include voting rights, access to residence rights and citizenship of 'heritage populations' and co-ethnic and co-religious groups, and measures to facilitate remittance transfers and stimulate investments by emigrants and diaspora populations (see Gamlen 2008). In destination countries, such policies can be controversial as they are sometimes seen as infringing on sovereignty and hindering the integration of immigrant groups.
	Policy area: Exit
Reintegration/return programmes	Policy measures – either as part of bilateral agreements or unilateral programmes – that aim at encouraging the voluntary return of migrants to origin countries through financial or institutional assistance. This can include both the actual assistance to return and subsequent measures established to foster their reintegration in the origin societies. An example is the 'departure bonuses' and financial aid packages developed for 'guestworkers' by destination countries such as Germany and the Netherlands in the 1970s and 1980s (see Entzinger 1985).
Deportation/expulsion	Policy measures to enforce the physical removal of migrants from the national territory, often in combination with detention policies. Irregular workers and rejected asylum seekers are the most common target groups of such policies. These policies are only effective if they are accompanied by readmission agreements with origin or transit countries.
Readmission agreements	Agreements between destination governments and governments of origin and transit countries for the readmission of undocumented migrants and/or rejected asylum seekers.
Exit visa/permit or exit ban	Measures that establish conditions for the exit of citizens, making the departure from the national territory subject to prior approval. This includes both rules on exit permits, as well as more absolute forms of exit bans. Particularly authoritarian states have often used systems of travel and emigration permits.

Source: Adapted from de Haas, Natter and Vezzoli 2014

12 Migrants and Minorities in the Labour Force

People migrate for many reasons. Although labour demand is a prime factor in driving migration, and government policies often focus on economic migration, a significant share of migration is not primarily for economic purposes. Family reunion, marriage, education, marriage, retirement and the search for new lifestyles are other important motives for migrating. About one out of every ten migrants cross borders to seek refuge from war and persecution. But many people do migrate for explicitly economic reasons: in search of higher incomes, better employment chances or professional advancement. Most international migration has some economic dimension: origin countries look to remittances, investments and technology transfer by migrants as resources for economic growth, while destination countries are concerned with the role of migrants in meeting demand for labour and skills.

Labour demand and economic motives also play an important secondary role in other forms of migration. For instance, family migration is often an indirect consequence of labour migration: once migrant workers decide to settle in destination countries this is often followed by family reunion, and upon arrival family members often join the migrant work force. Student migration also has a strong economic dimension, as studying abroad increases people's employability and income-earning power, and upon completing their degrees many students end up staying and taking up skilled jobs in destination countries, often aided by immigration policies that try to attract skilled workers and grant special job-seeking visas to foreign university graduates in an effort to retain them.

Destination country labour demand is arguably the most important direct and indirect determinant of migration. Given the prime importance of labour market dynamics in explaining processes and experiences of migration, this chapter will review the various ways in which fundamental changes in labour market structures have shaped contemporary migration, and how migrants have been incorporated into labour markets and destination country economies over time. It will particularly focus on the position of lower-skilled migrants, because unlike higher-skilled migration, this type of immigration is subject to considerable political controversy, but is often poorly understood. Because of data limitations, most of the data on migrant employment is on OECD countries, although many of the dynamics highlighted in this chapter, particularly on the role of precarious and informal work, also apply to non-OECD destinations, particularly in the Gulf region and South-East Asia, probably even more so.

The social construction of labour needs

It is often claimed that labour migration from poor to rich countries meets *mutual needs* (see CEC 2005; GCIM 2005). The underlying idea is that poor countries have more young labour market entrants than their weak economies can employ, so they 'need' to export surplus workers to rid themselves of problems like unemployment and poverty. Rich countries, by contrast, have declining numbers of young people entering their labour markets and who are able and willing to do certain jobs, so they 'need' to import labour. However, it is important to realize that such labour needs are socially constructed. The 'need' for low-skilled migrant labour in higher- and middle-income countries is the result of poor wages, conditions and social status in certain sectors, and the general unwillingness by native workers to take up these jobs. Such jobs generally include

> dirty, difficult and dangerous jobs, low-paid household service jobs, low-skilled jobs in the informal sector of the economy, jobs in sectors with strong seasonal fluctuation, for example farming, road repairs and construction, hotel, restaurant and other tourism-related services. (Münz *et al.* 2007: 7)

If the conditions and status of such '3D' (dirty, difficult and dangerous) jobs were improved, local workers might be more willing to take them although marginal employers might go out of business, while families would be more inclined to do the cleaning, cooking and caretaking themselves if they had more free time. The result might be that certain types of work would be relocated to lower-wage economies. Such 'offshoring' or 'outsourcing' has in fact become common since the 1970s in the manufacturing sector, where much of the production has been moved to new industrial economies. In many cases, 'global commodity chains' (Gereffi 1996) or 'global value chains' (Ponte and Gibbon 2005) have evolved, in which routine manufacturing tasks are carried out by low-paid workers in low-income countries. Higher-paid activities such as design, management and marketing initially remained in the rich countries. More recently, the 'back office' jobs of banks, insurance firms and IT firms has also been outsourced, and this process has also affected research and development work, which can be done more cheaply in emerging hi-tech areas within low-wage economies like India.

Agriculture also seems an obvious choice for outsourcing, since productivity is low, especially in fruit and vegetable production, and climatic conditions and labour costs may allow for higher productivity in tropical and subtropical countries. To some extent this has happened, for instance through European investments in vegetable- and flower-growing in Ethiopia, Kenya and Senegal (see Schewel 2019). However, in many cases local farmers – as well as unions of farm workers – would be hurt by moving production offshore, and they have had the political clout to prevent this happening. This explains the persistence of trade protection and US farm subsidies and the EU's Common Agricultural Policy (CAP), both of which are costly to taxpayers, disadvantageous to consumers and damaging to agriculture in poor countries, particularly through agricultural commodity dumping (Murphy and Hansen-Kuhn 2019). In this way, US farm subsidies for exported crops, which compete with Mexican products, have tilted the level playing field NAFTA was supposed to create heavily in favour of the US (Wise 2009).

This is why Martin and Taylor (1996) argued that unequal terms of trade, government subsidies, structural technological advantages and economies of scale in wealthy countries may harm the competitiveness of the South even in the production of labour-intensive goods (see also Krugman 1995). Under such circumstances, trade liberalization can paradoxically reinforce the concentrations of highly productive economic activities (such as intensive agriculture) in the North *along with more immigration of low-wage labourers to support them* (Martin and Taylor 1996). Sustained migration of Mexican and Central

American agricultural workers to the US and of North- and West-African workers to Southern Europe are cases in point.

Employers tend to talk about labour demand in terms of 'needs' because it creates an air of urgency and objectivity whereas in reality it is their interest to allow more migrant workers in, as it helps them to reduce labour costs or, as in the case of domestic work, because it allows native women to enter the labour market and thus become more 'productive'. But the social construction of the 'need' for migrant workers goes beyond such pecuniary cost-benefit calculations, but is also closely tied to social status. The prime reason why native workers, even when unemployed, often refuse to do certain jobs is because they are seen as 'below' their status. In addition, immigrants fulfil an important function in enabling natives (and high-skilled migrants of the 'expat' type) to afford particular luxurious lifestyles that enhance their status. This is particularly the case for domestic work.

Anderson (2001) argued that domestic work is on the one hand necessary, because humanity need to accommodate the child-rearing, cooking and basic hygiene to survive. On the other hand, it is also concerned with the reproduction of lifestyle, and, crucially, of social status. Anderson argued that 'nobody has to have stripped pine floorboards, hand-wash only silk shirts, dust-gathering ornaments, they all create domestic work, but they affirm the status of the household, its class, its access to resources of finance and personnel' (Anderson 2001: 22). Even employing a domestic worker in itself can be a status symbol, particularly if they are educated and English-speaking: for that reason, for instance, Filipina workers are often preferred in Arab countries above Arab or African workers.

For instance, while most people across cultures would agree that stinking clothes are an offence to human dignity, then exactly how often they are washed, how they need to be washed, whether they are ironed and so on, can quickly become issues of status (Anderson 2001). The organization of our homes and their accessories demonstrates our social status, and the richer people get, the bigger their houses, gardens, cars and swimming pools, the more 'stuff' they accumulate, and the greater their perceived needs in quality of maintenance, food, clothing, personal care, service and leisure. The reproduction of middle- and higher-class lifestyles thus creates a perceived need for personnel to carry out these tasks, although we objectively do not need such lifestyles to live our lives in basic human dignity. As Anderson quoted a Filipina domestic worker in Paris:

> Every day I am cleaning for my madam, one riding shoes, two walking shoes, house shoes, that is every day, just for one person ... plus the children, that is one rubber and one shoes for everyday school, that is another two. Fourteen shoes every day. My time is already finished ... You will be wondering why she has so many bathrobes, one silk and two cotton. I say, 'Why madam has so many bathrobes?' Every day you have to hang up. Every day you have to press the back because it is crumpled. (Anderson 2001: 21)

Narratives about the 'need' for migrant workers conceal subjective economic interests and social status motives, and the power of dominant classes to push through immigration policies – or the absence of enforcement – that reflect their interests. It also expresses certain political choices about how to organize societies and economies. For instance, because of facilities for part-time work, parental leave and subsidized childcare, North west European welfare states generate less demand for private domestic workers and nannies than Southern European economies. In the same vein, there may be good or bad reasons to maintain a subsidized and state-protected agriculture, but such political choices will inevitably sustain a demand for agricultural workers, who are likely to be recruited from abroad.

Rather than a *need*, it is therefore more appropriate to speak of a *demand* for migrant labour, although it may be cast as an objective need by powerful economic and political interests. The subjective nature of such needs does not mean that the resulting labour demand is not real. The fact that natives are no longer willing or available to do certain jobs does create a real demand for foreign workers. Governments in destination countries have often responded to this demand by actively recruiting foreign labour (Castles *et al.* 2012). However, if such demand is not mirrored by immigration policies, this can generate other responses, such as mechanization and automation, relocation of production, increasing wages, or indeed irregular immigration and informal employment.

The changing structure of labour demand

The structure of the demand for migrant labour underwent important shifts over the post-WWII period. As described in Chapter 6, foreign labour employment in Europe stagnated or declined after 1973 in a period of recession and economic restructuring. After the Oil Shock, many West European countries stopped recruitment, but were unable to prevent family reunion and permanent settlement. The US changed its immigration rules in 1965, but did not expect a significant increase in immigration. However, in recent decades there has been a major upsurge of lower- *and* higher-skilled migration to industrialized countries. This defied earlier expectations that the era of large-scale labour migration had come to an end, because the industries, mines, shipyards and farming sectors that employed most migrant workers no longer seemed viable. Views that lower-skilled workers were no longer 'needed' became increasingly popular over the 1980s and 1990s, particularly in West European societies that struggled with large-scale unemployment and social marginalization of former guestworkers from Turkey and North Africa.

What led to this change? Why did lower-skilled immigration continue? A major *economic factor* was the realization that industrialized countries could not export all low-skilled work to low-wage countries. The manufacture of cars, computers and clothing could be shifted to countries such as China, Brazil or Malaysia, but the construction industry, hotels and restaurants, hospitals and elderly care had to be where their consumers were. The same applies to all sorts of domestic jobs (often done by female migrant workers) such as childcare and cleaning. The demand for migrants to do such jobs increased because of increased formal labour market participation of native women and rising education, which depleted two main former sources of such labour: women and early 'school leavers'. At the same time, new sources of demand for such labour had appeared in the booming economies of the Gulf, East Asia and Southern Europe.

Local workers often prefer to not work or be unemployed rather than do unattractive jobs with very low social status while the poor cannot afford to be unemployed. This particularly applies to people from middle- and upper-class families who can afford to stay at home and not work – this applies to both industrialized and developing countries (Elder *et al.* 2015). Although students are often willing to do such jobs because their social status does not yet depend on it, upon graduation few would be willing to do the same jobs, and expect 'decent' jobs commensurate with their social status. Demand for foreign domestic workers is huge in Saudi Arabia and other Gulf countries, but most Arab women whose traditional tasks in cleaning, cooking and childcare are taken over by Asian and African workers do not enter the labour force, but instead use this time for leisure. This illustrates the extent to which the low social status of such work – and not just low pay – are important reasons why natives tend to shun such jobs (see Piore 1979).

The growing demand for migrant labour prompted increased recruitment and emigration of migrant workers from poorer countries. A major *demographic factor* behind this change in views and policies was the realization that fertility rates had fallen sharply and that the proportion of older people in the populations of highly developed countries was rising fast (see also Chapter 5). If there are fewer young people, they will expect better opportunities. In addition, in many industrialized countries a decreasing share of youngsters were willing to do vocational training in lower- and mid-level skilled jobs, such as in construction, catering and healthcare, which has led to real shortages in labour markets in terms of unavailability of personnel with the right skill sets.

In advanced economies, demand for manual jobs in manufacturing and perhaps also agriculture may further decline because of automatization and outsourcing, but, as a result of ageing and increasing education of native workers, there is likely to be a growth in demand for lower-skilled service workers in household and care jobs. In Italy alone, there are an estimated 1.5 million migrants working in households, and one Italian household out of ten employs a migrant domestic worker for care of children or the elderly (Ambrosini 2013). New industrial countries in East Asia also have low fertility and ageing populations, which will lead to major future challenges with regard to welfare and elderly care. One of the most striking findings of the regional overview chapters is that the Great Recession barely affected migration to industrialized countries. This confirms Piore's (1979) hypothesis that the demand for lower-skilled migrant labour is an inherent and therefore unavoidable feature of industrialized economies (see Chapter 3). The crucial questions are whether and how this demand will be met by policies, and where future migrant workers will come from.

Migrants in the labour market

The contribution of migrant workers to economic expansion

Some economists have argued that migrant labour made a crucial contribution to the post-WWII boom (see Kindleberger 1967). Migrants replaced local workers, who were therefore able to obtain more productive, higher-skilled jobs. Without the flexibility provided by immigration, bottlenecks in production and inflationary tendencies would have developed. However, other economists have argued that immigration can reduce efficiency, keeping low-productivity firms viable and holding back the shift to more capital-intensive forms of production such as through mechanization and automation. Such observers also claim that capital expenditure on housing and social services for immigrants reduced the capital available for productive investment (see for instance, Borjas 2000; Jones and Smith 1970).

There is a close correlation between economic growth and immigration levels. For instance, high immigration countries, like West Germany, Switzerland, France and Australia, had the highest economic growth rates in the 1945–73 period. Countries with relatively low net immigration (like the UK and the US at that time) had much lower growth rates (see Castles *et al.* 1984: Chapter 2; Castles and Kosack 1973: Chapter 9). However, such correlations do not prove causality, and in reality, the dominant direction of causality is rather the other way around: fast-growing economies attract more migrants because of high labour demand and skills shortages. Immigration can further boost such growth through boosted growth and profits by keeping labour costs down. In this way, economic growth and immigration can reinforce each other.

Economic impacts of low-skilled immigration: Winners and losers

In order to understand the ongoing political controversy on the question of 'do we need low skilled workers?', a crucial observation is that the socioeconomic benefits (and potential costs) are not evenly distributed across destination country populations. Some benefit much more than others, and some may actually lose out. Immigration *primarily* benefits businesses and higher income groups, while non-migrant low-skilled workers may feel more threatened and trade unions may feel it undermines their power. Most statistical studies suggest that the effect of immigration on wages of local workers, whether positive or negative, is small (Dustmann *et al.* 2016; Ottaviano and Peri 2012; UNDP 2009).

The most consistent finding of studies on the income effects of migration seems to be that the higher income earners benefit more than the low-income earners (Smith and Edmonston 1997; UNDP 2009). There are several reasons why higher income groups are likely to be the prime beneficiaries of low-skilled immigration. First, migrant workers do not compete for the same jobs as high-skilled native workers. Second, the availability of low-cost migrant labour is allowing the middle- and higher-income classes to buy in personal services like home-cleaning, childcare, gardening, dry cleaning, dog-walking, hairdressing and personal care, ironing and home maintenance at lower prices. The outsourcing of such services to migrant workers also allows higher-skilled groups, including high-skilled immigrants, to work longer hours and thus increase their income. Third, the supply of cheap migrant labour increases the productivity of businesses, the profits of which flow disproportionally to higher income groups either through direct ownership of these companies or indirectly through the ownership of stocks, pension savings and other capital investments. Fourth, the already well-to-do real-estate owning groups in society are also likely to benefit disproportionally from increases in house prices resulting from high economic growth and immigration, while (native and migrant) tenants tend to lose out from associated hikes in rent prices.

Whether lower-skilled native workers benefit from immigration is a more complex matter. If the skills of immigrant workers complement those of non-migrant workers, both groups can benefit. For instance, the availability of migrant domestic workers and childcarers enables non-migrant women to work (UNDP 2009), although these are unlikely to be among the poorest. Because lower-skilled migrants often do work that locals are no longer prepared to undertake, the view that migrants 'take away' jobs from local workers is often not correct. Negative effects on wages are most likely if new migrants compete

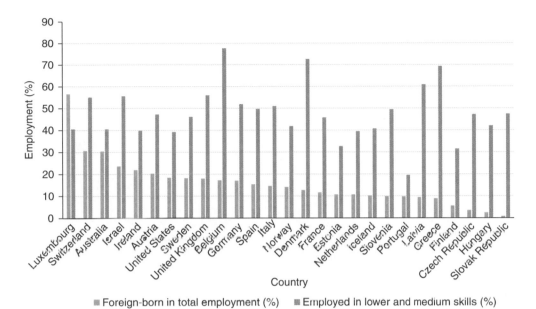

Figure 12.1　Employment patterns of foreign-born in OECD countries, 2017
Source: OECD 2018

with the existing workforce. Therefore, the workers most negatively affected by the entry of new migrants tend to be earlier migrants (UNDP 2009). However, it is important to remember that the most important finding from the cumulated evidence is that the overall magnitude of the wage effects of immigration on employment is *very small*. This should preclude exaggerated claims about the positive- or negative- effects of migration.

Migrant employment patterns: The first generation

Notwithstanding oscillations in immigration policies, the number of foreign workers in OECD countries has increased over the past decades. By 2017, foreign-born workers made up a substantial share of the labour force: around 30 per cent in Switzerland and Australia, 24 per cent in Israel, around 21 per cent in Ireland and Austria, 18 per cent in the US, the UK and Sweden, 17 per cent in Belgium and Germany, and 15 per cent in Spain and Italy, but only 11 per cent in the Netherlands and France (see Figure 12.1).

New immigrants increasingly bring skills with them: the old stereotype of the unskilled migrant coming in to take the least-qualified positions is no longer valid (Collins 2006; Portes and Rumbaut 2006). This reflects increasing education in origin countries as well as the changing structure of labour markets in destination countries. In the 1970s, high concentrations of male migrant workers were found in factories, building sites or services such as garbage collection and street cleaning. Women were also to be found in factories (especially textiles and clothing, engineering and food processing) and in services such as cleaning and health.

By the 2000s, migrants could be found right across the economy. In the service sector, the foreign-born were overrepresented in hotels and restaurants, and the health-care and social services sector (OECD 2007: 72–73). In 2005, more than 50 per cent

Source: Alamy/John Angerson

Photo 12.1   Spanish Nurses take care of patients in an NHS hospital in Blackburn, United Kingdom

of cleaning jobs were held by migrants in Switzerland, and more than 30 per cent in Austria, Germany, Sweden, Italy, Greece and the US. Migrant employment profiles have further diversified since, and foreign workers can now be found across the skills spectrum, although they remain comparatively over-represented in lower-skilled jobs (see Figure 12.1). In 2017, about 65 per cent of employed migrants in the OECD area worked in low- or medium-skilled jobs (OECD 2018). The share for native-born workers was almost 10 percentage points lower. Foreign-born workers remain over-represented in mining, manufacturing, hospitality catering and in domestic services (OECD 2018).

Migrant workers are also over-represented in specific service occupations requiring high-skill levels, such as teachers (such as in Switzerland and Ireland), doctors and nurses (in the UK) and computer experts (US). A striking example is the high dependence of the British National Health Service (NHS) on immigrants. In 2003, 29.4 per cent of NHS doctors were foreign born and 43.5 per cent of nurses recruited to the NHS after 1999 were born outside the UK (Simpson *et al.* 2010). Migrant employment in the tertiary sector seems increasingly 'dualistic' with concentration at low and high skill levels and a gap in between (Bauder 2006; see also OECD 2007, 2018). Despite the increasing importance of skilled migration, overall migrants still tend to have lower occupational status and higher unemployment rates than non-migrant workers.

The labour market advantage of 'new' immigrants compared with the 'old' should not be overstated. In general, migrants experience a 'poor return on education' (Reyneri and Fullin 2010: 45). The employment situation of migrants depends on national factors such as labour market policies and social security arrangements, but also on ethnicity, gender and legal status. A comparison of six European countries highlighted the diverse experiences of migrants.

They had lower unemployment levels but less qualified jobs in Southern European countries, while they had more qualified jobs but a higher risk of unemployment in Northwestern Europe (Reyneri and Fullin 2010: 43). In the UK, the labour force is stratified by ethnicity and gender. Generally, people of Indian, Chinese or Irish background have employment situations as good as or sometimes better than the average for white British (Dustmann and Fabbri 2005). By contrast, other groups are worse off, with a descending hierarchy of black Africans, black Caribbeans, Pakistanis and – at the very bottom – Bangladeshis. Gender distinctions varied: young women of black African and black Caribbean background performed better in education and employment than men of these groups, while the opposite appeared to be the case for Pakistanis and Bangladeshis (ONS 2004).

There is huge variation in unemployment levels among migrants in OECD countries (see Figure 12.2). What is particularly striking – and worrying – is the high gap in unemployment levels between foreign- and native-born in some countries. In Sweden, unemployment amongst the foreign-born is 3.5 times higher than among the native-born. Large gaps of 2 times or higher are also found in countries such as Norway, Austria, Belgium, Chechia, the Netherlands, Denmark and Germany. In other countries, the foreign labour force has *lower* unemployment levels than native workers, such as in Israel and the US. In other countries, such as the UK, Canada and Australia the differences are negligible.

Such differences may be related to differences in immigration policies. For instance, the legacy of guestworker policies, which led to mass unemployment after 1973, is still visible in countries that recruited such workers. It also reflects the fact that immigration into

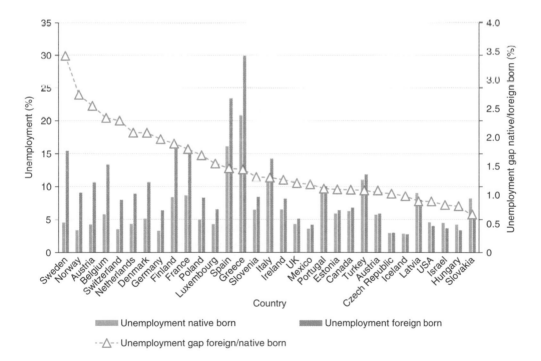

Figure 12.2   Unemployment of native- and foreign-born in OECD countries, 2017
Source: OECD 2018

some countries, such as the US, from regions such as Africa and the Middle East, is much more selective, mainly attracting the relatively higher skilled, who generally have lower unemployment levels. There also seems to be a correlation between levels of migrant unemployment and the level of social security. Although there is no evidence for the idea that immigrants are attracted by welfare state arrangements (the 'welfare magnet'; see UNDP 2009), low levels of social welfare leave migrants no choice but to work, while welfare provisions may be a 'retaining factor' motivating migrants to stay.

The second generation: Improvement and continued discrimination

Most labour migrants to OECD countries up to the 1970s were workers with low skill levels. From the 1980s and 1990s, a *second generation* (native-born persons with both parents foreign-born) has emerged. They generally received their education in the destination country. In most OECD countries *persons with a migration background* (foreign-born persons plus the second generation) made up a large proportion of young adults by 2005 – the highest share was in Australia (45 per cent of persons aged 20–29), followed by Switzerland and Canada (30–35 per cent), and then Sweden, the US, the Netherlands, Germany, France and the UK (20–30 per cent) (OECD 2007: 79).

In general, the second generation has much better average educational outcomes than the migrant parent generation. They also do better than young migrants of the same age group. However, despite significant progress, the outcomes of the second generation tend to lag behind those of native-born without migration background. This may be partly explained by the low educational and socioeconomic levels of their parents, since such factors tend to be transmitted across generations. The OECD's Programme for International Student Assessment (PISA) examined performance of 15-year-olds in mathematics, science, reading and cross-curricular competencies. The study showed that second-generation students remained at a substantial disadvantage. This applied particularly to former guest worker-recruiting countries, such as Germany, Belgium, Switzerland and Austria. However, second-generation education disadvantage was insignificant in Sweden, France, Australia and Canada (OECD 2007: 79–80).

This indicates that the original mode of labour market incorporation can have effects that cross generations (see also Portes and Rumbaut 2006: 92–101). In all OECD countries studied (except the US), second-generation women did better than their male counterparts at school (OECD 2007). Schooling thus seems to have an important emancipatory effect for second-generation women. However, there are also substantial differences across different ethnic groups. For instance, the children of Asian immigrants in the US tend to have much better educational outcomes than the children of other immigrant groups. In fact, they often perform better at school than many native and white groups.

The most important question for the second generation is whether they can get decent jobs. Again, the overall pattern found across many studies is one of strong intergenerational improvement. Second-generation members tend to have higher employment levels than immigrants in the same age group, but often still suffer significant disadvantages. There are large differences in labour market incorporation across immigrant groups. For instance, in Western Europe children of African immigrants seem to have the greatest difficulties. According to one study, they were up to twice as likely to be unemployed as non-migrant youngsters. Explanations include lack of access to informal networks that help in job-finding; lack of knowledge of the labour market; and racist discrimination (OECD

2007: 81–85). A meta-analysis of 43 studies conducted in OECD countries between 1990 and 2015 found that ethnic discrimination in hiring decisions has remained widespread. On average, equally qualified minority candidates need to send around 50 per cent more applications to be invited for an interview than majority candidates (Zschirnt and Ruedin 2016).

Migrant entrepreneurship

Migration has always been a selective process. Migrants tend to have more adventurous and entrepreneurial mindsets than people they leave behind, as migration entails considerable aspirations, creativity and determination (see Goldin *et al.* 2011). Up to the 1970s, particularly in Europe, migrants were generally seen as (largely temporary) wage-workers, and rarely as entrepreneurs. Although there have always been important exceptions, such as the historical prominence of Italians, ethnic Chinese and Lebanese in various catering businesses, the general pattern was that of the recruitment and deployment of migrants for jobs. In Germany, Switzerland and Austria their work permits even prohibited self-employment – this was linked to the dictum that migrants were only temporary and should not be encouraged to settle.

The situation was different in the US, Australia, the UK and France, where migrants began to run small shops and cafés early on. A historical example is the involvement of Jewish immigrants in the London clothing industry around the turn of the nineteenth century (see Chapter 5). The same has historically applied to the Indian, Chinese and Lebanese heritage populations in various parts of South-East Asia, Africa and Latin America, who often play a central role in various business sectors. Since the 1980s, migrant self-employment has become far more common everywhere. Particularly for immigrant and ethnic groups who face substantial discrimination on labour markets, self-entrepreneurship can be an important avenue for upward socioeconomic mobility.

A particular advantage enjoyed by migrant entrepreneurs is that they can often rely on the cheap or free labour of family members, who are often motivated to contribute to the family business out of strong feelings of solidarity to 'make it together'. The willingness to work in family business may decrease with the second and, particularly, third generation, as migrants' children and grandchildren start to aspire to different occupations as consequence of increasing education and cultural assimilation. However, in earlier phases of migrant settlement, entrepreneurship can be a crucial emancipatory activity, allowing migrant families to acquire assets, provide a good education for their children, move to better neighbourhoods and eventually achieve middle-class status.

In 2007–8, about 12.6 per cent of working-age migrants in OECD countries were involved in non-agricultural entrepreneurship, compared with 12 per cent among non-migrants (see Figure 12.3). Migrants were over-represented in self-employment in Australia, UK, France, Belgium, Denmark, Sweden, Norway, US and Central and Eastern Europe, but underrepresented in Southern Europe, Ireland, Israel, Germany, Austria and Switzerland (OECD 2011b: 142–143). Probable explanations include that past exclusion from entrepreneurship in former guest worker-recruiting countries has had a long-term effect, while in Southern Europe the relative recentness and the frequently irregular nature of migration have not been conducive to business formation. The over-representation of migrants in self-employment in Poland, the Slovak Republic, the Czech Republic and Hungary can be partly explained by relatively flexible visa regulations for migrant entrepreneurs (OECD 2011b: 143).

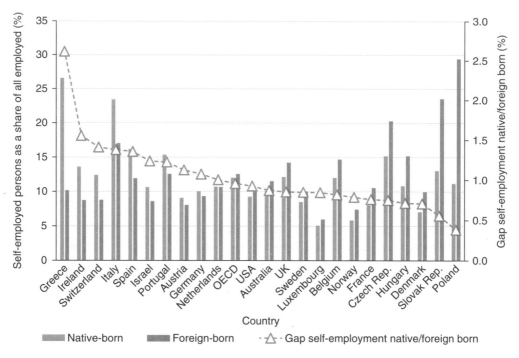

Figure 12.3   Self-employment of native and foreign-born in OECD countries, 2007–08
Source: OECD 2011b: 142

Although 'ethnic businesses' initially often cater for the needs of ethnic communities, they can rapidly expand across the economy. Typical initial migrant-owned businesses are ethnic restaurants, 'mom and pop' food stores and convenience stores (Waldinger *et al.* 1990). The fact that immigrant-owned businesses frequently employ family members from origin countries helps to hold costs low but also sustains the demand for migrant labour both through formal and informal channels. For instance, the origins of the Korean business community in Los Angeles can be traced back to the Korean War, which led to extensive migration from Korea to the US (see Chapter 7). Korean entrepreneurs came to dominate the New York greengrocery business, at first employing co-ethnics, then replacing them with Mexican workers at lower wages – and then re-employing Koreans when the Mexicans demanded better pay and conditions (Ness 2005: 58–95; see also Waldinger 1996). Migrant entrepreneurs often operate in particular niches, through which they can dominate particular sectors or create entirely new ones, thereby reshaping demand amongst urban populations.

The economic dynamism of immigrant entrepreneurs can have positive effects on economic growth and quality of life for consumers (Fix and Passel 1994: 53) and it can contribute to the revitalization and *gentrification* of run-down neighbourhoods. A more critical viewpoint stresses the suffering entailed by intense competition, long hours of work, and exploitation of family labour and of illegally employed migrants (Collins *et al.* 1995; Light and Bonacich 1988; Phan 2016). The growth of small businesses is partly linked to policies of economic deregulation, which have made it easier to start businesses and to employ workers on a casual basis. In many such businesses, both employers and workers are migrants or members of ethnic minorities. There is often no clear status distinction between the employed and the self-employed – the former can include high-status managers as well as low-paid service workers, while the latter range from medical professionals to food-stall operators, cab drivers, and food and parcel deliverers.

The contribution of migrant entrepreneurs to employment creation is significant. One study found that, on average, a self-employed migrant owning small or medium enterprises creates between 1.4 and 2.1 additional jobs, slightly less than native-born entrepreneurs. Their contributions to the host-country economy are not limited to job creation, but also include innovation and trade (OECD 2011b: 140). Kloosterman (2018) argued that 'superdiverse' cities across the globe offer ample opportunities for a diverse population of migrant entrepreneurs in various segments of urban economies. Migrant-run businesses generate employment and income for large groups of migrants, and migrant entrepreneurs may be seen as role models. Migrant entrepreneurs also provide a whole array of goods and services to the wider urban population (Kloosterman 2018).

The transformation of work

Subcontracting, temporary work and casualization

The neoliberal globalization of production since the late 1970s has reshaped the world of work, and this has had indirect, but crucial, effects on immigration. Global commodity chains, polarization of labour markets, devaluation of old skills and the decline of job security have affected workers everywhere, but migrants have often been most vulnerable to such changes. A key element of neoliberal employment practices has been the drive to turn wage-workers, who previously enjoyed the protection of labour law and collective agreements, into independent 'contractors', who have no guarantee of work, have to buy their own tools and equipment, and bear all the risks of accident, sickness or lack of jobs (Schierup *et al.* 2006: Chapter 9).

Increasingly, these 'contractors' are immigrants. The pressure to become independent contractors has affected occupations as diverse as building tradesmen, truck drivers, graphic designers and architects. A striking example from the US concerns the New York 'black cab drivers', who take executives and tourists to and from the airports. Once paid employees, they now have to buy the expensive luxury vehicles (on credit) and bear all operating costs, with no guarantee of work. The result is low income and extreme working hours. Most of the drivers are South Asians (Ness 2005: 130–180).

Employing migrants on a temporary basis is another way of enhancing employer control and reducing demands for better wages and conditions. Migrants are more likely to be employed in temporary jobs than natives (OECD 2007: 75–76). Economic deregulation has led to the removal of many legal controls on employment. This allowed a big expansion in *casual employment*: that is, hiring by the hour or for specific tasks, especially of migrants, young people and women. Casual jobs are typical for cleaning, catering and other service occupations, but also for the construction, textile and garment industries. Many big firms no longer engage directly in production, but subcontract it to smaller firms in sectors of the labour market, with a high degree of informality and scant regulation of working conditions. Through outsourcing to subcontractors, they strive for a maximum of flexibility. The frequently celebrated rise of 'ethnic entrepreneurship' needs also to be seen in the context of such trends, as it is partly a reaction to the lack of opportunities for security and upward mobility in formal labour markets.

Informal employment and irregular migration

Many migrants work in the informal sector. Although undocumented migrants are more likely to work in informal jobs, many migrants with residency status work in informal jobs, as many natives do. In fact, it has been estimated that more than 60 per cent of the world's employed population do informal jobs (Chacaltana *et al.* 2018). *Informal employment* refers to work without legal and social protection both inside and outside the *informal sector*, which can be defined as production and employment taking place in unregistered enterprises (see Chen 2009). About 93 per cent of the world's informal employment is in emerging and developing countries: in Africa, 85.8 per cent of employment is informal, 68.2 per cent in the Asia-Pacific region, 68.6 per cent in Arab countries, 40.0 per cent in the Americas and 25.1 per cent in Europe and Central Asia (Chacaltana *et al.* 2018). Poverty is both a cause and a consequence of informality – with the poor facing face higher rates of informal employment and poverty rates being higher among informal workers. Although the size of the informal economies generally decreases with income and education, there is significant variation.

In developed countries, informal jobs account for 18.3 per cent of all employed on average (Chacaltana *et al.* 2018). Some middle- and high-income countries have sizeable informal sectors, particularly in agriculture, which for instance account for 27.3 per cent of employment in Spain, 31.5 per cent in South Korea, 34.8 per cent in Turkey, 35.9 per cent in Russia, 53.4 per cent in Mexico and 79.9 per cent in Morocco. Irregular migrants enter well-rooted and flourishing local underground economies, the existence of which often predates irregular immigration (see Reyneri 2001). For instance, informal domestic, cleaning and agricultural work used to be done by lower-class native women and men. As domestic supply for such jobs has been drying up, migrant workers have increasingly taken their place. Informal employment of migrants is particularly common sectors such as agriculture, construction, catering, domestic work and other services.

In addition, the trends of subcontracting, temporary work and casualization have contributed to informalization, which can be defined as a 'redistribution of work from regulated sectors of the economy to new unregulated sectors of the underground or informal economy' (Ness 2005: 22). Some form of production has been reorganized into small-scale and decentralized economic units, with mass production giving way to 'flexible specialization', sometimes reverting to sweatshop production (Piore and Sabel 1984). The *informalization* of employment relations turned standard jobs into non-standard or atypical jobs with hourly wages but few benefits, or into piece-rate jobs with no benefits. In this way, the informal economy has become a permanent, subordinate feature of advanced economies (Portes *et al.* 1989; see Chen 2012).

Economic deregulation and employer practices have reinforced the informal sector and created new informal sector jobs, drawing in migrants (Reyneri 2003). Although the involvement of migrants in informal work is most obvious in the US, Southern Europe, the Gulf, and East and South-East Asia, informal work is widespread in other countries, for instance in British agriculture, cleaning and catering, but also in the cases of traffic wardens and security work – both services devolved by public authorities to subcontractors. The large-scale involvement of migrant women in domestic labour and cleaning has become an almost universal feature in the large city economies around the world. In Germany, for instance, the demand for care workers in private homes has been increasing fast, although the government has turned a blind eye to this issue. As Lutz and

Palenga-Möllenbeck (2010) argued, as a result of the mismatch between demand and restrictive policies, a large sector of undeclared migrant care work has emerged, which has become a *de facto* integral part of the German welfare state.

The feminization of precarious work

Labour market restructuring and the concomitant rise of *precarious work* has pushed certain categories of workers – and particularly migrants – into insecure and exploitative jobs. Migrant women are particularly susceptible to labour market exploitation. While surging demand for domestic workers around the world has increased opportunities for non-elite women in developing countries to earn a much higher income abroad, their gender and frequent lack of rights makes them vulnerable to abuse and exploitation. This particularly applies to countries with poor human right records, such as in the Gulf, but also in 'liberal democracies', where such workers have formed a new ethnic underclass. Morokvasic argued that migrant women from peripheral zones living in Western indus trial democracies represent:

> a readymade labour supply which is, at once, the most vulnerable, the most flexible and, at least in the beginning, the least demanding work force. They have been incorporated into sexually segregated labour markets at the lowest stratum in high technology industries or at the 'cheapest' sectors in those industries which are labour intensive and employ the cheapest labour to remain competitive. (Morokvasic 1984: 886)

Source: Getty Images/SOPA Images

Photo 12.2   Immigrants take selfies, make long distance calls and gather in Sham Shui Po on their Sunday day off in Hong Kong

Gender-segregated labour markets have become a crucial feature of new global divisions of labour. As the demand for manual male labour has remained stable or fallen because of declines in manufacturing employment, women have increasingly entered the migrant labour force. Migrant women are disadvantaged by at least three interlocking sets of mechanisms, also known as *intersectionality* (see Chapter 4). First, as migrants they are susceptible to abuse and discrimination, particularly if they have undocumented status and informal jobs. Second, they face gender-specific prejudices such as employers' assumptions that they are not primary breadwinners, but rather temporary workers who will leave to get married (Schrover *et al.* 2007). Third, migrant women are also disadvantaged by stereotypes of specific ethnic and racial groups, and often also by weak legal status (Browne and Misra 2003: 489). For instance, migrant women falling pregnant in Malaysia, Thailand and Taiwan risk deportation; such forms of 'triple discrimination' often make the position of migrant women often particularly precarious (Mendoza 2018).

Contrary to neoclassical assumptions that variations in employment status and pay reflect differing levels of objectifiable 'human capital', in practice citizenship, race, class, gender and sexual orientation often play a decisive role in determining labour market status. The disadvantaged position of migrant women is crucial to sectors such as the garment industry across Asia (see Mendoza 2018). Domestic workers form a category of gendered and racialized labour that has expanded rapidly in advanced economies around the world (Ambrosini 2013, Anderson 2000, 2007; Cox 2006; Parreñas 2000). Domestic work used to be the domain of lower-class native women, often rural-to-urban migrants. Over recent decades domestic work has increasingly become a niche for international migrant women from developing countries (Schrover *et al.* 2007: 536–537).

However bad the conditions, it does offer a chance of a job and an independent income, often combined with live-in conditions that are perceived by migrant women's families as sheltered. Care jobs are also less vulnerable to economic shocks than occupations such as construction where many male workers are concentrated. Female – but to a certain extent also male – workers are increasingly involved in providing care services for the elderly in Southern Europe, but to an increasing extent also in countries like Germany and Austria. Although they are often undocumented, the presence of such migrant workers is tolerated as they provide forms of 'invisible welfare' that have become a cornerstone of informal care systems (Ambrosini 2013, 2015).

Domestic work is marked by a hierarchy of work tasks, of formal and informal modes of employment, and of groups with varied statuses. For instance, Filipina domestic workers and nannies are preferred in some places due to their better education and English, but rejected in others because they are seen as too assertive in defending their rights. Domestic work by migrant women can be the result of increased opportunities of white-collar employment or leisure for majority-group women: hiring foreign maids can free women in Italy, the US or Singapore from housework and childcare (Ambrosini 2016; Huang *et al.* 2005; Iredale *et al.* 2002).

Such transnational care hierarchies sometimes go a stage further, when migrant domestic workers hire a maid in the home country to look after their own children, who may in turn rely on family members to take care of their children (Hochschild 2000; see also Chapter 3). Such 'global care chains' may mean higher living standards and a viable source of income for migrant women and better education for their children, but can come at a high emotional and social cost.

The chronic demand for migrant labour

Taken together, the various forms of labour force restructuring add up to a process of increased *labour market segmentation*. This implies that people's chances of getting secure jobs depend not only on their 'human capital' (their education and skills) but also on their gender, race, ethnicity and legal status. Labour market segmentation is anything but new. Piore's dual labour market theory (see Chapter 3) remains useful to understand why advanced economies generate a 'chronic' demand for low-skilled migrant labour. In Western Europe in the 1960s and 1970s, for instance, the discrimination inherent in guestworker policies funnelled immigrants into specific economic sectors and occupations (Castles and Kosack 1973). Neoliberal globalization seems to have reinforced the casualization of labour and growth of low-status, informal jobs that often only migrants want to do (Sassen 1988).

For instance, immigrants have played a crucial rule in the transformation of the social geography of New York City (Ness 2005: Chapter 2). In the early twentieth century, immigrant labour from Southern and Eastern Europe was crucial to the emergence of the garment, printing, meatpacking, construction and transportation industries. Industry was concentrated in 'ethnic neighbourhoods' and immigrants came to form the backbone of the city's strong labour movement. In the late twentieth century, these traditional industries were restructured, with most production jobs being moved to non-unionized 'sunbelt' states or offshore to the Caribbean, Latin America and Asia.

Many new jobs were created in retailing, personal services and business services (see also Waldinger 1996). The new economy was heavily stratified on the basis of ethnicity, with US-born white people getting high-skilled jobs in the services sector, African Americans and US-born Latinos getting public sector jobs, and immigrants often getting low-wage jobs in such areas private transportation, catering, delivery, security and building maintenance (Ness 2005: 17). Parallels to the changes in New York City can be found everywhere in advanced economies (see Goldring and Landolt 2011; Rath 2002).

The relatively small impact of the Great Recession on long-term migration trends illustrates the relevance of the dual labour market theory. As the regional overview chapters showed, migration to most crisis-affected countries in Europe and North America recovered remarkably quickly, despite significant rises in unemployment. The immediate impacts of the Great Recession were felt most strongly in the richer economies in Western Europe and North America (Phillips 2011b). Overall *unemployment* in the OECD grew by 55 per cent between December 2007 and January 2012. The largest increases were in Ireland, Spain, Greece, Iceland and Estonia (OECD 2012: 60). In most countries, migrants were more affected by unemployment than the native-born workers (OECD 2012: 63).

Yet the paradox was that foreign *employment* in European OECD countries actually increased by 5 per cent from early 2008 to the third quarter of 2010, while the employment of native-born persons declined by over 2 per cent (OECD 2011b: 74–75). In other words, foreign unemployment increased, but so did foreign employment! How could foreign unemployment and employment grow at the same time? To understand why this was the case requires looking at both short-term factors (what economists call 'conjunctural' issues like interest rates, consumer behaviour and entrepreneurs' propensity to invest along business cycles), and longer-term structural factors (like changes in underlying economic and demographic patterns). Analyses of migration often focus on short-term factors, such as economic ones, and fail to sufficiently recognize long-term structural factors such as labour market segmentation.

In the wake of the Great Recession, conjunctural factors led to a sharp decline in some types of production, especially of consumer goods and therefore also of the steel and plastics needed to make them. Yet the Recession did not affect much deeper, structural changes which have been going on for over half a century, like the shift away from manufacturing to the services, population ageing and the decreasing willingness of native workers to do dirty, difficult and dangerous jobs that lost even more appeal because of their increasing labelling as 'migrant jobs'.

Migrants working in sectors most adversely affected by the crisis often lost their jobs, but other migrants were able to gain or keep jobs in emerging sectors or sectors that were much less affected by the crisis. Migrant men were far more affected by job losses than migrant women. This is because migrant men tended to be employed in the sectors hardest hit by the downturn, especially manufacturing and construction, while migrant women were more concentrated in less-affected sectors, notably social services, child and elderly care work and domestic work. The Great Recession also reinforced the trend towards part-time, temporary and casual employment, with women more likely to enter such employment relationships than men (OECD 2011b: 78–81).

The continued importance of labour recruitment

As argued throughout this book, the control and recruitment of labour has been central in shaping global migration patterns. In the past, such recruitment was often organized by states and businesses, whether through indentured work in the era of European colonialism, state-assisted migration programmes to organize migration from Europe to the 'New World' or the recruitment of migrant workers. Although perhaps less visible than before, recruitment of migrant workers continues to be a vital mechanism for employers to access labour resources. Much of this recruitment has now been privatized and outsourced either to private recruitment agencies, ranging from large international employment and recruitment companies such as Randstad and Adecco, or to informal operators, agents, smugglers and various other 'migration intermediaries' (see Chapter 3). Economic deregulation has increased the scope for such private intermediaries. In the Netherlands alone, there are 3600 official employment agencies, which fulfil a crucial role in recruiting East European workers for the Dutch horticulture sector (Berghege *et al.* 2018).

Recruitment practices expose a fundamental dilemma. On the one hand, the foreign status of such workers makes them vulnerable to exploitation and abuses. The main reason why employers often *prefer* to employ migrants is that they are considered as a hardworking, malleable and docile workforce. On the other hand, recruitment can give access to international migration opportunities to relatively poor people who would otherwise not have had the means to migrate. For instance, the massive labour opportunities in Gulf countries represent the hope for a better future for millions of poor households in South and South-East Asia where migration to Europe or North America is often a prerogative of better-off groups.

While recruitment is important everywhere, recruitment through governments and agencies plays a particularly important role in facilitating Asian labour migration. Migration agents and labour brokers organize most recruitment of Asian migrant workers to the Gulf and within Asia, while matchmakers organize marriage migration. Annually, millions of Asian migrant workers across all skill levels leave their countries holding employment contracts issued or certified by government regulators (Agunias 2009, 2013). Licensed private recruitment agencies orchestrate much of the labour migration process from pre-departure, when the terms of conditions of the employment contract are negotiated, to migrants' eventual return. Origin and destination country governments have set up elaborate regulatory regimes governing recruitment, placement fees, recruiters' liability and employment conditions (Agunias 2013).

However, despite these rules and regulations, migrant workers remain vulnerable to exploitation and abuse from recruitment agencies and employers. For instance, despite regulations capping placement fees, recruitment agencies routinely overcharge migrant workers. *Contract substitution* – with migrants signing a new contract at the destination with inferior stipulations than the contract signed at the origin – is also common (Agunias 2009). Many workers are tied to exploitative employers, when reneging on their contracts and going home would incur large fees. Other common problems include underpayment or non-payment of wages, confiscation of passports, poor working and living conditions, and even physical abuse and sexual harassment (Agunias 2009, 2013). Because of this lack of enforcement, many international labour migrants leave their homes with contracts that look good on paper but do not adequately protect them on the ground (Agunias 2009, 2011). Problems include the proliferation of unlicensed subagents and brokers; insufficient capacity to identify unqualified employers; and a broken and inaccessible legal system for migrants.

Although countries like the Philippines and Vietnam have sought to regulate agencies, this has proved difficult in practice (Kim *et al.* 2007) (see Chapter 10). Sometimes, migrants become undocumented workers. For instance, labour migration from Indonesia to Malaysia and Myanmar to Thailand has been largely undocumented. Irregular migration is linked to the unwillingness of governments to allow legal entries and the desire of employers for easily available and exploitable workers. Spontaneous undocumented migration can meet labour needs effectively, but can create situations of insecurity and rightlessness for workers. Moreover, they can become easy scapegoats for social problems such as crime, disease and unemployment.

Migrant workers are generally more vulnerable to discrimination and harm compared to local workers, because they almost always lack the same legal and social protections. For instance, migrant workers play a key role in the ready-made garment sector manufacturing branded clothes for worldwide export in countries like Malaysia (involving workers from Nepal, Indonesia, Myanmar and the Philippines), Taiwan (involving workers from the Philippines and Vietnam) and Thailand (involving workers from Myanmar, Cambodia and Laos). These temporary migrant workers are typically young and female, and confront discriminatory workplace practices beyond those experienced by local workers (Mendoza 2018). For instance, workers routinely suffer from pregnancy discrimination. Migrant workers may be subjected to mandatory pregnancy-testing in their home country as part of the application process for a job overseas. While this type of pregnancy discrimination is legal in some countries, such as in Thailand and Malaysia, and banned in others, such as in Taiwan, pregnancy-testing happens nevertheless and if women become pregnant while working abroad this is often associated to loss of employment and income as well as deportation (Mendoza 2018).

Conclusion

This chapter highlighted the importance of lower- and higher-skilled migrant labour for advanced and emerging economies. Economic restructuring and demographic change increased the demand for immigrant labour, contributing to multilayered geographical patterns, with lower-skilled workers from low-income economies moving to middle-income economies, and lower-skilled workers from middle-income countries predominantly moving to high-income economies, while higher-skilled workers tend to be more mobile and migrate over larger distances following geographically more dispersed patterns.

Economic migration is vital for advanced and emerging economies. Lower- and higher-skilled workers provide *additional labour* at a time of high sector-specific demand resulting from economic, demographic and social shifts. They also provide *special types of labour* to fill gaps that native workers are incapable or unwilling to fill because of a lack of relevant qualifications and skills or because of their lack of motivation to do unattractive jobs. Migrants are willing to do unattractive jobs as they provide a major improvement in pay and prospects compared to opportunities back home. Migration thus helps to maintain labour market flexibility, encourage investment and economic growth. During times of recession, migrants also serve as a *buffer* which somewhat mitigates the effects of economic downturns on domestic unemployment forces, as migrants are generally the first to be dismissed.

Migrant workers often experience exploitation and abuse, particularly lower-skilled workers in the informal sectors. High unemployment rates – in some countries more than twice the average for native workers – reveal that migrant workers often still have a disadvantaged position. Many migrant workers now have service jobs, some of them in higher-skilled positions (such as doctors, nurses, engineers and teachers), but often in such sectors such as cleaning, catering, domestic work and care.

During the post-1945 boom, migrant workers in Western Europe and North America were steered into subordinate jobs: 'guestworkers' had strictly limited labour market rights, while colonial migrants were often subject to racial discrimination. In addition, many migrants lacked education and vocational training, and therefore entered the labour market at low levels. Since the 1980s and 1990s, migrants' work situation has become much more diverse, partly as a result of the shift to service-based economies.

The shift to neoliberal policies, the deregulation of labour markets and economic restructuring has been linked to a new international division of labour, in which migrant workers play important and increasingly varied roles. On the one hand, this has led to an increasing demand for higher-skilled workers in the service sector. On the other hand, this has made the position of lower-skilled workers more difficult. Stable employment within large-scale enterprises has in many cases been replaced by a variety of work arrangements that differentiate and separate workers along ethnic, class, education and gender lines. Temporary and casual employment, chains of subcontracting, informalization and new forms of labour market segmentation affect both native and migrant workers.

The most disadvantaged and vulnerable groups of workers – migrant women, undocumented workers, ethnic and racial minorities – tend to end up in the most precarious positions. Deprivation of human and worker rights for groups that lack legal status and market power seems to be an integral aspect of all advanced and emerging industrial economies today. The relevance of citizenship is particularly clear with regard to the right to cross national borders, and to work and take up residence in destination countries. Rich states compete with each other to attract highly skilled workers, fee-paying students and the wealthy, but take measures to prevent legal entry and access to rights of lower-skilled workers. Since there is a continuous demand for lower-skilled workers in agriculture, construction, manufacturing and the services, many migrants have irregular status. It is a question for debate whether such situations of illegality are willingly tolerated to create a vulnerable workforce that is easy to exploit, or whether this is the accidental by-product of misguided immigration policies led by short-term interests of employers.

Migrant workers have become essential to the economies of developed countries, especially in the Gulf, North America, Europe and various East and South-East Asian countries, where demographic change and increasing education is decreases local labour supply, particularly for lower-skilled jobs as well as certain specialized jobs for which not enough local workers have been trained. This illustrates the validity of Piore's (1979) argument that immigrants have become a structural feature of industrialized economies. As new industrial areas emerge in Asia, Latin America and Africa, migrant workers will play a part there too, and the differentiation of labour crucial to existing global commodity chains will become increasingly important in new contexts.

Perhaps the biggest challenge for policy makers is to reconcile the intrinsic demand for migrant workers generated by labour market dynamics with the political call to limit permanent settlement. 'We asked for workers, we got people instead', as the Swiss novelist Max Frisch summarized this conundrum, referring to the (for some, quite traumatic) 'guestworker' experience. Governments have never given up attempts to develop immigration policies that would ensure the return of workers (Castles 2006). However, as the experience with 'guestworkers' in Western Europe as well as with Mexican 'bracero' workers in the US has made clear, revolving door policies are notoriously difficult to enforce. Governments of destination countries in Africa and Asia and even the authoritarian governments of the Gulf region have also found it increasingly difficult to enforce temporariness. This highlights the powerful 'internal dynamics' such as network formation (see Chapter 3) that give migration processes their own momentum and that tend to turn migrants workers into increasingly permanent members of destination societies.

Guide to Further Reading

In this chapter, we were able to look only at a limited range of studies on the labour market experience of migrants and the economic effects of migration. It would be important to look at other indicators, such as wage levels (and how they change over time), poverty, employment rates and participation rates. Readers are recommended to use the further reading and to follow up the sources we cite for more on these issues. Anderson (2000) is a good introduction into the global politics of migrant domestic work. Parreñas (2001, 2005) are excellent studies on the role of migrant Filipina domestic workers in the global economy and the implication for their families, mothering and care at home. Piper (2008) gives an overview of the feminization of migration in Asia. Lutz (2016) is a useful collection on European

▶

migration and domestic work. Ambrosini (2013) discusses the crucial role of (female) migrant labour in care regimes and the ambiguous attitude of politics and society 'between rejection and practical tolerance'. Older but still highly relevant texts on the political economy of migrant labour include Munck (2011); Phillips (2011a); Piore (1979); and Sassen (1988, 2001). Gender and migrant labour are further examined in Browne and Misra (2003); Pessar and Mahler (2003); Phizacklea (1990, 1998); and Schrover *et al.* (2007). On irregular migration and the informal sector see Düvell (2006); Ness (2005); and Reyneri (2003).

Several issues of the OECD's annual *International Migration Outlook* contain good information on the employment situation, and the website of the International Labour Organization (ILO) also contains useful resources and publications on labour migration. *The Age of Migration* website includes an additional text on the educational and occupational success of the 'second generation' in Germany, as well as a summary of discussion between economists about whether migration is good or bad for destination-country economies and for specific groups of the population. The website also includes an analysis of labour market segmentation in the French car- and building-industries in the 1970s and 1980s.

Extra resources can be found at: **www.age-of-migration.com**

13 New Ethnic Minorities and Society

The migrations of the last half century have led to growing cultural diversity and the formation of new ethnic groups in many countries. Such groups are visible through the presence of different-looking people speaking their own languages, the development of ethnic neighbourhoods and the establishment of ethnic associations and businesses. This chapter examines the diverse experience of societies with the formation of migrant communities as well as their various modes of integration or 'incorporation' into destination societies. This chapter will review why ethnic group formation and growing diversity have been relatively easily accepted in some countries and for some immigrant groups, while in other cases this process has rather resulted in marginalization and exclusion. We will then go on to examine the consequences for the ethnic groups concerned and for society in general. The main argument is that the migratory process works in a rather similar way everywhere with respect to settlement, labour market segmentation, residential patterns and ethnic group formation. Racism and discrimination are also to be found in all societies, although their intensity varies. The main differences are to be found in public attitudes and government policies on immigration, settlement, education, housing, citizenship and cultural pluralism.

Immigration policies and minority formation

Three types of countries can be distinguished with regard to immigration and integration policies. The 'classical immigration' countries such as the United States, Canada, Australia and New Zealand encouraged family reunion and permanent settlement, and treated most legal immigrants as future citizens. The second group includes France, the Netherlands and the UK, where immigrants from former colonies were often citizens at the time of entry. Permanent immigration and family reunion have generally been permitted (though with some exceptions). Immigrants from other countries had a less favourable experience, although settlement and naturalization have often been permitted. The third group consists of countries which tried to cling to 'guestworker' models, above all Germany and Austria. Belgium and Switzerland are similar in many ways. Such countries tried (but failed) to prevent family reunion, were reluctant to grant secure residence status and had restrictive naturalization rules at least until the 1990s. Most immigration countries in East Asia, the Middle East and Africa also belong to the latter group, as they often oppose (or are in denial of) permanent settlement of migrants and have often very restrictive immigration and naturalization laws, except for co-ethnic groups.

The distinctions between these three categories are neither absolute nor static. Some countries do not fit the categories: for example, Sweden admitted both migrant workers and refugees, but always accepted family reunion. The Netherlands and Belgium had both colonial immigrants and 'guestworkers'. The openness of the US, Canada and Australia only applied to certain groups: all three countries had exclusionary policies towards Asians until the 1960s. The US tacitly permitted irregular farmworker migration from Mexico, but denied rights to such workers. France had very restrictive rules on family reunion until the 1970s. Germany and Switzerland gradually improved family reunion rules and access to long-term residence status. These countries could not completely deny the reality of settlement. There has been a gradual trend towards greater rights and security.

A major change has been the erosion of the privileged status of migrants from former colonies. Making colonized people into subjects of the Dutch or British crown, or citizens of France, was a way of legitimating colonialism. In the period of European labour short-age, it also seemed a convenient way of bringing in low-skilled labour. But citizenship for colonized peoples became to be seen as a liability when permanent settlement took place. All three countries removed citizenship from most former colonial subjects and put them on a par with foreigners. The Netherlands hastened Surinamese independence as it made Dutch citizens into foreigners, although the policy was counterproductive by triggering a massive 'now or never' migration wave (see Chapter 7).

As governments have to some degree been forced to adapt to migration realities on the ground, there has also been a degree of policy convergence: former colonial countries have become more restrictive with regard to citizenship, while former 'guestworker' countries have become less so (de Haas *et al.* 2018). But this has gone hand in hand with new forms of differentiation, such as through the creation of a privileged status for migrants in the European Community (and later EU) countries from 1968 onwards (see Chapter 10). At the same time, entry and residence have become more difficult for non-EU citizens, especially those from outside Europe. Political discourses which portray immigration as threatening to the nation, and which conflate legal and illegal migration, create problems for long-standing immigrants and even for their descendants. Anyone who looks differ-ent becomes suspect and may be forced to prove their identity as legal residents and their allegiance to the destination country culture.

Immigration policies have consequences for most other areas of policy towards immigrants, such as labour market rights, political participation and naturalization. Policies designed to keep migrants in the status of temporary mobile workers make it likely that settlement will take place under discriminatory conditions. Moreover, official ideologies of temporary migration create expectations among native popula-tions. When temporary sojourn eventually turns into settlement, and governments refuse to admit this, then it is often the immigrants who are blamed for the resulting problems.

Immigration policies also shape the consciousness of migrants themselves. In countries where permanent immigration is accepted and the settlers are granted secure residence status and civil rights, a long-term perspective is possible. Where the myth of short-term sojourn is maintained, as used to be the case in much of West Europe and still applies to many countries in Asia, Africa and the Gulf region, immigrants' perspectives are inevi-tably contradictory. Return to the country of origin may be difficult or impossible, but permanence in the immigration country is doubtful. Such immigrants settle and form ethnic groups, but they find it more difficult to see and plan a future as part of the wider society. Although discriminatory immigration policies cannot stop the completion of the

Box 13.1 Migrants and minorities in the US

US society is a complex ethnic mosaic deriving from five centuries of immigration. The white population is a mixture of the original mainly British colonists and later immigrants who came from all over Europe. Assimilation of newcomers is part of the 'American creed', but this process has always been racially selective. Native American societies were devastated, driven off their land and secluded in reservations by white expansion westwards, while millions of African slaves were brought to America to labour in the plantations of the South.

The US is becoming ever more culturally diverse. The foreign-born population grew by 13.4 million between 2000 and 2017, to reach an estimated 44.5 million. The foreign-born share in population rose from 4.7 per cent in 1970 to 13.7 per cent in 2017. By then, 52.2 per cent of the foreign-born residents were from the Americas, 31. per cent from Asia and only 10.8 per cent from Europe. Ethnic minorities make up over a quarter of the population – and are especially strongly represented in younger age cohorts. Hispanics are the descendants of Mexicans absorbed into the US through its south western expansion, as well as recent immigrants from Latin America. The Asian population is also growing fast.

The movement of European immigrants and African–Americans into low-skilled industrial jobs in the early and mid-twentieth century contributed to labour market segmentation and residential segregation. In the long run, many 'white ethnics' achieved upward mobility, while African–Americans became ghettoized. Despite the rise of a black middle class after the Civil Rights Movement of the 1960s, distinctions between the majority of African–Americans and whites in income, unemployment, social conditions and education are still extreme. Members of some recent immigrant groups, especially from Asia, have high educational and occupational levels.

US population by race and Hispanic origin, 1970, 1990, 2017 (millions)			
	1970 (millions)	1990 (millions)	2017 (millions)	Share of Population (%, 2017)
White	158.8	199.7	234.4	73.0
Black or African American	18.9	30.0	40.6	12.7
American Indian and Alaskan Native	0.6	2.0	2.6	0.8
Asian and Pacific Islander	1.0	7.3	17.8	5.5
Some other race	0.1	9.8	15.6	4.8
Two or more races			10.1	3.1
Total Population	**179.3**	**248.7**	**321.0**	
Hispanic (of any race)	5.8	22.4	56.5	17.6

Note: Data is for 'household population' (excluding people in institutions such as prisons). Race is based on self-identification.

Source: US Census Bureau (2017)

Incorporation of immigrants into the 'American dream' has been largely left to market forces. Nonetheless, government has played a role by making it easy to obtain US citizenship, and through compulsory public schooling. Undocumented migration has been a major political issues since the 1990s. The construction of fences and surveillance systems along the US–Mexico border did not cut migration but made it more dangerous and expensive: many migrants lost their lives trying to cross the desert. In view of the high risks and costs, many Mexican workers decided to stay on in the US and to bring their families. Thus, border control measures turned the temporary labour movement into permanent settlement (see Box 11.1).

Today immigration reform is a central issue in US politics. In 2016, 44.7 per cent of the 43.7 million foreign-born residents were naturalized citizens. Legal residents or legal temporary migrants (31.6 per cent) accounted for 31.6 per cent and undocumented immigrants made up 23.7 per cent of the total foreign-born population. However, federal reform efforts have stalled over the past decades. In response, there has been an increase in state initiatives. While some states (such as Arizona) have enacted tougher laws relating to immigration enforcement, other states (such as California) have adopted more immigrant-friendly laws while several cities have proclaimed to be safe havens for undocumented migrants.

Sources: Migration Information Source (2011); MPI Data Hub (2019); Pew research centre (2016); Krogstad *et al.* (2018); Portes and Rumbaut (2006); Massey and Pren (2012); Massey and Denton (1993); Portes and Zhou (1993)

migratory process, they can be the first step towards the marginalization of future settlers. This can contribute to their *downward assimilation* (see Portes and Zhou 1993) as they become part of low-status, racialized labour classes.

Labour market position: Segmented assimilation

As argued in Chapter 12, labour market segmentation based on ethnicity, race, legal status and gender has developed in all immigration countries. This was intrinsic in the type of labour migration practised until the mid-1970s in Western Europe, and which is still widespread in labour-importing countries of the Gulf and East and South-East Asia. However, today's migrants are much more diverse in educational and occupational status. Governments have increasingly encouraged the immigration of the higher skilled. Many refugees bring skills with them, although they are often not allowed to use them. Low-skilled migrants are officially less welcome, but often enter through family reunion, as asylum seekers or irregularly. Their contribution to low-skilled occupations, small business and households (childcare, cleaning) is of great economic and social importance, but is often officially unrecognized.

In any case, labour market segmentation is part of the migratory process. When people come from poor to rich countries, without local knowledge or networks, lacking proficiency in the language and unfamiliar with local ways of working, then their entry point into the labour market is likely to be at a low level. The question is whether there is a fair chance of later upward mobility. This partly depends on state policies. Some countries (such as Australia, Canada, Sweden and the UK) have active policies to improve the labour market position of immigrants and minorities through language courses, basic education, vocational training and anti-discrimination legislation. Other immigration countries (such as Malaysia, South Africa, Libya and the Gulf states) seem content to exploit the labour of immigrants, either through maintaining them in a situation of irregularity or by setting up discriminatory contract labour arrangements.

The former 'guestworker' countries form a third in-between category. Although they introduced education and training measures for foreign workers and foreign youth, they also maintained restrictions on labour market rights. However, the overwhelming majority of foreign workers would eventually gain long-term residency permits, which gave them virtual equality of labour rights with nationals. The US has a special position: there are equal opportunities, affirmative action and anti-discrimination legislation, but little in the way of language, education and training measures. This fits in with the laissez-faire model of social policy and with cuts in government intervention.

It is not clear which policies work better, and evidence seems mixed. Although levels of poverty and extreme marginalization of migrant groups may be lower in countries with extensive welfare, public education, health care and housing systems in Northwest Europe and Scandinavia, evidence reviewed in Chapter 12 shows that unemployment rates are often very high. In countries such as the US and to a lesser extent also the UK and Canada, labour force participation amongst migrants is much higher, although they may often have precarious, low paid and unattractive jobs with no prospect for promotion. National differences should not be exaggerated, as experiences of incorporation also differ widely across immigrant groups, depending on their ethnicity, education and migration history. In general, the experiences of high-skilled migrants are much better than those of low-skilled migrants.

Residential segregation

Some degree of residential segregation based on class, ethnicity and race can be found in all immigration countries, though nowhere is it as extreme as in the US, where in certain areas there is almost complete separation between blacks and whites, and sometimes Asians and Hispanics too. In other countries there are city neighbourhoods where minority and immigrant groups are highly concentrated, though they rarely form the majority of the population. Residential segregation arises partly from immigrants' situation as newcomers, lacking social networks and local knowledge. Equally important is their low social status and income.

Migrant workers often start work in low-income jobs and have few savings. Often they have to remit money home. Therefore, they tend to seek cheap housing in working-class areas. Proximity to work reinforces this choice of location. As a group becomes established, the earlier arrivals can assist the newcomers, which strengthens the tendency towards ethnic clustering. Another factor is discrimination by landlords: some refuse to let to immigrants, while others make a business of charging high rents for poor accommodation. In Germany in the 1980s, some landlords crowded migrants into poorly equipped apartments, in order to make conditions unbearable for long-standing German tenants who could not be legally evicted. When the German tenants left, the foreigners were evicted too, and the block could be demolished to make way for offices or luxury housing.

Source: Getty Images/Jonas Gratzer

Photo 13.1   A Pakistani asylum seeker, looks out over the ocean from the immigration detention centre on Manus Island, Papua New Guinea, 2018

Such practices increased racism towards the immigrants, who became the scapegoats of ruthless urban development speculation.

Institutional practices may also encourage residential segregation. Many migrant workers were initially housed by employers or public authorities. There were migrant hostels and camps in Australia, barracks provided by employers in Germany and Switzerland, and hostels managed by the government *Fonds d'Action Sociale* (FAS, or Social Action Fund) in France. In the Netherlands, in 1951 many former colonial soldiers from the Moluccas (now part of Indonesia) (see Chapter 6) and their families were initially housed in former concentration camps. Also in Gulf countries migrants often live physically separated from native populations in barracks and compounds. Many governments initially house asylum seekers separately in refugee camps or asylum seekers' centres. One the one hand, such isolation may facilitate the processing of asylum requests. On the other hand, this is often part of official policies to prevent their integration and the formation of social ties with local populations, who frequently resist their deportation in case asylum requests are rejected.

Migrant hostels and barracks often provide better conditions than private rented accommodation, but also increase control and isolation. Hostels also encourage clustering: upon leaving their initial accommodation, workers tend to seek housing in the vicinity. Where institutional racism is relatively weak, immigrants often move out of inner-city areas to better suburbs as their economic position improves, to be replaced by newer groups. Thus, ethnic clustering can be understood in such contexts as a transitory phenomenon. However, where racism and social exclusion are strong, concentration persists or may even increase.

Box 13.2 Migrants and minorities in the United Kingdom

The UK uses three classifications for its population of immigrant origin:

- In 2017, there were 6.2 million *foreign citizens* (9.6 per cent of the total population of about 62 million) – twice the 2004 figure;
- The *foreign-born population* numbered 9.4 million in 2017 (14.4 per cent of total population), compared to 4 million in 1993 (7 per cent). The main origin countries of the *foreign-born* residents were: Poland (9.8 per cent), India (8.8 per cent), Pakistan (5.6 per cent), Ireland (4.1 per cent) and Romania (4.1 per cent);
- The *ethnic minority population* are mostly British-born descendants of New Commonwealth immigrants who arrived from the 1950s to the 1970s. In 2010–11 about 7 million people (11 per cent of the population) identified themselves as having a 'non-white ethnicity'. Nearly 90 per cent identified themselves as white; 5 per cent as Asian; and 3 per cent as black.

Commonwealth immigrants who came before 1971 were British subjects and enjoyed full citizenship rights. The 1971 Immigration Act and the 1981 British Nationality Act put them on a par with foreigners.

The *race relations approach* of the late 1960s and the 1970s recognized ethnic groups. Acceptance of cultural and religious diversity was officially labelled as *multiculturalism*. Race Relations Acts outlawed discrimination in public places, employment and housing. However, racism remained a major problem. Minority youth rioted in inner-city areas in the 1980s and 1990s. The government responded with measures to reduce youth unemployment, improve education, rehabilitate urban areas and change police practices. But the 1999 Stephen Lawrence Inquiry (analysing the poor police response after the murder of a young black man by a white gang) revealed the strength of institutional racism. In 2001, riots broke out involving youth of Asian origin in northern cities.

In the 1990s and early 2000s the main immigration issue was asylum. Five new asylum laws were introduced between 1993 and 2006, successively tightening up entry rules, and introducing detention and restrictions on welfare. By the mid-2000s public attention had shifted to the perceived threat of Islam. The 2005 London bombings precipitated concern about the loyalty of young Muslims. Policies under the Labour Government emphasized 'social cohesion'. Citizenship tests for immigrants were introduced, based on ideas of 'Britishness' and 'core values', and this policy was continued by Conservative governments after 2010.

The UK experience highlights the contrast between formal equality enjoyed by ethnic minorities, and frequent experiences of inequality and racism. It shows that citizenship is not necessarily a protection against social disadvantage. Yet it also shows that in the long run, groups that have initial negative experiences do become incorporated into society and can experience upward mobility.

A more recent issue was the rapid and unexpected growth in migration from Poland and other Eastern European countries since joining the EU in 2004.

▶

Immigrants often found work in agriculture and food-processing, leading to concentrations in areas with little experience of immigration. Politicians blamed migrants for problems such as deteriorating health care, public housing and education.

Fears about a 'foreign invasion' stirred up by the UK Independence Party (UKIP) and the tabloid press played a major role in the 2016 Brexit Vote, as it was believed this would enable the UK to 'take back control' on immigration.

Sources: Anderson *et al.* (2006). Cohen and Bains (1988); Kubal (2016); Migrationobservatory.org (2019); Peach (1968); Solomos (2003); and Favell (1998)

Migrants typically concentrate in the largest cities. In 2011, London contained 58 per cent of Britain's black population and about a third of the population of South Asian origin. In 2018, one quarter of the population of Amsterdam belong to ethnic minorities, such as Surinamese, Antillean, Turkish and Moroccan origin groups, and these are mainly concentrated in certain neighbourhoods, where 'non-Western' population groups can constitute up to half or more of the population. Segregation is reinforced when members of the majority population move out of inner-city areas to the suburbs (a phenomenon also known as 'white flight'). A typical pattern is that of immigrants initially clustering in particular 'immigrant neighbourhoods', but after one or two generations they increasingly move out as their children achieve upward socioeconomic mobility, to be replaced by new immigrant groups. Newly arriving groups often settle in rather homogeneous enclaves, but after a few generations such groups would move out as they got better jobs and largely assimilated into mainstream culture.

Many immigrant neighbourhoods have seen such a succession of migrants coming and going. The East End of London, which from the seventeenth century onwards received successive immigration waves of Huguenot refugees, Irish weavers, Jews and Bangladeshis is a typical example of such a 'transitory immigrant neighbourhood' (see Chapter 5). Another example is the Lower East Side in New York City, which received successive waves of newly arrived immigrants, first of Germans, then of Italians and Eastern European Jews, as well as Greeks, Hungarians, Poles, Romanians, Russians, Slovaks and Ukrainians. The replacement of Greeks by Vietnamese in Richmond (Melbourne) and Marrickville (Sydney) is another, more recent example.

Residential segregation is a contradictory phenomenon. In terms of the theory of ethnic minority formation (see Chapter 4), it contains elements of both other-definition and self-definition. The relative weight of the two sets of factors varies from country to country and from group to group. Immigrants cluster together for economic and social reasons, and are often kept out of certain areas by racism. But they also frequently *want* to be together, in order to provide mutual support and protection, to develop family and neighbourhood networks, and to maintain their languages and cultures. Clustering of migrants in ethnic neighbourhoods allow the establishment of small businesses and agencies which cater for immigrants' needs, as well as the formation of associations, schools and religious institutions of all kinds. Residential segregation can thus facilitate community formation and the long-term emancipation of migrant groups.

It is perhaps useful to make a distinction between *ethnic enclaves* and 'ghettoes' (see Peach 2005), with the former indicating largely voluntary concentrated settlement in immigrant neighbourhoods (which initially offer many benefits such as mutual support, cultural facilities and ethnic businesses), and the latter indicating the largely involuntary and multigenerational spatial concentrations of particular ethnic groups resulting from their economic marginalization. While 'ethnic enclaves' can be vital 'emancipation machines', 'ghettoes' tend to be 'dead-end' areas typically characterized by grinding poverty, unemployment and sometimes crime, where people of particular ethnic groups get trapped because of a lack of opportunities to move out. Whereas ethnic enclaves mainly result from *internal closure*, ghettoes predominantly result from *external closure* because of exclusion.

Levels of residential segregation in Europe are more moderate compared to the US, although there are big differences across immigrant groups, cities and countries (Musterd 2005). Contrary to popular belief, overall segregation levels do not appear to increase,

---

Box 13.3 Migrants and minorities in France

France, like Britain, has a confusing array of statistical classifications for its immigrant populations. The category of *immigrés* (*immigrants*) has been introduced to give information on people born abroad of non-French parents. In 2015, *immigrés* numbered about 5.9 million (9.3 per cent of total population). In addition, about half a million *French citizens of Overseas Departments and Territories* are mostly of African, Caribbean and Pacific Island origin. France's immigrant population has changed from mainly Southern European in the 1970s to a majority of non-Europeans today: predominantly North and West Africans but with a substantial Asian component.

Reticence about the use of ethnic categories is based on France's 'republican model', which lays down principles of civic citizenship and equal individual rights for all. Recognition of cultural difference or ethnic communities is ideologically unacceptable. The idea is that immigrants should become citizens, and will then enjoy equal opportunities. The reality is often different. People of non-European origin (whether citizens or not) face considerable social exclusion and discrimination. Minorities have become concentrated in inner-city areas and in *banlieues*. The situation of several minorities is marked by low-status, insecure jobs and high unemployment, especially for youth: in 2010, 43 per cent of young men in 'sensitive urban zones' (ZUS, *zones urbaines sensibles*) were jobless.

The position of immigrants in French society has become highly politicized, with the emergence of an anti-immigrant right-wing party, the *Front National* (renamed in 2018 into *Rassemblement National*) and movements of citizens of North African origin. Major riots, notably in 2005 and 2007, concentrated political attention on the long-term effects of immigration.

President Sarkozy's government from 2007 to 2012 used anti-immigrant measures as an electoral tool. These included restriction on family reunion, deportations of Roma, attacks on religious slaughter, and public claims about the incompatibility of Islam and French identity. Migrants were obliged to sign a

▶

France: immigrant population 2014 by place of birth		
	Numbers (thousands)	Share of Immigrant Population (%)
Europe	2,185	35.4
European Union (28 states)	1,897	30.8
Spain	249	4.0
Italy	286	4.6
Portugal	622	10.1
UK	148	2.4
Other EU 28	591	9.6
Other Europe	288	4.7
Africa	2,754	44.6
Algeria	791	12.8
Morocco	741	12.0
Tunisia	270	4.4
Other Africa	952	15.4
Asia	883	14.3
Turkey	249	4.0
Cambodia, Laos, Vietnam	160	2.6
Other Asia	475	7.7
Americas, Oceania	346	5.6
Total	6,169	100

Note: Figures are for Metropolitan France (excluding Overseas Departments)
Source: INSEE (2019)

mandatory integration contract and a 'Charter of rights and duties of French citizens' upon naturalization.

In the run-up to the Presidential Election of 2012, Sarkozy claimed that France was threatened by Islamist terrorists, but was defeated, indicating perhaps that immigration was less important to electors than other issues. The Hollande government liberalized immigration and asylum policies, but counteracted undocumented migration and a renewed version of the integration contract was adopted. From 2017, President Emmanuel Macron maintained these policies and proposed expansion of the 'Talent Passport' to widened categories of migrants.

Sources: Bertossi (2007); Body-Gendrot and Wihtol de Wenden (2007); Cross (1983); Favell (1998); Miller 1978; OECD 2012; OECD 2018; Scott (2005); Weil (1991)

partly because the effects of new immigration are counterbalanced by the upward residential mobility of long-term settlers (Musterd and Van Kempen 2009). Particularly in welfare states such as Germany, Sweden and the Netherlands, immigrants often live in social housing of relatively high quality. Upward steps in migrants' residential careers (implying a move to a non-immigrant neighbourhood) reflect broader integration processes (Musterd and Vos 2007). This suggests that promoting education and labour market access are more effective integration policies than 'spatial social engineering projects' such as the promotion of mixed neighbourhoods (Musterd and Ostendorf 2009).

Migration and the global city

Immigration and ethnic minority formation are transforming post-industrial cities in contradictory ways. Global reorganization of finance, production and distribution have reinforced the dominant position of 'global cities' (Sassen 2001) which attract immigrants, both for highly specialized services and for low-skilled service jobs which enable the luxurious lifestyles of the (native and migrant) elites (see Hochschild 2000; Parreñas 2000). The classic examples are cities such as New York, London, Paris, Sydney and Tokyo, although many new global cities have emerged in Europe (Moscow, Berlin, Milan), the Middle East (Istanbul, Dubai, Tel Aviv), Asia (Singapore, Hong Kong, Mumbai), Latin America (São Paulo, Mexico City) and Africa (Johannesburg, Lagos, Cairo).

Ethnic clustering and community formation may be seen as necessary products of migration to global cities. This may lead to tensions and conflicts, but they can also lead to renewal and enrichment of urban life and culture. Much of the energy and innovative capacity within the cities lies in the cultural syncretism of the multi-ethnic populations. In some cases, the vibrant mix of cultures and ethnic businesses can make such neighbourhoods attractive for cultural elites and middle class. Ethnic neighbourhoods, such as the many 'Chinatowns' in the West, can be transformed into places of leisure and consumption (Rath *et al.* 2018). Such processes can lead to gentrification – the 'upgrading' of deteriorated neighbourhoods through urban renewal and the influx of more affluent residents – and subsequent displacement of lower-income groups including migrants and ethnic minorities. In other cases, ethnic neighbourhoods become places that reinforce marginalization.

Integration policies: A controversial issue

As migrants moved into inner cities and industrial towns, social conflicts with lower-income groups of majority populations developed. Immigrants are often blamed for rising housing costs, declining housing quality and deteriorating social amenities. In response, a whole set of social policies developed in many immigration countries. Sometimes policies developed to reduce segregation and ease social tension achieve the opposite.

Nowhere were the problems more severe than in France. In the 1960s *bidonvilles* (shantytowns) occupied by migrant workers and their families emerged. After 1968, measures were taken to eliminate *bidonvilles* and make public housing more accessible to immigrants. The concept of the *seuil de tolérance* (threshold of tolerance) was introduced, according to which the immigrant presence should be limited to a maximum of 10 or 15 per cent of residents in a housing estate or 25 per cent of students in a class (MacMaster 1991: 14–28; Verbunt 1985: 147–155). The assumption was that immigrant concentrations presented a problem, and that dispersal was the precondition for assimilation. Subsidies to public housing societies (*habitations à loyer modéré*, or HLMs) were coupled to quotas for immigrants. To minimize the perceived risk of conflicts with local residents, however, immigrant families were concentrated in specific estates. The HLMs could claim that they had adhered to the quotas – on an average of all their dwellings – while in fact creating new ghettoes (Weil 1991: 249–258).

By the 1980s, *banlieues*, high-rise estates on the periphery of the cities, were rapidly turning into areas of social problems and ethnic conflicts. Social policies focused on urban youth, and governments developed a range of programmes to improve housing and social conditions, and to boost educational outcomes and unemployment, paying special attention to youth of North African background. Such policies had little effect: the *banlieues* remained hotspots of segregation, youth unemployment and violence. They were designed to achieve integration into French society, but in fact they 'linked all the problems of these towns and neighbourhoods to immigration'. Thus, social policy has encouraged concentration of minorities, which obstructs integration and strengthens group religious and cultural affiliations (Weil 1991: 176–9). The ethnic youth uprisings of 2005 and 2007 reflected the deep feelings of discrimination and exclusion felt by inhabitants of the *banlieues* (Body-Gendrot and Wihtol de Wenden 2007; Roux and Roché 2016).

This is clearly a complex issue which requires careful analysis. The extent to which the state should introduce special social policies to facilitate immigrant integration is a controversial issue, pitting those arguing that active government intervention is needed to help migrants overcome disadvantage, against those arguing that that special policies may stereotype migrants, set them apart as 'problem groups' and that government assistance may discourage their self-emancipation through work and business. Special policies for immigrants have often reinforced tendencies to segregation. For instance, up to the 1980s, German education authorities pursued a 'dual strategy', designed to provide skills needed for life in Germany while at the same time maintaining homeland cultures to facilitate return. This led to special classes for foreign children, contributing to social isolation and poor educational performance (Castles *et al.* 1984). Housing policies in the UK were intended to be non-discriminatory, yet as in France they sometimes led to the emergence of 'black' and 'white' housing estates. In the Netherlands, critics of the Minority Policy argued that culturally specific integration policies may actually increase socioeconomic marginalization (Entzinger 2003). Sweden's special public housing schemes for immigrants led to a high degree of ethnic concentration and segregation from the Swedish population (Andersson 2007).

On the other hand, one can argue that immigrants need services that address their special needs with regard to education, language and housing. The absence of such measures and the denial of discrimination (particularly on the job market, see Chapter 12) can put immigrants and their children at a disadvantage and deny them opportunities for upward mobility. In the US, for instance, several migrant groups such as Puerto Ricans and Mexicans experience severe disadvantage, labour discrimination and downward assimilation (see Portes and Zhou 1993).

It is possible to suggest a rough classification of policy responses to immigration and minority formation. From the 1970s, Australia, Canada, the UK, Sweden and the Netherlands pursued active social policies targeting immigrants and minorities. In the first three, the label 'multicultural' was used. Britain also spoke of 'race relations policy',

Box 13.4 Migrants and minorities in Germany

Until the late 1990s, politicians declared that Germany was 'not a country of immigration'. Yet, with over 20 million newcomers since 1945 (many of them 'ethnic Germans' from the former Soviet Union, Poland and Romania), it has in fact had more immigration than any other European country, and has the second largest foreign-born population of all countries in the world, after the US. In 2017, Germany had 12.7 million *foreign-born persons* (up from 0.7 million in 1961) – 15.4 per cent of its total population of 82.8 million. The largest immigrant groups were from Poland (1.46 million), Turkey (1.32 million), Russian Federation (960,000), Kazakhstan (737,000) and Romania (657,000).

Children born to foreign citizen parents remain foreign citizens, though they can opt for German citizenship at maturity. Roughly one of every five foreigners living in Germany was born there.

Foreign born population in Germany by country of birth (2007 and 2017)			
Country of birth	**2007 (thousands)**	**2017 (thousands)**	**Share of immigrant population (%, 2017)**
Poland	797	1,468	11.5
Turkey	1,478	1,324	10.4
Russian Federation	947	960	7.5
Kazakhstan	358	737	5.8
Romania	330	657	5.2
Italy	432	508	4.0
Syria	.	479	3.8
Croatia	256	306	2.4
Greece	229	282	2.2
Ukraine	206	224	1.8
Bulgaria	50	215	1.7
Serbia	.	208	1.6
Hungary	81	207	1.6
Former USSR	.	201	1.6
Austria	192	190	1.5
Other	5,054	4,772	37.5
Total	**10,410**	**12,738**	**100**

Notes: Serbia includes persons recorded under the former country names 'Yugoslavia', 'Serbia-Montenegro' and 'Serbia or Kosovo'

Sources: IMO (2018)

▶

Most of the 'guestworkers' who came from Southern Europe and Turkey between the 1950s and the early 1970s were manual workers in manufacturing industries, leading to residential concentration in industrial areas and central city districts. Later economic restructuring eliminated many of the jobs held by immigrants, leading to unemployment rates of 20 per cent or more – nearly twice the national average. Lack of integration programmes meant that immigrants' children too tended to have poor labour market chances. Following German reunification in 1990, there was a wave of racist violence against immigrants and asylum seekers. The reality of permanent settlement and the dangers of creating an underclass became obvious.

Germany is an important example of the unforeseen effects of migration. Labour recruitment was designed to bring in temporary workers who would not stay, but in the long run it led to permanent settlement and the emergence of a multi-ethnic society. Official denial of Germany's status as an immigration country made things worse, as it exacerbated the exclusion of migrants from society. In the long run, public attitudes and policy approaches had to change.

Changes set in around the turn of the century, but proved a difficult and lengthy process. The Citizenship Law of 1999 marked a major change. It was designed to make it easier for immigrants and their children to become Germans by granting German citizenship based on place of birth instead of German heritage. Germany's first Immigration Law was passed in 2004. It was designed to establish a modern system for planning and managing migration intakes. It also established compulsory integration and language courses for new entrants and existing foreign residents. To counter ageing, successive governments have been keen to attract highly skilled immigrants. Although restrictions on low-skilled migration from outside the EU were maintained, labour migration from East-European states increased after the introduction of the EU Freedom of Movement Act in 2004.

During the 2015–2017 'refugee crisis' almost 1.4 million people, particularly Syrians, applied for asylum. While the German business sector and parts of civil society initially welcomed refugees, nationalist movements spurred anti-immigration sentiments. Worried about electoral repercussions, the Merkel government played a leading role in European efforts to decrease asylum flows (see Box 6.1). Facing the challenge of integrating refugees, the 2016 'Demand and Support' act connects the right to permanent residence to achieving integration goals and obtaining employment.

In sum, Germany has significantly loosened its migration policies but is still negotiating their response to the permanent settlement of immigrants.

Sources: Brubaker 1992; Eule 2016; Lutz and Palenga-Möllenbeck (2010); Rietig and Mülle (2016); OECD 2011b; OECD 2018; Schierup *et al.* (2006)

while Sweden used the term 'immigrant policy' and the Netherlands 'minorities policy', which generally went along with easy access to naturalization and tolerance of dual citizenship. In all these countries, social policies that specifically target immigrants and minorities have been heavily criticized from the 1990s onwards a part of a broader backlash against 'multiculturalism'. This has led to an increasing emphasis on 'integration', 'social cohesion' and 'shared citizenship values' (Entzinger 2003).

A second group of countries rejects special social policies for immigrants. US authorities regard special policies for immigrants as unnecessary government intervention. Nonetheless, the equal opportunities, anti-discrimination and affirmative action measures introduced after the Civil Rights Movement of the 1960s also benefited immigrants, and special social and educational measures are found at the local level. However, access to social benefits and education by non-citizens (especially irregular immigrants) has been under attack since the 1980s. French governments have rejected special social policies on the principle that immigrants should become citizens, and that any special treatment would hinder this. Yet despite this there have been programmes, such as Educational Priority Zones or, more generally, *la politique de la ville* (urban policy), which target areas of disadvantage, although without explicit mention of immigrants.

The third group of countries consists of those that have recruited migrant labour, particularly through 'guestworker' systems. Germany has pursued rather contradictory policies concerning the access of migrants to the welfare system. In the 1960s, the government commissioned charitable organizations (linked to the churches and the labour movement) to provide special social services for foreign workers. Although foreign workers were guaranteed equal rights to work-related health and pension benefits, they were excluded from some welfare rights. For example, application for social security payments on the grounds of long-term unemployment or disability could lead to deportation.

After recruitment stopped in 1973, migrants (supported by labour unions and NGOs) won landmark court cases on welfare rights and family reunion. As settlement became more permanent, welfare, health and education agencies began to take account of the needs of immigrants, despite the official claim that 'Germany was not a country of immigration'. However, anti-discrimination legislation or affirmative action programmes have little place in either Germany or Switzerland, while restrictive naturalization laws made it difficult for foreign residents to become citizens. Yet racist attacks in the 1990s prompted German authorities to seek ways of overcoming social exclusion of immigrants.

Since the 2000s, considerable convergence in integration policy seems to have occurred, fuelled by concerns across Western Europe about social exclusion, extremist violence or what a report on the 2001 riots in Northern England referred to as 'parallel lives' (Cantle 2001). Political leaders have questioned multicultural approaches, and have introduced such measures as citizenship tests and integration contracts. Special programmes to combat the social disadvantages faced by immigrants and their descendants can also be found almost everywhere, despite differences in rhetoric.

While there is a vast literature on official models and policies of integration, empirical evidence on the effectiveness of these policies is surprisingly limited (Ersanilli 2010). One study found that access to citizenship has positive effects on sociocultural integration of Turkish immigrants in France and Germany, countries which require some degree of assimilation, but not in the Netherlands (Ersanilli and Koopmans 2010). Based on a comparison of eight European countries, Koopmans (2010) argued that the effects of multicultural policies have been generally negative. However, others have questioned such views (see Banting and Kymlicka 2012; EFFNATIS 2001).

Yet the most salient insight is that the effects of various 'integration policies', whether positive or negative, are generally rather small (Ersanilli and Koopmans 2011), and that their importance should therefore not be exaggerated. To a large extent, modes of migrant incorporation are determined by skills, class and the social ties of migrants as well as structural factors such as access to education and labour markets (Fokkema and de Haas 2011; van Tubergen *et al.* 2004), which often lie beyond the scope of specific 'integration policies'.

There is no singular path towards the incorporation of immigrant groups. Portes and Zhou (1993) introduced the term *segmented assimilation* to describe the diverse possible outcomes of adaptation processes. Based on a review of the 'second generation' in the US, Portes *et al.* stressed that while the second generation assimilates in the sense of learning English and American culture, 'it makes a great deal of difference whether they do so by joining the mainstream middle class or the marginalized, and largely racialized, population at the bottom' (Portes *et al.* 2005: 1000). While most immigrant groups achieve high levels of upward socioeconomic mobility, others, such as some Latino groups in the US and North African migrant groups in Europe, face structural exclusion, discrimination and poverty, sometimes causing resentment and violent reactions.

Racism and minority formation

Three categories of settlers may be distinguished in terms of their trajectories of incorporation into destination societies. First, some settlers have merged into the general population and *do not constitute separate ethnic groups*. These are generally people who are culturally and socioeconomically similar to the majority of the receiving population: for instance, British settlers in Australia, Austrians in Germany, Paraguayans in Argentina, Algerians in Morocco or Syrians in Lebanon. Also, many higher-skilled migrants with dispersed settlement patterns tend to assimilate quickly, although others may form isolated 'expat' communities, living in luxurious neighbourhoods and refusing to learn the language of the destination country.

Second, some settlers form *ethnic communities*: they tend to cluster in certain neighbourhoods or 'ethnic enclaves' and to maintain their original languages and cultures, but they are not excluded from political participation and opportunities for economic and social mobility. Third, some settlers form *ethnic minorities*. Like ethnic communities, they tend to live in certain neighbourhoods and to maintain their languages and cultures. But, in addition, they may have a disadvantaged socioeconomic position and be partially excluded from the wider society by factors as weak legal status, labour market discrimination, and racist violence and harassment. Examples are some Middle Eastern and Asian immigrants in Australia, Canada and the US; certain Hispanic groups in the US; several Afro-Caribbean and Asian groups in the UK; North Africans and Turks in Western European countries; and asylum seekers just about everywhere.

Most immigration countries have all three categories, but the main concern is with the second and third categories. It is important to understand why some immigrants take on the character of ethnic communities, while others remain ethnic minorities. A further important question is why more immigrants take on minority status in some countries than in others. Two groups of factors appear relevant: those connected with the characteristics of the immigrants and minorities themselves, and those connected with the social structures, cultural practices and ideologies of destination societies.

Box 13.5 Migrants and minorities in Australia

Australia has pursued a programme of planned immigration since 1947: 7 million new settlers have arrived. Immigration has helped treble the population from 7.6 million in 1947 to 25.3 million by 2019. Australia is home to nearly 7 million overseas-born people, over a quarter of the population. A similar proportion is Australian-born people with at least one parent born overseas. In 2016 there were 649,000 Aboriginal and Torres Strait Islander people (2.7 per cent of total population).

Foreign born population in Australia by country of birth (2007 and 2017)			
Country of birth	2007 (thousands)	2017 (thousands)	Share of immigrant population (%, 2017)
United Kingdom	1,134	1,198	17.4
New Zealand	438	607	8.8
China	252	526	7.7
India	170	469	6.8
Philippines	142	246	3.6
Vietnam	178	237	3.4
Italy	218	195	2.8
South Africa	120	182	2.6
Malaysia	106	166	2.4
Germany	125	124	1.8
Sri Lanka	74	118	1.7
Greece	129	117	1.7
Korea	56	107	1.6
United States	75	104	1.5
Hong Kong, China	81	97	1.4
Other countries	1,736	2,381	34.6
Total	5,032	6,873	100

Historically, Australians have been fearful of migration from Asia, and a 'white Australia policy' was introduced in 1901. Post-1947 migration was designed to be mainly from Britain, with a gradual broadening to the rest of Europe. But the white Australia policy proved unsustainable, and Asian entries grew rapidly from the 1970s. By 2017, an estimated 71 per cent of the total population were born in Australia, while 29 per cent were born overseas. The large majority of origin of Australia's foreign-born populations originate from the British Commonwealth of

▶

Nations: United Kingdom (17.4 per cent), New Zealand (8.8 per cent) and India (6.8 per cent), closely followed by China (7.7 per cent) which accounts for an increasing share in Australian immigration.

Australia – like the US and Canada – has seen immigration as vital for nation-building. Family migration has been the norm. In the 1950s and 1960s, immigrants were expected to quickly assimilate. However, non-British immigrants (especially Eastern and Southern Europeans) tended to get low-paid manual jobs. This in turn meant clustering in low-income areas, providing the basis for ethnic community formation.

By the 1970s, a policy of multiculturalism was adopted. The Australian approach emphasizes the duty of the state to combat racism and to ensure that minorities have equal access to government services, education and jobs. Public support for immigration and multiculturalism waned in the 1990s. Disaffection with multiculturalism increased following 9/11 and the Bali bombing of 2002, in which 88 Australians were killed. Between 1996 and 2007, the Liberal–National Coalition promoted principles of social cohesion around 'core cultural values'. This was linked to a tough line on asylum, including mandatory detention of irregular entrants and the 'Pacific Solution' with asylum seekers being sent to camps in Nauru and Papua New Guinea. At the same time the Coalition encouraged immigration, particularly of high-skilled workers and students.

'Border protection' was a central issue in the 2010 federal election, with both major parties treating boat arrivals as a threat to national sovereignty and security, although asylum inflows have been small compared to soaring numbers of economic immigrants and students. Despite increasing emphasis on temporary migration, Australian immigration policy provides clear pathways to permanent residency. While Australia's points system and economic skilled migration program encourages skilled migration, asylum policies have become tougher. After heated elections in 2013, the newly elected Prime Minister, Toby Abbot, introduced Operation Sovereign Borders to stop maritime asylum arrivals by 'turnbacks' and 'towbacks' and announced that asylum seekers reaching Australia by boat would not be admitted through the resettlement program.

Sources: Castles *et al.* (2013); Collins (1991); DIAC (2012); Jupp (2002); Markus *et al.* (2009); OECD (2018); Vasta (1999)

Looking at the settlers, it is inescapable that visible or phenotypical difference (skin colour, dress, appearance) is a main marker for minority status. This applies even more to non-immigrant minorities, such as native or 'aboriginal' peoples in the Americas or Australia, African–Americans in the US, and Jews and Roma in Europe. Although they are generally not migrants, such ethnic minorities often make up the most marginalized groups. Visible difference is also a marker for exclusion in non-Western societies, for example of native black populations and sub-Saharan African immigrants in North Africa; Asian and, particularly, African workers in the Gulf oil countries; and South-East Asians in Korea and Japan.

There are four possible explanations for this: visible difference may coincide with recent arrival, with cultural distance or with socioeconomic position, or, finally, it may serve as a target for racism.

The first explanation is partly correct: in many cases, black, Asian or Hispanic settlers are among the more recently arrived groups. Historical studies reveal examples of past racism and discrimination against white immigrants (such as the Irish, Italians, Greeks and *Ashkenazi* Jews) quite as virulent as against non-whites today – or indeed, at the time these groups were seen as 'less white' than the original 'Anglo-Saxon' settlers. Recent arrival tends to make new, unknown groups appear more threatening, and new groups tend to compete more with local low-income groups for jobs and housing. But recent arrival cannot explain why indigenous populations are victims of exclusionary practices, nor discrimination against African–Americans and other long-standing minorities. Neither can it explain why racism against white immigrant groups tends to disappear in time, while that against non-white groups often continues over generations.

What about cultural distance? Some non-European settlers come from rural areas with pre-industrial cultures and may find it hard to adapt to industrial or post-industrial cultures, such as illiterate workers from rural areas of North Africa living in European metropolises have experienced. But many Asian settlers in North America and Australia are of urban background and highly educated. This does not protect them from racism and discrimination. Many people perceive culture mainly in terms of language, religion and values, and see non-European migrants as very different. This applies particularly to Muslims. Fear of Islam has a tradition going back to the medieval crusades. In recent decades, fear of religious extremism, and loss of modernity and secularity have played a role. In recent decades, concerns about terrorism have led to increased Islamophobia (fear of and hostility to Islam and Muslims), even though only a very small minority of Muslims actually support extremist ideologies (see Chapter 10). It could be argued that such fears are therefore based on racist ideologies rather than social realities. The strengthening of Muslim affiliations can be a protective reaction of discriminated groups ('internal closure') to exclusion by majority groups ('external closure), although it is important not to confound religious conservatism or fundamentalism with violent *jihadism*.

As for the third explanation, phenotypical difference does frequently coincide with socioeconomic status. Some immigrants from less developed countries lack the education and vocational training necessary for upward mobility in industrial economies. But even highly skilled immigrants may encounter discrimination. Many immigrants discover that they can only enter the labour market at the bottom, and that it is hard to move up the ladder subsequently. Thus, low socioeconomic status is as much a result of processes of marginalization as it is a cause of minority status.

We may therefore conclude that recentness of arrival is only a partial and temporary explanation of minority status, and that cultural difference and socioeconomic status are not adequate explanations on their own. The most significant explanation of minority formation lies therefore in practices of exclusion by the majority populations and the states of immigration countries. We refer to these practices as racism and to their results as the racialization of minorities (see Chapter 4). Traditions and cultures of racism are strong in all European countries and former European settler colonies (Essed 1991; Goldberg 1993; Murji and Solomos 2005). The increased salience of racism and racist violence seems to be linked to the growing socioeconomic insecurity of many people resulting from neoliberal globalization, with migrants and minorities receiving the blame for problems not of their making such as defunding of government services, decreased labour protection or growing inequality.

Minority formation is more likely to happen in countries that discriminate against migrants, particularly if they do not provide facilities for permanent residence and naturalization. Such structural exclusion of migrant workers is particularly prevalent in various destination countries in the Gulf, East and South-East Asia, and Africa. Despite discrimination and exclusion, ethnic minorities often succeed in improving their socio-economic situation over two or three generations, enabling them to become more fully incorporated into society. So, ethnic minorities may often become ethnic communities in the longer run. In some cases, however, sustained racism and exclusion lead to situations of structural and multigenerational marginalization of minorities, such as has been historically the case for Jews in Europe and African–Americans in the US, and the native populations of European settler societies in the Americas and the Pacific.

Citizenship acquisition

Citizenship rules: Ius sanguinis and ius soli

Varying historical experiences and models of nation states lead to different concepts of citizenship (see Chapter 4). Some countries make it very difficult for immigrants to become citizens, others grant citizenship but only at the price of cultural assimilation, while a third group make it possible for migrants to become citizens while maintaining distinct cultural identities.

Becoming a citizen is a crucial part of the integration process. *Citizenship* is a formal legal status (often referred to as *nationality*), designating membership of a nation-state. But it is also important to consider the content of citizenship. This is usually defined in terms of civil, political and social rights, but linguistic and cultural rights are also very important for immigrants. The rules for becoming a citizen are complex and have undergone considerable change over recent decades.

Historically, laws on citizenship or nationality derive from two competing principles: *ius sanguinis* (literally: law of the blood), which is based on descent from a national of the country concerned, and *ius soli* (law of the soil), which is based on birth in the territory of the country. *Ius sanguinis* is often linked to an ethnic or folk model of the nation-state (typical of Germany and Austria, but also common in Asia, Africa and the Middle East), while *ius soli* generally relates to a nation-state built through incorporation of diverse groups on a single territory (such as France and the UK) or through immigration (the US, Canada, Australia and New Zealand) (see Chapter 4). In practice, many modern states have citizenship rules based on a combination of *ius sanguinis* and *ius soli*, although one or the other may be predominant.

The rules for becoming a citizen in various countries are complex and have undergone considerable change in recent years (see Aleinikoff and Klusmeyer 2000, 2001). In Europe, there has been a trend of convergence. The distinction between *ius soli* and *ius sanguinis* countries was eroded by a trend towards more liberal rules in the 1990s (Bauböck *et al.* 2006). However, after 2000 citizenship rules again became more restrictive, especially in Denmark, France, Greece, the Netherlands, the UK and Austria. Rules have remained or become relatively less restrictive in Belgium, Finland, Germany, Luxembourg and Sweden (Bauböck *et al.* 2006: 23). Southern European countries such as Italy and Spain have *ius sanguinis* laws that provide easy residence and citizenship rights to migrants from Latin America who can prove ancestry. Several other 'ethno-states', such as Germany, Hungary, Turkey, Israel and Russia offer (instant) citizenship for co-ethnics living abroad.

Legal requirements for naturalization (such as 'good character', regular employment, language proficiency and other evidence of integration) are quite similar in various countries, but actual practices vary sharply. Switzerland, Austria and (until recently) Germany impose long waiting periods and complex bureaucratic practices, and treat naturalization as an act of grace by the state. Conversely, classical immigration countries encourage newcomers to become citizens. The act of becoming American (or Australian or Canadian) is seen as an occasion for celebration of the national myth. The introduction of citizenship ceremonies in some European countries such as the Netherlands is motivated by the desire to pass on 'national values', yet could have positive effects by providing a symbolic welcome into the nation.

Status of the second generation

The transmission of citizenship to the children of immigrants and subsequent generations is a key factor affecting patterns of immigrant incorporation. National variations parallel those found with regard to naturalization. In principle, *ius soli* countries confer *birthright citizenship* on all children born in their territory. *Ius sanguinis* countries confer citizenship only on children of existing citizens. However, most countries actually apply models based on a mixture of the two principles. Increasingly, entitlement to citizenship grows out of long-term residence in the country: the *ius domicili*.

Ius soli is applied most consistently in Australia, Canada, New Zealand and the US. A child born to immigrant parents in the US or Canada becomes a citizen, even if the parents are visitors or undocumented residents. In Australia, New Zealand and the UK, the child obtains citizenship if at least one of the parents is a citizen or a legal permanent resident. Destination countries in Latin America facilitate the acquisition of citizenship. Such countries use the *ius sanguinis* principle to confer citizenship on children born to their citizens while abroad (Çinar 1994: 58–60; Guimezanes 1995: 159). A combination of *ius soli* and *ius domicili* emerged in France, Italy, Belgium and the Netherlands in the 1990s. Children born to foreign parents in the territory obtained citizenship, providing they had been resident for a certain period and fulfilled other conditions. Since 2000, Germany, Finland and Spain have adopted similar arrangements. France, Belgium, the Netherlands and Spain also introduced the so-called *double ius soli*: children born to foreign parents, at least one of whom was also born in the country, acquire citizenship at birth. This means that members of the 'third generation' automatically become citizens, unless they specifically renounce this right upon reaching the age of majority (Bauböck *et al.* 2006; Çinar 1994: 61).

Where *ius sanguinis* is still applied (Austria, Switzerland, Japan and South Korea, as well as most Middle Eastern and African countries), children who have been born and grown up in a country may be denied not only security of residence, but also a clear national identity. They are formally citizens of a foreign country they may never have seen, and can even be deported there in certain circumstances. Other *ius sanguinis* countries (notably Germany) have taken steps towards *ius domicili*. This means giving an option of facilitated naturalization to young people of immigrant origin. There seems to be a pattern here: in the longer term, the reality of large-scale permanent migrant settlement tends to compel states which have long opposed and denied this permanency to eventually come to terms with these new realities by giving migrants and their descendants access to permanent residency and citizenship, although this is not always the case.

Most countries in East Asia, the Middle East and Africa are still characterized by a weak protection of migrant rights and a strict application of *ius sanguinis* principles. In East Asian countries such as Japan and Korea, this is linked to strong myths of ethnic homogeneity. In many other countries this seems linked to post-colonial efforts at nation state building. In Africa, for instance, newly established states endeavoured to instil a sense of national unity in ethnically diverse societies (see Davidson 1992). This increased the urge to assert national sovereignty by introducing severe immigration restrictions and to portray immigrants as a threat to sovereignty, security and ethnic homogeneity (see Vigneswaran and Quirk 2015). Only few African countries provide migrants' right to nationality, even for stateless children born on their territory (Manby 2016).

Dual citizenship

Trends are rather different with regard to dual or multiple citizenship (acquiring the nationality of a host country without renouncing the nationality of the origin country). This is a way of recognizing the multiple or transnational identities of migrants and their descendants and is also seen as an effective policy to encourage naturalization. Dual citizenship can be seen as a form of 'internal globalization' through which 'nation-state regulations implicitly or explicitly respond to ties of citizens across states' (Faist 2007: 3). This represents a major shift, since the idea of singular national loyalties has been historically central to state sovereignty. One reason for change is the trend towards gender equality. In the past, nationality in binational marriages used only to be transmitted through the father. Nationality rules in European countries were changed in the 1970s and 1980s. Once mothers obtained the same right to transmit their nationality as fathers, binational marriages automatically led to dual citizenship.

Countries such as Australia and Canada have long permitted dual citizenship for immigrants. Also in Europe, attitudes and laws have changed: by 2004 only five of the EU15 states required renunciation of the previous nationality (Bauböck *et al.* 2006: 24). The Netherlands introduced the right to dual citizenship in 1991, but withdrew it again in 1997 (Entzinger 2003) as part of a backlash against multiculturalism. Germany introduced measures to facilitate acquisition of nationality for immigrants and their children in 2000, but maintained its ban on dual citizenship. In both countries, however, there are important exceptions and many people do hold dual citizenship. In addition, many emigration countries have changed their nationality rules to allow emigrants to hold dual citizenship, as a way of maintaining links with their diasporas. Laws imposing singular citizenship are also difficult to enforce, particularly if emigration countries do not give their citizens the right to relinquish their nationality.

Acquisition of nationality by immigrants

In OECD countries, acquisitions of nationality have shown a slightly increasing trend over time, reaching levels of around 2 million per year in 2016. Acquisition includes naturalizations and other procedures, such as declarations on the part of descendants of foreign immigrant parents or conferral of nationality through marriage. Table 13.1 shows the absolute numbers of acquisitions of nationality in OECD countries as well as Russia. Over the 1996–2015 period, at least 40 million foreigners living in OECD countries and Russia obtained destination country citizenship. Real numbers are higher, because older data from several countries is missing (see Table 13.1). It shows high numbers of

Table 13.1　Acquisition of nationality in OECD countries, 1996–2015, average annual rates

Country	1996–2000	2001–2005	2006–2010	2011–2015
Australia	480,643	437,534	567,674	599,969
Austria	98,203	188,957	64,127	36,801
Belgium	176,657	209,374	173,035	148,996
Canada	818,075	852,939	937,291	929,262
Chile	–	–	3,383	4,668
Czech Republic	16,442	21,909	9,176	16,525
Denmark	54,254	60,958	26,924	25,642
Estonia	3,425	24,482	13,966	6,699
Finland	14,144	22,858	23,686	38,756
France	297,548	723,933	606,171	527,116
Germany	605,417	717,770	529,758	547,201
Greece	–	–	54,134	101,963
Hungary	40,963	32,474	34,606	60,904
Iceland	286	2,568	3,583	2,776
Ireland	1,143	17,116	27,750	94,706
Italy	51,526	46,407	259,754	530,170
Japan	76,267	78,850	69,863	48,373
Korea	–	39,533	77,781	73,018
Latvia	–	–	38,411	13,372
Luxembourg	3,392	3,842	11,912	22,793
Mexico	0,000	24,183	19,755	14,881
Netherlands	313,758	175,448	144,000	145,890
New Zealand	100,009	107,783	116,334	132,815
Norway	51,023	48,555	60,489	68,012
Poland	2,846	8,389	9,000	18,145
Portugal	4,704	6,483	77,987	111,053
Russia	–	1,498,123	1,536,813	672,601
Slovak Republic	–	8,901	3,582	1,277
Slovenia	–	–	5,844	6,735
Spain	60,314	146,274	421,637	663,165
Sweden	182,172	173,537	176,102	226,873
Switzerland	108,888	173,647	217,719	179,423
Turkey	–	59,950	33,028	9,216
United Kingdom	270,716	638,440	846,867	823,689
United States	3,834,706	2,786,548	3,773,233	3,615,231
Total	**7,676,601**	**9,337,765**	**11,065,375**	**10,518,716**

Source: Calculations based on data compiled from OECD reports

Table 13.2   Acquisition of nationality in OECD countries, as percentage
of foreign population, average annual rates

Country	1996–2000	2001–2005	2006–2010	2011–2015	Average
Austria	2.9	5.0	1.6	0.8	2.6
Belgium	3.9	4.9	3.7	2.6	3.8
Canada	–	–	11.4	5.7	8.5
Czech Republic	3.7	1.9	0.6	0.8	1.7
Denmark	4.4	4.6	1.9	1.4	3.1
Estonia	–	1.8	–	0.6	1.2
Finland	3.3	4.3	3.9	4.3	3.9
France	4.6	–	3.7	2.6	3.7
Germany	1.6	2.0	1.6	1.6	1.7
Greece	–	–	2.2	2.6	2.4
Hungary	5.6	5.2	4.3	6.9	5.5
Iceland	1.4	2.4	4.5	2.6	2.7
Ireland	–	1.8	1.0	3.4	2.1
Italy	1.0	0.7	1.7	2.5	1.5
Japan	1.1	0.9	0.7	0.5	0.8
Korea	–	2.2	2.0	1.5	1.9
Latvia	–	–	1.7	0.8	1.2
Luxembourg	0.5	0.5	1.2	2.0	1.0
Mexico	–	–	–	1.1	1.1
Netherlands	9.2	5.1	4.1	3.8	5.5
Norway	6.2	4.9	5.0	3.4	4.9
Poland	–	3.3	3.7	5.5	4.2
Portugal	0.6	0.4	3.6	5.2	2.4
Russia	–	3.1	–	21.7	12.4
Slovak Republic	–	10.6	2.7	0.4	4.6
Slovenia	–	–	–	1.3	1.3
Spain	1.9	2.1	1.8	2.4	2.0
Sweden	7.1	7.3	7.0	7.0	7.1
Switzerland	1.6	2.4	2.8	2.0	2.2
Turkey	–	–	8.7	5.5	7.1
United Kingdom	2.6	4.8	4.9	3.5	4.0
United States	–	–	3.6	3.3	3.4

Source: Calculations based on data compiled from OECD reports

naturalization in the *ius soli* countries Australia, Canada and the US. As mentioned before, these 'classical immigration countries' see citizenship for newcomers as essential for national identity. In the 2011–2015 period, the US plus Canada granted citizenship to 4.5 million immigrants, compared with 4.3 million for 24 EU countries.

However, these countries also have large migrant populations, so it is important to look at relative numbers. Table 13.2 displays average yearly rates of acquisition of nationality, expressed as a proportion of the total foreign population. Unfortunately, data from New Zealand and Australia is missing. It is difficult to detect clear trends based on citizenship models. It shows that acquisition rates in European countries such as Belgium, France or Denmark are on a par with the US. Russia, Canada, Sweden, Turkey, the Netherlands and Hungary had the highest acquisition rates of all countries, a very diverse group.

Trends are hard to interpret, because special factors (such as legal changes) may play a part. Also, a saturation effect may apply, where most immigrants have already become citizens. Some earlier trends are not visible here. For instance, between 1988 and 1995 there was an upward trend in acquisitions in several European countries such as Sweden and the Netherlands, reflecting efforts to encourage immigrants to become citizens. This trend continued from 1995 to 2005 for Sweden, but not for the Netherlands, where policies became more restrictive, although saturation may also have been a factor, as acquisition rates were extremely high in the 1996–2000 period. Germany moved away from its traditionally restrictive approach to citizenship for immigrants in the late 1990s, so acquisitions increased, although rates are still much lower than in other major destination countries. Southern European countries remain restrictive, but this may also reflect the relative recentness of most immigration. Japan and Korea have maintained their restrictive regimes, while naturalization rates vary across Central and Eastern European countries.

Linguistic and cultural rights

Many of the associations typically set up by migrants in the process of ethnic community formation are concerned with language and culture: they teach the mother tongue to the second generation, organize festivals and religious events, and carry out rituals. Language and culture not only serve as means of communication, but also take on a symbolic meaning which is central to ethnic group cohesion. In most cases, language maintenance applies in the first two to three generations, after which there is a rapid decline, exemplifying the general long-term trend towards assimilation. The significance of cultural symbols, religion and rituals may last much longer.

Many non-migrants see cultural difference as a threat to a supposed cultural homogeneity and to national identity. Migrant languages, religions and cultures become symbols of otherness and markers for discrimination, as shown by contemporary hostility to Islam and its visible symbols – such as women's clothing. Majority groups often see renouncing such practices as essential for success in the country of immigration. Failure to do so is regarded as indicative of a desire for separatism. Hostility to different languages and cultures is sometimes rationalized with the assertion that the official language is essential for economic success, and that migrant cultures are inadequate for a modern secular society. The alternative view is that migrant communities need their own languages and cultures to develop identity and maintain self-esteem.

Source: Getty Images/Thomas Imo

Photo 13.2   Pupils outside the Bastos Bilingual Primary School in October 2012 in Yaounda, Cameroon

Policies and attitudes on cultural and linguistic maintenance vary considerably. Some countries have histories of multilingualism. Canada's policy of bilingualism is based on two 'official languages', English and French; the same applies to Cameroon (English and French), Belgium (mainly Dutch and French, but also German) and to some extent also the UK (Welsh besides English). Large, multi-ethnic states (which often follow federal modes of government) such as India, Pakistan, Nigeria, South Africa, Kenya, Tanzania, Indonesia and Malaysia recognize a multitude of languages while often maintaining a common *lingua franca*, such as English, French, Hindi, Urdu, Swahili, Malay or Bahasa Indonesia. In general, as multicultural states 'by design', these have fewer difficulties with cultural diversity caused by immigration, although in times of nationalistic fervour and tension between ethnic groups this can change rapidly.

Multicultural policies can lead to limited recognition of – and support for – immigrant languages, but they have hardly penetrated into mainstream contexts, such as broadcasting. Switzerland has a multilingual policy for its four founding languages (German, French, Italian and Romansh), but does not recognize immigrant languages. Australia and Sweden both accept the principle of linguistic and cultural maintenance, and have multicultural education policies. They provide language services (interpreting, translating and mother-tongue classes), funding for ethnic media and support for ethnic community cultural organizations. However, such measures have been reduced in recent years as part of the general backlash against multiculturalism.

In the US, language has become a contentious issue. The tradition of English monolingualism is being eroded by the growth of the Hispanic community: in major cities such as Los Angeles and Miami, the number of Spanish speakers is overtaking that of English speakers. This led to a backlash in the 1980s, in the form of the 'English-only movement', which called for a constitutional amendment to declare English the official language. Most states passed legislation to introduce this measure, but it proved extremely hard to implement, and public agencies and private companies continued to provide multilingual material and services. *Monolingualism* is the basic principle applied to immigrants in France, the UK, Germany and the Netherlands (even where regional languages are officially recognized, such as Welsh in the UK and Frisian in the Netherlands). Nonetheless, all these countries have been forced to introduce at least some language services to take account of migrant needs in communicating with bureaucracies, health services and courts. Double standards are common. In the Netherlands, for instance, there is widespread tolerance of English-only speaking 'expats', while migrant workers and refugees not speaking fluent Dutch are sometimes accused of refusing to integrate.

Immigrants and nation

A crucial question is how immigrants and their descendants become part of receiving societies and nations. A second question is how states and civil society can and should facilitate this process. Answers have varied in different countries. The process is most commonly referred to as 'integration', but because this implies a specific idea of where the process should lead, others prefer the more neutral term 'incorporation'. In newer immigration countries, for instance the Gulf oil states and East Asia, the belief that immigrants are only temporary and should not be integrated at all is still widespread. Such ideas also prevailed a few decades ago in countries such as Germany, the Netherlands and Switzerland, where migrants were officially seen as 'guestworkers'. However, attitudes have shifted as those 'reluctant immigration countries' have gradually come to terms with the new reality of the formation of large immigrant communities.

The starting point for understanding incorporation is historical experiences of nation-state formation: in Europe and its settler colonies (such as Canada, Argentina or Australia) this refers to the ways in which emerging states handled difference when dealing with internal ethnic or religious minorities, conquering new territories, ruling subjugated peoples and incorporating immigrants. Differing ideas about citizenship developed from these experiences (see also Chapter 4). 'National models' for dealing with ethnicity and cultural difference emerged (see Bertossi 2007; Brubaker 1992; Favell 1998) and these ideologies affected how states and the public later reacted to immigrants (Castles and Davidson 2000).

For example, the British history of conquering Wales, Scotland and Ireland and of dealing with religious diversity led to a politically integrated state that accepted difference: the United Kingdom required political loyalty, but a person's group identity could be Welsh or Scottish, Protestant or Catholic while still identifying as 'British'. In France, the 1789 Revolution established principles of equality and the rights of man that rejected group cultural identity, and aimed to include individuals as equal political subjects. In both Britain and France, however, it was the expansion of the state that created the nation – political belonging came before national identity. In Russia, nation building and empire building happened concurrently (Pipes 1997).

Germany was different: it was not united as a state until 1871, and the nation came before the state. This led to a form of ethnic or folk belonging that was not consistent

Box 13.6 Migrants and minorities in Italy

Since the 1980s, Italy has experienced a dramatic migration transition. From 1945 to 1975, 7 million Italians emigrated to escape economic stagnation and poverty. Large Italian communities remain in the US, Argentina, Brazil, Australia, France, Germany, Belgium and Switzerland. But rapid economic growth and declining fertility have reversed former patterns. Italy's foreign-born population increased from just 0.4 million in 1984, to 4.6 million in 2011, to 6.1 million in 2017. Foreign-born residents made up about 10 per cent of Italy's total population of 60 million in the same year.

Italy: foreign-born population by country of birth in 2009 and 2017

Country of birth	2009 (thousands)	2017 (thousands)	Share immigrant (%, 2017)
Romania	1,021	1,036	17.1
Albania	443	458	7.6
Morocco	419	435	7.2
Ukraine	215	238	3.9
China	196	220	3.6
Germany	224	210	3.5
Switzerland	196	192	3.2
Moldova	161	182	3.0
India	130	156	2.6
Philippines	139	148	2.4
France	138	128	2.1
Bangladesh	90	120	2.0
Egypt	107	118	1.9
Poland	125	114	1.9
Peru	116	113	1.9
Other countries	2,096	2,187	36.1
Total	**5,814**	**6,054**	**100**

Source: OECD (2018)

Immigrants are important in sustaining agriculture, industry, care and domestic work. Irregular workers are concentrated in the 'underground economy', which comprises about a quarter of Italy's economic activity. Legal immigrants are important as workers – both skilled and unskilled – in the industries of Northern Italy and also in services throughout the country. Many female migrants work as caregivers and domestic workers for families and despite their often undocumented status their presence is widely tolerated. Several indicators show a trend to permanent settlement: increased family reunion and increasing numbers of children entering Italian schools.

Right-wing parties have regularly campaigned against immigration as a threat to law and order. Trade unions, left-wing parties, church organizations and advocacy groups support migrant rights and call for interculturalism, while employers' associations campaign for increased labour migration.

Italy had no immigration law until 1986, and it was not until 1998 that the Centre–Left Government tried to create a regulatory system, including a long-term residence permit (*carta di soggiorno*). Arguing that immigrants were a threat to the country, in 2002 the Centre–Right Berlusconi coalition repealed many of the 1998 measures, introducing tough measures against irregular immigration. But the Berlusconi government also introduced a legalization campaign, which led to a big increase in the legal foreign population.

In 2008, the third Berlusconi coalition government introduced the 'Security Package' (*Pachetto Sicurezza*), which included high fines and expulsion for irregular migrants. The government even expelled Roma and Sinti despite many being EU citizens. In this hostile climate, racist attacks on immigrants ensued, including the burning down of migrant shacks by mobs in Naples.

During the Great Recession, immigration was a relatively minor topic. However, the 'refugee crisis' of 2015 brought immigration back on top of the political agenda (see Box 6.1). As Italy received many asylum seekers crossing from Africa, Italian governments pushed for the resettlement of refugees in other EU-member states, but other states were reluctant to collaborate. As a result, the political discourse on boat migration has hardened across the political spectrum. In 2018 the right-wing nationalist Deputy Prime Minister Matteo Salvini took a hard stance and closed Italy's harbours to migrant boats and non-Italian humanitarian ships.

Sources: Ambrosini 2013; King *et al.* (2000); Scotto (2017); OECD (2018); Pastore *et al.* (2006); Paoletti (2011); Strozza and Venturini (2002)

with incorporation of minorities as citizens. These differing approaches imply different relationships between society and nation, and between civic belonging and national identity. In Britain a person could be a full member of the society and political nation, and yet belong to a distinct cultural or religious group. In France, civic identity required a unitary national identity. In Germany, national identity came first, and was the precondition for belonging as a citizen.

By contrast, the settler societies of the New World were built through the dispossession and murder of indigenous peoples, slavery, recruitment of indentured workers, and through immigration from Europe. Incorporation of immigrants as citizens was part of their national myths. This led to ideologies of assimilation, such as the US image of the 'melting pot'. In the settler societies, civic belonging was thought to lead to national identity, so that differing identities were acceptable as a passing phase on the way to 'Americanization' (or the equivalent). Of course, it was thought that only white people could be assimilated: Australia, New Zealand, Canada and the US all had racially selective immigration laws.

Other countries, and particularly states relying on myths of common ancestry, have struggled to come to terms with the realities of immigration. When immigration to industrialized countries started to gain ground in the post-1945 boom, incorporation of the newcomers was not a major issue. The numbers were not expected to be large, and there was a strong belief in the 'controllability of difference'. The 'classical immigration countries' (the US, Canada, Australia etc.) only wanted white settlers from their 'mother countries' or other Northwestern European countries, and saw no problem in assimilating them. Britain, France and the Netherlands also expected to be able to assimilate fairly small groups of immigrants from their former colonies and from other European countries such as Spain and Italy. Germany and other importers of 'guestworkers', such as Austria, Belgium, the Netherlands and Switzerland, did not anticipate family reunion or settlement, and therefore pursued polices of temporary admission to the labour market.

In Northeast Asia, new immigration countries such as Japan and South Korea have strong beliefs in myths of ethnic homogeneity, and these 'ethno-states' therefore find it very hard to incorporate people of different backgrounds, let alone accept them as fully part of the nation. South-East Asian countries such as Malaysia have culturally mixed populations, but the public and politicians often fear that incorporating newcomers could upset existing ethnic balances. Similarly, in post-apartheid South Africa, immigration from other African countries is often seen as a threat, and has led to violent clashes. In much of Africa and the Middle East, the main concern has often been with building a new national identity following decolonization and how ethnic and religious minorities and immigrants do – or do not – fit into this identity. Such contested notions of nationhood and the homogenizing tendencies and intolerance towards difference this involves have sparked significant violence and displacement.

Models of immigrant integration

It is possible to make a rough distinction between three approaches or 'models' towards immigrant integration. Chapter 4 distinguished four 'ideal-types' of citizenship: the (1) imperial; (2) folk or ethnic; (3) republican and (4) multicultural models. These provide important contexts for the various incorporation models, although these different models can often coexist in the same national context, or dominant ideas and ideologies can change over time.

Assimilation meant that immigrants were to be incorporated into society through a one-sided process of adaptation. They were to give up their distinctive linguistic, cultural or social characteristics and become indistinguishable from the majority population. The 'guestworker' model can be described as *differential exclusion*: migrants were to be temporarily incorporated into certain areas of society (above all the labour market) but initially denied access to others (especially citizenship and political participation) (Castles 1995). This model is still strictly applied by the Gulf oil states.

But the belief in the controllability of difference proved misplaced in all these cases. In the post-war boom, labour migration grew in volume and quickly became a structural feature of Western economies. Racially selective immigration rules broke down, and migrants increasingly came from more distant or culturally different countries. When the economic boom faltered in the 1970s, family reunion took place – even in 'guestworker' countries. Then the end of the Cold War and globalization brought new migrations from ever more diverse origins.

Since the 1970s, several governments replaced ideologies and discourses around assimilation (initially at least) with the principle of *integration*, which implied a recognition that adaptation is a gradual process that required some degree of mutual accommodation. Although acceptance of cultural maintenance and community formation might be a necessary stage, the final goal was implicitly still absorption into the dominant culture – integration was often simply a slower and gentler form of assimilation. Today, of all the major immigration countries, France comes closest to the assimilationist model. Elsewhere, however, there was a shift to approaches that recognized the long-term persistence of group difference. In some countries, such models were referred to as 'multiculturalism'; in other places terms such as 'minorities policy' and 'equality and freedom of choice' were used.

Multiculturalism implies that immigrants (and sometimes non-migrant minority groups) should be able to participate as equals in all spheres of society, without being expected to give up their own culture, religion and language, although usually with an expectation of conformity to certain key values. There have been two main variants of multicultural ideologies. In the laissez-faire approach typical of the US, cultural diversity and the existence of ethnic communities are officially accepted, but it is not seen as the role of the state to work for social justice or to support the maintenance of ethnic cultures. The second variant is multiculturalism as a public policy. Here, multiculturalism implies both the willingness of the majority group to accept cultural difference and state action to secure equal rights for minorities. Multiculturalism originated in Canada, and was taken up under various labels between the 1970s and the 1990s in Australia, the UK, the Netherlands, Sweden and elsewhere. Multiculturalism has always been controversial. To many members of the dominant ethnic group, multiculturalism appears as a threat to their culture and identity. Others criticize multiculturalism for leading to a superficial acceptance, without bringing real institutional change (see Vasta and Castles 1996).

All of the different approaches to incorporation have proved problematic in one way or another, so that by the early twenty-first century there appeared to be a widespread 'crisis of integration'. In recent years, debates on immigrant integration have been overshadowed by concerns about security (see Chapter 10), national identity and a lack of perceived integration, particularly of lower-skilled immigrant groups and refugees. As part of a general backlash against multiculturalism, the pendulum has swung back from celebrating diversity to insisting on forms of 'civic integration' and assimilation based on often rather unclear ideas about national values.

Scholars have questioned the validity of the different 'models' of integration and argued that they are partly a product of ideological imagination. For instance, Duyvendak and Scholten (2012) argued that although Dutch integration policies have often been labelled 'multiculturalist', in reality they were not driven by a single, coherent and consistent model or vision, but were rather the outcome of pragmatic concerns and political compromise. Based on extensive analysis of policy documents and public debate, they claim that Dutch policies were not that multicultural at all, and have always emphasized uniformity and the need to adapt to mainstream norms and values (Duyvendak and Scholten 2012).

Conclusion

The comparison of the situation in various immigration countries showed that ethnic group formation take place everywhere, but the conditions under which this happens varies considerably. On the one hand, heated ideological debates on integration and diversity tend to overlook the considerable extent to which the arrival, settlement and incorporation of migrants are autonomous social processes that will continue partly irrespective of state policies. On the other hand, the labour, educational and civic rights given to migrants as well as housing policies can be decisive in processes of settlement and incorporation, while state policies have sometimes been inflexible, unrealistic and sometimes counterproductive. This leads to different outcomes: in some countries immigrants and ethnic groups become marginalized and excluded minorities, whereas in others they blend in quickly or take the form of ethnic communities which are accepted as parts of pluralist societies. Outcomes also tend to be different for different migrant groups in the same country.

Comparison of national experiences and policies provides some useful conclusions. The first is that temporary migrant labour recruitment almost always leads to permanent settlement of at least a proportion of migrants; settlement in turn leads to the formation of ethnic groups. The second is that the character of future ethnic groups is influenced but not wholly determined by what the state does in the early stages of migration. Policies which deny the reality of immigration by tacitly tolerating large-scale illegal movements and exploitation of migrant workers lead to social marginalization, minority formation and racism. Third, in order to cope with the difficult experience of settlement in a new society, immigrants initially need their own organizations and social networks, as well as their own languages and cultures. Policies which deny legitimacy to these can lead to isolation and separatism. Fourth, the best way to prevent marginalization and social conflicts is to grant permanent immigrants full rights in all social spheres, including citizenship rights. Although instant access to social welfare is often seen as undesirable (and potentially undermining welfare states), the successful incorporation of immigrants including refugees is generally fostered by policies that provide a clear pathway to permanent residency and citizens, for instance after a number of years of work and residency.

Starting in the 1990s, policies on incorporation of immigrants and minorities have been questioned and revised. The inescapable reality of permanent settlement has led to the abandonment of the differential exclusionary approach in Germany and other former 'guestworker' countries. Immigration and citizenship laws have been reformed. While multiculturalist ideologies are increasingly rejected at the national level, local provision of special social and educational services for minorities is widespread in practice. However, there are limits to change: Germany still rejects dual citizenship and has introduced compulsory integration measures. Austria and Switzerland still cling to exclusionary policies, although these are modified by local integration efforts. Exclusion remains the dominant approach to foreign workers in most destination countries in Asia, the Gulf and Africa.

By the early 1990s, assimilation ideologies seemed to be on the way out everywhere, except in France. Democratic civil societies were thought to have an inherent trend towards multiculturalism (Bauböck 1996). That is no longer the case: there has been a widespread backlash against multiculturalism. Canada has maintained its multicultural principles, but watered down their implementation, and Australia has gone even further in this direction. Sweden, the Netherlands and the UK have all relabelled policies with much greater emphasis on 'integration', 'social cohesion' and 'core national values'.

According this interpretation, recognition of cultural diversity has had the perverse effect of encouraging ethnic separatism and the development of 'parallel lives'. A model of individual integration – based if necessary on compulsory integration contracts and citizenship tests – is thus seen as a way of achieving greater equality for immigrants and their children. The problem for such views, however, is that the one country that has maintained its model of individual assimilation is also experiencing serious problems. The experience with racism, marginalization and segregation in France showed that the republican model of individual integration has largely failed to overcome inequality and racism. More in general, this calls into question the validity, relevance and impact of different narratives, ideologies and 'national models' of integration, and shows the dangers of reducing the diverse array of policies national and local governments use to one particular label (see Alba and Foner 2014; Duyvendak and Scholten 2012).

All the varying policy approaches to incorporation of immigrants thus seem problematic: differential exclusion is useless once settlement takes place; multiculturalism may lead to separatism, and assimilation can perpetuate and deny marginalization and conflict. This situation actually reflects the inability or unwillingness of destination societies to deal with two issues. The first is the deep-seated cultures of racism that are a legacy of centuries of colonialism, imperialism, anti-Semitism and other forms of intolerance. In times of stress, such as economic restructuring, recession or international conflict, racism can reinforce social exclusion, discrimination and violence against minorities.

The second issue is the trend to increased inequality resulting from neoliberal globalization, economic restructuring, deregulation and privatization. Increased international competition has put pressure on job security, working conditions and welfare systems around the world. At the same time, neoliberal economic policies encourage greater pay differences and reduce the willingness of states to tax the wealthy in an effort to redistribute income to alleviate poverty and social disadvantage. In Western societies, job security has declined, welfare states have been partly dismantled, and real incomes and purchasing power of lower and middle-income groups have barely increased or actually declined over the past decades. Economic inequality has been on the rise while intergenerational social mobility is declining (see Piketty 2014). In this context, politicians can be easily tempted to blame migrants for these problems and by doing so deflecting the attention away from their own inability or unwillingness to address these problems.

Taken together, these factors have contributed to the racialization of ethnic difference (see Chapter 4). While many migrant groups are successful, other migrants and minorities may have poor employment situations, low incomes and high rates of impoverishment. This in turn leads to concentration in low-income neighbourhoods and residential segregation. The existence of separate and marginal communities is then taken as evidence of failure to integrate, and this in turn is perceived as a threat to the host society. Attempts to resolve the crisis through discrimination against minorities do not provide a solution. Rather, they threaten the fundamental values upon which democratic societies are based.

Guide to Further Reading

In earlier editions of *The Age of Migration*, we provided a detailed comparison of two very different immigration countries, Australia and Germany. For reasons of space, this chapter could not be included in this edition, but it is available on *The Age of Migration* website. The website also includes short accounts of the situation of migrants and minorities in Canada, the Netherlands, Sweden and South Korea. It would take up too much space to give further reading for individual countries here – instead we refer readers to the sources used for the country boxes. The annual OECD *International Migration Outlook* provides up-to-date statistics and policy information for many countries, while the Migration Information Source (www. migrationinformation.org) contains data and useful short country studies.

Useful comparative studies include Reitz (1998), which covers Canada, the US and Australia, Hollifield *et al.* (2013), which covers North America, Europe, Australia and some Asian countries, as well as Heckmann and Schnapper (2016) for comparison of French, German, British, Swedish, Swiss, Dutch, Finnish and Spanish experiences with immigrant integration. King *et al.* (2000) provides studies of Southern European immigration countries. Good comparative studies on citizenship include Aleinikoff and Klusmeyer (2000, 2001) and Bauböck *et al.* (2006). Useful works on multiculturalism include Parekh (2000, 2008), Modood (2007) and Kymlicka (1995). Kymlicka has also co-edited a book on *Multiculturalism in Asia* (Kymlicka and He 2005). Joppke (2007) gives an overview of the changes in citizenship policies.

14 Migration and Development in Origin Societies

Many migrants maintain strong transnational ties with origin communities even several decades after permanent settlement. In many cases, their main motive for migrating is to improve the living conditions of the families left behind. The money migrants remit often enables families to decrease livelihood insecurity, improve housing, nutrition, clothing, medical care and education. Many migrants aspire to return once they have saved enough money to invest in an enterprise, such as the family farm, a store, a truck or minibus, a workshop or a small factory. Migration may also serve cultural or religious purposes, for instance to perform the *hadj*, the religious pilgrimage to Mecca.

While the preceding chapters focused on the impacts of migration for destination societies, this chapter turns the analysis around and assesses how migration affects social transformation and development in origin societies. Migration research has traditionally focused on the settlement and integration of migrants and, more generally, the implications of migration for destination societies. This 'receiving country bias' is unfortunate, because the impacts on origin societies are equally, if not more, profound. Emigration arises through development and social transformation, but in turn brings further social, cultural, economic and political change in communities and societies of origin. While money remittances can upset traditional social, economic and ethnic hierarchies, and increase migration aspirations in origin communities, 'social remittances' often affect processes of sociocultural change, such as norms and practices around family, gender and religion. Because the causes and consequences of migration are interlinked, some of the theories discussed in Chapter 3 are therefore relevant here too.

The migration and development debate

While migration generally stimulates growth and innovation in destination countries, a much more contested question is *whether migration encourages development of the countries of origin or, conversely, hinders such development*. Over the past decades, this issue has sparked heated debate in policy and research, opposing 'migration optimists', who argue that migration brings growth and prosperity to origin countries, to 'migration pessimists', who argue that migration undermines development through draining origin countries of their scarce human and financial resources. While the pessimists predict a 'brain drain', the optimists predict a 'brain gain'. This reflects the more general division between historical–structural and functionalist theories that have rather opposed views on migration as either an *exploitation* or an *optimization* mechanism, respectively (see Chapter 3). Table 14.1 summarizes the main hypotheses of functionalist and historical–structural theories on migration and development. It shows how both paradigms reach quite opposite conclusions. As this chapter will argue, research evidence shows the need for more nuanced, in-between, positions.

Table 14.1   Brain drain versus brain gain: Opposing views on migration and development

Migration optimists		Migration pessimists
Functionalist	↔	Historical–structural
Neoclassical	↔	Neo-Marxist
Modernization	↔	Dependency
Net North–South transfer	↔	Net South–North transfer
Brain gain	↔	Brain drain
More equality	↔	More inequality
Remittance investment	↔	Consumption
Development	↔	Disintegration
Less migration	↔	More migration

However, it is important first to ask: what is development? This is a contested concept, and there are many definitions. Although there is no room in this book for a full discussion, it seems useful to adopt a broad definition of 'human development', which goes beyond narrow definitions that focus on income growth, and that therefore includes dimensions such as social security, welfare, education, health care and people's overall well-being. In this spirit, Amartya Sen defined development as the *process of expanding the substantive freedoms that people enjoy* (Sen 1999). In order to make this concept of freedoms more concrete, it is useful to draw on Sen's concept *of* human *capability*, which is the 'ability of human beings to lead lives they have reason to value and to enhance the substantive choices they have' (see Chapter 3). The capabilities concept is useful for achieving a comprehensive understanding of the development implications of migration that goes beyond narrow visions focused on economic impacts.

Sen (1999) argued that income growth – which can hide huge inequalities – should not be the focus of analysis, but rather the question of whether the real capabilities of people to control their own lives have expanded. Capabilities imply access to resources, particularly access to (1) social capital (useful connections to other people); (2) cultural or human capital (ideas, knowledge and skills) as well as (3) economic capital (money and assets). Capabilities and migration are connected in a reciprocal, two-way relationship, and can reinforce each other. First, in order to migrate, people need capabilities in the form of access to money, social networks, knowledge, skills and diplomas. This explains why the poorest tend to migrate less often and if they migrate, why they tend to do so over shorter distances and under more difficult circumstances.

Migration is thus shaped by people's capabilities but also affects such capabilities in its own right, as migration itself is a vital resource allowing migrant and their families to tap higher and more stable sources of income, which allows them to improve living standards, invest in education or to set up businesses. Because migration creates real opportunities and hope for the future, it is often the most attractive way 'out' of situations of stagnation, inequality and hopelessness for ambitious young adults. In fact, the real chance to change one's life and to access new opportunities is what motivates most people to migrate, irrespective of governments' views of migration policies.

Official views on 'migration and development' are diverse and often rather ambiguous. For instance, authoritarian governments may welcome migration as a 'safety valve' but fear migrants' political activism (see Chapter 10). Over the past decades, policy and academic debates about migration and development have swung back and forth, from predominant optimism in the post-1945 period, to predominant pessimism since the 1970s, to renewed optimism since the 2000s, although there have always been dissenting visions (see de Haas 2012). This section will review how the migration and development debate evolved. It will examine the theoretical foundations of optimistic and pessimistic views and will then outline more nuanced visions and summarize the main insights from the research literature on this topic.

Migration and development optimism: A 'win-win-win' scenario

During the 1950s and 1960s, many development economists stressed that labour migration was an integral and positive part of modernization. The reduction of labour surpluses (and hence unemployment) in areas of origin and the inflow of capital through remittances was expected to improve productivity and incomes in origin societies (Massey *et al.* 1998: 223). Many governments of 'labour exporting' countries shared this view, and considered migration as a key instrument to promote development (Adler 1981; Heinemeijer *et al.* 1977; Penninx 1982). 'Sending' and 'receiving' states signed recruitment agreements as they both saw mutual benefits in temporary or 'guestworker' migration. While West European countries recruited guestworkers, on the other side of the Atlantic the US recruited Mexican workers through the *Bracero* programme (see Chapters 6 and 7).

Migration optimists saw such labour migration as a process that stimulated growth in origin and destination countries (Adler 1981; Kindleberger 1967; Penninx 1982). It was widely thought that 'large-scale emigration can contribute to the best of both worlds: rapid growth in the country of immigration .. and rapid growth in the country of origin' (Kindleberger 1967: 253). While migration provided the industries of destination countries with much needed labour, *remittances* and the modern ideas and entrepreneurial attitudes migrants would acquire were expected to help origin countries in their modernization and economic take-off. The idea that migration can be a 'win-win-win' strategy for origin and destination countries as well as migrants themselves also underpins efforts to liberalize migration within regional economic blocs such as the European Union (EU), the Association of Southeast Asian Nations (ASEAN) and the Economic Community of West African States (ECOWAS).

Such views largely fit within neoclassical migration theory (see Chapter 3), which perceives unconstrained migration as a process contributing to a more optimal allocation of the factors of production labour and capital and, hence, to greater productivity in both sending and receiving areas. This would eventually lead to convergence of economic conditions between origin and destination countries (Massey *et al.* 1998). Such optimism fitted neatly in modernization theory (Rostow 1960): freshly de-colonialized nations in Africa and Asia were expected to quickly follow the same path of modernization, industrialization and economic growth as Western countries before them. As the *Marshall plan* had worked to spur economic development in post-WWII Europe, it was argued that large-scale capital transfer (whether argued loans, aid or remittances) would enable poor countries to jump on the bandwagon of rapid industrialization. Migration and remittances were ascribed an important and positive role in this process.

However, such views were not uncontested. In contrast to Western-aligned countries, 'non-aligned' and communist countries such as India, China, Algeria, Egypt (until Sadat's ascent to power in 1970) and Cuba saw emigration as a potential threat to independence and development. As part of a more general protectionist and anti-colonial policy, they tried to reduce emigration, particularly of the higher-skilled. Also in Mediterranean 'guestworker' countries, the long-term results of labour recruitment schemes were often disappointing, as they brought limited macro-economic benefits for origin countries and did little to resolve structural development problems (Abadan-Unat 1988; De Mas 1978; Heinemeijer *et al.* 1977; Martin 1991). At the same time, in Western Europe there was disillusionment with policies linking return migration and development through departure bonuses and investment programmes for former guestworkers (Entzinger 1985; Penninx 1982). Few migrants signed up for such programmes because of uncertain economic and political conditions in origin countries.

Migration and development pessimism: The 'migrant syndrome'

In the same period, awareness was growing that many temporary migrants would settle permanently. After the 1973 Oil Shock, the Western world entered a period of recessions, industrial restructuring and increasing unemployment. Because of the relocation of industrial production to low-wage countries and automation, it was widely thought the guestworker era had come to an end. Disillusionment with guestworker schemes also left its mark on dominant views on migration and development. From the 1970s, pessimistic views became predominant according to which 'migration undermines the prospects for local economic development and yields a state of stagnation and dependency' (Massey *et al.* 1998: 272).

The *migration pessimists* did not just argue that migration had negative effects, but saw it as one of the *very causes* of underdevelopment of origin countries. As Papademetriou (1985: 111–112) argued, migration contributed to 'an uncontrolled depletion of their already meagre supplies of skilled manpower – and the most healthy, dynamic, and productive members of their populations'. Although during this period many Asian and Middle Eastern governments started to stimulate migration to the Gulf region in the hope of alleviating unemployment and relieving political discontent, pessimistic or at least sceptical views prevailed amongst researchers about the long-term development impacts of migration.

Migration was believed to go along with a *brain drain*, which was seen as depriving poor countries of their scarce professional resources and draining their investments in education (GCIM 2005: 23–25; IOM 2005: 173). In what is often framed as a global competition to attract the 'best and brightest', OECD countries have drawn in the largest proportion of high-skilled migrants (Czaika and Parsons 2017). Increasing student migration can be seen as part of this brain drain. Foreign students tend to pay higher fees compared to domestic students, and thus provide an alternative stream of funding for the education systems of rich countries (Beine *et al.* 2014; Khadria 2008), which have been put under stress because of government cutbacks. Destination country governments increasingly encourage foreign graduates to stay and work. Indian and Chinese PhDs form the scientific backbone of Silicon Valley and other high-tech production areas. Another potential loss is known as *brain waste*: many migrants are unable to get their qualifications recognized or fail to find employment commensurate with their skills. The image of surgeons working as waiters or engineers driving taxis reflects reality for quite a few migrants and refugees.

From the 1980s, the dominant view increasingly became that remittances fuelled consumption and that migrants rarely invested their money productively. What seemed to make matter worse is that the 'lost labour' (the 'brawn drain' – see Penninx 1982) and dependency on remittance income seemed to encourage rural families to abandon their farms or to cease other local economic activities. Empirical studies suggested that remittances were primarily spent on (often conspicuous) consumption of goods and on 'non-productive' enterprises such as housing (Entzinger 1985; Lewis 1986; Lipton 1980; Rhoades 1979). Since many of these goods (for example, television sets, household appliances, stylish clothing, fertilizers, manufactured foodstuffs) had to be imported, this had the double effect of crowding out local production while strengthening the economies of wealthy countries, thus increasing economic gaps between origin and destination countries, and thus increasing the dependency of the 'Third World' on the 'First World'.

Because international migrants tend to come from relatively well-off groups, remittances were believed to increase inequality in origin communities (Lipton 1980; Zachariah *et al.* 2001). At the same time, the migration- and remittance-related decline of traditional agriculture and crafts deprived poorer non-migrant community members of work and income as farm workers, or from crafts such as tailoring, furniture making and various forms of traditional food and dairy processing. Remittance-fuelled consumption and land purchase by migrants were increasingly blamed for inflation (Russell 1992) and soaring land prices in origin areas (Appleyard 1989; Rubenstein 1992), from which the non-migrant poor would suffer most. In additional, large influxes of capital in the form of remittances were often believed to lead to an over-appreciation of the national currency, also known as the 'Dutch disease' (Taylor 1999). A related concern is that such remittances can 'crowd out' public finance by reducing government spending incentives in the presence of private substitutes for public goods (see Ambrosius 2019).

Pessimistic views also looked negatively at the sociocultural impacts of migration. Exposure to the relative wealth of (return) migrants and the goods and ideas they bring would lead to changes in rural tastes (Lipton 1980: 12) and an increase in consumerist attitudes. This further increased demand for imported goods (such as processed food and wheat instead of fresh vegetables and traditional grains; or various mass-produced consumer items as clothes, television sets and cars), further undermining local production. Increasing demand of consumer goods led to an increasing need for cash, which in turn reinforced the perceived need to migrate to generate income. Migration is also held responsible for the disruption of traditional kinship systems and care structures (King and Vullnetari 2006) and the breakdown of traditional institutions regulating village life and agriculture (de Haas 1998). Together, these changes make agrarian lifestyles less appealing, and demotivated youth to work in agriculture or other traditional occupations. This could give further rise to a 'culture of migration' (Heering *et al.* 2004; Massey *et al.* 1993), in which young people can only imagine a future through migrating (see Chapter 3).

Through these mutually reinforcing social, economic and cultural effects, migration would foster the development of remittance-dependent, non-productive and migration-obsessed communities. These negative views can be amalgamated into what Reichert (1981) called the 'migrant syndrome', or the vicious circle of: migration, leading to more underdevelopment, leading to more migration, and so on.

These pessimistic views became rather dominant in academic and development circles over the 1980s and 1990s. As an ILO official commented: 'migration and development – nobody believes that anymore' (quoted in Massey *et al.* 1998: 260). Although origin country governments in North Africa, South-East Asia and the Pacific (see Bertram and Watters 1985; Stahl and Habib 1991) continued to see migration as a welcome political-economic 'safety valve' to relieve unemployment and discontent, the idea that migration would boost national development was largely abandoned.

Remittances as development aid?

Remittance euphoria in a neoliberal era

Yet in the early twenty-first century, there was a rather sudden swing in assessment. After years of predominant pessimism and relative neglect of the migration issues amongst development agencies and organizations such as the World Bank, policy makers redis-covered the potential of migration to foster economic and social development in origin countries. This renewed optimism about migration and development was related to a spectacular increase in North–South remittances.

The publication in the World Bank's yearly World Development Finance of a chapter entitled *Workers' remittances: an important and stable source of external development finance* (Ratha 2003) drew the attention to the economic significant of remittances for origin countries and marked a major shift in the debate. A certain level of 'remittance eupho-ria' coincided with growing optimism about the development contributions of migrants through counter-flows of money, investment, trade relations, innovations, entrepreneur-ial and innovative attitudes and information (de Haas 2006b; Lowell and Findlay 2002). In a time that the effectiveness of development aid was increasingly questioned, remit-tances became a new 'development mantra' (Kapur 2003), and were increasingly seen as an ideal 'bottom-up' source of development finance, and an effective safety net for poor areas and countries (Jones 1998b).

This led to several policy initiatives by governments and international organizations to enhance the development impact of migration and remittances (DFID 2007; UNDP 2009; World Bank 2006). The World Bank in particular has been very active in finding ways to reduce the costs of sending remittances. Such rather sudden swings in thinking about migration and development partly reflect ideological changes. For instance, the shift to optimistic views partly reflected the growing influence of neoliberal ideologies. This is also evident in the fact that governments of liberal-capitalist states have gener-ally been much more positive about the development potential of emigration than social-democratic governments, whose often saw migration as a development undermin-ing leading to a brain drain.

Kapur pointed to the ideological roots of remittance euphoria by arguing that remit-tances exemplify the principle of self-help, in which 'Immigrants, rather than govern-ments, then become the biggest provider of "foreign aid"' (Kapur 2003: 10). This also fits within the 'Washington consensus' – the World Bank and IMF-led development ide-ology which advocates market liberalization, privatization and deregulation (Gore 2000; Stiglitz 2002) (see Chapter 3). Or, to put it less positively, the idea is that some of the world's most exploited workers should provide the capital for development, where aid pro-grammes have failed.

This raises the following question: to what extent are optimistic and pessimistic claims justified? What do research findings tell us about the development impacts of migration? Fortunately, the rapid increase in of empirical studies over recent decades has improved insight into this issue, and has given rise to more nuanced perspectives. The following sections will review the mixed evidence on this issue.

Remittance trends

According to World Bank data, in 1990 migrants sent back the equivalent of US$29 billion to lower- and middle-income countries: this amount had more than doubled to US$74 billion in 2000, after which it quadrupled to US$302 in 2010, to reach levels of 442 US$ billion in 2017 (see Figure 14.1). Although this fast increase partly reflects better measurement of remittances – which is in itself a result of the growing interest in the issue – there is little doubt that remittances have increased. Also high-income countries receive substantial remittances from their citizens living abroad, which accounted for 24 per cent of global remittances in 2017. However, the bulk of remittances (46 per cent) goes to lower middle-income countries with upper middle-income countries receiving another 25 per cent of global remittances. Middle-income countries thus receive almost three quarters of global remittances. The explanation is simple: middle-income countries tend to have the highest emigration rates. Low-income countries, by contrast, only receive 5 per cent of global remittances. This reflects their weak integration in global migration systems. This fact alone should be a reason for scepticism about the idea that migration and remittances could be a major force for reducing global inequality.

Still, remittances represent a major resource flow to developing countries. For instance, in 2017 the amount of remittances sent to lower- and middle-income countries (US$ 442 billion) was almost three times higher than the US$ 163 billion in official development assistance (ODA) and 82 per cent of the US$ 537 billion in Foreign Direct Investment

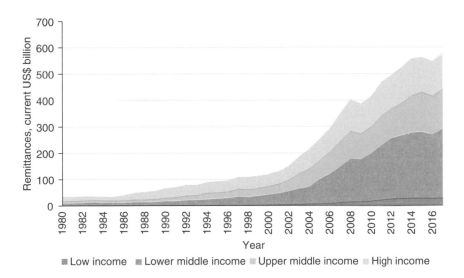

Figure 14.1   Remittances, current US$ billion, 1980–2017

Source: World Development Indicators database, accessed 26 January 2019

Photo 14.1   Multi-lingual Western Union sign advertising money transfer services in Paramaribo, Suriname

(FDI). Moreover, according to some estimates, unrecorded remittances through informal channels may add 50 per cent or more to recorded flows (World Bank 2007). In addition, remittances are sent directly by migrants to their families back home. Although banks and money transfer agencies often charge hefty fees for remittance transfers, remittances flow directly to those who need them, and no money is being siphoned off by state bureaucrats, expensive consultants or corrupt politicians (Kapur 2003).

Remittance dependency

In middle-income countries, remittances have increasingly dwarfed ODA, and have been 7 to 10 times higher over the 2010s. For low-income countries, remittances have also grown in importance, representing between 60 and 70 per cent of ODA in recent years. In addition to the vital importance of remittances for families and communities of origin countries, governments see remittances as an important source of hard currency, allowing them to cover a significant part of their trade deficits. The strategic importance of remittances for middle-income countries as a political-economic 'safety valve' and source of hard currency helps us to understand why their governments have often little real interest in curbing emigration despite the lip service they may pay to the migration control agendas of destination countries in international forums (de Haas 2008).

Table 14.2 shows the 20 most important remittance-receiving countries over the 2008–2017 period both absolute and relative to GDP terms. It also shows the most important countries from which remittances are sent. In dollar terms, India, the Philippines and Mexico are the largest remittance receivers, and next come Nigeria, Egypt and Germany. However, as the data on remittances as a share of GDP show, it is generally smaller countries with weak economies and high emigration rates that are most dependent on remittances, such as Tajikistan, the Kyrgyz Republic, Tonga, Nepal, Moldova and Haiti. Unsurprisingly, most remittances are sent from major migrant destination countries with the US, Saudi Arabia, Russia, Switzerland and Germany coming out firmly on top.

The data in Figure 14.2 reveals an interesting paradox: although the poorest countries in the world only receive 5 per cent of global remittances, their relative economic dependency on remittances is higher. The relative remittances dependency – expressed as a share of GDP – is highest in lower-middle and, particularly, low-income countries. Although low-income countries generally have lower long-distance emigration rates, this high

Table 14.2   Top 20 remittance-receiving and -sending countries, absolute and relative to GDP, average 2008–2017

Remittances received US$ billion		Remittances received as % of GDP		Remittances, paid US$ billion	
India	62.5	Tajikistan	37.1%	United States	57.0
Philippines	25.7	Kyrgyz Rep.	27.8%	Saudi Arabia	31.8
Mexico	25.2	Tonga	27.2%	Russia	25.6
France	22.8	Nepal	26.3%	Switzerland	22.3
China	22.2	Moldova	25.8%	Germany	17.7
Nigeria	20.3	Haiti	23.6%	Kuwait	14.2
Egypt	16.0	Bermuda	22.1%	France	12.9
Germany	14.9	Lesotho	21.2%	Italy	11.9
Pakistan	14.2	Samoa	19.1%	Luxembourg	11.6
Bangladesh	12.8	El Salvador	19.0%	Qatar	11.5
Belgium	10.7	Honduras	17.4%	Netherlands	10.4
Vietnam	10.2	Lebanon	17.0%	UK	10.3
Italy	8.9	Armenia	16.5%	South Korea	9.3
Ukraine	8.3	South Sudan	15.9%	Oman	8.2
Indonesia	7.8	Kosovo	15.9%	China	7.2
Poland	7.6	Jamaica	15.8%	Australia	5.8
Lebanon	7.2	Liberia	14.2%	Canada	5.3
Morocco	6.9	Marshall Isl.	14.1%	India	5.0
Russia	6.4	Jordan	14.0%	Norway	4.8
South Korea	6.4	W. Bank/Gaza	12.9%	Lebanon	4.5

Source: Calculations based on World Development Indicators database, accessed 26 January 2019

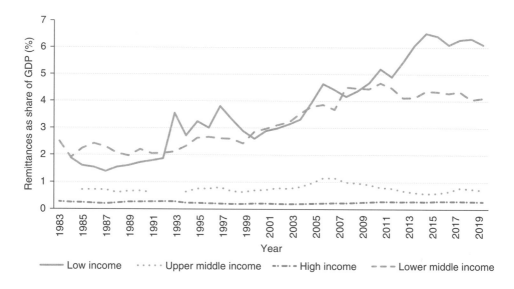

Figure 14.2   Remittances as a percentage of GDP, by country income group
Source: World Development Indicators database, accessed 26 January 2019

remittance dependency reflects the overall weak state of their economies. Many countries with high-emigration rates to the high-income countries (where most income gains can be made) are middle-income countries with comparatively stronger and more diversified economies, such as Mexico and Turkey, and to a lesser extent also the Philippines and Morocco.

The economic impacts of migration and remittances

The pessimistic and optimistic views on migration and development reviewed above represent rather extreme views, and in most cases the reality is somewhere in between. Since the 1990s a growing number of empirical studies have highlighted the diversity of migration impacts; and showed that this impact primarily depends on the specific conditions under which migration takes place. This shift towards more nuanced views partly reflect the growing influence of the *new economics of labour migration (NELM)* (Stark 1991; Taylor 1999) and livelihood approaches (see Chapter 3). These perspectives highlight the limited, but real, agency migrants have, and emphasize how internal and international migration is often a deliberate strategy used by marginalized and oppressed groups of people to overcome structural constraints (such as economic insecurity, inequality and violence) and improve the long-term wellbeing of families and communities in origin countries.

Research insights based on qualitative fieldwork and quantitative surveys emphasize that the impacts of migration primarily depend on the more general development context in origin societies. Several reviews of the rich empirical evidence (Agunias 2006; Clemens and McKenzie 2018; de Haas 2007c; Docquier and Rapoport 2012; Özden and Schiff 2005; Papademetriou and Martin 1991; Taylor *et al.* 1996b, 1996a) indicate that, despite its significant poverty-reducing, income-increasing and welfare-enhancing benefits for individuals and communities, migration and remittances alone cannot remove structural development obstacles at the national level.

Most studies support the view that migration within and from developing countries is part of deliberate attempts by families to improve their social and economic status rather than a 'desperate flight from poverty and misery' as it is often portrayed. Remittances are a relatively stable source of income which often help to diversify, stabilize and raise household income, to improve living conditions, nutrition, health and education, as well as to finance weddings, funerals and other ceremonies. On the national level, remittances provide a less volatile and more reliable source of foreign currency than other capital flows to developing countries. Remittances also tend to be relatively stable and can under certain circumstances be *anticyclical*, with remittances going up if needs increase in origin communities, for instance as a result of economic downturns or environmental shocks.

Effects of remittances on income, poverty and inequality

Migrants and their families have good reasons to expect considerable financial gains from migrating across borders, particularly for those moving from low- to high-income countries. For instance, migrant workers in the US earn on average about four times as much as they would in developing countries of origin, while Pacific Islanders in New Zealand increase their net real wages by a factor of three by migrating (Clemens *et al.* 2008; McKenzie *et al.* 2010). Low-skilled migrants from Thailand working in Hong Kong (China) and Taiwan are paid at least four times as much as they would earn at home (Sciortino and Punpuing 2009). Income gains can be large for the high-skilled as well as the low-skilled. A doctor from Côte d'Ivoire can raise her or his real earnings six times by working in France (Clemens 2009). Such income gains tend to increase over time, with acquisition of language skills leading to better improved labour market integration (Chiswick and Miller 1995; for further evidence see UNDP 2009: 50). Apart from salaries, migrants are also often motivated by factors such as better prospects for their children, better education and health care, improved security and better working conditions. This highlights that migration is generally a 'rational' strategy despite the significant costs and risks it may may entail.

However, this does not mean that remittances necessarily alleviate poverty in origin countries. Because migration tends to be a selective process requiring significant resources, *most* remittances flow neither to the poorest countries nor to the poorest members of societies and communities. These 'selection effects' become even stronger for long-distance migration to rich countries, which generally entails much higher costs and risks. For instance, in the Central Plateau region of Burkina Faso, poorer families embark on short-distance migration (such as to Mali) primarily for the purpose of income diversification (consistent with NELM theory), whereas overseas, intercontinental migration (such as to Europe) is only accessible for comparatively wealthy households, which allows them accumulate more wealth through remittances (Wouterse and Taylor 2008; Wouterse and Van den Berg 2011).

Large-scale statistical studies have nevertheless shown that international migration and remittances do reduce poverty in origin countries (Adams and Page 2005). In this context, Shaw and Ratha (2016) drew attention to the importance of 'South–South' migration. Although migration between developing countries generally goes along with smaller income gains, such migration is often more accessible for the relatively poor, and even small increases in income can have highly positive welfare implications for the poor (Shaw and Ratha 2016). However, the costs of South–South remittances are even higher than those of North–South remittances, because of factors such as lack of competition in the remittance market and high foreign exchange costs (Shaw and Ratha 2016).

Internal migration is even more accessible for the relatively poor, and much more massive in scale than international migration (see Chapter 1). *Internal* or *domestic remittances* can therefore play a vital key role in securing and improving livelihoods in rural areas of developing countries (De Brauw *et al.* 2013; Oucho 1996). In most countries, internal migration leads to significant income gains (Harttgen and Klasen 2009). Studies across a range of countries suggest that internal migration has enabled many households to lift themselves out of poverty (Housen *et al.* 2013; UNDP 2009: 50). The poverty-reducing potential of internal migration is generally larger because internal migration involves lower costs and is therefore less selective than migration across international borders. For instance, internal migrants in Bolivia experienced significant real income gains, with more than fourfold increases accruing to low-educated workers (Del Popolo *et al.* 2008; Molina and Yañez 2009).

In 2013, the roughly 250 million internal Chinese migrants were estimated to remit home around 160 billion US$ (Li and Wang 2015). This would mean that Chinese internal remittances alone were equal to 41 per cent of global international remittances (393 billion US$) in the same year. In China, internal remittances comprise between 20 and 50 per cent of the income of remittance-receiving households in rural areas and are crucial in sustaining household expenses and rural development (Cheng and Xu 2005, cited in Li and Wang 2015). In India, in 2008 about 58 per cent of domestic migrants have been estimated to send money home with an average remittance size of about 200 US$ per year (Wang 2018). Internal remittances can thus be important safety nets for relatively poor areas, even though the poorest of the poorest of households seem to receive fewer internal remittances, as even internal migration requires significant costs and risks, although less so than international migration.

The effects of migration on inequality in origin communities thus primarily depend on *who* migrates (the 'selectivity' of migration). For instance, if migrants are from relatively wealthy families, migration and remittances tend to reinforce socioeconomic inequality in origin communities (Jones 1998a). If migrants are from less wealthy sections of the population, or if 'selectivity' decreases because the relatively poor are increasingly able to migrate through the establishment of migrant networks, the effects can be neutral or become positive over time. One of the potential advantages of labour recruitment is that it can make migration accessible for poor and oppressed groups. In South Moroccan oases, internal migration and labour recruitment by French, Dutch and Belgian businesses have been important avenues for upward socioeconomic mobility for the *haratin*, a low-status ethnic group mainly consisting of black peasants and sharecroppers (de Haas 2006c).

Remittance investments

Remittances potentially enable households to invest in enterprises such as the family farm or small industry. There is also reason to cast migrants' inclination to spend money on consumption in a more positive light than is often done. From a capabilities point of view on development as proposed by Sen (see above), it is difficult to dismiss expenditure on education, health, food, medicines and housing as 'non-developmental' as such expenditures tend to significantly improve people's well-being. Moreover, remittance

Source: Hein de Haas

Photo 14.2   Tinghir, a migrant boomtown in Southern Morocco, where remittance investments have fuelled rapid urban expansion

expenditures on housing, services and local products can have significant economic multiplier effects by generating employment and income for non-migrants (de Haas 2006c; Taylor 1992).

In this way, also the non-migrant poor often benefit *indirectly* from migration. In emigration regions, for instance, construction of houses often stimulates economic growth and creates significant employment and income for non-migrants. Besides their intrinsic well-being-enhancing function, expenditure on lavish feasts, weddings and funerals can have economically beneficial multiplier effects by stimulating local demand (Mazzucato *et al.* 2006). A typical phenomenon in emigration regions is the large-scale construction of houses funded with remittances, which tends to generate considerable employment in the construction sectors. The employment and income created by such expenditures and investments by migrants helps to explain the poverty-reducing effect of remittances.

Remittance-fuelled regional growth partly explains why emigration regions in countries such as Mexico, Morocco and Turkey have become destinations for internal migrants from more peripheral regions in their own right, in a pattern reminiscent of 'replacement migration'. In the South-Moroccan Todgha valley, for instance, remittance expenditure and investments by migrants have stimulated the urbanizing regional economy and the rapid growth of Tinghir, its main urban center, which has generated employment and drawn in increasing numbers of migrants from its hinterland in the Saghro and Atlas mountains (de Haas 2006c). More generally, in rural Morocco, the preference of migrants to build houses, invest and resettle (upon return) in towns in their native regions (rather than in origin villages) has spurred the development of new urban centres around crossroads, markets and administrative centres, in a process of what (Berriane 1997) has called a 'micro-urbanization' of rural areas.

Besides this role of individual remittances in encouraging local economic dynamism, *hometown associations* and other migrant groups send *collective remittances* to be used for development projects for the benefit of the community of origin (Goldring 2004; Lacroix 2005). These usually take the form of basic infrastructure and communication projects such as the construction of roads, bridges, potable water systems, drainage, wells, electrification, telephones and so on. Nonetheless, collective remittances are only a fraction of those sent back individually to families (Orozco and Rouse 2007) and their potential development impact should therefore not be overestimated.

Criticism of migrants' ways of spending money often reveals an inability to comprehend the difficult conditions that prevail in countries of origin. For instance, policy makers and researchers have often castigated migrants for building large, richly ornamented houses which they consider 'exaggerated' (Ben Ali 1996: 354), reflecting an 'irrational' (Aït Hamza 1988) use of money on unnecessary status symbols. This is typically accompanied by a call for policies to divert remittances to 'productive' sectors of the economy by guiding migrants towards more 'rational' investment behaviour. However, besides the direct wellbeing-enhancing value of decent housing, this misses the point that houses are often a relatively secure and, hence, rational investment in generally insecure investment environments, particularly if legal systems are dysfunctional and corrupt and do therefore not adequately protect ownership rights.

In his study on remittance use in rural Egypt, Adams concluded that 'from the standpoint of the individual, housing expenses should be classified as an "investment", since new and improved housing offers possible future economic returns to the individual' (Adams 1991: 705–706). In addition, better housing can add to the safety and quality of life and the social status of families. Dismissive views about migrants' supposedly irrational consumption patterns also tend to be patronizing. For instance, by suggesting that migrants' families should stay in their traditional houses, policy makers apply different standards to 'other', poor and rural people, than they would probably do to themselves.

Brain drain or brain gain?

Debates on the 'brain drain' have also become more nuanced (see Docquier and Rapoport 2012; Gibson and McKenzie 2011; Özden and Phillips 2015). Only in a limited number of smaller countries, primarily in the Caribbean, the Pacific and some smaller African states have a large share of the high-skilled workers emigrated (see Adams 2003; Gibson and McKenzie 2011). Furthermore, if large-scale emigration of the highly skilled occurs, this is rather a symptom of structural problems rather than its main cause. For instance, low health staffing levels in some sub-Saharan African countries are primarily the result of unattractive working conditions and the failure to provide basic health services, which do not require highly trained personnel (Clemens 2007), and research has indicated that most elite health workers would not have provided basic health care to those most in need if they had stayed (DRC 2006).

In addition, many high-skilled workers have received at least part of their training abroad, which can potentially benefit origin countries. This casts doubt on the assumption that emigration *automatically* represents a permanent loss. What initially seems to be a loss, may become a gain over the longer term. Established migrants may support families over sustained periods, and migrants' ability to engage socially, politically and economically in countries of origin actually *increases* as their occupational situation in destination societies improves (Guarnizo *et al.* 2003). In fact, many governments deliberately create surpluses of certain categories of the highly skilled as part of a 'labour export' strategy. For instance, the Philippines government encourages education of nurses, for the explicit purpose of generating remittances.

A brain drain can also be accompanied by *brain gain*, which occurs when the prospect of moving abroad motivates those staying behind to continue education, through which the net effect of emigration on origin country education levels may actually be positive (Lowell and Findlay 2002; Stark *et al.* 1997). However, this only seems to occur if the opportunity to migrate increases the economic returns to education. Migration can also create *negative* incentives for education in cases of low-skilled, often irregular migration, where few positive returns on education can be expected (McKenzie 2006), and a significant *brain waste* occurs (Pires 2015). For instance, if Mexican or Moroccan university graduates end up working as undocumented migrants picking tomatoes in Californian or Spanish greenhouses, this provides little incentive for prospective migrants to continue education.

Social remittances

Besides the effects of migration on incomes and economic structures, migration can have profound impacts on social and cultural change in communities and countries of origin. Levitt (1998) coined the term 'social remittances' to describe ideas, behaviours, identities and social capital flowing from destination from origin communities. Migration and the associated confrontation with other values, norms and practices, as well as increasing awareness of opportunities and lifestyles elsewhere, can have profound effects on identity, aspirations and behaviour of people living in origin communities.

If migration becomes strongly associated with social and material success, migrating can become the norm rather than the exception, and staying home can become associated with failure, which can give rise to a 'culture of migration' (Massey *et al.* 1993). Such migration-fuelled cultural change can further strengthen migration aspirations along established pathways in communities and societies that can become obsessed with migration. In countries characterized by economic stagnation, inequality and corruption, securing a way to migrate overseas – such as by marrying a migrant or securing a work permit – opens unprecedented opportunities (see Ballard 2003; de Haas 2003). However, particularly in a context of restrictive immigration policies, the failure to do so can give rise to widespread frustrations among young generations of 'involuntarily immobile' (Carling 2002; de Haas 2003).

Migration is often held responsible for the disruption of traditional kinship systems and care structures (see Hayes 1991; King and Vullnetari 2006). However, on the other hand remittances often enable families to substantially improve their material well-being and overall livelihood security. Large-scale migration inevitably accelerates social and cultural change, and this has often positive as well as negative sides, also depending to which groups people belong, as some people benefit disproportionally from migration, while others may not benefit, or actually lose out, such as through remittance-fuelled increases in local prices, the disintegration of community care structure or traditional systems for collective land and water management.

The effect of migration on gender roles

Migration also affects the position of women and gender roles in origin communities. Such impacts can be more positive or negative, depending on the specific context, and particularly the question whether women have access to remittances or can migrate themselves (de Haas and van Rooij 2010; Gammage 2004; Taylor 1984). For instance, the out-migration of men has affected the position of women left behind in Moroccan rural areas, but does not seem to have fundamentally changed traditional gender roles. Although their husbands' migration temporarily increased their tasks and responsibilities, most women perceived this as a burden and this should therefore not be equated with emancipation in the meaning of making independent choices against prevailing gender norms, as women preferred to avoid overt rule-breaking in order to secure their social 'fall-back' position (de Haas 2003; de Haas and van Rooij 2010).

Menjívar and Agadjanian (2007) made similar observations based on their research among women who stayed behind in Armenia and Guatemala when their husbands migrated abroad. Along with remittances came strong pressures on the part of the husband for his wife to cease working outside the home, eliminating a source of power and independence. For Mexico, Amuedo-Dorantes and Pozo (2006) observed that remittances indeed reduce the female labour supply.

However, if women are able migrate themselves, it does often enable them to improve their economic position and power in origin communities. For instance, in rural Ethiopia women see temporary migration as domestic workers to the Gulf countries as an important avenue towards financial improvement, the ability to start a small enterprise and social independence. Their generally quite positive experiences challenge dominant narratives of trafficking, deception and victimization around this type of migration (Schewel 2018) (see also Chapter 2). In a similar vein, a study amongst migrant women from eight sub-Saharan African countries found that female remittances are substantial, and substantially increase their leverage in household decisions, remittance use and sponsoring of foster children from their side of the family (Eloundou-Enyegue and Calves 2006).

Demographic and political change

Another example of social remittances is the effect of migration on demographic behaviour. For instance, Fargues (2011b) argued that migration from Turkey and Morocco to West-European countries has contributed to the diffusion and adoption of European marriage patterns and small family norms, and so has played a significant role in accelerating the demographic transition. In the case of Egyptian migration to conservative Gulf countries, the effect of migration was the reverse, as it slowed down the transition to lower birth rates (see also Beine *et al.* 2013).

Migrants can also affect political processes in origin countries, either through activism from abroad or through direct participation in local, regional or national politics. Under certain circumstance, migrants can be a potential force for structural political change, for the better or worse. Throughout history, exile has been an important route for religious reformers, political opponents and revolutionaries to develop a power base from abroad. There also evidence that foreign-educated students can promote democracy in origin countries, but only if the foreign education is acquired in democratic countries (Spilimbergo 2009). Migrants and refugees can play a significant role in political reforms and post-conflict reconstruction in countries of origin. However, migrants may also contribute to violent conflicts, for instance by providing support for warring parties in countries such as Sri Lanka and Somalia (Nyberg-Sorensen *et al.* 2002; Van Hear 2004).

The political impacts of migration are thus ambiguous. Because migrants can have all possible political colours as well as religious, ethnic and class backgrounds, the political impacts of migration cannot be easily predicted. Largely depending on who migrates, *political remittances* can either challenge or reinforce the status quo. Migrants are often from middle-class or elite groups (Guarnizo *et al.* 2003), and may therefore not necessarily represent the views of the poor and the oppressed. As much as contributing to reform and greater equality, and the education and knowledge acquired abroad by elite youth can also be used for 'improving' internal oppression upon return.

Reform as a condition for migration and development

To summarize a rich and diverse literature, we can conclude that migration has enabled numerous families and communities around the world to secure and enhance their liveli-hoods. This is a testimony to the willpower of migrants to improve their own destinies, and the agency they deploy to achieve these goals. Internal and international migration has enabled millions of households to improve their living standards, health, housing, education and future prospects. Yet this does not mean that migration *alone* can set in motion processes of national economic and human development at the national level. As Papademetriou and Martin (1991) argued, there is no automatic mechanism through which migration and remittances result in development.

Migration and remittances have clear positive effects on reducing poverty in origin areas, on the welfare of migrants and their families, and on global GDP through increasing efficiency and productivity, but statistical studies have failed to find a direct effect of remit-tances on national economic growth in origin societies (Clemens and McKenzie 2018).

Source: Getty Images/Bloomberg

Photo 14.3   Tiny Owl employees work at computers inside the company's head office in Mumbai, India, in March 2015. Tiny Owl is a smartphone application that helps hungry city-dwellers scour nearby eateries for deliveries

To expect that migration alone can overcome structural development problems would be to overestimate the transformative potential of migration: after all, over the past decades remittances only represented about 4 per cent of the GDP of all developing countries. Although their importance for families and communities of origin can be huge and life-changing, and explains why migrants are often determined to succeed, the macro-economic importance of remittances is thus limited. Furthermore, a one-sided dependency on remittances is generally a sign of the weak state of national economies rather than a cause for celebration, although migration should not be identified as the culprit of such problems.

The crucial issue is that the extent to which migration can play a positive role in change in origin countries depends on more general development conditions, which individual migrants generally cannot change fundamentally. Under unfavourable conditions migration may actually undermine prospects for development. In the poorest countries international migration often remains a prerogative of the relatively better-off and may therefore deepen existing socioeconomic inequalities. Governments of middle-income countries such as Mexico, Morocco, Egypt and the Philippines have used migration as a political 'safety valve' to reduce unemployment, poverty and political unrest (Castles and Delgado Wise 2008; de Haas and Vezzoli 2010; Gammage 2006). Migration can therefore become an instrument to diminish pressure for structural reform and to sustain the position of elite groups, although authoritarian governments typically remain nervous that emigrants may raise their voice from abroad, and will do everything to oppress such voices.

The broader conditions under which migration occurs largely determine the extent to which the considerable development *potential* of migration can be fulfilled. Migration is a resource that tends to give people enhanced access to social, human and financial resources. It is this very capabilities-enhancing potential of migration that also increases the freedom of migrants to choose, either to invest in, or to withdraw from, origin countries, reunify their families and settle permanently at the destinatino - in which case they 'vote with their feet'. Whether they will invest or not and whether migrants will return or not, largely depends on (1) the political and economic situation in origin countries and the (2) position of migrants in destination countries. If origin countries offer little hope of progress, experiences are more likely to reflect the pessimistic paradigm. Under unfavourable conditions, emigration may also undermine the growth of a critical mass necessary to enforce political reform, which is exactly why ruling elites have often encouraged emigration of potential troublemakers.

However, if development in origin countries takes a positive turn, if fundamental political reform happens, if trust in governments increases and economic growth starts to take off, migration is more likely to play the positive role predicted by the migration optimists. Migrants are likely to be among the first to recognize such new opportunities, and reinforce and accelerate these positive trends through investing, trading, circulating and returning. In this way the vicious cycle of the 'migrant syndrome' (see above) can be transformed into a virtuous cycle in which migrants reinforce positive development trends. However, migrants alone cannot be expected to make that change.

Over the past decades, such mutually reinforcing interactions between migration and development have occurred in countries as diverse as Spain, Ireland, Taiwan, South Korea, India and Turkey. For instance, Taiwan experienced substantial loss of high-skilled people in the 1960s and 1970s, but when Taiwan's high-tech sector took off, the government was able to attract back experienced nationals from the US (see Newland 2007). Similarly, India set up Institutes of Technology from the 1950s on to support national development, but many of the graduates emigrated to the US and other rich countries. However, many IT experts later returned which boosted growth of India's own fast-growing IT sector (Khadria 2008; Rao and Balasubrahmanya 2017).

The important point to stress here is that migration was not the factor that triggered national development but, rather, that development enabled by structural political and economic reform unleashed the development potential of migration. So, it is essential to get the causality right. As Heinemeijer *et al.* (1977) already observed more than four decades ago, development is a *condition* for investment and return by migrants, rather than a *consequence* of migration.

Policy implications

Since the 1960s, governments have attempted to stimulate remittances and enhance the development impacts of migration, but such policies seem to have marginal effects as long as general investment environments remain unattractive. Policies to 'channel remittances into productive uses' are often based on rather condescending views that migrants behave irrationally, and miss the point that migrants have good reasons *not* to invest in risky enterprises if investment conditions are unfavourable and property rights not adequately protected. Such discourses also contain the rather problematic presumption that remittances can be 'tapped' by governments; while in fact remittances are private money – the result of migrants' hard work and frugal lifestyles under often difficult conditions. In addition, many migrants have a deep-seated and justified distrust of governments. It would therefore be naïve to assume that governments can deploy this money as if it were their own.

It is also important to be aware of the normative assumptions underlying some policy discourses which celebrate migrants as development pioneers. The idea that the transfer of the 'right' – that is, Western – attitudes and forms of behaviour ('social remittances') would bring about positive change goes back to the nineteenth-century idea of the 'civilizing mission' of Europe in the colonies. It was also central to the modernization theories of the 1950s and 1960s, according to which: '[d]evelopment was a question of instilling the "right" orientations – values and norms – in the cultures of the non-Western world so as to enable its people to partake in the modern wealth-creating economic and political institutions of the advanced West' (Rostow 1960: 230). More recently, neoliberal ideologies also assume that Western models of privatization and deregulation are the key to development. A certain scepticism about the assumed benefits of importing 'Western' attitudes therefore seems justified. More fundamentally, perhaps, it is important to realize that 'many ideas that are taken to be quintessentially Western have also flourished in other civilizations [and that these] are not as culture-specific as is sometimes claimed' (Sen 2000).

Governments and international organizations often advocate temporary migration as a 'win-win-win' strategy benefiting origin and destination countries as well as the migrants themselves. However, the stated development intentions of such programmes can camouflage an agenda of expelling undocumented immigrants or rejected asylum seekers after providing them some modest financial assistance or rapid and often ineffective vocational training (Weil 2002). Return is notoriously difficult to enforce, and as we have seen, restrictive immigration policies tend to paradoxically *reduce* return by pushing migrants into permanent settlement (Castles 2004; Czaika and de Haas 2017; see Chapter 11). The challenge is to design policies that encourage circulation: understanding how circular migration works when it develops spontaneously — and tailoring policies to fit those patterns of mobility. A key element of more successful programmes such as the Canadian Seasonal Agricultural Workers Program (SAWP, see Chapter 7) is guaranteeing repeat access for workers who comply with the programme's terms (to prevent 'overstaying') and to ensure the portability of social security and pension benefits (Newland *et al.* 2008). However, such programmes could only work for a limited range of occupations and require a genuine will to collaborate amongst origin and destination country governments.

The idea that remittances can be an alternative to government policies is equally naïve. Development is a complex and multifaceted process, requiring structural social, political and institutional reform, which cannot be achieved by individual migrants or remittances alone, and requires active state intervention. Notwithstanding their considerable blessings for individuals, households and communities, migration and remittances are no panacea to solve more structural development problems. If states fail to implement general social and economic reform, migration and remittances are unlikely to contribute in nationwide sustainable development (Gammage 2006; Taylor *et al.* 2006).

While stressing the developmental potential of migration, the empirical evidence also highlights the complexity, diversity and socially differentiated nature of migration-development interactions. This also provides a warning against naively optimistic views on migration and development, showing the need to be aware of the real but fundamentally limited ability of individual migrants to overcome structural constraints and, hence, the paramount importance of the more general development context in determining the extent to which the significant development potential of migration can be realized.

From this, some policy lessons can be drawn. First, significant gains are still to be made by making remitting money cheaper, particularly by breaking the monopolies or oligopolies of remittance transfer firms. Second, however, policies to facilitate remittances, to engage 'diasporas' and to encourage investment and return by migrants will have only limited effects if they are not accompanied by general reform in origin countries. Discourses celebrating migration, remittances and transnational engagement as self-help development 'from below' seem partly driven by neoliberal ideologies that shift the attention away from structural constraints and the limited ability of individual migrants to overcome these. Policies that improve infrastructure, legal security, governmental accountability and macroeconomic stability, while countering corruption and improving access to public education, health and credit, are crucial not only for creating positive conditions for development, but *also* for encouraging migrants to return and invest. This exemplifies the crucial role that states continue to play in shaping favourable conditions for human development to occur.

Third, destination country governments can increase the development potential of migration by lowering thresholds for immigration, particularly of the lower-skilled, and through favouring the socioeconomic integration of migrants. By deterring the relatively poor from migrating or, rather, forcing them into vulnerable or irregular positions, current immigration policies limit the potential of migration to alleviate poverty and stimulate development in origin countries. This also exposes one of the main problems standing in the way of achieving enhanced international governance of migration issues, since, for various reasons, destination country governments are wary of giving more rights to lower-skilled migrants.

The latter argument shows the importance of linking theories on immigrant settlement and incorporation to theories on migration and development. As much as conditions in origin societies affect people's decisions to stay or migrate, conditions in destination societies will affect migrants' capabilities and aspirations to engage with – and return to – to origin societies. Integration and transnational engagement are not necessarily substitutes but can reinforce each other, since a better position of migrants in destination societies will also increase their capacity to maintain ties with origin societies.

Conclusion

It would be an illusion to think that migration and remittances can reduce international inequality or can solve structural development problems in origin countries. The evidence is clear: most economic benefits of migration go to destination countries, primarily through the increase of labour force, innovation and productivity (Boubtane *et al.* 2013; Ortega and Peri 2013). Most of these benefits accrue to the middle- and higher-income groups. Lower-skilled local workers see few, if any, economic benefits from migration. Particularly, previous migrants may see their wages decreased (Ottaviano and Peri 2012), essentially because they compete for the same jobs as recent migrants (see Chapter 12).

While the macroeconomic benefits of migration for destination countries are considerable, there is no evidence that emigration stimulates macroeconomic growth in origin countries. In addition, most remittances flow to middle-income rather than low-income countries. Also, within countries and communities, it is not the poorest, but the relatively better-off middle groups that benefit most from migration. However, it is important to realize that for the migrants and their families and communities, the gains of migration are often huge. In fact, for countless people in developing countries, migration has made the key difference in creating new opportunities and improving the long-term well-being of their families.

This chapter has discussed how migration engenders social, economic, cultural and political change in communities and societies of origin, and has particularly reviewed the potential of migration and remittances to stimulate development in origin countries. A first central argument is that migration is no panacea to solve more structural development problems. Migrants and remittances can neither be blamed for a lack of development ('brain drain'), nor be expected to boost sustainable development ('brain gain') in generally unattractive investment environments. Despite its obvious benefits for families and communities at the micro- and meso-level, migration alone cannot generally remove structural development obstacles at the macro-level, such as corruption, nepotism and failing institutions. Migration is unlikely to *reverse* structural development trends, whether negative or positive. If states fail to implement reform, migration is unlikely to fuel national development – and can actually sustain situations of dependency, underdevelopment and authoritarianism. In that case, migrants are also more likely to settle permanently and to reunify their families at the destination. However, if development in origin countries takes a positive turn, migrants often reinforce these positive trends through investing and returning.

A second central argument is that migration represents a vital contribution to securing and improving the livelihoods of millions of families and communities in origin countries. Particularly, remittances fulfil a crucial function in improving conditions in origin countries and migration opens vital avenues to improved living conditions, housing, nutrition, healthcare and education for migrants and their families. Particularly in contexts of high inequality and corruption, people have good reasons to pin their hopes upon migration. What may appear as unattractive jobs for native workers in destination countries is often a 'game changer' for migrant workers and their families, often allowing them to increase their incomes. Migration is often a family investment in a better future, which often opens unprecedented opportunities, and this also explains why migrants are willing to endure considerable suffering and hardship in order to achieve their long-term goals, because it is socially and psychologically impossible to return empty-handed. This insight is crucial for understanding why states often find it difficult to control migration and why even migrants who are deported often try to return.

Guide to Further Reading

The 2009 Human Development Report *Overcoming Barriers: Human Mobility and Development* (UNDP 2009) gives a comprehensive overview on how migration affects development and growth in destination and origin societies, available in several languages online (www.undp.org). Bastia and Skeldon (2020) is a comprehensive collection of papers giving up-to-date insights into the links between migration and various dimensions of development. De Haas (2010b, 2012) provides an overview of theories on migration and development and critically discusses policy approaches. Taylor *et al.* (1996b, 1996a) (both appeared later as chapters in Massey *et al.* (1998)), Agunias (2006) and Katseli *et al.* (2006) provide reviews of the development impacts of migration and remittances. Docquier and Rapoport (2012) discuss evidence on the 'brain drain' from an economic perspective. Newland, Agunias and Terrazas (2008) give useful insights into factors that make (and unmake) effective circulation migration programmes. Castles and Delgado Wise (2008) examines the experiences of five major emigration countries. Stark (1991) is a collection of papers on the new economics of labour migration (NELM), while Taylor (1999) provides a useful introduction. The Migration and Remittances website of the World Bank contains useful data and research on remittances (www.worldbank.org). Adams (2011), Chami *et al.* (2008), Stark (2009) and Yang (2011) provide good discussions of the determinants and impacts of remittances. The Web Anthology on Migrant Remittances and Development (www.ssrc.org) is a useful online resource giving overviews of research insights.

Extra resources can be found at: **www.age-of-migration.com**

15 Conclusion

This book has argued that international migration is a constant, not an aberration, in human history. Population movements have always accompanied economic transformations, technological change, demographic transitions, geopolitical shifts, conflict and warfare. Over the last five centuries, migration has played a major role in colonialism, industrialization, nation state formation and the development of the capitalist world market. However, international migration has never been as pervasive, or as politically significant, as it is today. Never before have political leaders accorded such priority to migration concerns.

The hallmark of the age of migration is the global character of international migration: the way it affects more and more countries and regions, and its linkages with political, economic and cultural transformations affecting the entire world. This book has endeavoured to investigate the principal causes, processes and effects of international migration. Contemporary migration patterns, as discussed in Chapters 6, 7, 8 and 9, are rooted in historical relationships between societies highlighted in Chapter 5, and shaped by a multitude of political, demographic, socio-economic, geographical and cultural processes. Migration has contributed to increased ethnic diversity in countries and deepening transnational linkages between states and societies. International migrations are greatly affected by governmental policies as migrations are often the direct or indirect consequence of decisions to invade countries, to recruit workers or to admit refugees.

Yet international migrations may also possess a relative autonomy and be impervious to governmental policies. Official policies often fail to achieve their objectives, or even bring about the opposite of what is intended. People as well as governments shape international migration. Decisions made by individuals, families and communities play a vital role in determining patterns of migration and settlement. The social networks which arise through the migratory process tend to lower the costs and risks of migration, and tend to stimulate more people to migrate from particular origin areas to particular destinations. If recruitment is suspended or borders closed, networks then provide a vital social infrastructure facilitating the continuation of migration, for instance through family or unauthorized migration. In this way, once a certain number of migrants have settled at the destination, migration processes tend to gain their own momentum. Such network dynamics are powerful meso-level forces, explaining the continuation of migration partly independent from its initial structural causes and often defying attempts by governments to regulate migration. At the same time, significant sections of the populations of receiving countries may oppose immigration. As seen in Chapters 10 and 11, governments sometimes react by adopting strategies of denial, hoping that the problems will go away if they are ignored. In other instances, mass deportations and repatriations have been carried out. Governments vary greatly in their capacities to regulate international migration and in the credibility of their efforts to regulate unauthorized migration.

In Chapter 3, we discussed theoretical perspectives on the causes of migration, and in Chapter 4 we discussed how migration tends to lead to permanent settlement and the formation of distinct ethnic groups in destination societies, while Chapter 14 reviewed how migration affects development and social transformation in origin societies. We argued that migration should be understood as an intrinsic part of broader processes of development and social transformation (instead of a 'problem to be solved', or, conversely, as a solution to fundamental problems such as development failure), and to see people's mobility decisions as function of their capacities and aspirations to move, instead of a passive reaction to 'push' and 'pull' factors or a mere function of international inequalities.

Processes of socioeconomic development, increasing education and improvements in transport and infrastructure initially tend to increase levels of internal and international mobility because they increase people's capabilities and aspirations to migrate. The poorest of the poor are generally unable to migrate over large distances, which is why warfare, natural disasters and economic crises often have an immobilizing effect on the most vulnerable populations. This is also the main reason why the idea that climate change will in the future lead to massive international migration is unrealistic. Because migration requires considerable resources, it is no coincidence that countries with the highest levels of international out-migration are typically middle-income countries, such as Mexico, Morocco, Turkey and the Philippines over the past decades. It is therefore likely that future development in the poorest countries, such as in sub-Saharan Africa or South Asia will lead to more, not less, migration. At the same time, labour market dynamics in high- and middle-income countries will continue to create a structural and inevitable demand for lower- and higher-skilled migrant labour.

This defies outdated push–pull and neoclassical models, and shows the need for better theories that understand the nature of social transformations and how these are associated with migration patterns. These insights also show why migration is inevitable, irrespective of political preferences. The migratory process needs to be understood in its totality as a complex system of social interactions with a wide range of institutional structures and informal networks in origin, transit and destination countries. Certainly in democratic settings, legal admission of migrants will almost inevitably result in some settlement, even when migrants are admitted temporarily. But even for authoritarian states such as those of the Gulf it is increasingly difficult to prevent settlement from taking place.

Acceptance of the inevitability of *some* level of migration, as well as of permanent settlement and formation of ethnic groups, as part of the way contemporary societies are changing is the necessary starting point for any meaningful

Photo 15.1   Advertisement for a bus company offering direct services to various European countries, Todgha valley, Morocco 2000

Source: Hein de Haas

consideration of desirable public policies. The key to adaptive and effective policymaking in this realm (as in others) is a real understanding of the causes and dynamics of migration. Policies based on misunderstanding, politics of denial or mere wishful thinking are virtually condemned to fail. Knee-jerk policy responses to 'migration crises' are often motivated by a political desire to show decisiveness to constituencies but often reflect a failure or refusal to understand migration as a *social process*.

Chapter 11 showed that ill-conceived policies can have unintended, counterproductive consequences, such as border closures, which often interrupt circulation, thereby stimulating permanent settlement and consequently triggering large-scale family migration. Hence, if governments decide to admit foreign workers, they should refrain from creating 'guestworker illusions', but instead they should from the outset make provision for the legal settlement of that proportion of the entrants that is almost sure to remain permanently: a consideration that needs to be taken to heart by the governments of countries as diverse as the US, Germany, Japan, Malaysia, Thailand, South Korea, Mexico, Turkey, Saudi Arabia, Poland, South Africa and Côte d'Ivoire at present. The long-term consequences of the failure to do so can lead to severe problems such as the marginalization of migrant populations and ethnic minorities as well as unemployment, poverty, racist tensions and violence.

Today governments and peoples have to face up to some very serious dilemmas. The answers they choose will help shape the future of their societies, as well as the relations between the rich countries of the 'North' and the developing countries of the 'South'. Central issues include:

- Future perspectives for global migration and mobility;
- Improving international cooperation and governance in the migration arena;
- Policies towards irregular migration;
- Regulating legal immigration and integrating settlers;
- The role of ethnic diversity in social and cultural change, and the consequences for the nation state.

Future perspectives for global migration and mobility

When the first edition of *The Age of Migration* was published in 1993, its central concern was with immigration and its effects on advanced industrial economies. We showed how the labour migrations of the post-1945 period had led to (often unexpected, and often unwanted) settlement and minority formation processes, which were challenging ideas on national identity and citizenship. We also showed how migration within and from Africa, the Middle East, Latin America and Asia was growing in volume and significance. Although global migration levels have remained rather constant at around 3 per cent of the world population, the dominant direction of global migrations has undergone profound changes, notably through the transformation of Europe from the world's main sources of settlers and migrants, to a global migration destination.

This fundamental shift has triggered increasing migration from Asia, Latin America and the Caribbean to North America, and from North Africa, the Middle East and former colonies in Asia and the Caribbean to Europe. This has led to an increasing ethnic, cultural and religious diversity of immigrant populations in Europe, North America and other

Source: Getty Images/NurPhoto

Photo 15.2   Bolivian migrants in the streets of Villaverde Alto in Madris, Spain, in June 2017, celebrate the feast of the Jesus of Great Power, a tradition that takes place in the city of La Paz, Bolivia

industrialized countries. Particularly since the 1973 Oil Shock, the Gulf has emerged as a prime destination for migrants from the Middle East, South and South-East Asia as well as North and East Africa. Since 1989, the globalization of migration has progressed rapidly with the integration of an increasing number of countries into international migration systems as a result of the fall of communism, growing international links and, paradoxically perhaps, development in origin countries which has enabled and motivated young cohorts to search for better education, jobs and lifestyles elsewhere, primarily in cities, but also increasingly abroad.

It is impossible to predict exactly how these migration patterns will evolve over the coming decades. However, a number of key geopolitical and economic develop-ments are likely to play a preponderant role in shaping future global migrations. As the historical and regional overviews in chapter 4–9 showed, changes in the structure of global labour demand have been the dominant driving force of international migra-tion, and this is unlikely to change. Migration has played a crucial role in labour force growth in industrialized countries. Student, family and refugee migration often also have a strong economic element. Since the 1970s, much manufacturing employment has been outsourced to low-wage economies. Other types of work, such as in agri-culture and diverse services, cannot be outsourced. As a partial result of neoliberal economic policies, today's advanced economies have also been characterized by the resurgence of exploitative and poorly regulated, often unauthorized work in agricul-ture, services and manufacturing.

While we have seen a notable increase in the migration of the higher-skilled over the past decades, the demand for lower-skilled migrant labour has persisted. Patterns of labour market segmentation by gender, ethnicity, race, origins and legal status force many migrants into precarious forms of employment, characterized by subcontracting, spurious self-employment, temporary and casual work, and informalization, with immigrants typically picking up jobs that natives shun. In parallel, higher-skilled migrants have been increasingly welcomed. Today, such labour market segmentation is proliferating all around the world, and this partly explains why industrializing societies in Africa and Asia also attract significant numbers of labour immigrants, particularly from low-income countries.

Demand for higher- and lower-skilled migrant workers is likely to persist in high-income countries and to increase in fast growing middle-income economies. While economic transformations in middle-income societies such as China, Turkey and Mexico may generate an increasing demand for immigrant labour, their emigration rates are likely to decrease, provided that economic growth continues and political stability is maintained. Such fundamental economic transformations are likely to causes structural shifts in the global migration geography. As several of the top emigration countries of the 1945–2010 period are increasingly becoming destination countries, labour migrants are being recruited from poorer countries. This is likely to trigger a further shift of the international 'labour frontier' (see Skeldon 1997) – the imaginary line separating countries of net immigration from net emigration – into more peripheral zones, particularly in sub-Saharan Africa, South and South-East Asia and the poorer parts of Latin America. This is already visible in fast-increasing emigration from countries such as Ethiopia, Bolivia, Myanmar and Nepal, where increasing education and processes of economic and cultural change have been increasing people's capabilities and aspirations to migrate.

A second, closely connected shift has been the rapid demographic transition to low mortality and fertility and greater longevity in many countries around the world. Declining cohorts of young labour market entrants and increasing age dependency ratios make future labour demand at all skill levels seem likely. According to the medium variant of the UN's *2017 Revision* of *World Population Prospects*, the world population is expected to increase from 7.6 billion in 2017 to 9.8 billion in 2050 and to then grow more slowly to reach 11.1 billion by 2100 (UNDESA 2017). Virtually all of the expected 2.2 billion additional human beings in the coming 20–30 years will be born in the developing world, particularly in Africa. Although this book has shown that demographic factors by themselves do not 'lead' to migration (which would amount to naïve and dangerous Malthusian determinism), ageing in combination with economic growth is likely to sustain the demand for lower- and higher-skilled migrant labour particularly in service, manufacturing and agrarian sectors.

Citizens of wealthy and emerging countries in Europe, North America, East and South-East Asia and the Gulf are therefore likely to have to continue to rely on newcomers from Asia, Latin America and, particularly, Africa, especially in times of strong economic growth. An increasing group of industrialized and post-industrial Asian countries such as Japan, South Korea, Malaysia, Thailand, Singapore and also China are undergoing fast demographic transitions, with fertility rates now well under replacement level. This is linked to a third major shift resulting from the emergence of a multipolar world of regions, characterized by distinctive regionalization patterns. At the same time, the growing political and economic influence of emerging powers such as China, India, Turkey, South Africa, Brazil and Mexico will change the global landscape of migration as they may increasingly attract and compete for lower- and higher and perhaps also lower-skilled migrant labour.

Given the size of its economy, its population and rising global power, developments in China are particularly important to watch. On the one hand, decreasing domestic labour supply and increasing wage costs have already started a relocation of outsourcing of manufacturing to countries such as Bangladesh, Indonesia and Vietnam. On the other hand, this is likely to increase the demand for migrant labour, a process that is already visible in increasing migration from the Philippines, Vietnam, Laos and Myanmar to China while African migration to China is also increasing.

Some other areas that constituted important zones of emigration in the twentieth century, such as Mexico and the Maghreb countries, may become zones of immigration if future reforms stimulate growth and prosperity. At the same time, fertility rates in almost all middle- and low-income countries are decreasing fast. Ageing is rapidly becoming a global phenomenon. UN projections of demographic trends estimate that by 2050, all regions of the world except Africa will have nearly a quarter or more of their populations at ages 60 and above, and that the number of older persons in the world will increase to 1.4 billion in 2030 and 2.1 billion in 2050 (UNDESA 2017), and could rise to 3.1 billion in 2100.

Although the full labour market implications of this transformation will only be felt in one or two generations, particularly in Africa, it may in the longer term decrease the global supply of low-skilled workers that are willing to accept unattractive jobs that natives tend to shun. This might undermine the assumption that there is a 'quasi-unlimited' supply of cheap labour which is underpinning contemporary understandings of migration. As more and more previously isolated and low-income countries such as Ethiopia, Nepal, Lao, Myanmar, Bolivia and Peru are entering the global migration stage, this raises the question of where future workers will come from. Although we cannot predict the future, one scenario is that, by the middle of the twenty-first century, an increasing number of prosperous and ageing countries may be competing not just for highly skilled personnel – as they already do today – but also for low-skilled workers to build their houses, run their services, and look after children and the elderly.

A fourth possible shift is the emergence of more flexible types of international mobility: changes in transportation, technology and culture are making it normal for people to think beyond national borders and to cross them frequently for all types of reasons. While the share of international migrants in the world population has not significantly increased over the past half century, short-term mobility has soared. Mobility for study, tourism, marriage and retirement is assuming greater significance and is also affecting ideas on migration. Mobility implies an opening of borders for at least some kinds of movement, as well as more flexible types of movement, for a variety of purposes, which do not necessarily lead to long-term stay. For the foreseeable future, the world will experience both migrations in the traditional sense and new types of mobility, which will further intensify cultural exchange.

Improving international cooperation and governance

The increased economic importance of migration for many countries combined with public concerns about the social and cultural changes brought about by the inevitable long-term settlement of migrants may further increase the political saliency of migration. This raises the question of whether this might give rise to intensified international cooperation and governance, as has happened with finance (IMF and World Bank), trade (WTO) and many other forms of global connectivity (Held *et al.* 1999). International migration constitutes the most important facet of the international political economy not covered by a global regime for cooperation and governance. The only exception is refugees, which fall under the mandate of the United Nations High Commissioner for Refugees (UNHCR) and the Geneva Convention. However, besides the fact that UNHCR's powers and effectiveness are limited, refugees comprise less than 10 per cent of the international migrant population.

This means that the vast majority of international migrants do not fall under some form of comprehensive international regulation. Achievement of real change through international cooperation remains elusive despite several attempts to achieve this, including the creation of the Global Commission on International Migration and publication of its influential report (GCIM 2005), the convening of a High Level Dialogue on Migration and Development at the UN in 2006, its follow-up through an annual Global Forum on Migration and Development (GFMD) since 2007, and the UN Global Compact for Safe, Orderly and Regular Migration (GCM) which was adopted in Marrakesh in 2018.

For all their merits, these consultative démarches have not resulted in concrete measures towards a regime for improved international cooperation on migration. The unwillingness of states to move forward in this area is also seen in the very poor ratification record of the 1990 *International Convention on the Protection of the Rights of All Migrant Workers and Members of their Families,* passed by the UN General Assembly on 18 December 1990. As mentioned in Chapter 1, only 54 states (out of 193 UN member states) had ratified it by 2018. These were mainly countries of emigration; immigration countries have not been willing to support measures designed to protect migrants. Although the Global Compact for Migration (GCM) is championed as the first ever UN global agreement on migration, it is non-legally binding. It is very difficult to see how the lofty aims of the GCM to optimize the benefits of migration, while addressing its risks and challenges, and respecting human rights, non-discrimination, state sovereignty and responsibility-sharing can be met in practice.

At its core, the main obstacle to achieving cooperation and to effectively implementing agreements is that origin and destination countries often have opposed interests, with the former typically arguing in favour of more liberal immigration regimes particularly for the lower-skilled, while the political pressures in wealthy destination societies tend towards maintaining some level of control of immigration, to favour the immigration of the skilled while discriminating against lower-skilled workers. Such opposed interests also exist within the different world regions. For instance, within the African Union (AU) important destination countries such as South Africa and Côte d'Ivoire generally oppose agreements that would force them to give more rights to immigrants. This is why most progress on international regulation of migration has been achieved on the subregional level, as part of broader processes of regional unification, such as the EU or ECOWAS. The EU is the only major example of a regional union which successfully implemented legally binding instruments not only regulating travel and migration, but also the extension of the full right to work and establishment in member countries.

There are at least four reasons not to expect a global migration regime to emerge anytime soon. First, at least for the next few decades, there will remain a significant supply of foreign labour at the global level. This creates a disincentive to multilateral cooperation between origin and destination countries, as individual states can sign bilateral agreements to recruit foreign labour or (tacitly or openly) tolerate unauthorized entry and stay of foreign workers. However, this situation may change in the long term with the predicted coming stabilization of population and ageing in many world regions.

Second, there is no inherent reciprocity of interests between workers in high-income societies and workers in developing countries. The rich countries perceive little benefit in reciprocity. Their workers generally will not benefit from facilitated entry to less-developed states. Labour movements would be largely unidirectional, from less-developed areas to the more developed areas. Why would the most developed states cede sovereign prerogatives to regulate international migration to establish an international regime?

Third, as Koslowski (2008) has argued, leadership is vital to regime formation. US leadership since 1945 has helped forge liberal international trade regimes in many areas. Neither the US nor any of the other most powerful states have shown much leadership in forging a global regime concerning international migration. To the contrary, the US has been very sceptical about international forums on international migration, something already clearly evident in 1986 when the OECD convened the first major multilateral conference on international migration since the Second World War (Miller and Gabriel 2008). This became even more evident when in 2017 the Trump administration pulled out of the Global Compact for Migration (GCM), arguing that it would interfere with American sovereignty and run counter to US immigration policies. More generally, the influence of the US is still large but declining in an increasingly multipolar world, and it is difficult to envisage which country would be willing to take up this role.

Fourth, political leaders and public debates in immigration countries still generally treat migration as something fundamentally abnormal and problematic. The overwhelming concern seems to be to stop or reduce migration, as if it were inherently bad. Even well-meaning initiatives, like attempts to address the 'root causes' of emigration from poor countries through efforts to achieve 'durable solutions' to impoverishment and violence, are driven by the idea that migration should be reduced (Castles and Van Hear 2012). Political leaders still seem to cling to the naïve belief that development will curtail migration, despite overwhelming evidence that development in poor countries leads to more, instead of less, migration.

As we have shown in this book, migration has taken place throughout history and is caused by – and is an intrinsic part of – fundamental economic change, social transformation and globalization. One sign of this is the growth of highly skilled mobility between advanced economies. Another is the realization that rich countries like the US, UK, Germany and Australia also have large diasporas, which make important contributions to both destination and origin countries. Rather than necessarily reducing migration, a more realistic aim might be to work for greater economic and social equality between and within richer and poorer societies and better protection of migrant rights, so that migration will take place under better conditions and will enrich the welfare and experiences of migrants and communities. This will also increase the capability of migrants to contribute to development in origin countries. Thus, reducing 'unwanted migration' is a credible aim only if it is coupled with the understanding that this may well mean greater mobility overall – but mobility of a different and more positive kind. This would require measures that go well beyond the usual range of migration-related policies focused on addressing short-term political concerns rather than being part of a longer-term strategy.

Genuine reform of trade policies, for instance, could encourage economic growth in less-developed countries. A key issue is the level of prices for primary commodities as compared with industrial products. This is linked to constraints on world trade through tariffs and subsidies. Reforms could bring important benefits for less-developed countries. But trade policies generally operate within tight political constraints: few politicians are willing to confront their own farmers, workers or industrialists, particularly in times of economic recession. Reforms favourable to the economies of the less-developed countries will only come gradually, if at all.

Regional integration – the creation of free-trade areas and regional political communities – is sometimes seen as a way of diminishing 'unwanted' migration by reducing trade barriers and spurring economic growth, as well as by legalizing international movement of labour. As shown in the EU, freeing up migration can be a potentially effective strategy to stimulate circular mobility. After all, the evidence on past and contemporary migration dynamics reveals a crucial paradox and a fundamental policy dilemma: the more migration is subject to regulation and restrictions, the less circular migration tends to take place. This is because where migration entails high costs and risks, migrants find it hard to return home for fear they will not be able to migrate again; they therefore tend to remain for longer periods or even permanently in destination countries.

Both in the US and in the EU, immigration restrictions have pushed supposedly temporary immigrants inadvertently into permanent settlement. So, while restrictive immigration policy encourages more permanent types of migration, free migration policies tend to encourage mobility and circular movement. The effect of free mobility regimes on migration dynamics depends on specific economic, political and historical conditions. For instance, when Spain joined the EEC in 1986, the predicted mass emigration did not occur. Instead, many Spanish migrants returned home. The EU enlargements of 2004 and 2007 stimulated migration from Eastern Europe, but this migration is generally more temporary and circular compared to migration from outside the EU.

Restrictive policies and border controls have failed to significantly curb immigration to wealthy countries. Rather, they have led to greater reliance on increasingly risky and costly unauthorized migration. A commonly presented 'smart solution' to curb immigration is to address the perceived root causes of migration through increasing development aid. But, as we know, economic and human development tend to increase rather than decrease levels of emigration and overall mobility. Trade, aid and remittances tend to be complements to, rather than substitutes for, migration. At the same time, demand for both skilled and unskilled migrant labour is likely to persist. In fact, the focus on origin country development sustains the false suggestion that migration is mainly driven by poverty and inequality. This diverts the attention away from the demand-driven nature of much labour migration, and the fundamental fact that most migrants would not have migrated if there were no jobs and opportunities available for them.

The initial effect of economic and human development of less-developed countries and their integration into world markets is to boost internal and international migration. This is because the early stages of development lead to rural–urban migration and the acquisition by many people of the financial and cultural resources needed for international migration. Reaching the advanced stage of the 'migration transition' – through which emigration declines and immigration increases – requires demographic and economic conditions which generally take generations to develop. Neither restrictive measures nor development strategies can stop international migration, because there are such powerful forces stimulating population movement. These include the increasing pervasiveness of a global culture and the growth of cross-border movements of ideas, capital and commodities. This globalization can perhaps be mitigated or transformed by international cooperation, but it will be almost impossible to reverse. The world community will have to learn to live with large-scale migration for the foreseeable future.

Responding to irregular migration

In recent decades, there has been the emergence of a new generation of temporary foreign worker policies often touted as a way to better manage and substitute for irregular migration. Recently, international organizations have used the more positive label of 'circular migration'. There are many reasons to doubt that such policies will succeed.

Particularly in European countries such as Germany, some observers have suggested a need for increased immigration to deal with labour market shortages caused by low birth rates and an ageing population: foreign workers might provide the labour for elderly care and other services as well as the construction industry. But studies have shown that immigration cannot effectively counteract the demographic ageing of Western societies unless immigration reaches undesirable and unrealistic levels. Political constraints will not permit this. Public opinion may accept entry programmes for highly skilled labour, family reunification and refugees, but not a resumption of massive recruitment of foreign labour for low-level jobs. Most industrial democracies have to struggle to provide adequate employment for existing populations of low-skilled citizens and resident foreign workers.

One of the most pressing challenges for many countries today, therefore, is to find ways of responding to irregular and other forms of 'unwanted' migration. Irregular immigration is a somewhat vague blanket term, which embraces:

- Legal entrants who overstay their entry visas or who work without permission, which is the most important source of undocumented migration;
- Illegal border-crossers, which excludes refugees, because they have the right to cross borders to seek protection, according to international refugee law;
- Family members of migrant workers, prevented from entering legally by restrictions on family reunion;
- Asylum seekers not regarded as genuine refugees.

Many of such migrants come from poor countries and seek employment, but often lack recognized work qualifications. They may compete – or at least be perceived to compete – with lower-qualified local people for unskilled jobs, and for housing and social amenities. Many regions throughout the world have experienced an increase in such immigration over the past decades. Of course, the migration is not always as 'unwanted' as politicians proclaim in public: employers often benefit from cheap workers who lack rights, and often tacitly permit such movements, particularly in times of fast economic growth. There is a significant gap between politicians' 'tough talk', the actual policies and their implementation on the ground.

Appearing to 'crack down on 'illegal immigration' through deporting undocumented migrants, building walls and stricter border checks is increasingly regarded by governments as essential for safeguarding support and winning elections. In several regions of the world, the result has been a series of agreements designed to secure international cooperation in curbing irregular entries, to increase border control and to speed up the processing of applications for asylum. Several countries around the world have carried out quite draconian measures, such as mass expulsions of foreign workers (for example, US, Nigeria, Libya, Malaysia and South Africa), building fences and walls along borders (South Africa, Israel, Hungary, Turkey, India and the US), severe punishments for illegal entrants (even corporal punishment in Singapore) and sanctions against employers (South Africa, Japan and other countries). In addition, non-official punishments such as beatings by police are routinely meted out in some countries.

As long as labour markets continue to generate a demand for migrant workers, the effectiveness of such controls seems to be limited. To understand why these policies have remained popular, it is crucial to understand their important symbolical function. As Massey and his colleagues argued, 'elected leaders and bureaucrats increasingly have turned to *symbolic* policy instruments to create an *appearance* of control' (Massey *et al.* 1998: 288). Yet harsh political discourse on immigration which obscures and denies the real demand for migrant labour can be a catalyst for the very xenophobia and apocalyptic representations of a massive influx of migrants to which they claim to be a political-electoral response.

The difficulty in achieving effective migration control is not hard to understand. Barriers to mobility contradict the powerful forces of globalization which are leading towards greater economic and cultural interchange. In an increasingly international economy, it is difficult to open borders for movements of information, commodities and capital and yet close them to people. Global circulation of investment and know-how always means movements of people too. Moreover, as we have seen in the book, flows of highly skilled workers tend to also encourage flows of less-skilled workers. The instruments of border surveillance cannot be sufficiently fine-tuned to let through all those whose presence is officially wanted, but to stop all those who are not.

Nevertheless, there should be no mistaking that measures like enforcement of employer sanctions have a deterrent effect where there exists the genuine political will to punish unlawful employment of undocumented foreign workers. Because of technological advances in military, surveillance, identification and information technology, states have a higher-than-ever theoretical capacity to control migration. However, such political will is often absent despite the usually tough rhetoric through which politicians proclaim their desire to 'crack down' on undocumented migration and unlawful unemployment. Another problem is that the effective control on many forms of irregular migration and employment would require measures – such as raiding private homes to check whether domestic workers have residence or work permits – that come close to the practices of totalitarian states, are generally in conflict with fundamental human rights, and would not be supported by most people.

The matter is further complicated by a number of factors (see Castles *et al.* 2012): the contradiction between state and market – or more precisely between government rules that officially refuse entry to low-skilled workers and labour market demand for them; the eagerness of employers to hire foreign workers (whether documented or not) for menial jobs when nationals are unwilling to take such positions; the difficulty of adjudicating asylum claims and of distinguishing economically motivated migrants from those deserving of refugee status; and the inadequacies and inflexibility of immigration law. The trend towards economic and labour market deregulation since the 1980s and the weakening of organized labour and declining trade union membership in many Western democracies has also tended to increase irregular foreign employment.

'Neo-liberal' policies aimed at reducing labour market rigidities and enhancing competitiveness may result in expanded employer hiring of authorized and undocumented foreign workers (see Chapter 12). However, higher taxation associated with social welfare policies may also have unintended consequences, encouraging employment of undocumented foreign workers, since this allows employers to avoid paying social insurance contributions. Thus, despite the claimed desire of governments to stop irregular migration, many of the causes of such officially undesirable migrations are to be found in the very political, economic and social structures of destination countries. This explains why governments often cede to employer lobbies to let more legal workers in or turn a blind eye towards irregular migration, and why workplace enforcement is typically very low.

Legal migration and integration

Virtually all democratic states and many not-so-democratic states have growing foreign populations. As shown in Chapters 5 to 9, the presence of these immigrants is either the result of deliberate labour recruitment or immigration policies, or occurs more spontaneously because of the existence of various historical, political, military and economic linkages between origin and destination countries. In some cases (especially in Canada, Australia and New Zealand), policies of large-scale, planned immigration still exist. They are selective: economic migrants, family members and refugees are admitted according to quotas which are politically determined. In other regions, especially Europe, the Gulf and East Asia, migration tends to be driven mainly by employer demand, with governments often playing a facilitating role. In many cases, such demand-driven migration policies favour temporary labour migration over permanent settlement migration. In recent years, 'classical immigration countries' like Australia, New Zealand, Canada and (to a lesser extent) the US have increased their temporary intakes, although often with built-in pathways to permanent status. This is evidence of a growing international convergence in immigration policies.

Planned immigration seems conducive to acceptable social conditions for migrants as well as to relative social peace between migrants and local people. Countries with immigration quota systems generally decide on them through political processes which permit public discussion and the balancing of the interests of different social groups. Participation in decision-making increases the acceptability of immigration programmes. At the same time, this approach facilitates the introduction of measures to prevent discrimination and exploitation of immigrants, and to provide social services to support successful settlement. Although this seems to be an argument in favour of planned immigration policies, the actual trend has been towards the growth of employers' demand-driven migration programmes which primarily respond to an economic logic. Yet more and more societies in Europe, North America, Asia, the Middle East and Africa have been confronted by the increasingly permanent settlement of migrant populations, even though governments may still deny this. This shows the need for measures to safeguard the rights of temporary migrants, and to create policies and legal instruments that provide pathways to permanence and citizenship.

Chapters 10, 11 and 13 highlighted that governmental obligations towards immigrant populations are shaped by the nature of the political system in the receiving society, as well as the mode of entry of the newcomers. Governments possess an internationally recognized right to regulate entry of foreigners, a right that may be voluntarily limited through governmental signature of bilateral or multilateral agreements (for example, in the case of refugees and family migrants). Clearly it makes a difference whether or not a foreigner has arrived on a territory through legal means of entry. In principle, the proper course for action with regard to legally admitted foreign residents in a democracy is straightforward. They should be rapidly afforded equality of socioeconomic rights and a large measure of political freedom, for their status would otherwise diminish the quality of democratic life in the society. However, this principle is frequently ignored in practice. Unauthorized immigration, undocumented residence and employment make immigrants especially vulnerable to exploitation. The perceived illegitimacy of their presence can foster conflict, racism and anti-immigrant violence.

'Guestworker'-style restrictions on the employment and residential mobility of legally admitted foreigners appear difficult to reconcile with prevailing market principles, to say nothing of democratic norms. The same goes for restrictions on political rights. Freedom of speech, association and assembly should be unquestionable. The only restriction on the rights of legally admitted foreigners which seems compatible with democratic principles is the reservation of the right to vote and to stand for public office to citizens. This is only justifiable if resident foreigners are given the opportunity of naturalization, without daunting procedures or high fees. But, even then, some foreign residents are likely to decide not to become citizens for various reasons. A democratic system needs to secure their political participation too. This can mean setting up special representative bodies for resident non-citizens or extending local voting rights to non-citizens who fulfil certain criteria of length of stay (as in the EU and, in the nineteenth century, much of the US).

The global character of international migration results in the intermingling and cohabitation of people from various cultural backgrounds. Older immigration countries have developed approaches to incorporate newcomers into their societies, with a view to making them into citizens in the long run. Many newer immigration countries, for instance in East and South-East Asia (see Chapter 8) as well as the Middle East and Africa (see Chapter 9), reject the idea of permanent settlement and continue to treat migrants as temporary sojourners, however long they stay, and despite the fact that many of these countries have become de facto settlement countries. The long-term presence of new ethnic minorities may eventually compel those countries to come to terms with these new realities, and to accommodate permanent settlement, for instance through facilitating naturalization and developing integration policies.

Chapter 13 analysed the various models of immigrant integration, showing that there are important variations, ranging from 'exclusionary' approaches that keep migrants as a separate (and usually disadvantaged) part of the population, through 'assimilationist' approaches that offer full membership but at the price of abandoning migrants' original languages and cultures, to 'multicultural' approaches that offer both full membership and recognition of cultural difference. The book showed that in the post-1945 decades until the 1990s particularly in Western Europe there was a general trend away from exclusionary and assimilationist models towards multicultural approaches. Changes in citizenship laws to offer easier naturalization for migrants and birthright citizenship for their children were an important sign of change.

However, this trend has been questioned and to some extent reversed since the 1990s. Critics have argued that 'multiculturalism' is detrimental to the economic integration and success of migrants and minority populations, and that it can lead to permanent cultural and political divisions. This has prompted a renewed emphasis on 'national values', loyalty and a call to replace multicultural policies with measures to strengthen 'social cohesion' and national identity. Symptomatic of this trend has been the tightening-up of naturalization rules, restrictions on dual citizenship in some places and the introduction of citizenship tests in several Western countries. Yet, at the same time, many national governments and local councils have maintained the multilingual services and anti-discrimination rules typical of multicultural societies. In some places the rhetoric on multiculturalism seems to have changed more than the reality. At the time of writing, the picture is confused, indicating the persistence of important struggles in the public arena, and the universal difficulties in reconciling the on-the-ground realities of migration and settlement with official ideologies.

Ethnic diversity, social change and the nation state

The age of migration has already changed the world and many societies. Particularly countries in Western Europe, North America and the Gulf have become far more diverse than they were even a generation ago. In fact, few modern nations have ever been ethnically homogeneous. However, the nationalism of the last two centuries strove to create myths of homogeneity. In its extreme forms, nationalism even tried to bring about such homogeneity through expulsion of minorities, ethnic cleansing and genocide. But the reality for most industrialized countries today is that they have to contend with a new type of pluralism brought by about by the immigration and settlement of increasingly diverse populations. This has challenged assimilationist models of migrant integration, as well as the modern political construct of the nation state and the more ethnocentric models of citizenship. Clear-cut dichotomies of 'origin' or 'destination' and categories such as 'permanent', 'temporary' and 'return' migration are increasingly difficult to sustain in a world in which the lives of many migrants are characterized by circulation and simultaneous commitment to two or more societies (see de Haas 2005).

One reason why immigration and settlement of new ethnic groups have become the subject of heated debate is that these trends have coincided with the crisis of modernity and the transition to post-industrial societies. The labour migration of the pre-1973 period appeared at the time to be reinforcing the global economic dominance of the industrial nations in Western Europe and North America. Since then, growing international mobility of capital, the electronic revolution, economies policies focused on deregulation, the decline of old industrial areas and the rise of new ones, and international outsourcing of manufacturing and service jobs, are all factors which have led to rapid change in advanced economies.

The erosion of the old blue-collar working class, labour market segmentation, the increased polarization of the labour force between well-paid workers with secure jobs and low-paid workers with insecure jobs have led to a social crisis in which immigrants find themselves doubly at risk: many of them suffer unemployment and social marginalization, yet at the same time they are often inaccurately portrayed as the cause of the problems by politicians eager to manufacture immigration or integration 'crises' in an effort to deflect the attention away from the fact that declining job security, stagnating real wages for lower-skilled workers and deteriorating public services primarily result from *political* decisions and economic ideologies advocating deregulation and austerity.

That is why the emergence of societies in which the top strata are increasingly afflu-ent while the native working classes have become marginalized and the middle classes are increasingly under pressure, is often accompanied by ghettoization of the disadvantaged and the rise of racism. Nowhere is this more evident than in today's global cities: New York, London, Paris, Istanbul, Dubai, Tokyo, Singapore, Sydney, Shanghai, Hong Kong – to name just a few – are crucibles of social change, political conflict and cultural innovation. Yet they are marked by great gulfs: between the corporate elite and the informal sector workers who service them; between the well-guarded suburbs of the rich and the decaying inner cities or peripheral slum areas of the poor; between citizens of democratic states and 'illegal' non-citizens; between 'expats' and migrant workers; between dominant cul-tures and minority cultures.

The gulf may be summed up as that between inclusion and exclusion. The included are those who fit into the self-image of a prosperous, technologically innovative and demo-cratic society. The excluded are the shadow side: those who are needed to do the menial jobs in industry and the services, but who do not fit into the ideology of the model. Both groups include citizens and immigrants, though the immigrants are more likely to belong to the excluded and to be made into scapegoats. But the groups are more closely bound together than they might like to think: the corporate elite need the unauthorized immi-grants – for instance, to clean and maintain their houses, to cook their meals, to tend their gardens and to take care of their children – as much as the prosperous suburbanites need the slum-dwellers they find so threatening. It is out of this contradictory character of global cities that their enormous energy, cultural dynamism and its innovative capabil-ity emerge. But these coexist with potential for social breakdown, conflict, repression, vio-lence and deep socioeconomic divisions often drawn along ethnic, cultural and racial lines.

These new forms of ethnic diversity affect societies in many ways. Amongst the most important are issues of political participation, cultural pluralism and national identity. Immigration and formation of ethnic groups have already had major effects on politics in most developed countries. The increasing importance of minority voting in the US Presidential Elections (see Box 4.1) has been widely discussed. Immigration and growing diversity have also given impetus to the rise of anti-immigration movements and a resur-gence of nationalism and protectionism, as exemplified by the Brexit vote and the election of Donald Trump as US president in 2016.

The effects of growing immigrant populations and increased ethnic diversity are potentially destabilizing, if long-term residents find themselves excluded from politics. The only solution appears to lie in broadening political participation to embrace immi-grant groups, which in turn may mean rethinking the form and content of citizenship and decoupling it from ideas of ethnic homogeneity or cultural assimilation. The latter notion has turned out to be unrealistic even in the self-proclaimed 'ethno-states' of Europe, Africa, the Middle East and Asia, which have historically had a much higher degree of cul-tural and ethnic diversity than official state ideologies are willing to admit.

This leads on to the issue of cultural pluralism. Processes of marginalization and isola-tion of ethnic groups have gone so far in many countries that culture has become a marker for exclusion on the part of some sections of the majority population, and a mechanism of resistance by minorities. Even if serious attempts were made to end all forms of dis-crimination and racism, cultural and linguistic difference would persist for generations, especially if migration continues, as seems probable. That means that majority popula-tions will have to learn to live with cultural pluralism, even if it means modifying their own expectations of acceptable standards of behaviour and social conformity. In order to preserve social cohesion and prevent racist tensions and violence, majorities as well as minorities need to be incorporated in diverse societies.

If ideas of belonging to a nation have been based on myths of ethnic purity or of cultural superiority, then they are threatened by the growth of ethnic diversity. Whether the community of the nation has been based on belonging to an ethnic group (as in Germany or Japan) or on a unitary culture (as in France), ethnic diversity inevitably requires major political and psychological adjustments, which can seem deeply threatening to majority populations, particularly if politicians portray migrants as a fundamental threat to the prosperity and cultural integrity of the nation.

The shift and need for adjustment may seem smaller for countries that have seen themselves as nations of immigrants, for their political structures and models of citizenship are geared to incorporating newcomers. However, these countries too have historical traditions of racial exclusion and cultural homogenization which still need to be worked through. Citizens of immigration countries may have to re-examine their understanding of what it means to belong to their societies. Monocultural and assimilationist models of national identity are no longer adequate for the new situation, and probably have never really been.

Immigrants may be able to make a special contribution to the development of new forms of identity. It has always been part of the migrant condition to develop multiple identities, at least for one or two generations, which are linked to the cultures both of the country of origin and of the destination. Such personal identities possess complex new transcultural elements, manifest in growing transnationalism and expanding diasporic populations around the world. The strength of transnational social, economic and cultural ties also reinforces the potential of migrants to affect social transformation processes in origin countries through sending money, investing, cultural exchange and political participation.

Immigrants are not unique in this: multiple identities appear to have become an increasingly widespread characteristic of contemporary societies, but multiple belongings have probably always been part of the human condition (see Sen 2007). However, it is above all migrants who are compelled by their situation to have multiple identities, which are constantly in a state of transition and renegotiation. Moreover, migrants and ethnic minorities frequently develop a consciousness of their transcultural position, and the suffering and pain, but also joy and inspiration this can cause, which is reflected not only in their artistic and cultural work, but also in social and political action.

Despite current conflicts about the effects of ethnic diversity on national cultures and identity, immigration does offer perspectives for change. New principles of identity may emerge, which may be neither exclusionary nor discriminatory, and may provide the basis for better intergroup cooperation and understanding, such as through the rise of an increasingly global 'youth culture' over the past 70 years, which has made it arguably easier for young people all over the world to find common reference points around food, music, cinema, literature and other forms of artistic expression. While jazz and other Afro-American music crossed the Atlantic to Europe and elsewhere to amalgamate with local musical styles contributing to the rise of increasingly globalized popular music cultures, Italian, Chinese and Middle Eastern cuisines have transformed eating worldwide.

Diverse and transnational identities are likely to affect political structures. The democratic nation state is a fairly young political form, which came into being with the

American and French revolutions and achieved dominance in the nineteenth century. It is characterized by principles defining the relationship between people and government which are mediated through the institution of citizenship. The nation state was an innovative and progressive force at its birth, because it was inclusive and defined the citizens as free political subjects, linked together through democratic structures. But the nationalism of the nineteenth and twentieth centuries turned citizenship on its head by equating it with membership of a dominant ethnic group, defined on biological, religious or cultural lines. In many cases the nation state became an instrument of exclusion and repression.

National states, for better or worse, are likely to endure. But global economic and cultural integration and the establishment of regional agreements on economic and political cooperation are undermining the exclusiveness of national loyalties. The age of migration could be marked by the erosion or at least the challenging of ethno-nationalism and the concomitant weakening of divisions between peoples. Admittedly there are countervailing tendencies, such as racism, the rise of extreme-right organizations and the resurgence of nationalism in many countries. Coming transformations are likely to be uneven, and setbacks are possible, especially in the event of economic or political crises. But the inescapable central trends are the increasing ethnic and cultural diversity of most societies and the growth of cultural interchange. The age of migration may yet be a period of greater unity in tackling the pressing problems that beset our small planet.

Glossary

3D jobs	'Dirty, difficult and dangerous' jobs, typically low-skilled jobs in the informal sector or in sectors with strong seasonal fluctuation, such as domestic work, farming, road repairs, construction, catering and cleaning (Münz *et al.* 2007) (see *dual labour market theory*).
9/11	2001 terrorist attacks on the World Trade Center in New York and the Pentagon in Washington, DC.
Agency	People's capacity to make their own choices.
Aliens	Term used for foreigners (non-citizens) in the United States.
Anticyclical	Going against trends. Applied to remittances, it refers to the phenomenon of migrants sending *more* in remittances in reaction to economic crisis or environmental shocks in origin areas (complies with *new economics of labour migration*).
Anti-Semitism	Dislike of and prejudice against Jews.
Apartheid	Institutionalized system of racial segregation in South Africa (1948–1994) based on an ideology of white supremacy.
Assimilation	Process by minority groups fully adopting the culture and identity of majority groups, implying abandonment of their ancestors' identities and allegiances. Applies to immigrants and ethnic and religious minorities. Assimilation can be resisted through *social closure*, the development of *ethnic enclaves* and *diaspora* formation (see *incorporation* and *integration*).
Asylum seeker	A person who has applied for refugee status and is still awaiting a decision on her or his recognition as a *refugee*.
Aussiedler	Ethnic Germans living in Eastern Europe and the former Soviet Union; many of them migrated to Germany after the fall of the Berlin Wall.
Banlieues	High-rise estates on the periphery of cities in France, often associated with high concentrations of populations of migrant origin
Bidonvilles	Shantytowns (French) where many migrant workers settled in the post-WWII decades.
Bracero	Manual labourer; also name of the 1942–1964 *guestworker* programme between Mexico and the US.
Brain drain	Loss of knowledge and skills because of emigration of the highly educated.
Brain gain	The positive effect of emigration on educational achievements in origin countries, occurring when the prospect of moving abroad motivates those staying behind to continue education (see Stark *et al.* 1997). More broadly applied to refer to the potential gains of migration for origin countries through remittances, knowledge and ideas brought back by migrants.

Brain waste	Skill downgrading, usually applied to situations where migrants do jobs below their skill levels because of lack of diploma recognition and other discriminatory practices.
Brexit	Withdrawal of the United Kingdom from the European Union.
Business migrant	A person migrating primarily for business purposes.
Capabilities	The ability of human beings to lead lives they have reason to value and to enhance the substantive choices ('freedoms') they have (Sen 1999).
Carrier sanctions	Fines and other sanctions imposed on airlines and other transportation companies to prevent undocumented migrants from boarding airplanes, ships, trains and buses.
Casual employment	Hiring by the hour or for specific tasks, usually without labour contract (see *precarious work*).
Categorical substitution	The use of other legal (or illegal) entry channels in response to a change in migration policy, for instance the switch from labour to family or asylum channels in response to restrictions on labour immigration (see *substitution effects*) (de Haas 2011).
Chain migration	The phenomenon through which initial migration of a few persons leads to more migration, either through informal recruitment of workers or through *primary or secondary family reunion* (see also *migrant networks* and *migration multipliers*).
Circular migration	Forms of migration that involve multiple entries and returns, usually associated with free migration regimes, to be distinguished from *temporary migration*, which does not presume re-emigration.
Citizenship	Legal membership of a sovereign state laying down rights and duties.
Class	A group of people of similar social status based on their education, occupation and culture, as well as ownership and access to means of production and other resources.
Client politics	The process by which small interest groups gain disproportionate influence on political processes benefits at the expense of the larger public (Hollifield 1992).
Cold War	A period of tension between the Soviet Union and the United States and their allies through political pressure, the nuclear arms race and proxy warfare, lasting from 1946 until the fall of the Berlin Wall in 1989.
Collective remittances	Remittances sent by immigrant organizations for development purposes in their origin areas (see *hometown associations*).
Colons	Descendants of French and other European settlers in the Maghreb, particularly Algeria (see *pieds noirs*).
'Coolie'	Pejorative term for an indentured contract labourer, usually from South Asia.
Contract substitution	A practice faced by migrant workers in which original employment terms are changed on arrival at their destination.

Creolization	The emergence of new identities and forms of cultural expression as a result of the mixing of cultures (see Cohen 2007).
Culture of migration	The emergence of a collective mentality in which migration becomes the norm, and in which staying home is often associated with failure, often typical for high-emigration societies (see *involuntary immobility*).
Dehumanization	Extreme form of racism that denies the humanity of other people, or defines them as 'less human', usually serving to legitimize forced labour, enslavement and *genocide* (see Arendt 1951).
Denizenship	A person with rights between those of naturalized citizen and foreigner, generally coinciding with permanent residency.
Departure bonus	Financial incentives used to encourage voluntary return of migrants, for instance used for guestworkers in Europe and *Nikkeijin* in Japan.
Depolitization	Discursive strategies aimed at removing the political dimension from social issues.
Diaspora	Ethnic groups sharing a common experience of dispersal from an original homeland, often traumatically, characterized by a strong group consciousness, empathy, solidarity and the maintenance of 'transversal links' with co-ethnic members in other countries of settlement sustained over multiple generations (Cohen 1997).
Discourse (adj. discursive)	Thought systems and representations of the world as transmitted and reproduced through language, speech, concepts, art, attitudes, practices and actions.
Discursive gap	The gap between the stated objectives of politicians' (often 'tough') rhetoric or 'discourse', and the (often more watered-down) policies on the ground (de Haas *et al.* 2019a)
Downward assimilation	The assimilation of migrant workers and their offspring into precarious labour market segments and vulnerable classes of destination societies (see *segmented assimiliation*) (Portes & Zhou 1993).
Dual labour market theory	Theory explaining how the division of labour markets between protected workers in the primary sector and precarious workers in the secondary sector generates a structural demand for migrant workers (Piore 1979).
Efficacy gap	The degree to which implemented policies have the intended effect on the volume, timing, direction and 'selection' of migration (de Haas *et al.* 2019a).
Embedded legal constraints	The processes through which liberal democracies put legal constraints on the power of the executive to control immigration (Hollifield 1992) (see *external legal constraints*).
Essentialism (adj. essentialist)	The idea that every social unit has a number of fixed, objective and absolute characteristics that are inextricably linked to its identity.
Ethnic community	Community of people based on shared ethnic background and a distinctive culture, who generally blend in or assimilate within a few generations (see *social closure*, as opposed to *ethnic minority*).

Ethnic enclaves	Areas with high concentrations of particular migrant or diaspora groups that voluntarily choose to live there, offering benefits such as mutual support, cultural facilities and ethnic businesses (as opposed to *ghettoes*) (Peach 2005) (see *ethnic community, internal closure* and *social closure*).
Ethnic minority	An ethnic group with a subordinate position vis-à-vis more powerful groups in society, resulting from marginalization processes (as opposed to *ethnic community*, (see *external closure*).
Ethnicity	A sense of group belonging based on ideas of common origins, history, culture, experience and values (Fishman 1985: 4; Smith 1986: 27).
Ethno-states	States whose nationality ideology is constructed around ideas of ethnic homogeneity and common ancestry.
Exit revolution	Shift in the emphasis of migration policies from controlling exit to controlling entry from the mid-nineteenth century (Zolberg 2007).
External closure	Formation of ethnic groups or minorities as a result of exclusion (see also *ghettoes*), as opposed to *internal closure*.
External legal constraints	Legal constraints faced by states in the extent to which they can control migration as a consequence of their adherence to international agreements and human rights conventions (see *embedded legal constraints*) (see Joppke 1998).
Eugenics	The practice or idea of improving the human race through selective breeding (Wikler 1999).
Family migrant	Person who primarily migrates to join family members. Includes family reunion and international adoption.
Family migration/ reunion	Migration occurring when family members join migrants at the destination (see *primary* and *secondary family reunion*).
First Gulf War	War between Iraq and the United States after the Iraqi invasion of Kuwait (1990–1991).
Forced migration	Migration where staying is no option, either because of coercion (deportation, enslavement) or because of threat of abuse, violence, persecution or livelihood deprivation (see *refugee* and *asylum seeker*).
Frontex	The European Border and Coast Guard Agency, charged with controlling EU's external borders.
Frontier migration	Migration towards peripheral resource-rich areas suitable for colonization and resource extraction, such as through ranching, farming or mining, often stimulated by state-driven hegemonic expansion policies.
Genocide	The deliberate mass murder of ethnic groups (see Holocaust, *othering*, *racism*, and *dehumanization*).
Gentrification	The 'upgrading' of deteriorated neighbourhoods through urban renewal and the influx of more affluent residents.

Ghettoes	Areas typically characterized by poverty, unemployment, crime and high concentrations of particular ethnic groups that are 'trapped' there because of a lack of opportunities to move out (as opposed to *ethnic enclaves*) (Peach 2005).
Global care chains	'Series of personal links between people across the globe based on the paid or unpaid work of caring' (Hochschield 2000:131). Usually applied to domestic care networks in which migrant women supply domestic care labour in high-income societies, while women in origin countries care for the family members left behind.
Globalization	'The widening, deepening and speeding up of worldwide interconnectedness in all aspects of contemporary social life' (Held *et al.* 1999: 2).
Great Depression	A worldwide economic depression, lasting from 1929 to the late 1930s, prompting protectionist policies and severe immigration restrictions.
Great Migration	Period between 1914 to the 1950s during which African–Americans fled segregation and economic exploitation in the Southern states for better wages and rights in the North-East, Midwest and West.
Great Recession	A period of economic decline in the late 2000s and early 2010s, marked by financial and housing market crises, recessions and balance of payment problems, particularly affecting Western countries.
Guestworker	Migrants from Mediterranean countries who were recruited as temporary workers to work in the industries and mines of northwest European countries. Term is often seen as derogatory (see *bracero*).
Harkis	Algerians who served with the French army in the war of independence and fled to France after Algerian independence in 1962.
Holocaust	Systematic persecution and murder of around 6 million European Jews as well as Roma, Sinti and other groups by Nazi-Germany during the Second World War.
Hometown associations	Organizations that aim to maintain ties between immigrants and their hometowns often through supporting education, health care, infrastructure and development project in origin areas (see *collective remittances*).
Hukou system	Chinese policy to control internal migration introduced in 1958 (see *propiska*).
Human capital theory	Applied to migration: theory that sees migration as an investment in the productivity of knowledge and skills (human capital).
Human mobility	All forms of human movement outside of their direct living place and social environment (house, village or neighbourhood), irrespective of the distance and time-period implied, or whether this involved the crossing of administrative borders.

Human trafficking	'The recruitment, transportation, transfer, harbouring or receipt of persons, by means of the threat or use of force or other forms of coercion, of abduction, of fraud, of deception, of the abuse of power or of a position of vulnerability or of the giving or receiving of payments or benefits to achieve the consent of a person having control over another person, for the purpose of exploitation' (Protocol to Prevent, Suppress and Punish Trafficking in Persons).
Illegal entry	Arrival in a country without obtaining official permission to enter; does not apply to asylum seekers, who do have this right to cross borders in search of protection.
Illegal migration	Migration without authorization (see *irregular migration* and *undocumented migration*).
Illegal stay	Stay in a country without visa or residence permit.
Implementation gap	The gap between policies on paper and their actual implementation (de Haas *et al.* 2019a).
Incorporation	The ways in which immigrants settle and become part of destination societies (see *assimilation* and *integration*).
Indenture	Recruitment of migrant workers and their transportation to another area for work, on the basis of contracts that bind them to a particular employer for a fixed period of time.
Informal employment	Work without legal and social protection, both in formal and *informal sector*.
Informal sector	Production and employment taking place in unregistered enterprises, falling outside regulation and protection by the state.
Informalization	Shift of work from formal sector to informal sector employment.
Integration	Process in which members of a (migrant) group become part of the destination society and nation, without necessarily giving up their identity or certain cultural practices (see *multiculturalism*).
Internal closure	Formation of distinct ethnic groups or minorities as a result of free will (see also *ethnic enclaves* and *social closure*), as opposed to *external closure*.
Internal dynamics of migration processes	Feedback mechanisms through which the social, cultural and economic impacts of migration on origin and destination communities change the initial conditions under which migration takes place, often encouraging more migration along particular geographical pathways (see *chain migration, migration networks, culture of migration*).
Internal (or domestic) migrants	Migration that involves crossing of an administrative border within a country.
Internally displaced persons (IDPs)	Persons forced or obliged to flee or to leave their homes as a result of or in order to avoid the effects of armed conflict, situations of violence, violations of human rights, or natural or human-made disasters, and who have not crossed an internationally recognized state border (UN Guiding Principles on Internal Displacement)

Internal (or domestic) remittances	Money sent back to origin areas by internal migrants.
International migration	Migration that involves the crossing of a border between states.
Intersectionality	The interrelationship of gender, class, race, ethnicity and other social divisions, through which various disadvantages can reinforce each other.
Inter-temporal substitution	Change of timing of migration in in anticipation of expected change in migration policy, such as 'now or never' or 'beat the ban' migration surges in anticipation of border closure or as a reaction on an (allegedly temporary) border opening (see *substitution effects*) (see de Haas 2011).
Intifada	Palestinian resistance against Israeli occupation of the Gaza Strip and the West Bank.
Involuntary immobility	The phenomenon of people aspiring to migrate but lacking the capability to do so (Carling 2002), often leading to an obsession with moving out (see *culture of migration*).
Irregular migration	Migration taking place outside the regulatory norms of states; does not include asylum (see *undocumented migration*, *illegal migration*).
Islamophobia	Fear of and hostility to Islam and Muslims.
Ius domicili	Entitlement to citizenship which grows out of long-term residence in the destination country.
Ius sanguinis	'Birthright' laws on citizenship which are based on descent from a national (citizen) of the country concerned.
Ius soli	Laws on citizenship which are based on birth in the territory of the country.
Jim Crow system	Set of state and local laws designed to enforced racial segregation in the Southern US.
Kafala	Immigration regime based on sponsorship and bonded labour in Gulf countries (see Box 9.1).
Labour frontier	Imaginary line separating labour importing countries and countries experiencing peak emigration, usually middle-income countries going through fast economic and demographic transitions (Skeldon 1997) (see *migration transition*).
Labour market segmentation	The division of labour markets in various separate clusters, driven by increasing specialization of education, skills and work, as well as the growth of *dual labour markets*.
Labour migrant	A person migrating primarily for employment purposes.
Legalization effect	Sudden jumps in migrant numbers caused by the registration or legalization of previously arrived undocumented migrants (workers) rather than new arrivals.

Liberal paradox	The inherent tension between the need for liberal states to remain open to trade, investment and (labour) migration and protect the rights of its citizens (Hollifield 1992).
Lingua franca	Language used for communication between people who have a different native language, particularly common in multi-ethnic states, such as Nigeria, India, Indonesia or creole languages in the Caribbean.
Long-term or permanent migration	Change of country of residence for a longer period, usually coinciding with family reunion and family formation.
Maastricht Treaty	Agreement between 12 European countries to extent European cooperation, introduce EU citizenship and freedom of movement among member states (1992).
Maghreb	The region including Morocco, Algeria and Tunisia, and according to some definitions also Mauritania and Libya.
Maquiladoras	US-controlled factories in Mexico near the US border zone.
Marshall plan	'European Recovery Plan' initiated by the United States after the Second World War to revive Western-European economies and halt the rise of communism.
Marxism	A theory and methodology that views class relations and social conflict around the ownership of the means of production as central to the analysis of social transformation.
Mestizaje	Mixing of ethnic groups.
Mestizo	A person with a mixed ethnic background.
Migrant	A person who is living in another country, state, province or municipality than she or he was born.
Migrant/migration networks	Sets of interpersonal ties that connect migrants, former migrants, and non-migrants in origin and destination areas through bonds of kinship, friendship and shared community origin (see *internal dynamics of migration* and *chain migration*) (Massey et al. 1993: 448).
Migration	Change of habitual residency across administrative borders (e.g., municipalities, provinces, departments, federal or national states).
Migration corridor	An established link or route connecting particular places and communities.
Migration hump	Short- to medium-term hikes in emigration in the wake of trade reforms and other economic shocks (to be distinguished from *migration transition*) (Martin and Taylor 1996).
Migration industry	Groups of actors consisting of employers, travel agents, recruiters, brokers, smugglers, humanitarian organizations, housing agents, immigration lawyers and other intermediaries who have a strong interest in the continuation of migration (see *migration intermediaries*).

Migration intermediaries	Actors that facilitate, and sometimes drive, migration within and across borders (see *migration industry*).
Migration multiplier	The phenomenon that initial migration often leads to more migration, with as *migrant networks* facilitating migration along the same migration corridor. For instance, recruitment of a limited number of workers often leads to a much larger migration through family reunion (see also *chain migration, family reunion*).
Migration plateau	Long period of sustained out-migration as a result of development stagnation ('halted' *migration transition*) (Martin and Taylor 1996).
Migration policies	Laws, regulations and measures that states enact and implement with the explicit objective of affecting the volume, origin, direction and internal composition (or selection) of migration.
Migration reversal	A structural change in the dominant direction of migration flows between regions or countries.
Migration system	Set of places (or countries) linked by flows and counter-flows of people, goods, services and information, which tend to facilitate further exchange, including migration, between the places (or countries) (see *internal dynamics of migration processes*) (Mabogunje 1970).
Migration transition	The process through which economic development and social transformations associated to industrialization and modernization initially lead to accelerating emigration, with emigration generally peaking when countries reach upper-middle income status, after which emigration decreases and countries gradually transform from net emigration to net immigration countries (see *migration plateau, mobility transition* and *labour frontier*).
Migratory mobility	All forms of mobility qualifying as *migration*.
Migratory process	The ways migration evolves over time: the entire social process of migration over time, including migration decision-making, arrival, settlement, incorporation in destination societies, return and circulation, as well as the effects of migration on communities and societies of origin.
Minority	See *ethnic minority*.
Mobility freedom	The freedom to decide where to live, *including the option to stay at home* (de Haas 2014).
Mobility transition	The general expansion of individual mobility and the changing character of patterns of internal and international mobility in modernizing societies (Zelinsky 1971) (see *migration transition*).
Multi-layered (tiered) migration hierarchies	Migration pattern in which high-income countries (regions) attract migrants from middle-income countries (regions) while those countries (regions) in turn attract migrants from low-income countries (regions) (see *replacement migration, global care chains*).
Multiculturalism	Ideology implying that immigrants should be able to participate as equals in all spheres of society, without being expected to give up their own culture, religion and language, provided that they conform to national laws and certain key values.

Nation	An imagined political community of people, whose members are bound together by a sense of solidarity, a common culture, and consciousness (Anderson 1983: 15; Seton-Watson 1977: 1).
Nativism	Intense opposition to an internal minority on the grounds of its foreign connections; nativists see minorities as an outright threat to national identity and security (see *securitization*).
Neoliberalism	Political ideology that aims at limiting the intervention of government in the market by economic de-regulation, flexibilization of labour markets, privatization of state companies and partial dismantling of welfare states.
New economics of labour migration (NELM)	Theory developed to explain migration in and from developing countries, arguing that migration is often part of a (1) household or family (rather than an individual) sharing strategy aimed at (2) reducing livelihood risks (rather than maximizing income or utility) by diversifying income sources through remittances, (3) serving to spread income risks and overcome market constraints by raising investment capital through *remittances* (Stark 1978; 1991).
Nikkeijin	Descendants of past Japanese emigrants in South America, now often admitted as labour migrants in Japan.
Non-migratory mobility	All forms of mobility not qualifying as migration (e.g. commuting, shopping, tourism, family visits and business-related mobility).
Non-refoulement	Part of international refugee law which protects asylum seekers from deportation to countries where they may fear persecution.
Oil Shock	Economic shock caused by sudden rise in oil prices, commonly associated with the geopolitical and economic transformations prompted by the 1973 OPEC oil embargo, causing an Oil Boom in oil-producing countries and an Oil Crisis in oil-importing countries.
Other-definition	Ascription of undesirable characteristics and assignment to inferior social positions by dominant groups.
Othering	Representing and treating people from other social groups as essentially different from and generally inferior to the own group, as a form of *essentialism* (see also *nativism*).
Overstaying	Migrants or visitors staying after their residency permit or travel visa expires; generally the most important cause of *undocumented stay* (as opposed to *undocumented entry*).
Paradigm	Approach to scientific inquiry – centred around the use of particular vocabularies and assumptions as well as evaluation criteria determining the legitimacy of problem definition and 'appropriate' tools in terms of methodology and analysis (Kuhn 1962).
Phenotype	Physical appearance (traditional understanding of 'race').
Pieds-noirs	Algerians of French or other European ancestry who were 'repatriated' to France after Algeria gained independence in 1962 (see *colons*).

Points-based system	A migration policy tool that determines non-citizens' eligibility for temporary or permanent residence based on factors as education, wealth, age and job offers, first introduced by Canada in 1967.
Political remittances	Political impact of emigration on origin countries, either through direct political activism of emigrants, exiles and return migrants, or through the diffusion of ideas (see *social remittances*).
Politicization	To make an issue the subject of explicit political debate, emphasizing the political dimension of a social issue.
Power	The ability to influence the behaviour and thoughts of other people.
Precarious work	Insecure and exploitative jobs, often, but not necessarily, in the *informal sector*.
Price dumping	The practice of exporting commodities at prices below the cost of production, generally facilitated by government subsidies, often driving peasants in poor countries out of business.
Prima facie refugee recognition	'At first sight' approach to refugee migration that grants people refugee status based on apparent, objective circumstances in the origin country, instead of individual-level characteristics, common in the large-scale arrival of refugee groups where individual status determination is impractical or impossible.
Primary family migration/reunion	Migration occurring when migrants are joined by spouses and children living in origin areas (see *secondary family migration, chain migration, migration multipliers, migrant networks*).
Primary sector	Agrarian sector.
Primordial attachment	Approach to ethnicity according to which ethnicity is not primarily a matter of choice, but is pre-social, almost instinctual, something one is born into (Geertz 1963).
Proletarianization	Replacement of self-employment by wage-labour in capitalist industrial societies; Marxist concept associated with undermining of traditional peasant livelihoods, urbanization and the *rural exodus*.
Propiska	Soviet system aimed at regulating internal migration and the redistribution of labour (see also *hukou*).
Race	Social construct produced by racism (Miles 1989); highly contested term with different meanings over history and across societies, but generally referring to discourses and practices ascribing essential, unchangeable differences in phenotypical and cultural characteristics to social groups and their supposed members (see *racism, essentialism*, and *othering*).
Racialization	Discourses implying that socioeconomic or political problems are a 'natural' consequence of certain ascribed physical or cultural characteristics of minority groups (see *essentialism, othering*) (see Murji and Solomos 2005).

Racism	Process whereby social groups categorize other groups as different, hostile, inferior and 'less human', on the basis of phenotypical or cultural markers, often as part of discourses serving to provide moral justification for colonization, genocide, slavery and exploitation of migrant workers (see *essentialism*, *othering*, *anti-Semitism*, *Islamophobia*). Racism also implies making predictions about people's character, abilities or behaviour on the basis of socially constructed markers of difference.
Realism	School of thought about international relations that traditionally assumed that only sovereign states are relevant to analysis of questions of war and peace.
Refugee	A person fleeing violence, persecution or discrimination, legally defined as a person who, 'owing to a well-founded fear of persecution for reasons of race, religion, nationality, membership of a particular social group or political opinions, is outside the country of his nationality and is unable or, owing to such fear, is unwilling to avail himself of the protection of that country' (UN definition).
Relative deprivation	The feeling of being deprived emanating from comparisons with other members of social groups being perceived as better-off.
Re-emigration	Migration to the same destination after return, often associated to *circular migration*.
Remittances	Money sent back home by migrants (see *social remittances*).
Replacement migration	The process by which immigrants from more peripheral places fill the places of jobs left vacant by migrants who left to more central places, associated with the emergence of *multi layered migration hierarchies*.
Reverse flow substitution	Substitution effect occurring if immigration restrictions interrupt circulation by discouraging return and encouraging long-term or permanent settlement (see *substitution effects*) (de Haas 2011).
Reverse migration transition	Process by which countries transform from net-immigration to net-emigration societies, usually as a consequence of economic and political decline (such as Argentina over the late twentieth century).
Rural exodus	Large-scale migration from rural areas to cities, typically associated to the transition from agrarian to industrial societies (see *migration transition*, *proletarization*).
Schengen agreement	Treaty between European countries that created a zone in which citizens can travel freely without passport controls, signed in 1985, and implemented from 1995; it has come to include almost all EU states (except the UK and Ireland) as well as Iceland, Liechtenstein, Norway and Switzerland.
Secondary family migration/reunion	Migration following new unions between migrants' offspring and new spouses and partners in origin areas (see *primary family migration*, *chain migration*, *migration multipliers*, *migrant networks*).

Second generation	Children of migrants who were largely socialized in destination societies, generally includes those who migrated as infants, toddlers and young children.
Secondary sector	Industrial sector.
Securitization of migration	A procession of social construction that linking migration to security issues in discourse and policy, either through the portrayal of migrants and ethnic minorities as potential criminals, rapist or terrorists, or the portrayal of migration more in general as a threat to the cultural and religious identity of destination societies (see *nativism* and *Islamophobia*).
Sedentarization	The process of settling down, gaining a fixed residence, usually of (semi-) nomadic or itinerant people, often part of state policies.
Segmented assimilation	The incorporation of immigrants in particular labour market segments, socioeconomic strata and classes in destination societies (see *downward assimilation*).
Self-definition	Consciousness of group members of belonging together on the basis of shared cultural, religious and social characteristics.
Sharecropping	A form of semi-bonded agricultural labour in which families rent small plots of land in exchange for a (small) share of the harvest.
Situational ethnicity	The invocation of ethnicity for self-identification by members of a specific group.
Smuggling	The use of paid or unpaid intermediaries to cross borders without authorization, either as part of a business transaction or humanitarian activism.
Social capital	The aggregate of resources that can be accessed based on membership of social groups (based on Bourdieu 1979).
Social closure	The process in which a social group establishes rules and practices to exclude others in order to gain a competitive advantage or to preserve their identity (Weber 1968: 342)
Social construct	The imagined nature of social reality: the fundamental idea that humans can understand and give meaning to the world, themselves and groups, only through shared symbols, categories, concepts and ideas, usually expressed through discourse, music, arts and other forms of social interaction and cultural expression.
Social remittances	The flow of ideas, behavioural repertoires, identities and social capital from receiving to sending communities (Levitt 1998), can contribute to a *culture of migration*.
Soft power	The ability of states to achieve foreign policy and security objectives through political and cultural relations without recourse to military or economic coercion.
Spatial substitution	Rerouting of migration, through a change in migration destinations or the use of other geographical routes in response to border controls or other changes in migration policy (see *substitution effects*) (de Haas 2011).

State	Human community that (successfully) claims the monopoly of the legitimate use of physical force within a given territory (Weber 1919).
Student migrant	A person migrating primarily to pursue (usually secondary or tertiary) education.
Structural Adjustment Programmes (SAP)	Policy by international financial and development institutions to impose neoliberal economic reforms in developing countries (see *neoliberalism* and *Washington Consensus*).
Substitution effects	Unintended effects of migration policies which can decrease or undermine their effectiveness (see *categorical substitution*, *spatial substitution*, *inter-temporal substitution*, and *reverse flow substitution*) (de Haas 2011).
Supremacism	Belief that one's own racial or ethnic group is superior to other groups (see *racism*).
Target-earners	Migrants aiming to stay as long as they have saved the desired amount of money needed to fulfil a desired social, cultural or economic purpose at home (for example, marrying, building a house, starting a business), with return signalling the successful completion of the migration cycle (see *new economics of labour migration*, *temporary migration*).
Tertiary sector	Service sector.
Third-country national	Non-EU citizen.
Transnationalism	The maintenance of multiple ties and social interactions linking people or institutions across the borders of nation-states (Vertovec 1999), Transnationalism can be temporary by-product of migration or lead to *diaspora* formation.
Unauthorized migration	See *undocumented migration*.
Undocumented entry	Entry into a country or place without the appropriate documentation.
Undocumented migration	Crossing and residing without official permission (see *irregular, unauthorized* and *illegal migration*).
Undocumented stay	Staying in country or place without official permission.
Villagization	State policy targeted at the compulsory resettlement of dispersed people in designated areas or villages, generally to maintain control and prevent insurrection.
Visa run	Process in which migrants without residence permits leave and re-enter a state by renewing their travel visa before it expires.
Washington consensus	Development ideology which stresses the importance of market liberalization, privatization and deregulation as development recipes (see *Structural Adjustment Programmes* and *neoliberalism*).
Xenophobia	Fear of strangers, either foreigners or co-citizens with different religions, cultures and habits (see *anti-Semitism* and *Islamophobia*).

Bibliography

Abadan-Unat, N. 1988. The socio-economic aspects of return migration to Turkey. *Revue Européenne des Migrations Internationales*, 3, 29–59.

Abashin, S. 2017. Migration policies in Russia: Laws and debates. *In:* Heusala, A.-L. & Aitamurto, K. (eds) *Migrant Workers in Russia. Global Challenges of the Shadow Economy in Societal Transformation*, London, Routledge.

Abella, M. 1995. Asian migrant and contract workers in the Middle East. *In:* Cohen, R. (ed.) *The Cambridge Survey of World Migration*, Cambridge, Cambridge University Press.

Abella, M. & Ducanes, G. 2009. *Technical Note: The Effect of the Global Economic Crisis on Asian Migrant Workers and Governments' Responses*, Bangkok, ILO Regional Office for Asia and the the Pacific.

Aber, S. & Small, M. 2013. Citizen or subordinate: Permutations of belonging in the United States and the Dominican Republic. *Journal on Migration and Human Security*, 1, 76–96.

Abye, T. 2004. *Parcours d'Éthiopiens en France et aux Etats-Unis: De nouvelles formes de migrations*, Paris, L'Harmattan.

Acosta Arcarazo, D. & Freier, L. F. 2015. Turning the immigration policy paradox upside down? Populist liberalism and discursive gaps in South America. *International Migration Review*, 49, 659–696.

Adams, R. H. 1991. The economic uses and impact of international remittances in rural Egypt. *Economic Development and Cultural Change*, 39, 695–722.

Adams, R. H. 2003. International migration, remittances, and the brain drain: A study of 24 labor-exporting countries, Washington, D.C., World Bank.

Adams, R. H. 2011. Evaluating the economic impact of international remittances on developing countries using household surveys: A literature review. *Journal of Development Studies*, 47, 809–828.

Adams, R. H. & Page, J. 2005. Do international migration and remittances reduce poverty in developing countries? *World Development*, 33(10), 1645–1669.

Adamson, F. B. 2006. Crossing borders: International migration and national security. *International Security*, 31, 165–199.

Adamson, F. B. & Demetriou, M. 2007. Remapping the boundaries of 'state' and 'national identity': Incorporating diasporas into IR theorizing. *European Journal of International Relations*, 13, 489–526.

Adepoju, A. 2001. Regional integration, continuity and changing patterns of intra-regional migration in sub-Saharan Africa. *In:* Siddique, M. A. B. (ed.) *International Migration into the 21st Century*, Cheltenham, Edward Elgar.

Adepoju, A. 2006. Leading issues in international migration in sub-Saharan Africa. *In:* Cross, C., Gelderblom, D., Roux, N. & Mafukidze, J. (eds.) *Views on Migration in Sub-Saharan Africa*, Cape Town, HSRC Press.

Adler, S. 1981. *A Turkish Conundrum: Emigration, Politics and Development, 1961–1980*, Geneva, ILO.

Agunias, D. R. 2006. *Remittances and Development: Trends, Impacts, and Policy Options*, Washington, D.C., Migration Policy Institute.

Agunias, D. R. 2008. *Managing Temporary Migration: Lessons from the Philippine Model*, Washington, D.C., Migration Policy Institute.

Agunias, D. R., 2009. *Guiding the Invisible Hand: Making Migration Intermediaries Work for Development*. Human Development Research Paper No.22, New York, United Nations Development Programme, Human Development Report Office.

Agunias, D. R. 2011. *Running in Circles: Progress and Challenges in Regulating Recruitment of Filipino and Sri Lankan Labor Migrants to Jordan*, Washington, D.C., Migration Policy Institute.

Agunias, D. R. 2012. *Regulating Private Recruitment in the Asia-Middle East Labour Migration Corridor*, Washington, D.C., Migration Policy Institute.

Agunias, D. R. 2013. What we know about regulating the recruitment of migrant workers.

Policy Brief, Washington, D.C., Migration Policy Institute.

Agunias, D. R. & Newland, K. 2012. *Developing a Road Map for Engaging Diasporas in Development: A Handbook for Policymakers and Practitioners in Home and Host Countries*, Geneva, International Organization for Migration.

Aït Hamza, M. 1988. L'émigration, Facteur d'Intégration ou de Désintégration des Régions d'Origine. *Le Maroc et La Holllande. Actes de la Première Rencontre Universitaire*, Rabat, Université Mohammed V.

Alba, F. 2010. *Mexico: A Crucial Crossroads* [Online]. Washington, D.C., Migration Policy Institute. http://www.migrationinformation.org/Profiles/display.cfm?ID=772 [Accessed 9 March 2010].

Alba, R. & Foner, N. 2014. Comparing immigrant integration in North America and Western Europe: How much do the grand narratives tell us? *International Migration Review*, 48, S263–S291.

Aleinikoff, T. A. & Klusmeyer, D. (eds.) 2000. *From Migrant to Citizens: Membership in a Changing World*, Washington, D.C., Carnegie Endowment for International Peace.

Aleinikoff, T. A. & Klusmeyer, D. (eds.) 2001. *Citizenship Today: Global Perspectives and Practices*, Washington, D.C., Carnegie Endowment for International Peace.

Al Sharmani, M. 2007. *Contemporary Migration and Transnational Families: The Case of Somali Diaspora*, Cairo, American University in Cairo, Forced Migration & Refugee Studies Program.

AlShehabi, O. H. 2019. Policing labour in empire: The modern origins of the Kafala sponsor-ship system in the Gulf Arab States. *British Journal of Middle Eastern Studies* https://doi.org/10.1080/13530194.2019.1580183

Altamirano Rúa, T. 2010. *Migration, Remittances and Development in Times of Crisis*, Lima, Pontificia Universidad Católica del Perú.

Álvarez de Flores, R. 2006–2007. Evolución Histórica de las Migraciones en Venezuela: Breve Recuento. *Aldea Mundo*, 11, 89–93.

Amaral, E. & Fusco, W. 2005. *Shaping Brazil: The Role of International Migration* [Online]. http://www.migrationinformation.org/Profiles/display.cfm?ID=311 [Accessed 19 September 2011].

Ambrosini, M. 2013. *Irregular Migration and Invisible Welfare*, Basingstoke, Palgrave Macmillan.

Ambrosini, M. 2015. Irregular but tolerated: Unauthorized immigration, elderly care recipients, and invisible welfare. *Migration Studies*, 3, 199–216.

Ambrosini, M. 2016. From "illegality" to tolerance and beyond: Irregular immigration as a selective and dynamic process. *International Migration*, 54, 144–159.

Ambrosius, C. 2019. Government reactions to private substitutes for public goods: Remittances and the crowding-out of public finance. *Journal of Comparative Economics*, 47, 396–415.

Amelina, A. & Lutz, H. 2018. *Gender and Migration: Transnational and Intersectional Prospects*, London, Routledge.

Amin, S. 1974. *Accumulation on a World Scale*, New York, Monthly Review Press.

Amnesty International. 2011. Amnesty International Annual Report 2011: The State of the World's Human Rights.

Amrith, S. S. 2011. *Migration and Diaspora in Modern Asia*, Cambridge, Cambridge University Press.

Amuedo-Dorantes, C. & Pozo, S. 2006. Migration, remittances, and male and female employment patterns. *American Economic Review*, 96(2), 222–226.

Ananta, A. & Arifin, E. N. 2004. *International Migration in Southeast Asia*, Institute of Southeast Asian Studies.

Andall, J. 2003. *Gender and Ethnicity in Contemporary Europe*, Oxford, Berg.

Anderson, B. 1983. *Imagined Communities: Reflections on the Origin and Spread of Nationalism*, London, Verso.

Anderson, B. 2000. *Doing the Dirty Work: The Global Politics of Domestic Labour*, London, Zed Books.

Anderson, B. 2001. Why madam has so many bathrobes?: Demand for migrant workers in the EU. *Tijdschrift voor Economische en Sociale Geografie*, 92, 18–26.

Anderson, B. 2007. A very private business: Exploring the demand for migrant domestic workers. *European Journal of Women's Studies*, 14, 247–264.

Anderson, B. & Andrijasevic, R. 2008. Sex, slaves and citizens: The politics of anti-trafficking. *Soundings*, 40, 135–145.

Anderson, B., Ruhs, M., Rogaly, B. & Spencer, S. 2006. *Fair Enough? Central and East European Migrants in Low-Wage Employment in the UK*, York, Joseph Rowntree Foundation.

Andersson, R. 2007. Ethnic residential segregation and integration processes in Sweden. *In:* Schönwälder, K. (ed.) *Residential Segregation and the Integration of Immigrants: Britain, The Netherlands and Sweden*, Berlin, Wissenschaftszentrum.

Andres Henao, L. 2009. African immigrants drift toward Latin America. *Reuters*, 15 November 2009.

Ángel Castillo, M. 2006. *Mexico: Caught Between the United States and Central America* [Online]. http://www.migrationinformation.org/Feature/display.cfm?ID=389 [Accessed 11 November 2011].

Angelucci, M. 2012. US border enforcement and the net flow of Mexican illegal migration. *Economic Development and Cultural Change*, 60 (2), 311–357.

Anthias, F. & Yuval-Davis, N. 1989. Introduction. *In:* Anthias, F. & Yuval-Davis, N. (eds.) *Woman-Nation-State*, London, Macmillan.

Appleyard, R. 1989. Migration and development: Myths and reality. *International Migration Review*, 23, 486–499.

Archdeacon, T. 1983. *Becoming American: An Ethnic History*, New York, The Free Press.

Arendt, H. 1951. *The Origins of Totalitarianism*, New York, Harcourt, Brace.

Aronson, G. 1990. *Israel, Palestinians and The Intifada: Creating Facts on the West Bank*, London, Kegan Paul International.

Arthur, J. A. 1991. International labor migration patterns in West Africa. *African Studies Review*, 34, 65–87.

Asis, M. M. B. 2006. *The Philippines' Culture of Migration* [Online], Washington, D.C, Migration Policy Institute http://www.migrationinformation.org/Feature/display.cfm?ID=364 [Accessed 1 June 2019]

Asis, M. M. B. 2008. How international migration can support development: A challenge for the Philippines. *In:* Castles, S. & Delgado Wise, R. (eds.) *Migration and Development: Perspectives from the South*, Geneva, International Organization for Migration.

Augoustinos, M. & De Garis, S. 2012. 'Too black or not black enough': Social identity complexity in the political rhetoric of Barack Obama. *European Journal of Social Psychology*, 42, 564–577.

Baban, F., Ilcan, S. & Rygiel, K. 2017. Syrian refugees in Turkey: Pathways to precarity, differential inclusion, and negotiated citizenship rights. *Journal of Ethnic and Migration Studies*, 43(1), 41–57.

Back, L. & Solomos, J. 2013. *Theories of Race and Racism: A Reader*, London, Routledge.

Bade, K. 2003. *Migration in European History*, Oxford, Blackwells.

Baeck, L. 1993. *Post-War Development Theories and Practice*, Paris, UNESCO and The International Social Science Council.

Baganha, M. I. B. & Fonseca, M. L. 2004. *New Waves: Migration from Eastern to Southern Europe*, Lisbon, Luso-American Foundation.

Bakewell, O. 2000. Repatriation and self-settled refugees in Zambia: Bringing solutions to the wrong problems. *Journal of Refugee Studies*, 13, 356–373.

Bakewell, O. 2008. 'Keeping them in their place': the ambivalent relationship between development and migration in Africa. *Third World Quarterly*, 29, 1341–1358.

Bakewell, O. 2009. *South-South Migration and Human Development: Reflections on African Experiences*, New York, UNDP.

Bakewell, O. & de Haas, H. 2007. African migrations: Continuities, discontinuities and recent transformations. *In:* De Haan, L., Engel, U. & Chabal, P. (eds.) *African Alternatives*, Leiden, Brill.

Bakewell, O. & Jónsson, G. 2011. *Migration, Mobility and the African City*. IMI working paper 50, Oxford, International Migration Institute, University of Oxford.

Baldwin-Edwards, M. 2005. *Migration in the Middle East and the Mediterranean*, Switzerland, Global Commission on International Migration Geneva.

Baldwin-Edwards, M. & Schain, M. A. (eds.) 1994. *The Politics of Immigration in Western Europe*, Ilford, Essex, Frank Cass.

Balibar, E. 1991. Racism and nationalism. *In:* Balibar, E. & Wallerstein, I. (eds.) *Race, Nation, Class: Ambiguous Identities*, London, Verso.

Balibar, E. & Wallerstein, I. (eds.) *Race, Nation, Class: Ambiguous Identities*, London, Verso.

Ballard, R. 2003. A case of capital-rich underdevelopment: The paradoxical consequences of successful transnational entrepreneurship from Mirpur. *Contributions to Indian Sociology*, 37, 25–57.

Banting, K. & Kymlicka, W. 2012. *Is There Really a Backlash Against Multiculturalism Policies? New Evidence from the Multiculturalism Policy Index.* GRITIM working paper 14, Barcelona, Universitat Pompeu Fabra.

Barlán, J. 1988. *A System Approach for Understanding International Population Movement: The Role of Policies and Migrant Community in the Southern Cone.* IUSSP Seminar, Malaysia, Genting Highlands.

Barta, T. 2008. "They appear actually to vanish from the face of the Earth." Aborigines and the European project in Australia Felix. *Journal of Genocide Research*, 10, 519–539.

Bartram, D. 1999. *Foreign Labour and Political Economy in Israel and Japan*, Dissertation, Madison, University of Wisconsin.

Bartram, D. 2005. *International Labor Migration: Foreign Workers and Public Policy*, New York, Palgrave Macmillan.

Basok, T. 2007. *Canada's Temporary Migration Program: A Model Despite Flaws* [Online]. http://www.migrationinformation.org/Feature/display.cfm?ID=650 [Accessed 30 September 2011].

Bastia, T. & Skeldon, R. (eds.) 2020. *Routledge Handbook of Migration and Development*, London, Routledge.

Batalova, J. & Zong, J. 2016. *Caribbean Immigrants in the United States* [Online], Washington, D.C., Migration Policy Institute. https://www.migrationpolicy.org/article/caribbean-immigrants-united-states [Accessed 17 November 2018].

Batalova, J. & Zong, J. 2017. *Cuban Immigrants in the United States*[Online], Washington, D.C., Migration Policy Institute. https://www.migrationpolicy.org/article/cuban-immigrants-united-states. [Accessed 30 May 2019].

Batalova, J. & Zong, J. 2018. *Cuban Immigrants in the United States* [Online], Washington, D.C., Migration Policy Institute. https://www.migrationpolicy.org/article/cuban-immigrants-united-states. [Accessed 17 November 2018].

Bauböck, R. 1991. Migration and Citizenship. *New Community*, 18(1).

Bauböck, R. (ed.) 1994a. *From Aliens to Citizens: Redefining the Status of Immigrants in Europe*, Aldershot, Avebury.

Bauböck, R. 1994b. *Transnational Citizenship: Membership and Rights in International Migration*, Aldershot, Edward Elgar.

Bauböck, R. 1996. Social and cultural integration in a civil society. In: Bauböck, R., Heller, A. & Zolberg, A. R. (eds.) *The Challenge of Diversity: Integration and Pluralism in Societies of Immigration*, Aldershot, Avebury.

Bauböck, R., Ersbøll, E., Groenendijk, K. & Waldrauch, H. 2006. *Acquisition and Loss of Nationality: Policies and Trends in 15 European states*, Amsterdam, Amsterdam University Press.

Bauböck, R. & Rundell, J. (eds.). 1998. *Blurred Boundaries: Migration, Ethnicity, Citizenship*, Aldershot, Ashgate.

Bauder, H. 2006. *Labor Movement: How Migration Regulates Labor*, Oxford, Oxford University Press.

Bauder, H. 2017. Sanctuary cities: Policies and practices in international perspective. *International Migration*, 55, 174–187.

Bauer, T. & Zimmermann, K. 1998. Causes of international migration: A survey. In: Gorter, P., Nijkamp, P. & Poot, J. (eds.) *Crossing Borders: Regional and Urban Perspectives on International Migration*, Aldershot, Ashgate.

BBC News. (26 November 2007) *China in Africa: Developing ties.* http://news.bbc.co.uk/1/hi/world/africa/7086777.stm [Accessed 26 November 2007].

BBC/MPI. 2010. *Migration and Immigrants Two Years after the Financial Collapse: Where Do We Stand?*, London and Washington, D.C., BBC World Service and Migration Policy Institute

Beasley, W. G. 1987. *Japanese Imperialism, 1894–1945*, Oxford, Oxford University Press.

Beaugé, G. 1986. La kafala: un système de gestion transitoire de la main-d'œuvre et du capital dans les pays du Golfe. *Revue Européenne des Migrations Internationales*, 2–1, 109–122.

Beine, M., Docquier, F. & Schiff, M. 2013. International migration, transfer of norms and home country fertility. *Canadian Journal of Economics-Revue Canadienne D' Economique*, 46, 1406–1430.

Beine, M., Noël, R. & Ragot, L. 2014. Determinants of the international mobility of students. *Economics of Education Review*, 41, 40–54.

Beine, M. & Parsons, C. 2015. Climatic factors as determinants of international migration. *The Scandinavian Journal of Economics*, 117, 2, 723–767.

Bélanger, D., Lee, H.-K. & Wang, H.-Z. 2010. Ethnic diversity and statistics in East Asia: 'foreign brides' surveys in Taiwan and South Korea. *Ethnic and Racial Studies*, 33, 1108–1130.

Bell, D. 1975. Ethnicity and social change. *In:* Glazer, N. & Moynihan, D. P. (eds.) *Ethnicity – Theory and Experience*, Cambridge, MA., Harvard University Press.

Ben Ali, D. L'Impact de Transferts des Résidents Marocains à l'Etranger (RME) sur l'Investissement Productif. Séminaire sur "La Migration Internationale", 6–7 juin 1996., 1996 Rabat. Centre d'Etudes et de Recherches Démographiques (CERED), 345–263.

Bencherifa, A. 1996. Is sedentarization of pastoral nomads causing desertification? The case of the Beni Guil in eastern Morocco. In Swearingen, W. D. & Bencherifa, A., *The North African Environment at Risk*, Boulder, Westview Press, 117–131.

Bensaad, A. 2003. Agadez, carrefour migratoire sahélo-maghrébin. *Revue Européenne Des Migrations Internationales*, 19.

Berghege, M., Boer, M. D., Braak, S. V. D., Hilhorst, S. & Peek, S. 2018. 'Ik had mijn eigen huid in mijn handen' [Online]. De Groene Amsterdammer. https://www.groene.nl/artikel/ik-had-mijn-eigen-huid-in-mijn-handen [Accessed 17 May 2019].

Berriane, J. 2012. Ahmad al-Tijani and his Neighbors. The Inhabitants of Fez and their Perceptions of the Zawiya. *In:* Desplat, P. & Schulz, D. (eds.) *Prayer in the City. The Making of Sacred Place and Urban Life*, Bielefeld, Transcript Verlag.

Berriane, J. 2015. Sub-Saharan students in Morocco: Determinants, everyday life, and future plans of a high-skilled migrant group. *The Journal of North African Studies*, 20, 573–589.

Berriane, M. 1997. Emigration Internationale du Travail et Micro-Urbanisation dans le Rif Oriental: Cas du Centre de Taouima (Région de Nador, Maroc). *Migration Internationale et Changements Sociaux dans le Maghreb. Actes du Colloque Internationale du Hammamet, Tunisie (21–25 juin 1993)*. Tunis, Université de Tunis.

Berriane, M. & de Haas, H. 2012. *African Migrations Research: Innovative Methods and Methodologies*, Trenton, NJ, Africa World Press.

Berriane, M., De Haas, H. & Natter, K. 2018. Introduction: Revisiting Moroccan migrations. *Revisiting Moroccan Migrations*, London, Routledge.

Bertossi, C. 2007. *French and British Models of Integration: Public Philosophies, Policies and State Institutions*. Working Paper 46, Oxford, ESRC Centre on Migration, Policy and Society, University of Oxford.

Bertram, I. G. & Watters, R. F. 1985. The MIRAB economy in South Pacific microstates. *Pacific Viewpoint*, 26, 497–519.

Bérubé, M. 2005. *Colombia: In the Crossfire* [Online]. http://www.migrationinformation.org/Profiles/display.cfm?ID=344 [Accessed 1 November 2011].

Bettini, G. 2013. Climate Barbarians at the Gate? A critique of apocalyptic narratives on 'climate refugees'. *Geoforum*, 45, 63–72.

Bhagwati, J. 2003. Borders beyond control. *Foreign Affairs*, 82, 98–104.

Björgo, T. & Witte, R. (eds.) 1993. *Racist Violence in Europe*, London, Macmillan.

Black, R. 2001. *Environmental Refugees: Myth or Reality?* UNHCR working paper 34, Sussex, University of Sussex.

Black, R. 2003. Breaking the convention: Researching the "illegal" migration of refugees to Europe. *Antipode*, 35, 34–54.

Black, R., Adger, W. N., Arnell, N. W., Dercon, S., Geddes, A. & Thomas, D. 2011. The effect of environmental change on human migration. *Global Environmental Change*, 21, S3–S11.

Blackburn, R. 1988. *The Overthrow of Colonial Slavery 1776–1848*, London, Verso.

Böcker, A. 1994. Chain migration over legally closed borders: Settled migrants as bridgeheads and gatekeepers. *Netherlands' Journal of Social Sciences*, 30, 87–106.

Bodomo, A. 2010. The African trading community in Guangzhou: An emerging bridge for Africa–China relations. *The China Quarterly*, 203, 693–707.

Body-Gendrot, S. & Wihtol de Wenden, C. 2007. *Sortir Des Banlieues. Pour En Finir Avec La Tyrannie Des Territoires*, Paris, Autrement.

Böhning, W. R. 1984. *Studies in International Labour Migration*, London and New York, Macmillan and St. Martin's.

Bonjour, S. 2011. The power and morals of policy makers: Reassessing the control gap debate. *International Migration Review*, 45, 89–122.

Bonjour, S. & Duyvendak, J. W. 2018. The "migrant with poor prospects": Racialized intersections of class and culture in Dutch civic integration debates. *Ethnic and Racial Studies*, 41, 882–900.

Borjas, G. J. 1989. Economic theory and international migration. *International Migration Review*, 23, 457–485.

Borjas, G. J. 1990. *Friends or Strangers: The Impact of Immigration on the US Economy*, New York, Basic Books.

Borjas, G. J. 2000. *Issues in the Economics of Immigration*, Chicago, University of Chicago Press.

Borjas, G. J. 2001. *Heaven's Door: Immigration Policy and the American Economy*, N.J. , Princeton University Press.

Boserup, E. 1965. *The Conditions of Agricultural Growth: The Economics of Agrarian Change under Population Pressure*, Chicago, Aldine.

Boswell, C. 2007a. Theorizing migration policy: Is there a third way? *International Migration Review*, 41, 75–100.

Boswell, C. 2007b. Migration control in Europe after 9/11: Explaining the absence of securitization. *JCMS-Journal of Common Market Studies*, 45, 589–610.

Boswell, C. & Geddes, A. 2011. *Migration and Mobility in the European Union*, Basingstoke, New York, Palgrave Macmillan.

Boubakri, H. 2004. Transit migration between Tunisia, Libya and Sub-Saharan Africa: Study based on Greater Tunis. Paper presented at the Regional Conference on 'Migrants in transit countries: Sharing responsibility for management and protection, Istanbul, 30 September – 1 October 2004, Strasbourg, Council of Europe.

Boubtane, E., Coulibaly, D. & Rault, C. 2013. Immigration, growth, and unemployment: Panel VAR evidence from OECD countries. *Labour*, 27, 399–420.

Bourdieu, P. 1979. Le Capital Social: Notes Provisoires. *Actes De La Recherche En Sciences Sociales*, 31, 2–3.

Bourdieu, P. 1985. The forms of capital. *In:* Richardson, J. G. (ed.) *Handbook of Theory and Research for the Sociology of Education*, New York, Greenwood.

Brachet, J. 2012. From one stage to the next: Transit and Transport in (Trans) Saharan Migrations. *In:* Berriane, M. & de Haas, H. (eds.) *African Migrations Research: Innovative Methods and Methodologies*, Trenton, NJ, Africa World Press.

Brachet, J. 2016. Policing the desert: The IOM in Libya beyond war and peace. *Antipode*, 48, 272–292.

Brachet, J. 2018. Manufacturing smugglers: From irregular to clandestine mobility in the Sahara. *The ANNALS of the American Academy of Political and Social Science*, 676, 16–35.

Bredeloup, S. & Pliez, O. 2005. Editorial: Migrations entre les deux rives du Sahara. *Autrepart*, 4, 3–20.

Breton, R., Isajiw, W. W., Kalbach, W. E. & Reitz, J. G. 1990. *Ethnic Identity and Equality*, Toronto, University of Toronto Press.

Brettell, C. B. 2016. *Gender and Migration (Immigration and Society)*, Cambridge, Malden, Polity Press.

Brettell, C. B. & Hollifield, J. F. 2014. *Migration Theory: Talking across Disciplines*, London, Routledge.

Briggs, V. M. 1984. *Immigration Policy and the American Labor Force*, Baltimore and London, Johns Hopkins University Press.

Brochmann, G. & Hammar, T. 1999. *Mechanisms of Immigration Control: A Comparative Analysis of European Regulation Policies*, London, Bloomsbury Academic.

Browne, I. & Misra, J. 2003. The intersection of gender and race in the labor market. *Annual Review of Sociology*, 29, 487–513.

Brubaker, R. 1992. *Citizenship and Nationhood in France and Germany*, Cambridge, MA., Harvard University Press.

Bruni, M. 2018. *New Evidence on Yemeni Return Migrants from the Kingdom of Saudi Arabia*, Sana'a, International Organization for Migration.

Cahill, D. 1990. *Intermarriages in International Contexts*, Quezon City, Scalabrini Migration Center.

Calmont, R. 1981. La communauté guyanaise en France. *Revue Guyanaise D'histoire Et De Géographie*, 21–43.

Cantle, T. 2001. *Community Cohesion: A Report of the Independent Review Team*, London, Home Office.

Cantor, D., Freier, L. F. & Gauci, J. P. 2013. *A Liberal Tide: Towards A Paradigm Shift in Latin American Migration and Asylum Policy-Making?* Conference Report. 18 March 2013, Senate House, University of London.

Capps, R., McCabe, K. & Fix, M. 2012. *Diverse streams: African Migration to the United States*, Washington, D.C., Migration Policy Institute.

Carling, J. 2002. Migration in the age of involuntary immobility: Theoretical reflections and Cape Verdean experiences. *Journal of Ethnic and Migration Studies*, 28, 5–42.

Carling, J. 2006. *Migration, Human Smuggling and Trafficking from Nigeria to Europe*, Geneva, International Organisation for Migration.

Carling, J. 2007. Migration control and migrant fatalities at the Spanish-African borders. *International Migration Review*, 41, 316–343.

Carling, J. & Collins, F. 2018. Aspiration, desire and drivers of migration. *Journal of Ethnic and Migration Studies*, 44, 909–926.

Castells, M. 1997. *The Power of Identity*, Oxford, Blackwells.

Castillo, M. Á. 2006. *Mexico: Caught Between the United States and Central America* [Online]. http://www.migrationinformation.org/Feature/display.cfm?id=389 [Accessed 30 September 2011].

Castles, S. 1985. The guests who stayed—the debate on 'foreigners policy'in the German Federal Republic. *International Migration Review*, 19, 517–534.

Castles, S. 1986. The Guest-Worker In Western-Europe – An Obituary. *International Migration Review*, 20(4), 761–778.

Castles, S. 1995. How Nation-states respond to immigration and ethnic diversity. *New Community*, 21(3), 293–308.

Castles, S. 2002. *Environmental Change and Forced Migration: Making Sense of the Debate. New Issues in Refugee Research.* Working paper 70, Geneva, UNHCR.

Castles, S. 2004a. The factors that make and unmake migration policy. *International Migration Review*, 38, 852–884.

Castles, S. 2004b. Why migration policies fail. *Ethnic and Racial Studies*, 27, 205–227.

Castles, S. 2006. Guestworkers in Europe: A resurrection? *International Migration Review*, 40, 741–766.

Castles, S., Arias Cubas, M., Kim, C. & Ozkul, D. 2012. Irregular migration: Causes, patterns and strategies. *In:* Omelaniuk, I. & National Institute for Migration Mexico (Inami) (eds.) *Reflections on Migration and Development*, Berlin and Geneva, Springer and International Organization for Migration.

Castles, S., Booth, H. & Wallace, T. 1984. *Here for Good: Western Europe's New Ethnic Minorities*, London, Pluto.

Castles, S. & Davidson, A. 2000. *Citizenship and Migration: Globalisation and the Politics of Belonging*, London, Macmillan.

Castles, S., de Haas, H., Van Hear, N. & Vasta, E. 2010. Special issue: Theories of Migration and Social Change. *Journal of Ethnic and Migration Studies*, 36.

Castles, S. & Delgado Wise, R. (eds.) 2008. *Migration and Development: Perspectives from the South*, Geneva, International Organization for Migration.

Castles, S. & Kosack, G. 1973. *Immigrant Workers and Class Structure in Western Europe*, London, Oxford University Press.

Castles, S. & Van Hear, N. 2012. Root causes. *In:* Betts, A. (ed.) *Global Migration Governance*, Oxford, Oxford University Press.

Castles, S., Vasta, E. & Ozkul, D. 2013. Australia: A classical immigration country in transition. *In:* Hollifield, J., Martin, P. & Orrenius, P. (eds.) *Controlling Immigration: A Global Perspective*, Stanford, CA, Stanford University Press.

CEC. 2005. *Green Paper on an EU Approach to Managing Economic Migration*, Brussels, Commission of the European Communities.

Cernea, M. M. & McDowell, C. (eds.) 2000. *Risks and Reconstruction: Experiences of Resettlers and Refugees*, Washington, D.C., World Bank.

Ceuppens, B. & Geschiere, P. 2005. Autochthony: Local or global? New modes in the struggle over citizenship and belonging in Africa and Europe. *Annual Review of Anthropology*, 34, 385–407.

Chacaltana, J., Bonnet, F. & Leung, V. 2018. *Women and Men in the Informal Economy: A Statistical Picture*. Third edition, Geneva, International Labour Organization.

Chaillou-Atrous, V. 2016. *Indentured Labour in European Colonies during the 19th Century* [Online]. Encyclopédie pour une histoire nouvelle de l'Europe. http://ehne.fr/en/node/1128 [Accessed 28 May 2019].

Chalcraft, J. 2011. Migration and popular protest in the Arabian Peninsula and the Gulf in the 1950s and 1960s. *International Labor and Working Class History*, 79 (1(S)), 28–47.

Chami, R., Barajas, A., Cosimano, T., Fullenkamp, C., Gapen, M. & Montiel, P. 2008.

Macroeconomic Consequences of Remittances, International Monetary Fund Occasional Paper, 259.

Chan, J. & Selden, M. 2017. The labour politics of China's rural migrant workers. *Globalizations*, 14, 259–271.

Chaney, E. M. 1989. The context of Caribbean migration. *Center for Migration Studies special issues*, 7, 1–14.

Chatty, D. & Colchester, M. (eds.). 2002. *Conservation and Mobile Indigenous Peoples: Displacement, Forced Settlement, and Sustainable Development*, New York, Oxford, Berghahn.

Chebel d'Appollonia, A. 2012. *Immigration and Insecurity in the United States and Europe*, Ithaca, NY, Cornell University Press.

Chen, M. A. 2009. Informalization of labour markets: Is formalization the answer? *The Gendered Impacts of Liberalization*, London, Routledge.

Chen, M. A. 2012. The informal economy: Definitions, theories and policies. *WIEGO Working Paper 1*, Cambridge, Women in Informal Employment, Globalizing and Organizing.

Cheng, E. & Xu, Z. 2005. Domestic Money Transfer Services for Migrant Workers in China the Consultative Group to Assist the Poor. Paper presented at Seminar on Issues of Migrant Workers' Remittance, supported by the World Bank, Beijing. 2005.

Cherti, M. & Collyer, M. 2015. Immigration and Pensée d'Etat: Moroccan migration policy changes as transformation of 'geopolitical culture'. *Journal of North African Studies*, 20, 590–604.

Chiffoleau, S. 2003. Un champ à explorer: le rôle des pèlerinages dans les mobilités nationales, régionales et internationales du Moyen Orient. *Revue Européenne Des Migrations Internationales*, 19, 285–289.

Chimni, B. S. 1998. The geo-politics of refugee studies: A view from the South. *Journal of Refugee Studies*, 11, 350–374.

Chin, A. 2005. Long-run labor market effects of Japanese American internment during World War II on working-age male internees. *Journal of Labor Economics*, 23(3), 491–525.

Chishti, M. & Bolter, J. 2019. *Remain in Mexico Plan Echoes Earlier U.S. Policy to Deter Haitian Migration*, Washington, D.C., Policy Institute.

Chishti, M., Pierce, S. & O'Connor, A. 2019. *Despite Flurry of Actions, Trump Administration Faces Constraints in Achieving Its Immigration Agenda*, Washington, D.C., Migration Policy Institute.

Chiswick, B. R. 2000. Are immigrants favorably self-selected? An economic analysis. *In*: Brettell, C. B. & Hollifield, J. F. (eds.) *Migration Theory: Talking Across Disciplines*, London, Routledge.

Chiswick, B. R. & Miller, P. W. 1995. The endogeneity between language and earnings: International analyses. *Journal of Labor Economics*, 13, 246–288.

Christian Aid. 2007. *Human Tide: The Real Migration Crisis*, London, Christian Aid.

Chudinovskikh, O. & Denisenko, M. 2017. *Russia: A Migration System with Soviet roots* [Online], Washington, D.C., Migration Policy Institute. https://www.migrationpolicy.org/article/russia-migration-system-soviet-roots [Accessed 30 May 2019].

Cinanni, P. 1968. *Emigrazione e Imperialismo*, Rome, Riuniti.

Çinar, D. 1994. From aliens to citizens: A comparative analysis of the rules of transition. *In*: Bauböck, R. (ed.) *From Aliens to Citizens*, Aldershot, Avebury.

Cissé, D. 2013. South-South migration and Sino-African small traders: A comparative study of Chinese in Senegal and Africans in China. *African Review of Economics and Finance*, 5, 17–28.

Clemens, M. A. 2007. Do visas kill? Health effects of African health professional emigration, Washington, D.C., Center for Global Development.

Clemens, M. A. 2009. Skill flow: A fundamental reconsideration of skilled worker mobility and development. *Human Development Research Paper No. 8*, New York, United Nations Development Programme, Human Development Report Office.

Clemens, M. A. 2014. *Does Development Reduce Migration?*, Washington, D.C., Center for Global Development.

Clemens, M. A. & McKenzie, D. 2018. Why don't remittances appear to affect growth? *The Economic Journal*, 128, 179–209.

Clemens, M. A., Montenegro, C. E. & Pritchett, L. 2008. *The Place Premium: Wage Differences for Identical Workers across the US Border*, Washington, D.C., World Bank.

Clinton, J. & Roush, C. 2016. *Poll: Persistent Partisan Divide Over 'Birther' Question* [Online]. NBC News. https://www.nbcnews.com/politics/2016-election/poll-persistent-partisan-divide-over-birther-question-n627446 [Accessed 20 April 2019].

Cohen, P. & Bains, H. S. 1988. *Multi-Racist Britain*, London, Macmillan.

Cohen, R. 1987. *The New Helots: Migrants in the International Division of Labour*, Oxford, Oxford University Press and Oxford Publishing Services.

Cohen, R. 1991. East-West and European migration in a global context. *New Community*, 18, 9–26.

Cohen, R. 1995. Asian indentured and Colonial migration. *In:* Cohen, R. (ed.) *The Cambridge Survey of World Migration*, Cambridge, Cambridge University Press.

Cohen, R. 1997. *Global Diasporas: An Introduction*, London, University College London Press.

Cohen, R. 2007. Creolization and cultural globalization: The soft sounds of fugitive power. *Globalizations*, 4, 369–384.

Coleman, J. S. 1988. Social capital in the creation of human-capital. *American Journal of Sociology*, 94, S95-S120.

Collier, J. L. 1978. *The Making of Jazz: A Comprehensive History*, London, Macmillan.

Collier, P. 2013. *Exodus: How Migration is Changing Our World*, Oxford, Oxford University Press.

Collins, J. 1991. *Migrant Hands in a Distant Land: Australia's Post-War Immigration*, Sydney, Pluto.

Collins, J. 2006. The changing political economy of Australian immigration. *Tijdschrift voor Economische en Sociale Geografie*, 97, 7–16.

Collins, J., Gibson, K., Alcorso, C., Castles, S. & Tait, D. 1995. *A Shop Full of Dreams: Ethnic Small Business in Australia*, Sydney, Pluto.

Collyer, M. 2003. *Explaining Change in Established Migration Systems: The Movement of Algerians to France and the UK*. Migration working paper 16, Sussex, Sussex Centre for Migration Research, University of Sussex.

Collyer, M. 2005. When do social networks fail to explain migration? Accounting for the movement of Algerian asylum-seekers to the UK. *Journal of Ethnic and Migration Studies*, 31, 699–718.

Collyer, M. & de Haas, H. 2012. Developing dynamic categorisations of transit migration. *Population Space and Place*, 18, 468–481.

Collyer, M., Duvell, F. & de Haas, H. 2012. Critical approaches to transit migration. *Population Space and Place*, 18, 407–414.

Commission on Filipino Overseas. 2013. Stock Estimates of Filipinos Overseas, Manila, Department of Foreign Affairs Philippine Overseas Employment Administration.

Comtat, E. 2009. *Les Pieds-Noirs Et La Politique. Quarante Ans Après Le Retour*, Paris, Presses de Sciences Po.

Condon, S. A. & Ogden, P. E. 1991. Emigration from the French Caribbean: The origins of an organized migration. *International Journal of Urban and Regional Research*, 15, 505–523.

Cordeiro, A. 2006. Portugal and the immigration challenge. *In:* Majtczak, O. (ed.) *The Fifth International Migration Conference*, Warsaw, Independent University of Business and Government.

Cornelius, W. A. 2001. Death at the border: Efficacy and unintended consequences of US immigration control policy, *Population and Development Review*, 27 (4), 661–685.

Cornelius, W. A. 2005. Controlling 'unwanted' immigration: Lessons from the United States, 1993–2004. *Journal of Ethnic and Migration Studies*, 31 (4), 775–794.

Council of Labor Affairs Taiwan. 2012. *Foreign Labor (in Chinese)* [Online]. http://www.cla.gov.tw/cgi-bin/siteMaker/SM_theme?page=4aadd21a [Accessed 22 January 2012].

Courtis, C. 2011. Marcos Institucionales, Normativos y de Políticas sobre Migración Internacional en Argentina, Chile y Ecuador. *In:* Martínez Pizarro, J. (ed.) *Migración Internacional en América Latina y el Caribe – Nuevas Tendencias, Nuevos Enfoques*, Santiago de Chile, CEPAL.

Cousens, S. 1960. The Regional Pattern of Emigration during the Great Irish Famine, 1846–51. *Transactions and Papers (Institute of British Geographers)*, 119–134.

Cox, R. 2006. *The Servant Problem: Domestic Employment in a Global Economy*, London and New York, I. B. Tauris.

Crenshaw, K. 1989. Demarginalizing the intersection of race and sex: A black feminist critique of antidiscrimination doctrine, feminist theory and antiracist politics. *University of Chicago Legal Forum*, 139–168.

Cross, C., Gelderblom, D., Roux, N. & Mafukidze, J. (eds.) 2006. *Views on Migration in Sub-Saharan Africa*, Cape Town, HSRC Press.

Cross, G. S. 1983. *Immigrant Workers in Industrial France: The Making of a New Laboring Class*, Philadelphia, PA, Temple University Press.

Crush, J. 2008. *South Africa: Policy in the Face of Xenophobia* [Online], Washington, D.C., Migration Policy Institute. http://www.migrationinformation.org/Profiles/display.cfm?ID=689 [Accessed 24 July 2012].

Cruz, J. M. 2010. Central American maras: From youth street gangs to transnational protection rackets. *Global Crime*, 11(4), 379–398.

Czaika, M. & de Haas, H. 2012. The role of internal and international relative deprivation in global migration. *Oxford Development Studies*, 40, 423–442.

Czaika, M. & de Haas, H. 2013. The effectiveness of immigration policies. *Population and Development Review*, 39, 487–508.

Czaika, M. & de Haas, H. 2014. The globalisation of migration: Has the world become more migratory? *International Migration Review*, 48, 283–323.

Czaika, M. & de Haas, H. 2017. The effect of visas on migration processes. *International Migration Review*, 51, 893–926.

Czaika, M., de Haas, H. & Villares-Varela, M. 2018. The global evolution of travel visa regimes. *Population and Development Review*, 44, 589.

Czaika, M. & Hobolth, M. 2016. Do restrictive asylum and visa policies increase irregular migration into Europe? *European Union Politics*, 17, 345–365.

Czaika, M. & Neumayer, E. 2017. Visa restrictions and economic globalisation. *Applied Geography*, 84, 75–82.

Czaika, M. & Parsons, C. R. 2017. The gravity of high-skilled migration policies. *Demography*, 54, 603–630.

Da Silva, S. A. 2013. Brazil, a new Eldorado for immigrants?: The case of Haitians and the Brazilian immigration policy. *Urbanities – Journal of Urban Ethnography*, 3, 3–18.

Daniels, R. 2004. *Guarding the Golden Door*, New York, Hill and Wang.

Danış, D. & Perouse, J. 2005. La politique migratoire turque: vers une normalisation? *Migrations et Société*, 19, 93–106.

Davidson, B. 1992. *The Black man's burden: Africa and the curse of the nation-state*, New York, Times Books.

Davis, A. 2012. *The Impact of Anti-Immigration Parties on Mainstream Parties' Immigration Positions in the Netherlands*, Flanders and the UK 1987–2010: Divided electorates, left-right politics and the pull towards restrictionism. PhD thesis, Florence, European University Institute.

Davis, D. K. 2005. Indigenous knowledge and the desertification debate: Problematising expert knowledge in North Africa. *Geoforum*, 36, 509–524.

de Brauw, A., Mueller, V. & Woldehanna, T. 2013. Motives to remit: Evidence from tracked internal migrants in Ethiopia. *World Development*, 50, 13–23.

de Bree, J. 2007. *Belonging, Transnationalism and Embedding: Dutch Moroccan Return Migrants In Northeast Morocco*, MA thesis, Radboud University, Nijmegen.

de Bree, J., Davids, T. & de Haas, H. 2010. Post-return experiences and transnational belonging of return migrants: A Dutch–Moroccan case study. *Global Networks*, 10, 489–509.

de Bruijn, M., Nyamnjoh, F. & Brinkman, I. (eds.) 2009. *Mobile Phones: The New Talking Drums of Everyday Africa*, Leiden and Bamenda, African Studies Centre/Langaa.

de Bruijn, M. & Van Dijk, H. 2003. Changing population mobility in West Africa: Fulbe pastoralists in Central and South Mali. *African Affairs*, 102, 285–307.

de Bruijn, M., van Dijk, R. & Foeken, D. 2001. *Mobile Africa: Changing Patterns of Movement in Africa and Beyond*, Leiden and Boston, Brill.

de Haan, A. 1999. Livelihoods and poverty: The role of migration. *Journal of Development Studies*, 36, 1–47.

de Haan, A., Brock, K., Carswell, G., Coulibaly, N., Seba, H. & Toufique, K. A. 2000. *Migration and Livelihoods: Case Studies in Bangladesh, Ethiopia and Mali*. IDS Research Report 46, Brighton and Sussex, Institute of Development Studies.

de Haas, H. 1998. Socio-Economic transformations and oasis agriculture in Southern Morocco. *In:* de Haan, L. & Blaikie, P. (eds.) *Looking at Maps in the Dark*, Utrecht/Amsterdam, KNAG/FRW UvA.

de Haas, H. 2003. *Migration and Development in Southern Morocco: The Disparate Socio-Economic Impacts of Out-Migration on the Todgha Oasis Valley*, PhD thesis, Radboud University, Nijmegen.

de Haas, H. 2005. International migration, remittances and development: Myths and facts. *Third World Quarterly*, 26, 1269–1284.

de Haas, H. 2006a. *Trans-Saharan Migration to North Africa and the EU: Historical Roots and Current Trends* [Online] Washington, D.C., Migration Policy Institute, https://www.migrationpolicy.org/article/trans-saharan-migration-north-africa-and-eu-historical-roots-and-current-trends. [Accessed 30 May 2019].

de Haas, H. 2006b. *Engaging Diasporas: How Governments and Development Agencies can Support Diasporas' Involvement in Development of Origin Countries*, A study for Oxfam Novib, Oxford, International Migration Institute, University of Oxford.

de Haas, H. 2006c. Migration, remittances and regional development in Southern Morocco. *Geoforum*, 37, 565–580.

de Haas, H. 2007a. *Between Courting and Controlling: The Moroccan State and 'Its' Emigrants*. COMPAS working paper 54, Oxford, Centre on Migration, Policy and Society, University of Oxford.

de Haas, H. 2007b. Morocco's migration experience: A transitional perspective. *International Migration*, 45, 39–70.

de Haas, H. 2007c. *Remittances and Social Development: A Conceptual Review of the Literature*, Geneva, UNRISD.

de Haas, H. 2008. The Myth of Invasion – The inconvenient realities of African migration to Europe. *Third World Quarterly*, 1305–1322.

de Haas, H. 2009. *Mobility and Human Development*, New York, UNDP.

de Haas, H. 2010a. The internal dynamics of migration processes: A theoretical inquiry. *Journal of Ethnic and Migration Studies*, 36, 1587–1617.

de Haas, H. 2010b. Migration and development: A theoretical perspective. *International Migration Review*, 44 227–264.

de Haas, H. 2010c. *Migration Transitions: A Theoretical and Empirical Inquiry into the Developmental Drivers of International Migration*. IMI working paper 24, International Migration Institute, University of Oxford.

de Haas, H. 2011. *The Determinants of International Migration*. IMI working paper 2, International Migration Institute, University of Oxford.

de Haas, H. 2012. The migration and development pendulum: A critical view on research and policy. *International Migration*, 50, 8–25.

de Haas, H. 2014a. *Migration theory: Quo vadis?* IMI working paper 100, International Migration Institute, University of Oxford.

de Haas, H. 2014b. Un siècle de migrations marocaines: Transformations, transitions et perspectives d'avenir. *In:* Berriane, M. (ed.) *Marocains de l'Extérieur*, Rabat, Fondation Hassan II pour les Marocains Résidant à l'Etranger.

de Haas, H. 2019 *Trends and Drivers of African Migration*. IMI Working paper, International Migration Institute, University of Amsterdam.

de Haas, H., Czaika, M., Flahaux, M.-L., Mahendra, E., Natter, K., Vezzoli, S. & Villares-Varela, M. 2019a. International migration: Trends, determinants and policy effects. *Population and Development Review* (forthcoming).

de Haas, H., Vezzoli, S. & Villares-Varela, M. 2019b. *Opening the Floodgates?: European Migration Under Restrictive and Liberal Border Regimes 1950–2010*. IMI working paper 150, International Migration Institute, University of Amsterdam.

de Haas, H. & Fokkema, T. 2010. Intra-household conflicts in migration decisionmaking: Return and pendulum migration in Morocco. *Population and Development Review*, 36, 541–561.

de Haas, H. & Fransen, S. 2018. *Social Transformation and Migration: An Empirical Inquiry*. IMI Working paper 141, International Migration Institute, University of Amsterdam.

de Haas, H. & Natter, K. 2014. *The Determinants of Migration Policies: Does the Political Orientation of Governments Matter?*, IMI working paper 117, International Migration Institute, University of Oxford.

de Haas, H., Natter, K. & Vezzoli, S. 2014. *Compiling and Coding Migration Policies: Insights from the DEMIG POLICY Database*. IMI working paper 87, International Migration Institute, University of Oxford.

de Haas, H., Natter, K. & Vezzoli, S. 2015. Conceptualizing and measuring migration policy change. *Comparative Migration Studies*, 3, 15.

de Haas, H., Natter, K. & Vezzoli, S. 2018. Growing restrictiveness or changing selection? The nature and evolution of migration policies. *International Migration Review*, 52, 324–367.

de Haas, H. & Sigona, N. 2012. Migration and revolution. *Forced Migration Review*, 39, 4–5.

de Haas, H. & van Rooij, A. 2010. Migration as emancipation? The impact of internal and international migration on the position of women left behind in rural Morocco. *Oxford Development Studies*, 38, 43–62.

de Haas, H. & Vezzoli, S. 2010. *Migration and Development: Lessons from the Mexico-US and Morocco-EU Experiences*, International Migration Institute, University of Oxford.

de Haas, H. & Vezzoli, S. 2011. *Leaving Matters: The Nature, Evolution and Effects of Emigration Policies*. IMI working paper 34, International Migration Institute, University of Oxford.

de Haas, H. & Vezzoli, S. 2013. Migration and development on the South–North frontier: A comparison of the Mexico–US and Morocco–EU cases. *Journal of Ethnic and Migration Studies*, 39(7), 1041–1065.

de Lepervanche, M. 1975. Australian immigrants 1788–1940: Desired and unwanted. *In:* Wheelwright, E. L. & Buckley, K. (eds.) *Essays in the Political Economy of Australian Capitalism*, Sydney, Australia and New Zealand Book Co.

De Mas, P. 1978. *Marges Marocaines: Limites De La Cooperation Au Développement Dans Une Région Périphérique: Le Cas Du Rif*, 'S-Gravenhage, NUFFIC/IMWOO/Projet Remplod.

De Mas, P. 1990. Overlevingsdynamiek in het Marokkaanse Rif-Gebergte. De Samenhang tussen Circulaire Migratie en Demografische Structuur van Huishoudens. *Geografisch Tijdschrift*, XXIV, 73–86.

de Regt, M. 2010. Ways to come, ways to leave: Gender, mobility, and il/legality among Ethiopian domestic workers in Yemen. *Gender & Society*, 24, 237–260.

De Soto, H. 2000. *The Mystery of Capital*, London, Bantam.

Debessay, H. 2003. Eritrea: transition to dictatorship, 1991–2003. *Review of African Political Economy*, 30(97), 435–444.

Decloîtres, R. 1967. *The Foreign Worker*, Paris, OECD.

Del Popolo, F., Oyarce, A. M., Ribotta, B. & Rodríguez, J. 2008. *Indigenous Peoples and Urban Settlements: Spatial Distribution, Internal Migration and Living Conditions*, Santiago, UNELAC.

Delano, A. 2011. *Mexico and Its Diaspora in the United States*, Cambridge, Cambridge University Press.

Delgado Wise, R. & Covarrubias, H. M. 2009. Capitalist restructuring, development and labour migration: The Mexico-US case. *In:* Munck, R. (ed.) *Globalisation and Migration: New Issues, New Politics*, London and New York, Routledge.

DeParle, J. 2007. Fearful of Restive Foreign Labor, Dubai Eyes Reforms, *New York Times*, 6 August 2007.

DeWind, J., Kim, E. M., Skeldon, R. & Yoon, I.-J. 2012. Korean development and migration. *Journal of Ethnic and Migration Studies*, 38, 371–388.

DFID. 2007. *Moving Out of Poverty - Making Migration Work Better for Poor People*, London, Department for International Development.

DHS. 2019. *Worksite Enforcement* [Online]. Department of Homeland Security. https://www.ice.gov/features/worksite-enforcement [Accessed 17 May 2019].

DIAC 2012. *2010–11 Migration Program Report*, Canberra, Department of Immigration and Citizenship.

Dietz, B. & Roll, H. 2017. Ethnic German and Jewish immigrants from post-Soviet countries in Germany: Identity formation and integration prospects. *In:* Isurin, L. & Riehl, C. M. (eds.) *Integration, Identity and Language Maintenance in Young Immigrants*, Amsterdam, John Benjamins.

Diop, A., Johnston, T. & Le, K. T. 2015. Reform of the Kafāla System: A Survey Experiment from Qatar. *Journal of Arabian Studies*, 5, 116–137.

Docquier, F. & Rapoport, H. 2012. Globalization, brain drain, and development. *Journal of Economic Literature*, 50, 681–730.

Dodson, B. & Crush, J. 2015. *Migration Governance and Migrant Rights in the Southern African Development Community (SADC): Attempts at Harmonization in a Disharmonious Region*, Geveva, UNRISD.

Dohse, K. 1981. *Ausländische Arbeiter und bürgerlicher Staat*, Konistein/Taunus, Hain.

Doña, C. & Levinson, A. 2004. *Chile: Moving Towards a Migration Policy* [Online]. http://www.

migrationinformation.org/Profiles/display. cfm?ID=199 [Accessed 5 December 2011].

Donoghue, J. 2013. Indentured servitude in the 17th century English Atlantic: A brief survey of the literature. History Compass, 11(10):893–902.

Doomernik, J. 2013. Does circular migration lead to "Guest Worker outcomes?" *International Migration*, 51(1), 24–39.

Douglass, M. & Roberts, G. 2000. *Japan and Global Migration: Foreign Workers and the Advent of a Multicultural Society*, London, Routledge.

Douki, C. 2007. The liberal Italian states and mass emigration. *In:* Green, N. & Weil, F. (eds.) *Citizenship and Those Who Leave Urbana*, Champaign, University of Illinois Press.

DRC. 2006. *Skilled Migration: Healthcare Policy Options*, Brighton, Development Research Centre (DRC) on Migration, Globalisation and Poverty, University of Sussex.

Duany, J. 2012. The Puerto Rican diaspora to the United States: A postcolonial migration?. *In:* Bosma, U., Lucassen, J. & Oostindie, G. (eds.) *Postcolonial Migrants and Identity Politics: Europe, Russia, Japan and the United States in Comparison*, Oxford, Berghahn.

Durand, J. 2004. *From Traitors to Heroes: 100 Years of Mexican Migration Policies* [Online]. Migration Policy Institute. http://www. migrationinformation.org/feature/display. cfm?ID=203 [Accessed 1 May 2011].

Dustmann, C. & Fabbri, F. 2005. Immigrants in the British labour market. *Fiscal Studies*, 26, 423–470.

Dustmann, C., Schönberg, U. & Stuhler, J. (2016). The impact of immigration: Why do studies reach such different results? *Journal of Economic Perspectives*, 30(4), 31–56.

Düvell, F. 2006a. *Crossing the fringes of Europe: Transit migration in the EU's neighbourhood.* Working paper 33, Centre on Migration, Policy and Society, University of Oxford.

Düvell, F. (ed.) 2006b. *Illegal Immigration in Europe: Beyond Control*, Basingstoke, Palgrave Macmillan.

Düvell, F. 2012. Transit migration: A blurred and politicised concept. *Population, Space and Place*, 18, 415–427.

Duyvendak, J. W. & Scholten, P. 2012. Deconstructing the Dutch multicultural model: A frame perspective on Dutch immigrant integration policymaking. *Comparative European Politics*, 10, 266–282.

Earnest, D. 2008. *Old Nations, New Voters*, Albany, NT, State University of New York Press.

EFFNATIS. 2001. Effectiveness of National Integration Strategies Towards Second Generation Migrant Youth in a Comparative European Perspective, final report of the EFFNATIS project. European Forum for Migration Studies, University of Bamberg.

Elder, S., de Haas, H., Principi, M. & Schewel, K. 2015. *Youth and Rural Development: Evidence from 25 School-To-Work Transition Surveys*, Geneva, ILO.

Ellermann, A. 2006. Street-level democracy: How immigration bureaucrats manage public opposition. *West European Politics*, 29, 293–309.

Ellman, M. 2007. Stalin and the Soviet famine of 1932–33 revisited. *Europe-Asia Studies*, 59, 663–693.

Eloundou-Enyegue, P. M. & Calves, A. E. 2006. Till marriage do us part: Education and remittances from married women in Africa. *Comparative Education Review*, 50, 1–20.

Engels, F. 1962. The condition of the working class in England. *Marx, Engels on Britain*, Moscow, Foreign Languages Publishing House.

Engstrom, D. W. 1997. *Presidential Decision Making Adrift: The Carter Administration and the Mariel Boatlift*, Lanham, Rowman & Littlefield.

Entzinger, H. 1985. Return migration in Western Europe: Current policy trends and their implications, in particular for the second generation. *International Migration*, XXIII, 263–290.

Entzinger, H. 2003. The rise and fall of multiculturalism: The case of the Netherlands. *In:* Joppke, C. & Morawaska, E. (eds.) *Towards Assimilation and Citizenship: Immigration in Liberal Nation-States*, Basingstoke, Palgrave Macmillan.

Ersanilli, E. 2010. *Comparing Integration. Host Culture Adoption and Ethnic Retention Among Turkish Immigrants and their Descendents in France, Germany and the Netherlands.* PhD thesis, Amsterdam, Vrije Universiteit.

Ersanilli, E. & Koopmans, R. 2010. Rewarding integration? Citizenship regulations and the Socio-Cultural integration of immigrants in the Netherlands, France and Germany. *Journal of Ethnic and Migration Studies*, 36, 773–791.

Ersanilli, E. & Koopmans, R. 2011. Do immigrant integration policies matter? A three-country comparison among Turkish immigrants. *West European Politics*, 34, 208–234.

Esposito, J. L. & Mogahed, D. 2007. *Who Speaks for Islam?: What a Billion Muslims Really think*, Simon and Schuster.

Essed, P. 1991. *Understanding Everyday Racism*, London and New Delhi, Sage

Eule, T. G. 2016. *Inside Immigration Law: Migration Management and Policy Application in Germany*, London, Routledge.

EUMC 2006. *The Annual Report on the Situation regarding Racism and Xenophobia in the Member States of the EU*, Vienna, EuMC on Racism and Xenophobia.

Faist, T. 2006. Extension du domaine de la lutte: International Migration and Security before and after 11 September 2011. *In:* Messina, A. & Lahav, G. (eds.) *The Migration Reader*, Boulder, CO and London, Lynne Rienner.

Faist, T. (ed.) 2007. *Dual Citizenship in Europe*, Aldershot, Ashgate.

Fargues, P. 2005. *How Many Migrants from, and to, Mediterranean Countries of the Middle East and North Africa?*, Florence, CARIM, RSCAS, EUI.

Fargues, P. (ed.) 2007. *Mediterranean Migration: 2006-2007 Report*, San Domenico Fiesole (FI), European University Institute, RSCAS.

Fargues, P. 2011a. Immigration without Inclusion: Non-Nationals in Nation-Building in the Gulf States. *Asian and Pacific Migration Journal*, 20, 273–292.

Fargues, P. 2011b. International migration and the demographic transition: A two-way interaction. *International Migration Review*, 45, 588–614.

Fargues, P. 2017. *Four Decades of Cross-Mediterranean Undocumented Migration to Europe: A Review of the Evidence*, Geneva, International Organisation for Migration.

Favell, A. 1998. *Philosophies of Integration: Immigration and the Idea of Citizenship in France and Britain*, London, Macmillan.

Fawcett, J. T. 1989. Networks, linkages, and migration systems. *International Migration Review*, 23, 671–680.

Feinsilver, J. M. 2010. Fifty years of Cuba's medical diplomacy: From idealism to pragmatism. *Cuban Studies*, 41, 85–104.

Fernandez, B. 2010. Cheap and disposable? The impact of the global economic crisis on the migration of Ethiopian women domestic workers to the Gulf. *Gender & Development*, 18, 249–262.

Ferrer, M. 2011. Marcos Institucionales, Normativos y de Políticas sobre Migración Internacional: el Caso de El Salvador y una Exploración en Costa Rica y el Caribe. *In:* Martínez Pizarro, J. (ed.) *Migración Internacional en América Latina y el Caribe - Nuevas Tendencias, Nuevos Enfoques*, Santiago de Chile, CEPAL.

Findlay, A. & Geddes, A., 2011. Critical views on the relationship between climate change and migration: some insights from the experience of Bangladesh. *In:* Pécoud A., Piguet, E. & de Guchteneire, P. (eds.), *Migration and Climate Change*, UNESCO/Cambridge University Press, Paris, 138–159.

Findley, S. E. 2004. *Mali: Seeking Opportunity Abroad* [Online], Washington, D.C., Migration Policy Institute. Available: https://www.migrationpolicy.org/article/mali-seeking-opportunity-abroad [Accessed 30 May 2019].

Finotelli, C. & Michalowski, I. 2012. The heuristic potential of models of citizenship and immigrant integration reviewed. *Journal of Immigrant & Refugee Studies*, 10, 231–240.

Fishman, J. A. 1985. *The Rise and Fall of the Ethnic Revival: Perspectives on Language and Ethnicity*, Berlin, New York and Amsterdam, Mouton.

FitzGerald, D. S. (2014) The sociology of international migration. *In:* Brettell, C. B. & Hollifield, J. F. (eds.), *Migration Theory: Talking across Disciplines*, London, Routledge, 115–147.

FitzGerald, D.S. (2019) *Refuge beyond Reach: How Rich Democracies Repel Asylum Seekers*. Oxford: Oxford University Press.

FitzGerald, D. S. & Cook-Martín, D. 2014. *Culling the Masses: The Democratic Origins of Racist Immigration Policy in the Americas*, Cambridge, MA., Harvard University Press.

Fix, M. & Passel, J. S. 1994. *Immigration and Immigrants: Setting the Record Straight*, Washington, D.C., The Urban Institute.

Flahaux, M.-L. 2017. The role of migration policy changes in Europe for return migration to Senegal. *International Migration Review*, 51 (4), 868–892.

Flahaux, M.-L. & de Haas, H. 2016. African migration: Trends, patterns, drivers. *Comparative Migration Studies*, 4, 1.

Flahaux, M.-L. & Vezzoli, S. 2016. How do post-colonial ties and migration regimes shape travel visa requirements? The case of Caribbean nationals. *Journal of Ethnic and Migration Studies*, 43, 1141–1163.

Flahaux, M.-L. & Vezzoli, S. 2017. Examining the role of border closure and post-colonial ties in Caribbean migration. *Migration Studies*, 6, 165–186.

Fokkema, T. & de Haas, H. 2011. Pre- and Post-Migration determinants of socio-cultural integration of African immigrants in Italy and Spain. *International Migration Review*, 53(6), 3–26.

Foot, P. 1965. *Immigration and Race in British Politics*, Harmondsworth, Penguin.

Ford, R. 2008. Is racial prejudice declining in Britain? *The British Journal of Sociology*, 59, 609–636.

Foresight. 2011. *Foresight: Migration and Global Environmental Change*, London, UK Government Office for Science.

Fox-Genovese, E. & Genovese, E. D. 1983. *Fruits of Merchant Capital: Slavery and Bourgeois Property in the Rise and Expansion of Capitalism*, New York, Oxford University Press.

Frank, A. G. 1966. The Development of Underdevelopment. *Monthly Review*, September.

Frank, A. G. 1969. *Capitalism and Underdevelopment in Latin America*, New York, Monthly Review Press.

Fransen, S. & de Haas, H. 2019. *The Volume and Geography of Global Refugee Migration*. IMI working paper, International Migration Institute, University of Amsterdam.

Franz, J. G. 1939. Review: Cityward Migration: Swedish Data. by Jane Moore. *Sociometry*, 2, 109.

Fratkin, E. & Roth, E. A. 2006. *As pastoralists settle: Social, health, and economic consequences of the pastoral sedentarization in Marsabit District, Kenya*, Berlin, Springer Science & Business Media.

Freeman, G. P. 1995. Modes of immigration politics in liberal democratic states. *International Migration Review*, 24, 881–902.

Freier, L. F. 2013. Open doors (for almost all): Visa policies and ethnic selectivity in Ecuador. Working Paper 188. San Diego, Center for Comparative Immigration Studies, UC San Diego.

Freier, L. F. 2018. *Understanding the Venezuelan Displacement Crisis* [Online]. E-International Relation. https://www.e-ir.info/2018/06/28/understanding-the-venezuelan-displacement-crises/[Accessed 28 May 2019].

Froebel, F., Heinrichs, J. & Kreye, O. 1980. *The New International Division of Labour*, Cambridge, Cambridge University Press.

Galeano, E. H. 1973. *Open Veins of Latin America: Five Centuries of the Pillage of a Continent*, New York, Monthly Review Press.

Galenson, D. W. 1984. The rise and fall of indentured servitude in the Americas: An economic analysis. *The Journal of Economic History*, 44, 1–26.

Gamlen, A. 2006. *Diaspora Engagement Policies: What are they, and What Kinds of States use them?*, Centre on Migration, Policy and Society (COMPAS), University of Oxford.

Gamlen, A. 2008. The emigration state and the modern geopolitical imagination. *Political Geography*, 27, 840–856.

Gammage, S. 2004. Exercising exit, voice and loyalty: A gender perspective on transnationalism in Haiti. *Development and Change*, 35, 743–771.

Gammage, S. 2006. Exporting people and recruiting remittances – A development strategy for El Salvador? *Latin American Perspectives*, 33, 75–100.

GAO. 2006. *Foreign Workers-Information on Selected Countries Experiences*, Washington, D.C., US Governmental Accountability Office.

García, M. C. 2006. *Seeking Refuge: Central American Migration to Mexico, the United States, and Canada*, Berkeley, University of California Press.

García-Calvo, C. & Reinares, F. 2016. Patterns of involvement among individuals arrested for Islamic State-related terrorist activities in Spain, 2013–2016. *Perspectives on Terrorism*, 10.

Garrard, J. A. 1971. *The English and Immigration: A Comparative Study of the Jewish Influx 1880–1910*, London, Oxford University Press.

GCIM. 2005. *Migration in an Interconnected World: New Directions for Action: Report of the*

Global Commission on International Migration, Geneva, Global Commission on International Migration.

Geddes, A. 2000. *Immigration and European Integration: Towards Fortress Europe?*, Manchester and New York, Manchester University Press.

Geddes, A. 2003. *The Politics of Migration and Immigration in Europe*, London, Sage.

Geertz, C. 1963. *Old Societies and New States - the Quest for Modernity in Asia and Africa*, Glencoe, IL, Free Press.

Gellman, I. F. 1971. The "St. Louis" Tragedy. *American Jewish History*, 61, 144.

Gellner, E. 1983. *Nations and Nationalism*, Oxford, Blackwell.

Gemenne, F. 2010. What's in a name: Social vulnerabilities and the refugee controversy in the wake of Hurricane Katrina. In Afifi, T. and J, Jäger (eds.) *Environment, Forced Migration and Social Vulnerability*. Heidelberg, Dordrecht, London, New York, Springer, 29–42.

Gemenne, F. 2011. Why the numbers don't add up: A review of estimates and predictions of people displaced by environmental changes. *Global Environmental Change*, 21, S41–S49.

Gemenne, F. & Blocher, J. 2017. How can migration serve adaptation to climate change? Challenges to fleshing out a policy ideal. *The Geographical Journal*, 183, 336–347.

Gereffi, G. 1996. Global commodity chains: New forms of coordination and control among nations and firms in international industries. *Competition and Change*, 1, 427–439.

Geschiere, P. 2005. Autochthony and citizenship: New modes in the struggle over belonging and exclusion in Africa. Forum for Development Studies, Abington, Taylor & Francis, 371–384.

Geschiere, P. 2009. *The perils of belonging: Autochthony, citizenship, and exclusion in Africa and Europe*, Chicago and London, University of Chicago Press.

Ghorashi, H. 2005. Layered meanings of community experiences of Iranian women exiles in 'Irangeles'. *In:* Davids, T. & Driel, F. V. (eds.) *The Gender Question in Globalization*, Aldershot, Ashgate.

Ghosh, P. R. 2010. African immigrants gravitating to China. *International Business Times* [Online], 16 August 2010.

Gibney, M. J. 2000. *Outside the Protection of the Law: The Situation of Irregular Migrants in Europe*, Refugee Studies Centre, University of Oxford.

Gibson, C. & Lennon, E. 1999. *Historical Census Statistics on the Foreign-born Population of the United States: 1850–1990. Population Division working paper 29*, Washington, D.C., U.S. Census Bureau.

Gibson, J. & McKenzie, D. 2011. Eight questions about brain drain. *Journal of Economic Perspectives*, 25, 107–128.

Giddens, A. 2002. *Runaway World: how Globalisation is Reshaping our Lives*, London, Profile.

Givens, T. E. & Maxwell, R. 2012. *Immigrant Politics: Race and Representation in Western Europe*, Boulder, CO, Lynne Rienner.

Glazer, N. & Moynihan, D. P. 1975. Introduction. *In:* Glazer, N. A. & Moynihan, D. P. (eds.) *Ethnicity: Theory and Experience*, Cambridge, MA., Harvard University Press.

Glennie, A. & Chappell, L. 2010. *Jamaica: From Diverse Beginning to Diaspora in the Developed World.* [Online] www.migrationinformation.org/Profiles/display.cfm?ID=787 [Accessed 30 September 2011].

Glick Schiller, N. 1999. Citizens in transnational nation-states: The Asian experience. *In:* Olds, K., Dicken, P., Kelly, P. F., Kong, L. & Yeang, H. W. (eds.). *Globalisation and the Asia-Pacific: Contested Territories*, London and New York, Routledge, 202–218.

Glick Schiller, N., & Salazar, N. B. (2013). Regimes of mobility across the globe. *Journal of ethnic and migration studies*, 39(2), 183–200.

Go, S. P. 2002. Detailed case study of Philippines. *In:* Iredale, R., Hawksley, C. & Lyon, K. (eds.) *Migration Research and Policy Landscape: Case Studies of Australia, the Philippines and Thailand*, Wollongong, Asia Pacific Migration Research Network.

Goldberg, D. 1993. *Racist Culture: Philosophy and the Politics of Meaning*, Oxford, Blackwell.

Goldhagen, D. J. 1996. *Hitler's Willing Executioners: Ordinary Germans and the Holocaust*, New York, Alfred A. Knopf.

Goldin, I., Cameron, G. & Balarajan, M. 2011. *Exceptional People: How Migration Shaped Our World and Will Define Our Future*, Princeton, NJ and Oxford, Princeton University Press.

Goldring, L. 2004. Family and collective remittances to Mexico: A multi-dimensional typology. *Development and Change*, 35, 799–840.

Goldring, L. & Landolt, P. 2011. Caught in the work-citizenship matrix: The lasting effects of precarious legal status on work for Toronto immigrants. *Globalizations*, 8, 325–341.

González Ferrer, A. 2012. ¿Se van los españoles? Sí. Y deberíamos preocuparnos. *El Diario*, 8 October 2012 [Online].

Gordon, S. L. 2016. Welcoming refugees in the rainbow nation: Contemporary attitudes towards refugees in South Africa. *African Geographical Review*, 35, 1–17.

Gore, C. 2000. The rise and fall of the Washington consensus as a paradigm for developing countries. *World Development*, 28, 789–804.

Grant, E. E. 1925. Scum from the melting-pot. *American Journal of Sociology*, 30(6), 641–651.

Grau, P. C. 1994. Italian Presence in Modern Venezuela: Socioeconomic Dimension and Geo-cultural Changes, 1926–1990. *Center for Migration Studies special issues*, 11(3) 152–172.

Green, N. & Weil, P. 2007. *Citizenship and Those Who Leave*, Urbana, University of Illinois Press.

Greenhill, K. M. 2010. *Weapons of Mass Migration: Forced Displacement, Coercion, and Foreign Policy*, Ithaca, NY, Cornell University Press.

Griffith, A. 2017. *Building a Mosaic: The Evolution of Canada's Approach to Immigrant Integration* [Online], Washington D.C., Migration Policy Institute. https://www.migrationpolicy.org/article/building-mosaic-evolution-canadas-approach-immigrant-integration. [Accessed 30 May 2019].

Guarnizo, L. E., Portes, A. & Haller, W. 2003. Assimilation and transnationalism: Determinants of transnational political action among contemporary migrants. *American Journal of Sociology*, 108, 1211–1248.

Guimezanes, N. 1995. Acquisition of nationality in OECD countries. In: OECD (ed.) *Trends in International Migration: Annual Report*, Paris, OECD.

Gurak, D. T. & Caces, F. 1992. Migration networks and the shaping of international migration systems. In: Kritz, M. M., Lim, L. L. & Zlotnik, H. (eds.) *International Migration Systems: A Global Approach*, Oxford, Clarendon.

Gutierrez, D., J., B. & Terrazas 2012. *The 2012 Mexican Presidential Election and Mexican Immigrants of Voting Age in the United States* [Online], Washington, D.C., Migration Policy Institute. https://www.migrationpolicy.org/article/2012-mexican-presidential-election-and-mexican-immigrants-voting-age-united-states [Accessed 30 May 2019].

Gzesh, S. 2006. *Central Americans and Asylum Policy in the Reagan Era* [Online]. http://www.migrationinformation.org/Feature/display.cfm?id=384 [Accessed 11 November 2011].

Habermas, J. & Pensky, M. 2001. *The Postnational Constellation: Political Essays*, Cambridge, Polity in association with Blackwell Publishers.

Hage, G. 1998. *White Nation: Fantasies of White Supremacy in a Multicultural Society*, Sydney and New York, Pluto Press and Routledge.

Hakimzadeh, S. 2006. *Iran: A Vast Diaspora Abroad and Millions of Refugees at Home* [Online], Washington D.C., Migration Policy Institute,. http://www.migrationinformation.org/Profiles/display.cfm?id=424. [Accessed 3 September 2006].

Halliday, F. 1985. Migrations de main d'oeuvre dans le monde arabe: l'envers du nouvel ordre économique. *Revue Tiers Monde*, 26, 103.

Hamilton, N. & Stoltz Chinchilla, N. 1991. Central American Migration: A Framework for Analysis. *Latin American Research Review*, 26, 75–110.

Hammar, T. (ed.) 1985. *European Immigration Policy: A Comparative Study*, Cambridge, Cambridge University Press.

Hammar, T. 1990. *Democracy and the Nation-State: Aliens, Denizens and Citizens in a World of International Migration*, Aldershot, Avebury.

Hamood, S. 2006. *African Transit Migration through Libya to Europe: The Human Cost*, Cairo, FMRS, AUC.

Handlin, O. 1951. *The Uprooted. The Epic Story of the Great Migrations that Made the American People*, Boston, MA, Little, Brown.

Hardt, M. & Negri, A. 2000. *Empire*, Cambridge, MA., Harvard University Press.

Harris, J. R. & Todaro, M. P. 1970. Migration, unemployment and development: A two-sector analysis. *American Economic Review*, 60, 126–142.

Harris, N. 2002. *Thinking the Unthinkable: The Immigration Myth Exposed*, London, I. B. Tauris.

Harttgen, K. & Klasen, S. 2009. A Human Development Index by Internal Migration Status. *Human Development Research Paper No. 54*, New York, United Nations Development Programme, Human Development Report Office.

Hashim, I. H. 2003. Cultural and gender differences in perceptions of stressors and coping skills: A study of Western and African college students in China. *School Psychology International*, 24, 182–203.

Hatton, T. J. 2009. The rise and fall of asylum: What happened and why? *Economic Journal*, 119, F183–F213.

Hatton, T. J. & Williamson, J. G. 1993. After the famine: Emigration from Ireland, 1850–1913. *The Journal of Economic History*, 53, 575–600.

Hatton, T. J. & Williamson, J. G. 1998. *The Age of Mass Migration: Causes and Economic Effects*, Oxford and New York, Oxford University Press.

Haugen, H. Ø. 2012. Nigerians in China: A second state of immobility. *International Migration*, 50, 65–80.

Haugen, H. Ø. 2013. China's recruitment of African university students: Policy efficacy and unintended outcomes. *Globalisation, Societies and Education*, 11, 315–334.

Hayes, G. 1991. Migration, metascience, and development policy in Island Polynesia. *The Contemporary Pacific*, 3, 1–58.

Hearing, L. & Erf, R. V. D. 2001. Why do people migrate. *Statistics in Focus. Population and Social Conditions. Eurostat/European Communities*, 1–7.

Heckmann, F. & Schnapper, D. 2016. *The Integration of Immigrants in European Societies: National Differences and Trends of Convergence*, Berlin, Walter de Gruyter.

Heering, L., van der Erf, R. & van Wissen, L. 2004. The role of family networks and migration culture in the continuation of Moroccan emigration: A gender perspective. *Journal of Ethnic and Migration Studies*, 30, 323–337.

Heinemeijer, W. F., van Amersfoort, J. A., Ettema, W., De Mas, P. & van der Wusten, H. 1977. *Partir pour rester, une enquête sur les incidences de l'émigration ouvrière à la campagne marocaine*, Den Haag, NUFFIC.

Held, D., McGrew, A., Goldblatt, D. & Perraton, J. 1999. *Global Transformations: Politics, Economics and Culture*, Cambridge, Polity.

Henry, S., Schoumaker, B. & Beauchemin, C. 2004. The impact of rainfall on the first out-migration: A multi-level event-history analysis in Burkina Faso. *Population and Environment*, 25, 423–460.

Heusala, A.-L. & Aitamurto, K. (eds.) 2016. *Migrant Workers in Russia: Global Challenges of the Shadow Economy in Societal Transformation*, London, Routledge.

Higham, J. 2002. *Strangers in the land: Patterns of American nativism, 1860–1925*, New Brunswick, Rutgers University Press.

Hirsh, M. 2017. Emerging infrastructures of low-cost aviation in Southeast Asia. *Mobilities*, 12, 259–276.

HKSARG. 2011. *Foreign Domest Helpers* [Online], Hong Kong, Immigration Department, The Government of the Hong Kong Special Administrative Region. http://www.immd.gov.hk/ehtml/faq_fdh.htm#9 [Accessed 22 January 2012].

Hochschild, A. R. 2000. Global care chains and emotional surplus value. In: Hutton, W. & Giddens, A. (eds.) *On the Edge: Living with Global Capitalism*, London, Jonathan Cape.

Hochschild, A. R. 2018. *Strangers in their Own Land: Anger and Mourning on the American Right*, New York, The New Press.

Hoerder, D. 2002. *Cultures in Contact: World Migrations in the Second Millennium*, Durham, Duke University Press.

Hollifield, J. 1992. *Immigrants, Markets and States: The Political Economy of Postwar Europe*, Cambridge, MA., Harvard University Press.

Hollifield, J., Martin, P. & Orrenius, P. (eds.) 2013. *Controlling Immigration: A Global Perspective*, Stanford, CA, Stanford University Press.

Homer-Dixon, T. & Percival, V. 1996. *Environmental Security and Violent Conflict: Briefing Book*, Toronto, University of Toronto and American Association for the Advancement of Science.

Homze, E. L. 1967. *Foreign Labor in Nazi Germany*, Princeton, NJ, Princeton University Press.

Horst, C. 2006. *Transnational Nomads: How Somalis Cope with Refugee Life in the Dadaab Camps of Kenya*, New York, Oxford, Berghahn.

Horwood, C. & Hooper, K. 2016. *Protection on the Move: Eritrean Refugee Flows through the Greater Horn of Africa*, Washington, D.C., Migration Policy Institute.

Housen, T., Hopkins, S. & Earnest, J. 2013. A systematic review on the impact of internal remittances on poverty and consumption in developing countries: Implications for policy. *Population, Space and Place*, 19(5), 610–632.

Huang, S., Yeoh, B. & Rahman, N. A. 2005. *Asian Women as Transnational Domestic Workers*, Singapore, Marshall Cavendish Academic.

Hugo, G. (ed.) 2013. *Migration and Climate Change*, Cheltenham, Edward Elgar.

Hugo, G. J. 2005. *Migration in the Asia-Pacific Region*, Geneva, Global Commission on International Migration.

Hugo, G. J. 2016. Internal and international migration in East and Southeast Asia: Exploring the linkages. *Population, Space and Place*, 22, 651–668.

Human Rights First. 2008. *2008 Hate Crime Survey [Online]*, Washington, D.C., Human Rights First.: https://www.humanrightsfirst.org/sites/./FD-081103-hate-crime-survey-2008.pdf. [Accessed 30 May 2019]

Human Rights Watch. 2011. *World Report 2011: Events of 2010*, New York, Human Rights Watch.

Hur, -J.-J. & Lee, K. 2008. Demographic Change and International Labor Mobility in Korea. *PECC-ABAC Conference on "Demographic Change and International Labor Mobility in the Asia Pacific Region: Implications for Business and Cooperation"*, Seoul, Korea, 25–26 March.

İçduygu, A. 2000. The Politics of International Migratory Regimes: Transit Migration Flows in Turkey. *International Social Science Journal*, 165, 357–366.

İçduygu, A. & Yükseker, D. 2012. Rethinking transit migration in Turkey: Reality and re-presentation in the creation of a migratory phenomenon. *Population, Space and Place*, 18, 441–456.

Ignatieff, M. 1994. *Blood and Belonging: Journeys into the New Nationalism*, London, Vintage

ILO. 2006. *Realizing Decent Work in Asia: Fourteenth Asian Regional Meeting: Report of the Director-General*, Geneva, International Labour Office.

ILO. 2007. *Labour and Social Trends in ASEAN 2007*, Bangkok, International Labour Office Regional Office for Asia and the Pacific.

Immigration Department of Malaysia. 2012. *Foreign Worker* [Online]. http://www.imi.gov.my/[Accessed 24 January 2012].

Infantino, F. 2010. La frontière au guichet. Politiques et pratiques des visas Schengen aux Consulat et à l'Ambassade d'Italie au Maroc. *Champ pénal/Penal field, Nouvelle Revue Internationale de Criminologie*, VII.

Infantino, F. 2014. Bordering 'fake' marriages? The everyday practices of control at the consulates of Belgium, France, and Italy in Casablanca. *Etnografia e Ricerca Qualitativa*, 1, 27–48.

INS 2002. *Statistical Yearbook of the Immigration and Naturalization Service, 1999*, Washington, D.C., US Government Printing Office.

IOM 2005. *World Migration 2005: Costs and Benefits of International Migration*, Geneva, International Organization for Migration.

Iredale, R., Guo, F. & Rozario, S. (eds.) 2002. *Return Skilled Migration and Business Migration and Social Transformation*, Wollongong, Centre for Asia Pacific Social Transformation Studies.

Ivakhnyuk, I. 2009. The Russian migration policy and its impact on human development: The historical perspective. *Human Development Research Paper No. 91*, New York, United Nations Development Programme, Human Development Report Office.

Jachimowicz, M. 2006. *Argentina: A New Era of Migration and Migration Policy* [Online]. http://www.migrationinformation.org/USfocus/display.cfm?ID=374 [Accessed 5 December 2011].

Jackson, J. A. 1963. *The Irish in Britain*, London, Routledge and Kegan Paul.

Janmaat, J. G. 2007. The ethnic 'other'in Ukrainian history textbooks: The case of Russia and the Russians. *Compare*, 37, 307–324.

Janmaat, J. G. & Vickers, E. 2007. Education and identity formation in post-cold war Eastern Europe and Asia. *Compare*, 37, 267–275.

Jimenez, M. 2009. *Humanitarian Crisis: Migrant Deaths at the U.S.–Mexico Border*. San Diego: American Civil Liberties Union of San Diego and Imperial Counties & Comisión Nacional de los Derechos Humanos.

Jokisch, B. 2007. *Ecuador: Diversity in Migration* [Online]. http://www.migrationinformation.org/USfocus/display.cfm?ID=575 [Accessed 1 November 2011].

Jones, K. & Smith, A. D. 1970. *The Economic Impact of Commonwealth Immigration*, Cambridge, Cambridge University Press.

Jones, R. C. 1998a. Remittances and inequality: A question of migration stage and geographical scale. *Economic Geography*, 74, 8–25.

Jones, R. C. 1998b. Introduction: The renewed role of remittances in the new world order. *Economic Geography*, 74, 1–7.

Jónsson, G. 2010. *The Environmental Factor in Migration Dynamics – A Review of African Case Studies*. IMI Working paper 21, International Migration Institute, University of Oxford.

Joppke, C. 1998. *The Challenge to the Nation-State: Immigration in Western Europe and the United States*, New York, Oxford University Press.

Joppke, C. 1999. *Immigration and the Nation-State: The United States, Germany and Britain*, Oxford, Oxford University Press.

Joppke, C. 2001. The legal-domestic sources of immigrant rights: The United States, Germany, and the European Union. *Comparative Political Studies*, 34, 339–366.

Joppke, C. 2007. Transformation of citizenship: Status, rights, identity. *Citizenship Studies*, 11, 37–48.

Jordan, B. & Düvell, F. 2002. *Irregular Migration: The Dilemmas of Transnational Mobility*, Cheltenham, Edward Elgar.

Jung Park, Y. 2009. *Chinese Migration in Africa*, Johannesburg, South African Institute of International Affairs (SAIIA).

Jupp, J. (ed.) 2001. *The Australian People: An Encyclopedia of the Nation, Its People and Their Origins*, Cambridge, Cambridge University Press.

Jupp, J. 2002. *From White Australia to Woomera: The History of Australian Immigration*, Melbourne, Cambridge University Press.

Jureidini, R. 2003. L'échec de la protection de l'État: les domestiques étrangers au Liban. *Revue Européenne des Migrations Internationales*, 19, 95–127.

Kanso, H. 2018. *Despite reforms, Qatar's migrant workers still fear exploitation* [Online]. Reuters. Despite reforms, Qatar's migrant workers still fear exploitation [Accessed 28 May 2019].

Kapur, D. 2003. Remittances: The new development mantra? *Paper prepared for the G-24 Technical Group Meeting, 15–16 September*, New York and Geneva, United Nations.

Kapur, D. & McHale, J. 2005. *Give us Your Best and Brightest: The Global Hunt for Talent and its Impact on the Developing World*, Washington, D.C., Center for Global Development.

Kashiwazaki, C. & Akaha, T. 2006. *Japanese Immigration Policy: Responding to Conflicting Pressures* [Online]. http://www.migrationinformation.org/Profiles/display.cfm?ID=487 [Accessed 19 September 2011].

Kassim, A. & Zin, R. H. M. 2011. Policy on Irregular Migrants in Malaysia: An Analysis of its implementation and Effectiveness. *Discussion Paper Series No. 2011–34*.

Katseli, L. T., Lucas, R. E. B. & Xenogiani, T. 2006. *Effects of Migration on Sending Countries: What do we know?*, Paris, OECD.

Kay, D. & Miles, R. 1992. *Refugees or Migrant Workers? European Volunteer Workers in Britain 1946–1951*, London, Routledge.

Keely, C. B. 2001. The international refugee regimes(s): The end of the Cold War matters. *International Migration Review*, 35, 303–314.

Kellner, D. 2017. Brexit plus, whitelash, and the ascendency of Donald J. Trump. *Cultural Politics*, 13, 135–149.

Kenbib, M. 1999. Les migrations des juifs marocains à l'époque contemporaine. *In*: Berriane, M. & Popp, H. (eds.) *Migrations Internationales entre le Maghreb et l'Europe*, Rabat, Université Mohammed V.

Kenny, K. 2006. Race, violence, and anti-Irish sentiment in the nineteenth century. *In*: Lee, J. J. & Casey, M. R. (eds.). *Making the Irish American*, London & New York, New York University Press, 364–378.

Kenny, M. 1962. Twentieth-Century Spanish expatriates in Mexico: An urban sub-culture. *Anthropological Quarterly*, 35, 169–180.

Kepel, G. 2002. *Jihad: The Trail of Political Islam*, Cambridge, MA, Belknap of Harvard University.

Khadria, B. 2008. India: Skilled migration to developed countries, labour migration to the Gulf. *In*: Castles, S. & Delgado Wise, R. (eds.) *Migration and Development: Perspectives from the South*, Geneva, International Organization for Migration.

Kim, G. & Kilkey, M. 2018. Marriage migration policy in South Korea: Social investment beyond the nation state. *International Migration*, 56, 23–38.

Kim, H. M., Kim, G.-D., Kim, M.-J., Kim, J. S. & Kim, C. 2007. Research on labour and marriage migration process from Mongolia and Vietnam to Korea and the impact on migrant rights (in Korean). *2007 Joint Project of Inter-Asian NIs on Current Human Rights Issues*, Seoul, National Human Rights Commission.

Kim, W. B. 1996. Economic interdependence and migration dynamics in Asia. *Asian and Pacific Migration Journal*, 5, 303–317.

Kindleberger, C. P. 1967. *Europe's Postwar Growth – the Role of Labor Supply*, Cambridge, MA., Harvard University Press.

King, R., Lazaridis, G. & Tsardanidis, C. (eds.) 2000. *Eldorado or Fortress? Migration in Southern Europe*, London, Macmillan.

King, R. & Skeldon, R. 2010. 'Mind the Gap!' Integrating approaches to internal and international migration. *Journal of Ethnic and Migration Studies*, 36, 1619–1646.

King, R., Thomson, M., Fielding, T. & Warnes, T. 2006. Time, generations and gender in migration and settlement. *In:* Penninx, R., Berger, M. & Kraal, K. (eds.) *The Dynamics of International Migration and Settlement in Europe*, Amsterdam, Amsterdam University Press.

King, R. & Vullnetari, J. 2006. Orphan pensioners and migrating grandparents: The impact of mass migration on older people in rural Albania. *Ageing & Society*, 26, 783–816.

Kirişci, K. 2006. National identity, asylum and immigration: The EU as a vehicle of postnational transformation in Turkey. *In:* Kieser, H. L. (ed.) *Turkey Beyond Nationalism: Toward Post-Nationalist Identities*, London, IB Tauris.

Kirişçi, K. 2007. Turkey: A country of transition from emigration to immigration. *Mediterranean Politics*, 12, 91–97.

Kiser, G. & Kiser, M. (eds.) 1979. *Mexican Workers in the United States*, Albuquerque, University of New Mexico Press.

Klepp, S. 2017. Climate change and migration. *Oxford Research Encyclopedia of Climate Science*, 1.

Klepp, S. E. & Smith, B. G. (eds.) 1992. *The Infortunate: The Voyage and Adventures of William Moraley an Indentured Servant*, University Park, Pennsylvania State University Press.

Kloosterman, R. C. 2018. Migrant entrepreneurs and cities: New opportunities, newcomers, new issues. *The Routledge Handbook of the Governance of Migration and Diversity in Cities*, London, Routledge.

Klug, F. 1989. "Oh to be in England": The British case study. *In:* Yuval-Davis, N. & Anthias, F. (eds.) *Woman–Nation–State*, London, Macmillan.

Komai, H. 1995. *Migrant Workers in Japan*, London, Kegan Paul International.

Komine, A. 2018. A closed immigration country: Revisiting Japan as a negative case. *International Migration*, 56, 106–122.

Koopmans, R. 2010. Trade-offs between equality and difference: Immigrant integration, multiculturalism and the welfare state in cross-national perspective. *Journal of Ethnic and Migration Studies*, 36, 1–26.

Korean Immigration Service. 2002. *Immigration Statistics 2001 (in Korean)* [Online], Korea, Ministry of Justice. http://www.immigration.go.kr [Accessed 20 January 2012].

Korean Immigration Service. 2012. *Immigration Statistics, December 2011 (in Korean)* [Online], Korea, Ministry of Justice. http://www.immigration.go.kr [Accessed 20 January 2012].

Koslowski, R. 2000. *Migrants and Citizens*, Ithaca, NY, Cornell University Press.

Koslowski, R. 2008. Global mobility and the quest for an international migration regime. *Conference on International Migration and Development: Continuing the Dialogue-Legal and Policy Perspectives*, New York, CMS and IOM.

Kothari, U. 2014. Political discourses of climate change and migration: Resettlement policies in the Maldives. *The Geographical Journal*, 180, 130–140.

Kramer, R. 1999. *Developments in International Migration to the United States*, Washington D.C., Department of Labor.

Kress, B. 2006. *Burkina Faso: Testing the Tradition of Circular Migration* [Online], Washington, D.C., Migration Policy Institute, Available: https://www.migrationpolicy.org/article/burkina-faso-testing-tradition-circular-migration. [Accessed 30 May 2019].

Krissman, F. 2005. Sin coyote ni patrón: Why the "migrant network" fails to explain international migration. *International Migration Review*, 39, 4–44.

Kritz, M. M. 1975. The impact of international migration on Venezuelan demographic and social structure. *International Migration Review*, 9(4), 513–543.

Kritz, M. M., Lim, L. L. & Zlotnik, H. (eds.) 1992. *International Migration System: A Global Approach*, Oxford, Clarendon.

Krogstad, J. M., Passel, J. S. & Cohn, D. V. 2018. *5 facts about illegal immigration in the U.S.* [Online]. Pew Research Center. https://www.pewresearch.org/fact-tank/2018/11/28/5-facts-about-illegal-immigration-in-the-u-s/ [Accessed 28 May 2019].

Krugman, P. 1995. *Development, Geography, and Economic Theory*, Cambridge, MA, MIT Press.

Kubal, A. 2013. Conceptualizing semi-legality in migration research. *Law and Society*, 47, 555–587.

Kubal, A. 2016. *Socio-Legal Integration: Polish Post-2004 EU Enlargement Migrants in the United Kingdom*, London, Routledge.

Kubal, A. 2019. *Immigration and Refugee Law in Russia. Socio-Legal Perspectives*, Cambridge, Cambridge University Press.

Kubal, A., Bakewell, O. & de Haas, H. 2011. The evolution of Brazilian migration to the UK: A scoping study report. *Theorizing the Evolution of European Migration Systems (THEMIS)*, International Migration Institute, University of Oxford.

Kubat, D. 1987. Asian immigrants to Canada. *In:* Fawcett, J. T. & Cariño, B. V. (eds.) *Pacific Bridges: The New Immigration from Asia and the Pacific Islands*, New York, Center for Migration Studies.

Kuhn, T. S. 1962 *The Structure of Scientific Revolutions*, Chicago, University of Chicago Press.

Kulischer, E. M. 1948. *Europe on the Move: War and Population Changes*, New York, Columbia University Press.

Kureková, L. 2011. *The Role of Welfare Systems in Affecting Out-Migration: The Case of Central and Eastern Europe*. IMI working paper 46, International Migration Institute, University of Oxford.

Kureková, L. 2013. Welfare systems as emigration factor: Evidence from the new accession states. *Journal of Common Market Studies*, 51, 721–739.

Kuzio, T. 2002. History, memory and nation building in the post-Soviet colonial space. *Nationalities Papers*, 30, 241–264.

Kyle, D. & Liang, Z. 2001. Migration merchants: Human smuggling from Ecuador and China. *In:* Guiraudon, V. & Joppke, C. (eds.) *Controlling a New Migration World*, London and New York, Routledge.

Kymlicka, W. 1995. *Multicultural Citizenship: A Liberal Theory of Minority Rights*, Oxford, Clarendon.

Kymlicka, W. & He, B. 2005. *Multiculturalism in Asia*, Oxford University Press on Demand.

Lacroix, T. 2005. *Les réseaux marocains du développement: Géographie du transnational et politiques du territorial*, PhD thesis, Paris, Presses de Sciences Po.

Lahav, G. 2004. *Immigration and Politics in the New Europe: Reinventing Borders*, New York, Cambridge University Press.

Landau, L. B. & Freemantle, I. 2009. Tactical cosmopolitanism and idioms of belonging: Insertion and self-exclusion in Johannesburg. *Journal of Ethnic and Migration Studies*, 36, 375–390.

Laub, Z. 2014. *The Taliban in Afghanistan* [Online]. Council on Foreign Relations. https://www.cfr.org/backgrounder/taliban-afghanistan. [Accessed 13 May 2019].

Laurens, H. 2005. Les migrations au Proche-Orient de l'empire ottoman aux etats-nations. Une perspective historique. *In:* Jaber, H. & Métrai, F. (eds.) *Mondes en mouvements. Migrants et migrations au Moyen-Orient au tournant du XXIe siècle*, Beyrouth, Institut Français du Proche Orient IFPO.

Lavergne, M. 2003. Golfe arabo-persique: un système migratoire de plus en plus tourné vers l'Asie. *Revue Européenne Des Migrations Internationales*, 19, 229–241.

Lee, E. S. 1966. A Theory of Migration. *Demography*, 3, 47–57.

Lee, H.-K. 2008. International marriage and the state in South Korea: Focusing on governmental policy. *Citizenship Studies*, 12, 107–123.

Lee, H. K. 2010. Family migration issues in North-East Asia. *Background Paper for the World Migration Report 2010*, Geneva, International Organization for Migration.

Lee, J. S. & Wang, S.-W. 1996. Recruiting and managing of foreign workers in Taiwan. *Asian and Pacific Migration Journal*, 5, 2–3.

Leichtman, M. A. 2005. The legacy of transnational lives: Beyond the first generation of Lebanese in Senegal. *Ethnic and Racial Studies*, 28, 663–686.

Lever-Tracey, C. & Quinlan, M. 1988. *A Divided Working Class*, London and New York, Routledge and Kegan Paul.

Levine, B. B. 1987. The Puerto Rican exodus: Development of the Puerto Rican circuit. *In:* Levine, B. B. (ed.) *The Caribbean Exodus*, West Port, CT, Praeger.

Levitt, P. 1998. Social remittances: Migration driven local-level forms of cultural diffusion. *International Migration Review*, 32, 926–948.

Levitt, P. & Glick Schiller, N. 2004. Conceptualising simultaneity: A transnational social field perspective on society. *International Migration Review*, 38, 1002–1039.

Levitt, P. & Lamba-Nieves, D. 2011. Social remittances revisited. *Journal of Ethnic and Migration Studies*, 37, 1–22.

Levy, D. 1999. Coming home? Ethnic Germans and the transformation of national identity in the Federal Republic of Germany. *In:* Geddes, A. & Adrian, F. (eds.) *The Politics of Belonging: Migrants and Minorities in Contemporary Europe*, Aldershot, Ashgate.

Lewin, P. A., Fisher, M. & Weber, B. 2012. Do rainfall conditions push or pull rural migrants: Evidence from Malawi. *Agricultural Economics*, 43, 2, 191–204.

Lewis, J. R. 1986. International labour migration and uneven regional development in labour exporting countries. *Tijdschrift voor Economische en Sociale Geografie*, 77, 27–41.

Lewis, W. A. 1954. Economic development with unlimited supplies of labour. *Manchester School of Economic and Social Studies*, 22, 139–191.

Li, X. & Wang, D. 2015. The impacts of rural–urban migrants' remittances on the urban economy. *Annals of Regional Science*, 54, 591–603.

Li, X. R., Harrill, R., Uysal, M., Burnett, T. & Zhan, X. 2010. Estimating the size of the Chinese outbound travel market: A demand-side approach. *Tourism Management*, 31, 250–259.

Lieten, G. K. & Nieuwenhuys, O. 1989. Introduction: Survival and emancipation. *In:* Lieten, G. K., Nieuwenhuys, O. & Schenk-Sandbergen, L. (eds.) *Women, Migrants and Tribals: Survival Strategies in Asia*, New Delhi, Manohar.

Light, I. & Bonacich, E. 1988. *Immigrant Entrepreneurs*, Berkeley, University of California Press.

Lightfoot, D. R. & Miller, J. A. 1996. Sijilmassa: The rise and fall of a walled oasis in medieval Morocco. *Annals of the Association of American Geographers*, 86, 78–101.

Lindley, A. 2012. *The Early Morning Phonecall: Somali Refugees' Remittances*, New York, Oxford, Berghahn.

Lindsay, C. 2001. The Caribbean community in Canada. *Profiles of Ethnic Communities in Canada*, Ottawa, Social and Aboriginal Statistics Division.

Lipton, M. 1980. Migration from the rural areas of poor countries: The impact on rural productivity and income distribution. *World Development*, 8, 1–24.

Lockard, C. A. 2013. Chinese migration and settlement in Southeast Asia before 1850: Making fields from the sea. *History Compass*, 11, 765–781.

Loescher, G. 2001. *The UNHCR and World Politics: A Perilous Path*, Oxford, Oxford University Press.

Lohrmann, R. 1987. Irregular migration – a rising issue in developing countries. *International Migration*, 25, 253–266.

Longva, A. N. 1999. Keeping migrant workers in check: The Kafala system in the Gulf. *Middle East Report*, 29, 20–22.

López, G., Bialik, K. & Radford, J. 2018. *Key Findings about U.S. Immigrants* [Online]. Pew Research Center. https://www.pewresearch.org/fact-tank/2018/11/30/key-findings-about-u-s-immigrants/[Accessed 28 May 2019].

Lopez, M. & Taylor, P. 2012. *Latino Voters in the 2012 Election* [Online]. Available: http://www.pewhispanic.org/2012/11/07/latino-voters-in-the [Accessed 7 November 2012].

Lovejoy, P. E. 1989. The impact of the Atlantic slave trade on Africa: A review of the literature. *The Journal of African History*, 30, 365–394.

Lowell, L. B. & Findlay, A. 2002. *Migration of Highly Skilled Persons from Developing Countries: Impact and Policy Responses*, Geneva and London, International Labour Organization and United Kingdom Department for International Development.

Luan, L. 2018. *Profiting from Enforcement: The Role of Private Prisons in U.S. Immigration Detention* [Online]. Migration Policy Institute. https://www.migrationpolicy.org/article/profiting-enforcement-role-private-prisons-us-immigration-detention [Accessed 17 May 2019].

Lubkemann, S. C. 2008. Involuntary immobility: On a theoretical invisibility in forced migration studies. *Journal of Refugee Studies*, 21, 454–475.

Lucas, R. E. B. & Stark, O. 1985. Motivations to remit: Evidence from Botswana. *Journal of Political Economy*, 93, 901–918.

Lucassen, J. 1995. Emigration to the Dutch colonies and the USA. In: Cohen, R. (ed.) *The Cambridge Survey of World Migration*, Cambridge, Cambridge University Press.

Lucassen, L. 2005. *The Immigrant Threat: The Integration of Old and New Migrants in Western Europe since 1890*, Urbana and Chicago, University of Illinois Press.

Lucassen, L., Feldman, D. & Oltmer, J. 2006. Immigrant integration in Western Europe, then and now. In: Lucassen, L., Feldman, D. & Oltmer, J. (eds.) *Paths of Integration: Migrants in Western Europe (1880–2004)*, Amsterdam, Amsterdam University Press.

Lustik, I. S. 2011. Israel's migration balance: Demography, politics and ideology. *Israel Studies Review*, 26, 33–65.

Lutz, H. 2016. *Migration and Domestic Work: A European Perspective on a Global Theme*, London, Routledge.

Lutz, H. 2018. Care migration: The connectivity between care chains, care circulation and transnational social inequality. *Current Sociology*, 66, 577–589.

Lutz, H. & Palenga-Möllenbeck, E. 2010. Care work migration in Germany: Semi-compliance and complicity. *Social Policy and Society*, 9, 419–430.

Lutz, W., Kritzinger, S. & Skirbekk, V. 2006. The demography of growing European identity. *Science*, 314, 425–425.

Mabogunje, A. L. 1970. Systems Approach to a Theory of Rural-Urban Migration. *Geographical Analysis*, 2, 1–18.

Mac Con Uladh, D. H. T. 2005. *Guests of the Socialist Nation? Foreign Students and Workers in the GDR, 1949–1990*. PhD thesis, University College London.

MacMaster, N. 1991. The "seuil de tolérance": The uses of a "scientific" racist concept. In: Silverman, M. (ed.) *Race Discourse and Power in France*, Aldershot, Avebury.

Mafukidze, J. 2006. A discussion of migration and migration patterns and flows in Africa. In: Cross, C., Gelderblom, D., Roux, N. & Mafukidze, J. (eds.) *Views on Migration in Sub-Saharan Africa*, Cape Town, HSRC Press.

Mahendra, E. 2014a. *Financial Constraints, Social Policy and Migration: Evidence from Indonesia*, IMI working paper 101, International Migration Institute, University of Oxford.

Mahendra, E. 2014b. *Trade Liberalisation and Migration Hump: NAFTA as a Quasi-Natural Experiment*. IMI working paper 98, International Migration Institute, University of Oxford.

Mahler, S. & Ugrina, D. 2006. *Central America: Crossroads of the Americas* [Online]. http://www.migrationinformation.org/Feature/display.cfm?ID=386 [Accessed 30 September 2011].

Maier, T. 2015. *Historical Developments of Immigration and Emigration* [Online]. Bundeszentrale für Politische Bildung. https://www.bpb.de/gesellschaft/migration/laenderprofile/203942/historical-developments [Accessed 28 May 2019].

Mallee, H. & Pieke, F. N. 2014. *Internal and International Migration: Chinese Perspectives*, London, Routledge.

Mamdani, M. 2014. *When Victims Become Killers: Colonialism, Nativism, and the Genocide in Rwanda*, Princeton, NJ, Princeton University Press.

Manby, B. 2016. *Citizenship Law in Africa*, New York, African Minds.

Manning, P. 2005. *Migration in World History*, New York, Routledge.

Manuh, T. (ed.) 2005. *At Home in the World? International Migration and Development in Contemporary Ghana and West Africa*, Accra, Ghana, Sub-Saharan Publishers.

Martin, P. L. 1991. *The Unfinished Story: Turkish Labour Migration to Western Europe*, Geneva, International Labour Office.

Martin, P. L. 1993. *Trade and Migration: NAFTA and Agriculture*, Washington, D.C., Institute for International Economics.

Martin, P. L. & Miller, M. J. 2000. *Employer Sanctions: French, German and US Experiences*, Geneva, ILO.

Martin, P. L. & Taylor, J. E. 1996. The anatomy of a migration hump. In: Taylor, J. E. E. (ed.) *Development Strategy, Employment, and Migration: Insights from Models*, Paris, OECD Development Centre.

Martínez Pizarro, J. (ed.) 2011. *Migración Internacional En América Latina Y El Caribe - Nuevas Tendencias, Nuevos Enfoques*, Santiago de Chile, CEPAL.

Martiniello, M. 1994. Citizenship of the European Union: A critical view. *In:* Bauböck, R. (ed.) *From Aliens to Citizens*, Aldershot, Avebury.

Marx, K. 1976. *Capital*, Harmondsworth, Penguin.

Massey, D. S. 1988. Economic development and international migration in comparative perspective. *Population and Development Review*, 14, 383–413.

Massey, D. S. 1990. Social structure, household strategies, and the cumulative causation of migration. *Population Index*, 56, 3–26.

Massey, D. S. 1999. International migration at the dawn of the twenty-first century: The role of the state. *Population and Development Review*, 25, 303–322.

Massey, D. S. 2000. To study migration today, look to a parallel era. *Chronicle of Higher Education*, 46, 5.

Massey, D. S. 2004. Social and economic aspects of immigration. *Annals of the New York Academy of Sciences*, 1038, 206–212.

Massey, D. S. 2007. *Categorically unequal: The American stratification system*, Russell Sage Foundation.

Massey, D. S., Arango, J., Hugo, G., Kouaouci, A., Pellegrino, A. & Taylor, J. E. 1993. Theories of international migration: A review and appraisal. *Population and Development Review*, 19, 431–466.

Massey, D. S., Arango, J., Hugo, G., Kouaouci, A., Pellegrino, A. & Taylor, J. E. 1998. *Worlds in Motion: Understanding International Migration at the End of the Millennium*, Oxford, Clarendon Press.

Massey, D. S. & Denton, N. A. 1993. *American Apartheid: Segregation and the Making of the Underclass*, Cambridge, Harvard University Press.

Massey, D. S., Durand, J. & Pren, K. A. 2016. Why border enforcement backfired. *American Journal of Sociology*, 121, 1557–1600.

Massey, D. S. & España, F. G. 1987. The social process of international migration. *Science*, 237, 733–738.

Massey, D. S. & Pren, K. A. 2012. Unintended consequences of US immigration policy: Explaining the post-1965 surge from Latin America. *Population and Development Review*, 38, 1–29.

Mazzucato, V., Kabki, M. & Smith, L. 2006. Locating a Ghanaian funeral: Remittances and practices in a transnational context. *Development and Change*, 37, 1047–1072.

Mbaye, L. M. 2014. "Barcelona or die": Understanding illegal migration from Senegal. *IZA Journal of Migration*, 3, 1–19.

McCabe, K. 2011. *Caribbean Immigrants in the United States* [Online]. http://www.migrationinformation.org/USfocus/display.cfm?ID=834 [Accessed 11 November 2011].

McCarthy, J. 1995. *Death and Exile: The Ethnic Cleansing of Ottoman Muslims 1821–1922*, Princeton, Darwin Press.

McDougall, J. & Scheele, J. 2012. *Saharan Frontiers: Space and Mobility in Northwest Africa*, Bloomington, IN, Indiana University Press.

McDowell, C. & de Haan, A. 1997. *Migration and Sustainable Livelihoods: A Critical Review of the Literature*, Sussex, Institute of Development Studies.

McKenzie, D., Stillman, S. & Gibson, J. 2010. How important is selection? Experimental vs. non-experimental measures of the income gains from migration. *Journal of the European Economic Association*, 8, 913–945.

McKenzie, D. J. 2006. Beyond remittances: The effects of migration on Mexican households. *In:* Özden, Ç. & Schiff, M. (eds.) *International Migration, Remittances, and the Brain Drain*, Washington, D.C., World Bank.

McKeown, A. 2004. Global migration, 1846–1940. *Journal of World History*, 15, 155–189.

McKeown, A. 2010. Chinese emigration in global context, 1850–1940. *Journal of Global History*, 5, 95–124.

McKinnon, M. 1996. *Immigrants and Citizens: New Zealanders and Asian Immigration in Historical Context*, Wellington, Institute of Policy Studies.

Meissner, D., Kerwin, D. M., Chishti, M. & Bergeron, C. 2013. *Immigration Enforcement in the United States: The Rise of a Formidable Machinery*, Washington, D.C., Migration Policy Institute.

Meissner, D., Papademetriou, D. & North, D. 1987. *Legalization of Undocumented Aliens: Lessons from Other Countries*, Washington, D.C., Carnegie Endowment for International Peace.

Meissner, F. & Vertovec, S. 2015. Comparing super-diversity. *Ethnic and Racial Studies*, 38, 541–555.

Mendoza, D. R. 2018. *Triple Discrimination: Woman, Pregnant, and Migrant*, Washington, D.C., Fair Labor Association.

Menjívar, C. & Agadjanian, V. 2007. Men's migration and women's lives: Views from rural Armenia and Guatemala. *Social Science Quarterly*, 88, 1243–1262.

Migration Dialogue. 2007. *Migration News: Latin America* [Online]. http://migration.ucdavis.edu/mn/more.php?id=3250_0_2_0 [Accessed 27 December 2011].

Migration Dialogue. 2011. *Migration News: Latin America*. http://migration.ucdavis.edu/mn/more.php?id=3674_0_2_0 [Accessed 27 December 2011].

Migration Information Source. 2011. *Heading into the 2012 Elections, Republican Presidential Candidates Walk the Immigration Policy Tightrope*. Washington DC, Migration Policy Institute. www.migrationinformation.org/Feature/print.cfm?ID=867 [Accessed 23 February 2012].

Miles, R. & Brown, M. 2003. *Racism*, London, Routledge.

Milia-Marie-Luce, M. 2007. La grande migration des Antillais en France ou les années BUMIDOM. *Dynamiques migratoires de la Caraïbe*, 6, 93.

Miller, M. J. 1978. *The Problem of Foreign Worker Participation and Representation in France, Switzerland and the Federal Republic of Germany*, Madison, University of Wisconsin.

Miller, M. J. 1981. *Foreign Workers in Western Europe: An Emerging Political Force*, New York, Praeger.

Miller, M. J. 1999. Prevention of unauthorized migration. *In:* Bernstein, A. & Weiner, M. (eds.) *Migration and Refugee Policies: An Overview*, London and New York, Pinter.

Miller, M. J. 2002. Continuity and change in postwar French legalization policy. *In:* Messina, A. (ed.) *West European Immigration and Immigrant Policy in the New Century*, Westport, CT and London, Praeger.

Miller, M. J. & Gabriel, C. 2008. The US-Mexico honeymoon of 2001: A retrospective. *In:* Gabriel, C. & Pellerin, H. (eds.) *Governing International Labour Migration: Current Issues, Challenges and Dilemmas*, New York, Routledge.

Miller, M. J. & Stefanova, B. 2006. NAFTA and the European referent: Labor mobility in European and North American regional integration. *In:* Messina, A. & Lahav, G. (eds.) *The Migration Reader: Exploring Politics and Policies*, Boulder, CO, Lynne Reiner.

Ministry of Employment and Labor. 2010. *Introduction of Industry, Employment Permit System* [Online]. Seoul, Ministry of Employment and Labor. http://www.eps.go.kr/en/index.html [Accessed 5 July 2011].

Ministry of Manpower Singapore. 2011. Singapore Yearbook on Manpower Statistics, 2011.

Ministry of Overseas Indians Affairs. 2011. *Facts on Indian Diaspora* [Online]. Overseas Indian Facilitation Centre. http://www.oifc.in/Facts/Facts-on-Indian-Diaspora [Accessed 26 Janaury 2012].

Mitchell, K. & Sparke, M. 2016. The new Washington consensus: Millennial philanthropy and the making of global market subjects. *Antipode*, 48, 724–749.

Mitchell, M. I. 2012. Migration, citizenship and autochthony: strategies and challenges for state building in Côte d'Ivoire. *Journal of Contemporary African Studies*, 30, 267–287.

Moch, L. P. 1992. *Moving Europeans. Migration in Western Europe since 1650*, Bloomington, IN, Indiana University Press.

Moch, L. P. 1995. Moving Europeans: Historical migration practices in Western Europe. *In:* Cohen, R. (ed.) *The Cambridge Survey of World Migration*, Cambridge, Cambridge University Press.

Modood, T. 2007. Multiculturalism: a civic idea. Cambridge: Polity.

Mohan, G. & Tan-Mullins, M. 2016. Chinese migrants in Africa as new agents of development? An analytical framework. *The Power of the Chinese Dragon*, Berlin, Springer.

Molina, G. G. & Yañez, E. 2009. The moving middle: Migration, place premiums and human development in Bolivia. *Human Development Research Paper No. 46*, New York, United Nations Development Programme, Human Development Office.

Morgan-Trostle, J., Zheng, K. & Lipscombe, C. 2016. *The State of Black Immigrants*, New York, New York School of Law and Black Alliance for Just Immigration (BAJI).

Mori, H. 1997. *Immigration Policy and Foreign Workers in Japan*, London, Macmillan.

Morokvasic, M. 1984. Birds of passage are also women. *International Migration Review*, 18, 886–907.

Muller, J. Z. 2008. Us and them: The enduring power of ethnic nationalism. *Foreign Affairs*, 87(2) 18–35.

Müller, K. & Schwarz, C. 2018. *Making America Hate Again? Twitter and Hate Crime Under Trump*. Available at SSRN: https://ssrn.com/abstract=3149103 or http://dx.doi.org/10.2139/ssrn.3149103 [Accessed 21 July 2019].

Munck, R., Schierup, C.-U. & Delgado Wise, R. (eds.) 2011. *Globalizations: Special Issue: Migration, Work and Citizenship in a Global Era*, London, Routledge.

Münz, R. 1996. A continent of migration: European mass migration in the twentieth century. *New Community*, 22, 201–226.

Münz, R., Straubhaar, T., Vadean, F. & Vadean, N. 2007. *What are the Migrants' Contributions to Employment and Growth? A European Approach*, Hamburg, Hamburg Institute of International Economics.

Murji, K. & Solomos, J. (eds.) 2005. *Racialization: Studies in Theory and Practice*, Oxford, Oxford University Press.

Murphy, S. & Hansen-Kuhn, K. 2019. The true costs of US agricultural dumping. *Renewable Agriculture and Food Systems*, 1–15.

Musterd, S. 2005. Social and ethnic segregation in Europe: Levels, causes, and effects. *Journal of Urban Affairs*, 27, 331–348.

Musterd, S. & Ostendorf, W. 2009. Residential segregation and integration in the Netherlands. *Journal of Ethnic and Migration Studies*, 35, 1515–1532.

Musterd, S. & Van Kempen, R. 2009. Segregation and housing of minority ethnic groups in Western European cities. *Tijdschrift voor economische en sociale geografie*, 100, 559–566.

Musterd, S. & Vos, S. D. 2007. Residential dynamics in ethnic concentrations. *Housing Studies*, 22, 333–353.

Mutluer, M. 2003. Les migrations irrégulières en Turquie. Traduit par Stéphane de Tapia. *Revue Européenne Des Migrations Internationales*, 19, 151–172.

MVA. 1938. *Memorie van Antwoord*, Rijksbegroting voor het dienstjaar 1938.2.IV.9: 16. Bijlage A. Den Haag: Tweede Kamer der Staten Generaal.

Myers, N. & Kent, J. 1995. *Environmental Exodus: An Emergent Crisis in the Global Arena*, Washington, D.C., Climate Institute.

Myrdal, G. 1957. *Rich Lands and Poor*, New York, Harper and Row.

National Immigration Agency Taiwan. 2012. *Statistics* [Online]. http://www.immigration.gov.tw/lp.asp?ctNode=29986&CtUnit=16677&BaseDSD=7&mp=2 [Accessed 22 January 2012].

Natter, K. 2014. *Fifty years of Maghreb emigration: How states shaped Algerian, Moroccan and Tunisian emigration*. IMI working paper 95, University of Oxford, International Migration Institute.

Natter, K. 2018. Rethinking immigration policy theory beyond 'Western liberal democracies'. *Comparative Migration Studies*, 6(4), 1–21.

Nayar, D. 1994. International labour movements, trade flows and migration transitions: A theoretical perspective. *Asian and Pacific Migration Journal*, 3, 31–47.

Ness, I. 2005. *Immigrants, Unions and the New U.S. Labor Market*, Philadelphia, PA, Temple University Press.

Neumayer, E. 2006. Unequal access to foreign spaces: How states use visa restrictions to regulate mobility in a globalized world. *Transactions of the Institute of British Geographers*, 31, 72–84.

Newland, K. 2007. *A New Surge of Interest in Migration and Development* [Online], Washington, D.C., Migration Policy Institute. www.migrationinformation.org [Accessed 6 February 2007].

Newland, K., Agunias, D. R. & Terrazas, A. 2008. *Learning by Doing: Experiences of Circular Migration*, Washington, D.C., Migration Policy Institute

Noiriel, G. 1988. *Le creuset français: Histoire de l'immigration XIXe-XXe siècles*, Paris, Seuil.

Noiriel, G. 2007. *Immigration, antisémitisme et racisme en France (XIXe-XXe siècle)*, Paris, Fayard.

Norris, P. & Inglehart, R. F. 2012. Muslim integration into Western cultures: Between origins and destinations. *Political Studies*, 60(2), 228–251.

Nyberg-Sorensen, N., Van Hear, N. & Engberg-Pedersen, P. 2002. The migration-development nexus evidence and policy options state-of-the-art overview. *International Migration*, 40, 3–47.

Nye, J. P. 2004. *Soft Power: The Means to Success in the World Politics*, New York, Public Affairs.

O'Connor, A. & Batalova, J. 2019. *Korean Immigrants in the United States* [Online],

Washington D.C., Migration Policy Institute,. https://www.migrationpolicy.org/article/korean-immigrants-united-states [Accessed 30 May 2019].

O'Neil, K., Hamilton, K. & Papademetriou, D. 2005. *Migration in the Americas*, Geneva, Global Commission on International Migration.

Odmalm, P. 2011. Political parties and 'the immigration issue': Issue ownership in Swedish parliamentary elections 1991–2010. *West European Politics*, 34, 1070–1091.

OECD. 1987. *The Future of Migration*, Paris, OECD.

OECD. 2004. *Trends in International Migration: Annual Report 2003*, Paris, OECD.

OECD. 2006. *International Migration Outlook: Annual Report 2006*, Paris, OECD.

OECD. 2007. *International Migration Outlook: Annual Report 2007*, Paris, OECD.

OECD. 2011a. *Education at a Glance 2011: OEDC Indicators*, Paris, OECD.

OECD. 2011b. *International Migration Outlook: SOPEMI 2011*, Paris, OECD.

OECD. 2012. *International Migration Outlook: SOPEMI 2012*, Paris, OECD.

OECD. 2018. *International Migration Outlook: SOPEMI 2018*, Paris, OECD.

Olayo-Méndez, J. A. 2018. *Migration, Poverty, and Violence in Mexico: The Role of Casas de Migrantes*, DPhil Thesis, University of Oxford.

ONS. 2004. *Focus on Ethnicity and Identity* [Online], London, Office for National Statistics. www.statistics.gov.uk [Accessed 15 March 2004].

Oostindie, G. J. 2009. Migration paradoxes of non-sovereignty; A comparative perspective on the Dutch Caribbean. *In:* Clegg, P. & Pantojas-Garcia, E. (eds.) *Governance in the Non-Independent Caribbean; Challenges and Opportunities in the 21st Century*, Kingston and Miami, Ian Randle Publishers.

Organization of American States. 2011. *International Migration in the Americas: First Report of the Continuous Reporting System on International Migration in the Americas [SICREMI]*, Washington, D.C., OAS.

Orozco, M. & Rouse, R. 2007. *Migrant Hometown Associations and Opportunities for Development: A Global Perspective* [Online], Washington, D.C., Migration Policy Institute.

http://migrationinformation.org [Accessed 6 February 2007].

Ortega, F. & Peri, G. 2013. The effect of income and immigration policies on international migration. *Migration Studies*, 1, 47–74.

Ottaviano, G. I. P. & Peri, G. 2012. Rethinking the effect of immigration on wages. *Journal of the European Economic Association*, 10, 152–197.

Oucho, J. O. 1996. *Urban Migrants and Rural Development in Kenya*, Nairobi, Nairobi University Press.

Özden, Ç. & Phillips, D. 2015. What really is brain drain? Location of birth, education and migration dynamics of African doctors, Washington, D.C., Global Knowledge Partnership on Migration and Development.

Özden, Ç. & Schiff, M. (eds.) 2005. *International Migration, Remittances, and The Brain Drain*, Washington, D.C., International Bank for Reconstruction and Development/ World Bank.

Paice, E. 2006. *Tip & Run: The Untold Tragedy of the Great War in Africa*, London, Weidenfeld and Nicolson.

Paoletti, E. 2011. *The Migration of Power and North-South Inequalities: The Case of Italy and Libya*, Basingstoke, Palgrave Macmillan.

Papademetriou, D. G. 1985. Illusions and reality in international migration: Migration and development in post World War II Greece. *International Migration*, XXIII, 211–223.

Papademetriou, D. G. & Martin, P. L. (eds.) 1991. *The Unsettled Relationship. Labor Migration and Economic Development*, New York, Greenwood Press.

Parekh, B. 2000. *Rethinking Multiculturalism: Cultural Diversity and Political Theory*, London, Macmillan.

Parekh, B. 2008. *A New Politics of Identity*, Basingstoke, Palgrave Macmillan.

Pargenter, A. 2008. *The New Frontiers of Jihad*, Philadelphia, PA, University of Pennsylvania Press.

Parreñas, R. S. 2000. Migrant Filipina domestic workers and the international division of reproductive labor. *Gender & Society*, 14, 560–580.

Parreñas, R. S. 2001. *Servants of Globalization: Migration and Domestic Work*, Redwood City, Stanford University Press.

Parreñas, R. S. 2005. *Children of Global Migration: Transnational Families and Gendered Woes*, Redwood City, Stanford University Press.

Passaris, C. 1989. Immigration and the evolution of economic theory. *International Migration*, 27, 525–542.

Pastore, F., Monzini, P. & Sciortino, G. 2006. Schengen's soft underbelly? Irregular migration and human smuggling across land and sea borders to Italy. *International Migration*, 44, 95–119.

Peach, C. 1968. *West Lndian Migration to Britain: A Social Geography*, London, Oxford University Press.

Peach, C. 1991. *The Caribbean in Europe: Contrasting Patterns of Migration and Settlement in Britain, France and the Netherlands*, Centre for Research in Ethnic Relations, University of Warwick.

Peach, C. 2005. The ghetto and the ethnic enclave. *In:* Varady, D. P. (ed.) *Desegregating the City: Ghettos Enclaves, and Inequality*, Albany, NT, SUNY Press.

Pedersen, M. H. 2003. *Between Homes: Post-war Return, Emplacement and the Negotiation of Belonging in Lebanon New Issues in Refugee Research*. Working paper 79, Denmark, United Nations High Commissioner for Refugees.

Pellegrino, A. 2000. Trends in international migration in Latin America and the Caribbean. *International Social Science Journal*, 52, 395–408.

Penninx, R. 1982. A critical review of theory and practice: The case of Turkey. *International Migration Review*, 16, 781–818.

Penninx, R., Berger, M. & Kraal, K. (eds.) 2006. *The Dynamics of International Migration and Settlement in Europe*, Amsterdam, Amsterdam University Press.

Perez-Lopez, J. & Diaz-Briquets, S. 1990. Labor migration and offshore assembly in the socialist world: The Cuban experience. *Population and Development Review*, 16, 273–299.

Pessar, P. R. & Mahler, S. J. 2003. Transnational migration: Bringing gender in. *International Migration Review*, 37, 812–846.

Petersen, W. 1958. A general typology of migration. *American Sociological Review*, 23, 256–266.

Petras, J. & Veltmayer, H. 2000. Globalisation or imperialism? *Cambridge Review of International Affairs*, 14, 1–15.

Pew. 2010. *Growing Number of Americans Say Obama Is a Muslim [Online]*, Washington, D.C., Pew Research Centre. https://www.pewforum.org/2010/08/18/growing-number-of-americans-say-obama-is-a-muslim/[Accessed 20 April 2019].

Pfahlmann, H. 1968. *Fremdarbeiter und Kriegsgefangene in der deutschen Kriegswirtschaft 1939–45*, Darmstadt, Wehr und Wissen.

Phillips, J., Klapdor, M. & Simon-Davies, J. 2010. Migration to Australia since federation: A guide to the statistics, Background note. October 2010, Canberra, Australian Parliamentary Library.

Phillips, N. (ed.) 2011a. *Migration in the Global Political Economy*, Boulder, CO, Lynne Rienner.

Phillips, N. 2011b. Migration and the global economic crisis. *In:* Phillips, N. (ed.) *Migration in the Global Politiical Economy*, Boulder, CO, Lynne Rienner.

Phizacklea, A. 1983. *One Way Ticket? Migration and Female Labour*, London, Routledge and Kegan Paul.

Phizacklea, A. 1990. *Unpacking the Fashion Industry: Gender Racism and Class in Production*, London, Routledge.

Phizacklea, A. 1998. Migration and globalisation: A feminist perspective. *In:* Koser, K. & Lutz, H. (eds.) *The New Migration in Europe*, London, Macmillan.

Picquet, M., Pelligrino, A. & Papail, J. 1986. L'immigration au Venezuela. *Revue Européenne des Migrations Internationales*, 2(2).

Pieke, F. N. 2011. Immigrant China. *Modern China*, 38, 40–77.

Piketty, T. 2014. *Capital in the Twenty-First Century*, Cambridge and London, Harvard University Press.

Pilkington, H. 1998. *Migration, Displacement, and Identity in Post-Soviet Russia*, London and New York, Routledge.

Piore, M. 1979. *Birds of Passage: Migrant Labor and Industrial Societies*, Cambridge, Cambridge University Press.

Piore, M. & Sabel, C. 1984. *The Second Industrial Divide*. New York, Basic Books.

Piper, N. 2008. Feminisation of migration and the social dimensions of development: The Asian case. *Third World Quarterly*, 29, 1287–1303.

Piper, N. & Lee, S. 2016. Marriage migration, migrant precarity, and social reproduction in Asia: An overview. *Critical Asian Studies*, 48, 473–493.

Pipes, R. 1997. Is Russia still an enemy? *Foreign Affairs*, 76(5), 65–78.

Pires, A. J. G. 2015. Brain drain and brain waste. *Journal of Economic Development*, 40, 1–34.

Plewa, P. 2006. How have regularization programs affected Spanish governmental efforts to integrate migrant populations. In: Majtczak, O. (ed.) *The Fifth International Migration Conference*, Warsaw, Independent University of Business and Government.

Pliez, O. 2005. Le Sahara libyen dans les nouvelles configurations migratoires. *Revue Européenne des Migrations Internationales*, 16, 165–181.

Polanyi, K. 1944. *The Great Transformation: The Political and Economic Origins of Our Time*, New York, Farrar and Rinehart.

Ponte, S. & Gibbon, P. 2005. Quality standards, conventions, and the governance of global value chains. *Economy and Society*, 34, 1–31.

Popp, M. 2018. Firing at Refugees: EU Money Helped Fortify Turkey's Border. *Spiegel Online*. 29 March 2018. [Accessed 16 July 2019].

Portes, A. 1998. Social capital: Its origins and applications in modern sociology. *Annual Review of Sociology*, 24, 1–24.

Portes, A. 1999. Conclusion: Towards a new world the origins and effects of transnational activities. *Ethnic and Racial Studies*, 22, 463–477.

Portes, A. 2010. Migration and social change: Some conceptual reflections. *Journal of Ethnic and Migration Studies*, 36, 1537–1563.

Portes, A. & Bach, R. L. 1985. *Latin Journey: Cuban and Mexican Immigrants in the United States*, Berkeley, University of Calfornia Press.

Portes, A. & Böröcz, J. 1989. Contemporary immigration: Theoretical perspectives on Its determinants and modes of incorporation. *The International Migration Review*, 23, 606–630.

Portes, A., Castells, M. & Benton, L. A. 1989. *The Informal Economy: Studies in Advanced and Less Developed Countries*, Baltimore, JHU Press.

Portes, A., Fernandez-Kelly, P. & Haller, W. 2005. Segmented assimilation on the ground: The new second generation in early adulthood. *Ethnic and Racial Studies*, 28, 1000–1040.

Portes, A. & Grosfoguel, R. 1994. Caribbean diasporas: Migration and ethnic communities. *Annals of the American Academy of Political and Social Science*, 533, 48–69.

Portes, A., Guarnizo, L. E. & Landolt, P. 1999. The study of transnationalism: Pitfalls and promise of an emergent research field. *Ethnic and Racial Studies*, 22, 217–237.

Portes, A. & Rumbaut, R. G. 2006. *Immigrant America: A Portrait*, Berkeley, University of California Press.

Portes, A. & Zhou, M. 1993. The new 2nd-generation – segmented assimilation and its variants. *Annals of the American Academy of Political and Social Science*, 530, 74–96.

Postel, H., Rathinasamy, C. & Clemens, M. 2015, *Europe's Refugee Crisis Is Not as Big as You've Heard, and Not Without Recent Precedent [Online]*, Washington, D.C., Center for Global Development. https://www.cgdev.org/blog/europes-refugee-crisis-not-big-youve-heard-and-not-without-recent-precedent [Accessed 31 May 2019].

Potts, L. 1990. *The World Labour Market: A History of Migration*, London, Zed Books.

Preibisch, K. 2010. Pick-Your-Own Labour: Migrant Workers and Flexibility in Canadian Agriculture. *International Migration Review*, 44, 404–441.

Price, C. (1963). *Southern Europeans in Australia*. Melbourne, Oxford University Press.

Pries, L. 2018. *Refugees, Civil Society and the State*, Cheltenham, Northampton, MA, Edward Elgar.

Productivity Commission 2010. *Population and Migration: Understanding the Numbers*. December ed., Melbourne, Productivity Commission.

Prost, A. 1966. L'immigration en France depuis cent ans. *Esprit*, vol. 34.

Putnam, R. D. 2000. *Bowling Alone: The Collapse and Revival of American Community*, New York, Simon & Schuster.

Ranis, G. & Fei, J. H. C. 1961. A theory of economic development. *American Economic Review*, 51, 533–565.

Rao, P. M. & Balasubrahmanya, M. 2017. The rise of IT services clusters in India: A case of growth by replication. *Telecommunications Policy*, 41, 90–105.

Rath, J. 2002. *Unravelling the Rag Trade: Immigrant Entrepreneurship in Seven World Cities*, Oxford, Berg.

Rath, J., Bodaar, A., Wagemaakers, T. & Wu, P. Y. 2018. Chinatown 2.0: The difficult flowering of an ethnically themed shopping area. *Journal of Ethnic and Migration Studies*, 44, 81–98.

Ratha, D. 2003. Workers' remittances: An important and stable source of external development finance. *Global Development Finance 2003*, Washington, D.C., World Bank.

Ravenstein, E. G. 1885. The laws of migration. *Journal of the Royal Statistical Society*, 48, 167–227.

Ravenstein, E. G. 1889. The laws of migration. *Journal of the Royal Statistical Society*, 52, 214–301.

Reichert, J. S. 1981. The migrant syndrome: Seasonal U.S. labor migration and rural development in central Mexico. *Human Organization*, 40, 56–66.

Reitz, J. G. 1998. *Warmth of the Welcome: The Social Causes of Economic Success for Immigrants in Different Nations and Cities*, Boulder, CO, Westview Press.

Rex, J. 1986. *Race and Ethnicity*, Maidenhead, Open University Press.

Rex, J. & Mason, D. (eds.) 1986. *Theories of Race and Ethnic Relations*, Cambridge, Cambridge University Press.

Reyneri, E. 2001. *Migrants' Involvement in Irregular Employment in the Mediterranean Countries of the European Union*, Geneva, International Labour Organization.

Reyneri, E. 2003. Immigration and the underground economy in new receiving South European countries: Manifold negative effects, manifold deep-rooted causes. *International Review of Sociology*, 13, 117–143.

Reyneri, E. & Fullin, G. 2010. Labour market penalties of new immigrants in new and old receiving West European countries. *International Migration*, 49, 31–57.

Rhoades, R. E. 1979. From caves to main street: Return migration and the transformations of a Spanish village. *Papers in Anthropology*, 20, 57–74.

Rhoda, R. 1983. Rural development and urban migration: Can we keep them down on the farm?. *International Migration Review*, 17, 34–64.

Rietig, V. & Mülle, A. 2016. *The New Reality: Germany Adapts to Its Role as a Major Migrant Magnet [Online]*, Washington, D.C., Migration Policy Institute. https://www.migration policy.org/article/new-reality-germany-adapts-its-role-major-migrant-magnet [Accessed 1 June 2019].

Rodrik, D. 2011. *The Globalization Paradox: Why Global Markets, States, and Democracy Can't Coexist*, Oxford, Oxford University Press.

Roman, H. 2006. *Transit Migration in Egypt*, Florence, CARIM, European University Institute.

Romero, F. 1993. Migration as an issue in European interdependence and integration: The case of Italy. *In:* Milward, A., Lynch, F., Renieri, R., Romero, F. & Sorensen, V. (eds.) *The Frontier of National Sovereignty*, London, Routledge.

Rosenberg, C. D. 2006. *Policing Paris: The Origins of Modern Immigration Control Between the Wars*, Ithaca, NY and London, Cornell University Press.

Rostow, W. W. 1960. *The Stages of Economic Growth: A Non-Communist Manifesto*, Cambridge, Cambridge University Press.

Roux, G. & Roché, S. 2016. Police et phénomènes identitaires dans les banlieues: entre ethnicité et territoire. *Revue française de science politique*, 66, 729–750.

Roy, O. 2003. Euroislam:The Jihad within? *The National Interest*, 63–73.

Rubenstein, H. 1992. Migration. Development and Remittances in Rural Mexico. *International Migration*, 30, 1992.

Ruhs, M. & Anderson, B. 2010. Semi-compliance and illegality in migrant labour markets: An analysis of migrants, employers and the state in the UK. *Population Space and Place*, 16, 195–211.

Ruhs, M. 2013. *The price of rights: regulating international labor migration*. Princeton: Princeton University Press.

Ruiz, I. & Vargas-Silva, C. 2009. Another consequence of the economic crisis: A decrease in migrants' remittances. *Applied Financial Economics*, 20, 171–182.

Rummel, R. J. 1998. *Statistics of Democide: Genocide and Mass Murder since 1900*, LIT Verlag Münster.

Russell, S. S. 1992. Migrant remittances and development. *International Migration*, 30.

Rycs, J. F. 2005. Le "sponsorship" peut-il encore canaliser les flux migratoires dan les pays du Golfe? Le cas de Emirats Arabes Unis. *In:* Jaber, H. & Métrai, F. (eds.) *Mondes En Mouvements. Migrants Et Migrations Au Moyen-Orient Au Tournant Du XXIe Siècle*, Beyrouth, Institut Français du Proche Orient IFPO.

Sabater, A. & Domingo, A. 2012. A new immigration regularization policy: The settlement program in Spain. *International Migration Review*, 46, 191–220.

Safran, W. 1991. Diasporas in modern societies: Myths of homeland and return. *Diaspora*, 1, 83.

Salazar, N. B. & Glick Schiller, N. (eds.). 2016. *Regimes of Mobility: Imaginaries and Relationalities of Power*, London and New York, Routledge.

Sarmah, S. 2007. *Is Obama Black Enough?* [Online]. https://archives.cjr.org/politics/is_obama_black_enough.php [Accessed 20 April 2019].

Sassen, S. 1988. *The Mobility of Labour and Capital*, Cambridge, Cambridge University Press.

Sassen, S. 2001. *The Global City: New York, London, Tokyo*, Princeton, NJ, Princeton University Press.

Sassen Koob, S. 1979. Economic growth and immigration in Venezuela. *International Migration Review*, 13(3), 455–474.

Sayad, A. 1999. Immigration et pensée d'etat. *Actes de la recherche en sciences sociales*, 129, 5–14.

Schaffner, B. F., MacWilliams, M. & Nteta, T. 2018. Understanding white polarization in the 2016 vote for president: The sobering role of racism and sexism. *Political Science Quarterly*, 133, 9–34.

Schain, M. A. 2008. Commentary: Why political parties matter. *Journal of European Public Policy*, 15, 465–470.

Schama, S. 2006. *Rough Crossings: Britain, the Slaves and the American Revolution*, London, BBC Books.

Schaub, M. L. 2012. Lines across the desert: Mobile phone use and mobility in the context of trans-Saharan migration. *Information Technology for Development*, 18, 126–144.

Scheele, J. 2010. Traders, saints, and irrigation: Reflections on Saharan connectivity. *The Journal of African History*, 51, 281–300.

Scheele, J. 2012. *Smugglers and Saints of the Sahara: Regional Connectivity in the Twentieth Century*, Cambridge, Cambridge University Press.

Schenk, C. 2018. *Why Control Immigration? Strategic Uses of Migration Management in Russia*, Toronto, Buffalo, University of Toronto Press.

Schewel, K. 2018. *Why Ethiopian Women Go To The Middle East: An Aspiration-Capability Analysis of Migration Decision-Making*. IMI working paper 148, International Migration Institute, University of Amsterdam.

Schewel, K. 2019a. Understanding Immobility: Moving Beyond the Mobility Bias in Migration Studies. *International Migration Review*, 1–28.

Schewel, K. 2019b. *Moved by Modernity*. PhD thesis, Amsterdam, University of Amsterdam.

Schewel, K. & Fransen, S. 2018. Formal education and migration aspirations in Ethiopia. *Population and Development Review*, 44, 555.

Schielke, S. & Graw, K. (eds.) 2012. *The Global Horizon: Migratory Expectations in Africa and the Middle East*, Leuven, Leuven University Press.

Schierup, C. U. & Alund, A. 1987. *Will They Still be Dancing?*, Stockholm, Almquist and Wiksell International.

Schierup, C. U., Hansen, P. & Castles, S. 2006. *Migration, Citizenship and the European Welfare State: A European Dilemma*, Oxford, Oxford University Press.

Schiff, M. 1994. *How Trade, Aid, and Remittances Affect International Migration*, Washington, D.C., World Bank.

Schnapper, D. 1991. A host country of immigrants that does not know itself. *Diaspora*, 1, 353–364.

Schnapper, D. 1994. *La Communauté des Citoyens*, Paris, Gallimard.

Schrover, M., Van der Leun, J. & Quispel, C. 2007. Niches, labour market segregation, ethnicity and gender. *Journal of Ethnic and Migration Studies*, 33, 529–540.

Schwenkel, C. 2014. Rethinking Asian mobilities: Socialist migration and post-socialist repatriation of Vietnamese contract workers in East Germany. *Critical Asian Studies*, 46, 235–258.

Sciortino, R. & Punpuing, S. 2009. *International Migration in Thailand*, Bangkok, International Organisation for Migration.

Scott, J. W. 2005. Symptomatic politics: The banning of Islamic head scarves in French public schools. *French Politics, Culture & Society*, 23(3), 106–127.

Scotto, A. 2017. *From Emigration to Asylum Destination, Italy Navigates Shifting Migration Tides* [Online], Washington, D.C., Migration Policy

Institute. https://www.migrationpolicy.org/article/emigration-asylum-destination-italy-navigates-shifting-migration-tides [Accessed 1 June 2019].

Seccombe, I. J. 1986. Immigrant Workers in an Emigrant Economy. *International Migration*, 24, 377–396.

Seilonen, J. 2016. *Fortress Europe – A Brief History of the European Migration and Asylum Policy*, PhD thesis, University of Helsinki.

Sell, R. R. 1988. Egyptian international labor migration and social processes: Toward regional integration. *International Migration*, 22, 87–108.

Semyonov, M. & Lewin-Epstein, N. 1987. *Hewers of Wood and Drawers of Water*, Ithaca, NY, ILR Press.

Sen, A. 1999. *Development as Freedom*, Oxford, Oxford University Press.

Sen, A. 2000. East and West: The reach of reason. *New York Review of Books*, 47.

Sen, A. 2006. *Identity and Violence: The Illusion of Destiny*, London, Penguin .

Seol, D.-H. 2001. Situation of and Measures on Undocumented Foreign Workers in Korea (in Korean). *Shinhak Sasang*, 113, 49–75.

Seton-Watson, H. 1977. *Nations and States*, London, Methuen.

Shaw, M. 2000. *Theory of the Global State: Globality as Unfinished Revolution*, Cambridge, Cambridge University Press.

Shaw, W. & Ratha, D. 2016. *South-South Migration and Remittances*, Washington, D.C., World Bank.

Shevel, O. 2009. The politics of citizenship policy in new states. *Comparative Politics*, 41(3), 273–291.

Shimpo, M. 1995. Indentured migrants from Japan. *In:* Cohen, R. (ed.) *The Cambridge Survey of World Migration*, Cambridge, Cambridge University Press.

Shulewitz, M. H. 2000. *Forgotten millions: The modern Jewish exodus from Arab lands*. London and New York, Continuum.

Siegelbaum, L. H. & Moch, L. P. 2014. *Broad is my Native Land: Repertoires and Regimes of Migration in Russia's Twentieth Century*, Ithaca, NY and London, Cornell University Press.

Simpson, J. M., Esmail, A., Kalra, V. S. & Snow, S. J. 2010. Writing migrants back into NHS history: Addressing a 'collective amnesia'and

its policy implications. *Journal of the Royal Society of Medicine*, 103, 392–396.

Sinn, E. (ed.) 1998. *The Last Half Century of Chinese Overseas*, Hong Kong, Hong Kong University Press.

Sjaastad, A. H. 1962. The costs and returns of human migration. *Journal of Political Economy*, 70, 80–93.

Skeldon, R. 1977. Evolution of migration patterns during urbanization in Peru. *Geographical Review*, 67, 394–411.

Skeldon, R. 1990. *Population Mobility in Developing Countries: A Reinterpretation*, London, Belhaven Press.

Skeldon, R. 1997. *Migration and Development: A Global Perspective*, Harlow, Essex, Addison Wesley Longman.

Skeldon, R. 2006. Interlinkages between internal and international migration and development in the Asian region. *Population, Space and Place*, 12, 15–30.

Skeldon, R. 2012. Migration Transitions Revisited: Their Continued Relevance for The Development of Migration Theory. *Population, Space and Place*, 18, 154–166.

Slater, R. B. 2000. African immigrants in the United States are the nation's most highly educated group. *The Journal of Blacks in Higher Education*, 26, 60–61.

Smith, A. D. 1986. *The Ethnic Origins of Nations*, Oxford, Blackwell.

Smith, A. D. 1991. *National Identity*, London, Penguin.

Smith, A. E. 1947. *Colonists in Bondage: White Servitude and Convict Labor in America, 1607–1776*, Chapel Hill, Chapel Hill Press.

Smith, D. P. & King, R. 2012. Special issue: Re-making migration theory: Transitions, intersections and cross-fertilisations. *Population, Space and Place*, 18, i–ii, 127–224.

Smith, H. W. (ed.) 2011. *The Oxford Handbook of Modern German History*, Oxford, Oxford University Press.

Smith, J. 2006. *Guatemala: Economic Migrants Replace Political Refugees* [Online]. http://www.migrationinformation.org/Profiles/display.cfm?id=392 [Accessed 11 November 2011].

Smith, J. P. & Edmonston, B. (eds.) 1997. *The New Americans: Economic, Demographic and Fiscal Effects of Immigration*, Washington, D.C., National Academy Press.

Smith, R. 2003. Migrant membership as an instituted process: Transnationalization, the state and extra-territorial conduct of Mexican politics. *International Migration Review*, 37, 297–343.

Snel, E., Engbersen, G. & Leerkes, A. 2006. Transnational involvement and social integration. *Global Networks-a Journal of Transnational Affairs*, 6, 285–308.

Solomos, J. 2003. *Race and Racism in Britain*, Basingstoke, Palgrave Macmillan.

Soysal, Y. N. 1994. *Limits of Citizenship: Migrants and Postnational Membership in Europe*, Chicago and London, University of Chicago Press.

Spilimbergo, A. 2009. Democracy and foreign education. *The American Economic Review*, 99, 528–543.

Squire, V. 2015. The securitisation of migration: An absent presence? *The Securitisation of Migration in the EU*, Berlin, Springer.

Stahl, C. W. & Habib, A. 1991. Emigration and development in South and Southeast Asia. In: Papademetriou, D. G. & Martin, P. L. (eds.) *The Unsettled Relationship: Labor Migration and Economic Development*, Connecticut, Greenwood Publishing Group.

Stalker, P. 2000. *Workers Without Frontiers: The Impact of Globalization on International Migration*, London and Boulder, CO, Lynne Rienner.

Stark, O. 1978. *Economic-Demographic Interactions in Agricultural Development: The Case of Rural-to-Urban Migration*, Rome, FAO.

Stark, O. 1980. On the role of urban-to-rural remittances in rural development. *Journal of Development Studies*, 16, 369–374.

Stark, O. 1991. *The Migration of Labor*, Cambridge and Oxford, Blackwell.

Stark, O. 2009. Reasons for Remitting. *World Economics*, 10, 147–157.

Stark, O. & Bloom, D. E. 1985. The new economics of labor migration. *American Economic Review*, 75, 173–178.

Stark, O., Helmenstein, C. & Prskawetz, A. 1997. A brain gain with a brain drain. *ECOLET*, 55, 227–234.

Stark, O. & Levhari, D. 1982. On migration and risk in LDCs. *Economic Development and Cultural Change*, 191–196.

Statista. 2019. Hate crime in the U.S [Online]. Available: https://www.statista.com/topics/4178/hate-crimes-in-the-united-states/ [Accessed 30 May 2019].

Steinberg, S. 1981. *The Ethnic Myth: Race Ethnicity and Class in America*, Boston, Beacon Press.

Stiglitz, J. 2002. *Globalization and Its Discontents*, London, Allen Lane, Penguin Press.

Stirn, H. 1964. *Ausländische Arbeiter im Betrieb*, Frechen/Cologne, Bartmann.

Strozza, S. & Venturini, A. 2002. Italy is no longer a country of emigration. Foreigners in Italy: How many, where do they come from?. In: Rotte, R. & Stein, P. (eds.) *Migration Policy and the Economy: International Experiences*, Munich, Hans Seidel Stiftung.

Sullivan, M. J. 1994. The 1988–89 Nanjing anti-African protests: Racial nationalism or national racism? *The China Quarterly*, 138, 438–457.

Surak, K. 2013. The migration industry and developmental states in East Asia. In: Gammeltoft-Hansen, T. & Sorensen, N. N. (eds.) *The Migration Industry and the Commercialization of International Migration*, London and New York, Routledge.

Surk, B. & Abbot, S. 2008. India wants oil-rich Emirates to pay workers better wages. *Sunday News Journal* (Wilmington, DE).

Sze, L.-S. 2007. *New Immigrant Labour from Mainland China in Hong Kong* [Online], Hong Kong, Asian Labour Update. http://www.amrc.org.hk/alu_article/discrimination_at_work/new_immigrant_labour_from_mainland_china_in_hong_kong [Accessed 23 March 2007].

Taeuber, I. B. 1951. Family, migration, and industrialization in Japan. *American Sociological Review*, 16, 149–157.

Tan, Y. 2020. Development induced displacement and resettlement: An overview of issues and interventions. In: Bastia, T. & Skeldon, R. (eds.) *Routledge Handbook of Migration and Development*, London, Routledge.

Tan-Mullins, M., Urban, F. & Mang, G. 2017. Evaluating the behaviour of Chinese stakeholders engaged in large hydropower projects in Asia and Africa. *The China Quarterly*, 230, 464–488.

Tapinos, G. P. 1990. *Development Assistance Strategies and Emigration Pressure in Europe and Africa*, Washington, D.C., Commission for the Study of International Migration and Co-operative Economic Development.

Taylor, E. 1984. Egyptian migration and peasant wives. *Merip Reports*, 124, 3–10.

Taylor, J. E. 1992. Remittances and inequality reconsidered – direct, indirect, and intertemporal effects. *Journal of Policy Modeling*, 14, 187–208.

Taylor, J. E. 1999. The new economics of labour migration and the role of remittances in the migration process. *International Migration*, 37, 63–88.

Taylor, J. E., Arango, J., Hugo, G., Kouaouci, A., Massey, D. S. & Pellegrino, A. 1996a. International migration and community development. *Population Index*, 62, 397–418.

Taylor, J. E., Arango, J., Hugo, G., Kouaouci, A., Massey, D. S. & Pellegrino, A. 1996b. International migration and national development. *Population Index*, 62, 181–212.

Taylor, M. J., Moran-Taylor, M. J. & Ruiz, D. R. 2006. Land, ethnic, and gender change: Transnational migration and its effects on Guatemalan lives and landscapes. *Geoforum*, 37, 41–61.

Taylor, P., Gonzalez-Barrera, A., Passel, J. & Lopez, M. H. 2012. *An Awakened Giant: The Hispanic Electorate is Likely to Double by 2030*, Washington, D.C., Pew Hispanic Center.

Tekeli, I. 1994. Involuntary displacement and the problem of the resettlement in Turkey from the Ottoman Empire to the present. *In:* Shami, S. (ed.) *Population Displacement and Resettlement: Development and Conflict in the Middle East*, New York, Center for Migration Studies.

Terrazas, A. M. 2007. *Beyond Regional Circularity: The Emergence of an Ethiopian Diaspora* [Online], Washington, D.C., Migration Policy Institute. Beyond Regional Circularity: The Emergence of an Ethiopian Diaspora [Accessed 30 May 2019].

Ther, P. 1996. The integration of expellees in Germany and Poland after World War II: A historical reassessment. *Slavic Review*, 55, 779–805.

Thibos, C. 2014. *Competitive Identity Formation in the Turkish Diaspora*. DPhil thesis, University of Oxford

Thiollet, H. 2011. Migration as diplomacy: Labor migrants, refugees, and Arab regional politics in the oil-rich countries. *International Labor and Working-Class History*, 103–121.

Thiollet, H. 2014. From migration hub to asylum crisis: The changing dynamics of contemporary migration in Yemen. *In:* Lackner, H. (ed.) *Why Yemen Matters*. Saqi Books.

Thiollet, H. 2015. Migration et (contre)révolution dans le Golfe: Politiques migratoires et politiques de l'emploi en Arabie saoudite. *Revue Européenne des Migrations Internationales*, 31(3), 121–143.

Thiollet, H. 2016. *Managing Migrant Labour in the Gulf: Transnational Dynamics of Migration Politics Since the 1930s*, IMI working paper 131. International Migration Institute, University of Oxford.

Thomas, L. 2015. *Puerto Ricans in the United States* [Online]. Oxford Research Encyclopedias. https://oxfordre.com/americanhistory/view/10.1093/acrefore/9780199329175.001.0001/acrefore-9780199329175-e-32 [Accessed 28 May 2019].

Thomas, W. I. & Znaniecki, F. 1918. *The Polish Peasant in Europe and America: Monograph of an Immigrant Group*, Chicago, University of Chicago Press.

Thomas-Hope, E. 1996. The Dynamic of Caribbean Migration Culture. *Journal of Social Sciences* 3(1+2): 3–19.

Thomas-Hope, E. M. 2000. *Trends and Patterns of Migration to and from Caribbean Countries*, Economic Commission for Latin America and the Caribbean (ECLAC/CEPAL).

Thränhardt, D. 1996. European migration from East to West: Present patterns and future directions. *New Community*, 22, 227–242.

Tilly, C. 1976. *Migration in modern European history*. CRSO working paper 145. Ann Arbor, Center for Research on Social Organization.

Tilly, C. 1990. *Coercion, capital, and European states, AD 990*, Cambridge, Basil Blackwell.

Timmer, A. S. & Williamson, J. G. 1998. Immigration policy prior to the 1930s: Labor markets, policy interactions, and globalization backlash. *Population and Development Review*, 24, 739–742.

Todaro, M. P. 1969. A model of labor migration and urban unemployment in less-developed countries. *American Economic Review*, 59, 138–148.

Todaro, M. P. & Maruszko, L. 1987. Illegal migration and US immigration reform: A conceptual framework. *Population and Development Review*, 13, 101–114.

Tolts, M. 2003. Mass Aliyah and Jewish emigration from Russia: Dynamics and factors. *East European Jewish Affairs*, 33, 71–96.

Toma, S. & Villares-Varela, M. 2019. The role of migration policies in the attraction and retention of international talent: The case of Indian researchers. *Sociology*, 53, 52–68.

Tomas, K. & Münz, R. 2006. *Labour Migrants Unbound? EU Enlargement, Transitional Measures and Labour Market Effects*, Stockholm, Institute for Futures Study.

Torpey, J. 1998. Coming and going: On the state monopolization of the legitimate "means of movement". *Sociological Theory*, 16, 239–259.

Torpey, J. C. 2000a. Mobility and modernity: Migration in Germany, 1820–1989. *Journal of Interdisciplinary History*, 31, 281–284.

Torpey, J. C. 2000b. *The Invention of the Passport: Surveillance, Citizenship and the State*, Cambridge, Cambridge University Press.

Torpey, J. 2007. Leaving: A comparative view. In: Green, N. & Weil, F. (eds.) *Citizenship and Those Who Leave*, Urbana, University of Illinois Press.

Triandafyllidou, A. & Lazarescu, D. 2009. *The Impacts of the Recent Global Economic Crisis on Migration. Preliminary Insights from the South-Eastern Borders of the EU (Greece)*, CARIM. Robert Schuman Centre for Advanced Studies, San Domenico di Fiesole, European University Institute.

Tribalat, M. 1995. *Faire France: une enquete sur les immigrés et leurs enfants*, Paris, La Découverte.

Uehling, G. 2004. *Beyond memory: The Crimean Tatars' deportation and return*, Houndmills, Basingstoke, Hampshire, Palgrave Macmillan.

UNDESA. 2009. *Trends in International Migrant Stock: The 2008 Revision*, New York, United Nations Department of Economic and Social Affairs, Population Division.

UNDESA. 2017. *2017 Revision of World Population Prospects*, New York, United Nations Population Division.

UNDP. 2009. *Overcoming Barriers: Human Mobility and Development*, New York City, Human Development Report 2009, New York, UNDP.

UNHCR. 1995. *The State of the World's Refugees: In Search of Solutions*, Oxford, Oxford University Press.

UNHCR. 2000. *The State of the World's Refugees: Fifty Years of Humanitarian Action*, Oxford, Oxford University Press.

UNHCR. 2011a. *Statistical Yearbook 2010*, Geneva, United Nations High Commissioner for Refugees.

UNHCR. 2011b. *2011 UNHCR Country Operations Profile - Colombia* [Online], Geneva, UNHCR. http://www.unhcr.org/cgi-bin/texis/vtx/page?page=49e492ad6&submit=GO [Accessed 13 December 2011].

UNHCR. 2011c. *UNHCR Statistical Database Online* [Online], Geneva, UNHCR. [Accessed 6 September 2011].

UNHCR. 2012a. *2012 UNHCR country operations profile – Sudan* [Online], Geneva, UNHCR. http://www.unhcr.org/cgi-bin/texis/vtx/page?page=49e483b76&submit=GO [Accessed 24 July 2012].

UNHCR. 2012b. *2012 UNHCR Country Operations Profile– Tanzania* [Online], Geneva, UNHCR. http://www.unhcr.org/pages/49e45c736.html [Accessed 25 July 2012].

UNHCR. 2017. Global trends: Forced displacement in 2017. In: Refugees, UNHCR. (ed.), Geneva, UHCR.

UNHCR. 2018. *Desperate Journeys* [Online], Geneva, UNHCR. https://www.unhcr.org/desperatejourneys/[Accessed 30 May 2019].

UNPD. 2017. *Migrant Stock by Origin and Destination*. New York, United Nations Population Division.

USCR. 2001. World Refugee Survey 2001, Washington, D.C., US Committee for Refugees.

USINS. 1999. *Statistical Yearbook of the Immigration and Naturalization Service 1997*, Washington, D.C., US Immigration and Naturalization Service.

Ustubici, A. 2016. Political activism between journey and settlement: Irregular migrant mobilisation in Morocco. *Geopolitics*, 21, 303–324.

van Amersfoort, H. 2011. *How the Dutch Government stimulated the unwanted migration from Suriname*. IMI working paper 47, International Migration Institute, University of Oxford.

Van Hear, N. 1998. *New Diasporas: The Mass Exodus, Dispersal and Regrouping of Migrant Communities*, London, UCL Press.

Van Hear, N. 2004. *Diasporas, Remittances, Development, and Conflict [Online]*, Washington, D.C., Migration Policy Institute. https://www.migrationpolicy.org/article/refugee-diasporas-remittances-development-and-conflict [Accessed 30 May 2019].

van Liempt, I. 2007. *Navigating Borders. An Inside Perspective into the Process of Human Smuggling*, Amsterdam, Amsterdam University Press.

van Liempt, I. & Sersli, S. 2013. State responses and migrant experiences with human smuggling: A reality check. *Antipode*, 45, 1029–1046.

van Tubergen, F., Maas, I. & Flap, H. 2004. The economic incorporation of immigrants in 18 western societies: Origin, destination, and community effects. *American Sociological Review*, 69, 704–727.

Vasta, E. 1993. Immigrant women and the politics of resistance. *Australian Feminist Studies*, 18, 5–23.

Vasta, E. 1999. Multicultural politics and resistance: Migrants unite? *In:* Hage, G. & Couch, R. (eds.) *The Future of Australian Multiculturalism*, Sydney, RIHSS Sydney University.

Vasta, E. & Castles, S. (eds.) 1996. *The Teeth are Smiling: The Persistence of Racism in Multicultural Australia*, Sydney, Allen and Unwin.

Vasta, E., Rando, G., Castles, S. & Alcorso, C. 1992. The Italo-Australian community on the Pacific Rim. *In:* Castles, S. E. A. (ed.) *Australia's Italians*, Sydney, Allen & Unwin.

Vecoli, R. J. 1964. Contadini in Chicago: A critique of the uprooted. *The Journal of American History*, 51, 404–417.

Verbunt, G. 1985. France. *In:* Hammar, T. (ed.) *European Immigration Policy: A Comparative Study*, Cambridge, Cambridge University Press.

Vertovec, S. 1999. Conceiving and researching transnationalism. *Ethnic and Racial Studies*, 22, 445–462.

Vertovec, S. 2007. Super-diversity and its implications. *Ethnic and Racial Studies*, 30, 1024–1054.

Vertovec, S. 2009. *Transnationalism*, London, Routledge.

Vezzoli, S. 2014. *The Effects of Independence, State Formation and Migration Policies on Guyanese Migration*. IMI working paper 94, International Migration Institute, University of Oxford.

Vezzoli, S. 2015. *Borders, Independence and Post-Colonial Ties: The Role of the State in Caribbean Migration*. PhD thesis, University of Maastricht.

Vezzoli, S. & Flahaux, M.-L. 2017. How do postcolonial ties and migration regimes shape travel visa requirements? The case of Caribbean nationals. *Journal of Ethnic and Migration Studies*, 43, 1141–1163.

Vigneswaran, D. 2012. Experimental data collection methods and migration governance. *In:* Berriane, M. & de Haas, H. (eds.) *African Migrations Research: Innovative Methods and Methodologies*, Trenton, NJ, Africa World Press.

Vigneswaran, D. 2016. *Weak state/tough territory: The South African mobility regime complex*. MMG working paper 16–03. Göttingen, Max Planck Institute for the Study of Religious and Ethnic Diversity.

Vigneswaran, D. & Quirk, J. 2015. *Mobility makes States: Migration and Power in Africa*, Philadelphia, University of Pennsylvania Press.

Villarreal, A. 2014. Explaining the decline in Mexico-US migration: The effect of the Great Recession. *Demography*, 51, 2203–2228.

Vogel, D. (2009) Size and development of irregular migration to the EU. Comparative Policy brief CLANDESTINO project). http://clandestino.eliamep.gr/wp-content/uploads/2009/12/clandestino_policy_brief_comparative_size-of-irregular-migration.pdf (Accessed: 17 July 2012).

Vono de Vilhena, D. 2011. Panorama migratorio en España, Ecuador y Colombia a partir de las estadísticas locales. *In:* Martínez Pizarro, J. (ed.) *Migración Internacional En América Latina Y El Caribe - Nuevas Tendencias, Nuevos Enfoques*, Santiago de Chile, CEPAL.

Vora, N. & Koch, N. 2015. Everyday inclusions: Rethinking ethnocracy, Kafala, and belonging in the Arabian Peninsula. *Studies in Ethnicity and Nationalism*, 15(3), 540–552.

Waithanji, E. M. 2008. *Gendered impacts of sedentarization of nomads on the Somali community in Mandera Central division of northeastern Kenya*, PhD dissertation, Worcester, MA, Clark University.

Waldinger, R. 1996. *Still the Promised City? African-Americans and New Immigrants in Postindustrial New York*, Cambridge, MA and London, Harvard University Press.

Waldinger, R., Aldrich, H. & Ward, R. 1990. *Ethnic Entrepreneurs: Immigrant Business in Industrial Societies*, Newbury Park, CA, London and New Delhi, Sage Publications.

Wallace, C. & Stola, D. 2001. *Patterns of Migration in Central Europe*, Springer.

Wallerstein, I. 1974. *The Modern World System I, Capitalist Agriculture and the Origins of the European World Economy in the Sixteenth Century*, New York, Academic Press.

Wallerstein, I. 1980. *The Modern World System II, Mercantilism and the Consolidation of the European World-Economy, 1600–1750*, New York, Academic Press.

Wallerstein, I. 1984. *The Politics of the World Economy: The States, the Movements, and the Civilisations*, Cambridge, Cambridge University Press.

Wallman, S. 1986. Ethnicity and boundary processes. *In:* Rex, J. & Mason, D. (eds.) *Theories of Race and Ethnic Relations*, Cambridge, Cambridge University Press.

Wang, D. 2018. Internal remittances, vocational training costs and rural-urban migration in developing countries. *The International Economy*, 21,15–26.

Wang, G. 1997. *Global History and Migrations*, Boulder, CO, Westview Press.

Weber, M. 1919. *Politik Als Beruf*, Munich, Duncker & Humblodt.

Weber, M. 1968. *Economy and Society*, New York, Bedminister Press.

Wei, Y.-L. & Chang, S. C. 2011. Taiwan, Philippines sign MOU on hiring laborers. *Focus Taiwan*, 26 July.

Weil, P. 1991. *La France et Ses étrangers*, Paris, Calmann-Levy.

Weil, P. 2002. Towards a coherent policy of co-development. *International Migration*, 40, 41–56.

Weiner, M. & Hanami, T. (eds.) 1998. *Temporary Workers or Future Citizens? Japanese and U.S. Migration Policies*, New York, New York University Press.

Weiss, L. 1997. Globalization and the myth of the powerless state. *New Left Review*, 3–27.

Wejsa, S. & Lesser, J. 2018. *Migration in Brazil: The Making of a Multicultural Society* [Online], Washington, D.C., Migration Policy Institute. https://www.migrationpolicy.org/article/migration-brazil-making-multicultural-society [Accessed 30 May 2019].

Werner, H. 1973. *Freizügigkeit der Arbeitskräfte und die Wanderungsbewegungen in den Ländern der Eurpäischen Gemeinschaft*, Nuremburg, Institut für Arbeitsmarkt- und Berufsforschung.

White, A. 2009. Internal migration, identity and livelihood strategies in contemporary Russia. *Journal of Ethnic and Migration Studies*, 35(4), 555–573.

Wieviorka, M. 1995. *The Arena of Racism*, London, Sage.

Wikler, D. 1999. Can we learn from eugenics? *Journal of Medical Ethics*, 25(2), 183–194

Willner-Reid, M. 2017. *Afghanistan: Displacement Challenges in a Country on the Move* [Online], Washington, D.C., Migration Policy Institute. https://www.migrationpolicy.org/article/afghanistan-displacement-challenges-country-move [Accessed 30 May 2019].

Wise, T. A. 2009. *Agricultural Dumping under NAFTA: Estimating the Costs of US Agricultural Policies to Mexican Producers*, Medford, Global Development and Environment Institute, Tufts University.

Wishnie, M. J. 2007. Prohibiting the employment of unauthorized immigrants: The experiment fails. *University of Chicago Legal Forum*, 193.

Witte, J. 2012. *Turks in Germany Fear Racially motivated Murders* [Online]. Hamburg: Spiegel Online International. Available: http://www.spiegel.de/international/germany/0,1518,808949,00.html [Accessed 9 March 2012].

World Bank 2006. *Global Economic Prospects 2006: Economic Implications of Remittances and Migration*, Washington, D.C., World Bank.

World Bank 2007. *Remittance Trends 2006*, Washington, D.C., Migration and Remittances Team, Development Prospects Group, World Bank.

World Bank 2011. *Migration and Remittances Factbook 2011*, Washington, D.C., World Bank.

Wouterse, F. & Taylor, J. E. 2008. Migration and income diversification: Evidence from Burkina Faso. *World Development*, 36, 625–640.

Wouterse, F. & Van Den Berg, M. 2011. Heterogeneous migration flows from the Central Plateau of Burkina Faso: The role of natural and social capital. *The Geographical Journal*, 177, 357–366.

Wunderlich, D. 2010. Differentiation and policy convergence against long odds: Lessons from Implementing EU Migration Policy In Morocco. *Mediterranean Politics*, 15, 249–272.

Xiang, B., Yeoh, B. S. & Toyota, M. 2013. *Return: Nationalizing Transnational Mobility in Asia*, Duke University Press.

Yang, D. 2011. Migrant remittances. *Journal of Economic Perspectives*, 25, 129–151.

Yanovich, L. 2015. *Children left behind: The impact of labor migration in Moldova and Ukraine* [Online]. https://www.migrationpolicy.org/article/children-left-behind-impact-labor-migration-moldova-and-ukraine [Accessed 9 May 2019].

Yeates, N. 2004. Global care chains. *International Feminist Journal of Politics*, 6, 369–391.

Yue, C. S. 2011. Foreign labor in Singapore: Trends, Policies, Impacts, and Challenges. *Discussion Paper Series*. Makati City, Philippine Institute for Development Studies http://dirp3.pids.gov.ph/ris/dps/pidsdps1124.pdf [Accessed 21 July 2019].

Yuval-Davis, N. 2006. Intersectionality and feminist politics. *European Journal of Women's Studies*, 13, 193–209.

Zachariah, K. C., Mathew, E. T. & Rajan, S. I. 2001. Impact of migration on Kerala's economy and society. *International Migration*, 39, 63–88.

Zelinsky, Z. 1971. The hypothesis of the mobility transition. *Geographical Review*, 61, 219–249.

Zhang, S. X., Sanchez, G. E. & Achilli, L. 2018. Crimes of solidarity in mobility: Alternative views on migrant smuggling. *The ANNALS of the American Academy of Political and Social Science*, 676, 6–15.

Zickel, R. E. (ed.) 1989. *Soviet Union: A Country Study Federal Research Division*, Washington, D.C., Library of Congress.

Zohry, A. & Harrell-Bond, B. 2003. *Contemporary Egyptian Migration: An Overview of Voluntary and Forced Migration*. Working paper C3, Development Research Centre on Migration, Globalisation and Poverty, University of Sussex.

Zolberg, A. R. 2006. *A Nation by Design: Immigration Policy in the Fashioning of America*, Cambridge and London, Harvard University Press.

Zolberg, A. R. 2007. The exit revolution. *In:* Green, N. L. & Weil, F. (eds.) *Citizenship and Those Who Leave: The Politics of Emigration and Expatriation*, Urbana and Chicago, University of Illinois Press.

Zolberg, A. R. & Benda, P. M. (eds.) 2001. *Global Migrants, Global Refugees: Problems and Solutions*, New York and Oxford, Berghahn Books.

Zolberg, A. R., Suhrke, A. & Aguayo, S. 1989. *Escape from Violence*, Oxford and New York, Oxford University Press.

Zong, K. & Batalova, J. 2017a. *Chinese Immigrants in the United States* [Online], Washington, D.C., Migration Policy Institute. https://www.migrationpolicy.org/article/chinese-immigrants-united-states. [Accessed 30 November 2018].

Zong, K. & Batalova, J. 2017b *Indian Immigrants in the United States* [Online], Washington, D.C., Migration Policy Institute. https://www.migrationpolicy.org/article/indian-immigrants-united-states. [Accessed 30 November 2018].

Zong, K. & Batalova, J. 2017c *Korean Immigrants in the United States* [Online], Washington, D.C., Migration Policy Institute. https://www.migrationpolicy.org/article/korean-immigrants-united-states. [Accessed 30 November 2018].

Zong, K. & Batalova, J. 2018. *Filipino Immigrants in the United States* [Online], Washington D.C., Migration Policy Institute. https://www.migrationpolicy.org/article/filipino-immigrants-united-states. [Accessed 29 November 2018].

Zschirnt, E. & Ruedin, D. 2016. Ethnic discrimination in hiring decisions: A meta-analysis of correspondence tests 1990–2015. *Journal of Ethnic and Migration Studies*, 42, 1115–1134.

Index